Optical Waveguides

From Theory to Applied Technologies

T0179233

OPTICAL SCIENCE AND ENGINEERING

Founding Editor
Brian J. Thompson
University of Rochester
Rochester, New York

1. Electron and Ion Microscopy and Microanalysis: Principles and Applications, *Lawrence E. Murr*
2. Acousto-Optic Signal Processing: Theory and Implementation, *edited by Norman J. Berg and John N. Lee*
3. Electro-Optic and Acousto-Optic Scanning and Deflection, *Milton Gottlieb, Clive L. M. Ireland, and John Martin Ley*
4. Single-Mode Fiber Optics: Principles and Applications, *Luc B. Jeunhomme*
5. Pulse Code Formats for Fiber Optical Data Communication: Basic Principles and Applications, *David J. Morris*
6. Optical Materials: An Introduction to Selection and Application, *Solomon Musikant*
7. Infrared Methods for Gaseous Measurements: Theory and Practice, *edited by Joda Wormhoudt*
8. Laser Beam Scanning: Opto-Mechanical Devices, Systems, and Data Storage Optics, *edited by Gerald F. Marshall*
9. Opto-Mechanical Systems Design, *Paul R. Yoder, Jr.*
10. Optical Fiber Splices and Connectors: Theory and Methods, *Calvin M. Miller with Stephen C. Mettler and Ian A. White*
11. Laser Spectroscopy and Its Applications, *edited by Leon J. Radziemski, Richard W. Solarz, and Jeffrey A. Paisner*
12. Infrared Optoelectronics: Devices and Applications, *William Nunley and J. Scott Bechtel*
13. Integrated Optical Circuits and Components: Design and Applications, *edited by Lynn D. Hutcheson*
14. Handbook of Molecular Lasers, *edited by Peter K. Cheo*
15. Handbook of Optical Fibers and Cables, *Hiroshi Murata*
16. Acousto-Optics, *Adrian Korpel*
17. Procedures in Applied Optics, *John Strong*
18. Handbook of Solid-State Lasers, *edited by Peter K. Cheo*
19. Optical Computing: Digital and Symbolic, *edited by Raymond Arrathoon*
20. Laser Applications in Physical Chemistry, *edited by D. K. Evans*
21. Laser-Induced Plasmas and Applications, *edited by Leon J. Radziemski and David A. Cremers*

22. Infrared Technology Fundamentals, *Irving J. Spiro and Monroe Schlessinger*
23. Single-Mode Fiber Optics: Principles and Applications, Second Edition, Revised and Expanded, *Luc B. Jeunhomme*
24. Image Analysis Applications, *edited by Rangachar Kasturi and Mohan M. Trivedi*
25. Photoconductivity: Art, Science, and Technology, *N. V. Joshi*
26. Principles of Optical Circuit Engineering, *Mark A. Mentzer*
27. Lens Design, *Milton Laikin*
28. Optical Components, Systems, and Measurement Techniques, *Rajpal S. Sirohi and M. P. Kothiyal*
29. Electron and Ion Microscopy and Microanalysis: Principles and Applications, Second Edition, Revised and Expanded, *Lawrence E. Murr*
30. Handbook of Infrared Optical Materials, *edited by Paul Klocek*
31. Optical Scanning, *edited by Gerald F. Marshall*
32. Polymers for Lightwave and Integrated Optics: Technology and Applications, *edited by Lawrence A. Hornak*
33. Electro-Optical Displays, *edited by Mohammad A. Karim*
34. Mathematical Morphology in Image Processing, *edited by Edward R. Dougherty*
35. Opto-Mechanical Systems Design: Second Edition, Revised and Expanded, *Paul R. Yoder, Jr.*
36. Polarized Light: Fundamentals and Applications, *Edward Collett*
37. Rare Earth Doped Fiber Lasers and Amplifiers, *edited by Michel J. F. Digonnet*
38. Speckle Metrology, *edited by Rajpal S. Sirohi*
39. Organic Photoreceptors for Imaging Systems, *Paul M. Borsenberger and David S. Weiss*
40. Photonic Switching and Interconnects, *edited by Abdellatif Marrakchi*
41. Design and Fabrication of Acousto-Optic Devices, *edited by Akis P. Goutzoulis and Dennis R. Pape*
42. Digital Image Processing Methods, *edited by Edward R. Dougherty*
43. Visual Science and Engineering: Models and Applications, *edited by D. H. Kelly*
44. Handbook of Lens Design, *Daniel Malacara and Zacarias Malacara*
45. Photonic Devices and Systems, *edited by Robert G. Hunsberger*
46. Infrared Technology Fundamentals: Second Edition, Revised and Expanded, *edited by Monroe Schlessinger*
47. Spatial Light Modulator Technology: Materials, Devices, and Applications, *edited by Uzi Efron*
48. Lens Design: Second Edition, Revised and Expanded, *Milton Laikin*
49. Thin Films for Optical Systems, *edited by Francoise R. Flory*
50. Tunable Laser Applications, *edited by F. J. Duarte*
51. Acousto-Optic Signal Processing: Theory and Implementation, Second Edition, *edited by Norman J. Berg and John M. Pellegrino*

52. Handbook of Nonlinear Optics, *Richard L. Sutherland*
53. Handbook of Optical Fibers and Cables: Second Edition, *Hiroshi Murata*
54. Optical Storage and Retrieval: Memory, Neural Networks, and Fractals, *edited by Francis T. S. Yu and Suganda Jutamulia*
55. Devices for Optoelectronics, *Wallace B. Leigh*
56. Practical Design and Production of Optical Thin Films, *Ronald R. Willey*
57. Acousto-Optics: Second Edition, *Adrian Korpel*
58. Diffraction Gratings and Applications, *Erwin G. Loewen and Evgeny Popov*
59. Organic Photoreceptors for Xerography, *Paul M. Borsenberger and David S. Weiss*
60. Characterization Techniques and Tabulations for Organic Nonlinear Optical Materials, *edited by Mark G. Kuzyk and Carl W. Dirk*
61. Interferogram Analysis for Optical Testing, *Daniel Malacara, Manuel Servin, and Zacarias Malacara*
62. Computational Modeling of Vision: The Role of Combination, *William R. Uttal, Ramakrishna Kakarala, Spiram Dayanand, Thomas Shepherd, Jagadeesh Kalki, Charles F. Lunskis, Jr., and Ning Liu*
63. Microoptics Technology: Fabrication and Applications of Lens Arrays and Devices, *Nicholas Borrelli*
64. Visual Information Representation, Communication, and Image Processing, *edited by Chang Wen Chen and Ya-Qin Zhang*
65. Optical Methods of Measurement, *Rajpal S. Sirohi and F. S. Chau*
66. Integrated Optical Circuits and Components: Design and Applications, *edited by Edmond J. Murphy*
67. Adaptive Optics Engineering Handbook, *edited by Robert K. Tyson*
68. Entropy and Information Optics, *Francis T. S. Yu*
69. Computational Methods for Electromagnetic and Optical Systems, *John M. Jarem and Partha P. Banerjee*
70. Laser Beam Shaping, *Fred M. Dickey and Scott C. Holswade*
71. Rare-Earth-Doped Fiber Lasers and Amplifiers: Second Edition, Revised and Expanded, *edited by Michel J. F. Digonnet*
72. Lens Design: Third Edition, Revised and Expanded, *Milton Laikin*
73. Handbook of Optical Engineering, *edited by Daniel Malacara and Brian J. Thompson*
74. Handbook of Imaging Materials: Second Edition, Revised and Expanded, *edited by Arthur S. Diamond and David S. Weiss*
75. Handbook of Image Quality: Characterization and Prediction, *Brian W. Keelan*
76. Fiber Optic Sensors, *edited by Francis T. S. Yu and Shizhuo Yin*
77. Optical Switching/Networking and Computing for Multimedia Systems, *edited by Mohsen Guizani and Abdella Battou*
78. Image Recognition and Classification: Algorithms, Systems, and Applications, *edited by Bahram Javidi*

79. Practical Design and Production of Optical Thin Films: Second Edition, Revised and Expanded, *Ronald R. Willey*
80. Ultrafast Lasers: Technology and Applications, *edited by Martin E. Fermann, Almantas Galvanauskas, and Gregg Sucha*
81. Light Propagation in Periodic Media: Differential Theory and Design, *Michel Nevière and Evgeny Popov*
82. Handbook of Nonlinear Optics, Second Edition, Revised and Expanded, *Richard L. Sutherland*
83. Polarized Light: Second Edition, Revised and Expanded, *Dennis Goldstein*
84. Optical Remote Sensing: Science and Technology, *Walter Egan*
85. Handbook of Optical Design: Second Edition, *Daniel Malacara and Zacarias Malacara*
86. Nonlinear Optics: Theory, Numerical Modeling, and Applications, *Partha P. Banerjee*
87. Semiconductor and Metal Nanocrystals: Synthesis and Electronic and Optical Properties, edited by *Victor I. Klimov*
88. High-Performance Backbone Network Technology, *edited by Naoaki Yamanaka*
89. Semiconductor Laser Fundamentals, *Toshiaki Suhara*
90. Handbook of Optical and Laser Scanning, *edited by Gerald F. Marshall*
91. Organic Light-Emitting Diodes: Principles, Characteristics, and Processes, *Jan Kalinowski*
92. Micro-Optomechatronics, *Hiroshi Hosaka, Yoshitada Katagiri, Terunao Hirota, and Kiyoshi Itao*
93. Microoptics Technology: Second Edition, *Nicholas F. Borrelli*
94. Organic Electroluminescence, *edited by Zakya Kafafi*
95. Engineering Thin Films and Nanostructures with Ion Beams, *Emile Knystautas*
96. Interferogram Analysis for Optical Testing, Second Edition, *Daniel Malacara, Manuel Sercin, and Zacarias Malacara*
97. Laser Remote Sensing, *edited by Takashi Fujii and Tetsuo Fukuchi*
98. Passive Micro-Optical Alignment Methods, *edited by Robert A. Boudreau and Sharon M. Boudreau*
99. Organic Photovoltaics: Mechanism, Materials, and Devices, *edited by Sam-Shajing Sun and Niyazi Serdar Saracftci*
100. Handbook of Optical Interconnects, *edited by Shigeru Kawai*
101. GMPLS Technologies: Broadband Backbone Networks and Systems, *Naoaki Yamanaka, Kohei Shiomoto, and Eiji Oki*
102. Laser Beam Shaping Applications, *edited by Fred M. Dickey, Scott C. Holswade and David L. Shealy*
103. Electromagnetic Theory and Applications for Photonic Crystals, *Kiyotoshi Yasumoto*
104. Physics of Optoelectronics, *Michael A. Parker*
105. Opto-Mechanical Systems Design: Third Edition, *Paul R. Yoder, Jr.*
106. Color Desktop Printer Technology, *edited by Mitchell Rosen and Noboru Ohta*
107. Laser Safety Management, *Ken Barat*
108. Optics in Magnetic Multilayers and Nanostructures, *Štefan Višňovský*
109. Optical Inspection of Microsystems, *edited by Wolfgang Osten*

110. Applied Microphotonics, *edited by Wes R. Jamroz, Roman Kruzelecky, and Emile I. Haddad*
111. Organic Light-Emitting Materials and Devices, *edited by Zhigang Li and Hong Meng*
112. Silicon Nanoelectronics, *edited by Shunri Oda and David Ferry*
113. Image Sensors and Signal Processor for Digital Still Cameras, *Junichi Nakamura*
114. Encyclopedic Handbook of Integrated Circuits, *edited by Kenichi Iga and Yasuo Kokubun*
115. Quantum Communications and Cryptography, *edited by Alexander V. Sergienko*
116. Optical Code Division Multiple Access: Fundamentals and Applications, *edited by Paul R. Prucnal*
117. Polymer Fiber Optics: Materials, Physics, and Applications, *Mark G. Kuzyk*
118. Smart Biosensor Technology, *edited by George K. Knopf and Amarjeet S. Bassi*
119. Solid-State Lasers and Applications, *edited by Alphan Sennaroglu*
120. Optical Waveguides: From Theory to Applied Technologies, *edited by María L. Calvo and Vasudevan Lakshminarayanan*
121. Gas Lasers, *edited by Masamori Endo and Robert F. Walker*
122. Lens Design, Fourth Edition, *Milton Laikin*
123. Photonics: Principles and Practices, *Abdul Al-Azzawi*
124. Microwave Photonics, *edited by Chi H. Lee*

Optical Waveguides

From Theory to Applied Technologies

Edited by
María L. Calvo
Vasudevan Lakshminarayanan

CRC Press
Taylor & Francis Group
Boca Raton London New York

CRC Press is an imprint of the
Taylor & Francis Group, an **informa** business

CRC Press
Taylor & Francis Group
6000 Broken Sound Parkway NW, Suite 300
Boca Raton, FL 33487-2742

First issued in paperback 2019

© 2007 by Taylor & Francis Group, LLC
CRC Press is an imprint of Taylor & Francis Group, an Informa business

No claim to original U.S. Government works

ISBN-13: 978-1-57444-698-2 (hbk)
ISBN-13: 978-0-367-38953-6 (pbk)

This book contains information obtained from authentic and highly regarded sources. Reasonable efforts have been made to publish reliable data and information, but the author and publisher cannot assume responsibility for the validity of all materials or the consequences of their use. The authors and publishers have attempted to trace the copyright holders of all material reproduced in this publication and apologize to copyright holders if permission to publish in this form has not been obtained. If any copyright material has not been acknowledged please write and let us know so we may rectify in any future reprint.

Except as permitted under U.S. Copyright Law, no part of this book may be reprinted, reproduced, transmitted, or utilized in any form by any electronic, mechanical, or other means, now known or hereafter invented, including photocopying, microfilming, and recording, or in any information storage or retrieval system, without written permission from the publishers.

For permission to photocopy or use material electronically from this work, please access www.copyright.com (http://www.copyright.com/) or contact the Copyright Clearance Center, Inc. (CCC), 222 Rosewood Drive, Danvers, MA 01923, 978-750-8400. CCC is a not-for-profit organization that provides licenses and registration for a variety of users. For organizations that have been granted a photocopy license by the CCC, a separate system of payment has been arranged.

Trademark Notice: Product or corporate names may be trademarks or registered trademarks, and are used only for identification and explanation without intent to infringe.

Library of Congress Cataloging-in-Publication Data

Optical waveguides : from theory to applied mechanics / edited by Maria L. Calvo, Vasudevan Lakshminarayanan.
p. cm. -- (Optical science and engineering series ; 120)
Includes bibliographical references and index.
ISBN 1-57444-698-3 (alk. paper)
1. Optical wave guides. 2. Integrated optics. I. Calvo, Maria L. II. Lakshminarayanan, Vasudevan.

TA1800.O6975 2007
621.36'9--dc22 2006050575

Visit the Taylor & Francis Web site at
http://www.taylorandfrancis.com

and the CRC Press Web site at
http://www.crcpress.com

Preface

Optical waveguide theory is based on the electromagnetic nature of light and its interaction with matter. The development of fiber optics technology and optical communications has impacted our daily life tremendously. Nowadays, new technologies combine the confinement of radiation with novel optical properties of new photomaterials. The possibilities offered by micro- and nanotechnologies offer great promise of revolutionary devices for optical communication and data storage. Moreover, optical waveguide technologies are applied in life sciences also, specifically in biomedical imaging instrumentation and in the design of new biooptical fibers that are applied in various medical areas. One may consider other new interdisciplinary areas closely related to optics, such as neutron optics. The latter, unfortunately, is not quite as well known to scientists and technologists, despite the development of quite new and emerging areas, namely that of neutron waveguides.

It is in this context that we offer a new book on optical waveguides combining the most relevant aspects of waveguide theory along with the study of very detailed current waveguiding technologies and specifically focused on photonic devices, telecommunication applications and biomedical optics. This is a very carefully designed edition, which includes original images to show the high performance of the most outstanding photonic technologies.

This text is offered to scientists, technologists and professionals who require new insights on the subject and also to students looking for more specialized subjects for study and research. It is our wish that the readers will enjoy it and we are sure they will be surprised by the sophistication of the new generation of photonic devices that play a leading role in the new era for the information society in this, the photonics century.

María L. Calvo
V. Lakshminarayanan

Preface

Introduction

The fundamentals of optical waveguides are based on the electromagnetic theory of light and its interaction with matter. These principles have been well known for more than a century and many popular applications of optical fibers have been known for more than four decades. The historical roots of the technology go back to the Victorian Era, but basic science and technology have given us important components of optical fiber telecommunications such as high speed electronics, solid state lasers, and low loss optical materials only in the past few decades.

Shortly after the development of the telephone, Alexander Graham Bell proposed a wireless communication technology using modulated sunlight. He called the device a photophone! In 1926, Clarence W. Hansell outlined the basic elements of a fiber optic communications system in a laboratory at RCA and patented a design for transmission of images over bundles of glass fibers from remote locations. The concept of transmitting multiple wavelengths (wavelength division multiplexing) had its origin in the 1930s. However, it was not until the late 1970s that the first low-loss single-mode optical fibers were fabricated.

One reason for the tremendous impact of light waveguiding-based optical devices in daily life has been the remarkable development in the fields of fiber optics technology and optical communications. The importance and potential of the confined propagation of radiation go far beyond the telecommunications area. The advancement of theoretical formalisms for studying the confinement of light, the availability of very fast and efficient numerical algorithms for computational simulations, and the related advancements in optical technologies in material and laser science have made scientists and technologists well aware of the enormous impacts of waveguiding phenomena. The development of fiber optic technology for myriad applications ranging from communications, sensors, and imaging to medicine represents a great convergence of applied mathematics, physics, electronics, materials science, engineering, and manufacturing technologies.

Fiber optics technology also serves as the backbone of the Internet and the worldwide telecommunications infrastructure. Other advancements include the study of nonlinear behaviors of certain materials that created the emerging field of nonlinear optical waveguides and a promising future for communication and data transfer through the use of nonconventional signals (solitons) generated in these media.

Solitons are ultrafast short pulses that propagate through fibers without changing their spectral and temporal pulse shapes. Solitons derived as solutions to the nonlinear wave equation balance dispersion and nonlinearity, resulting in little or no degradation of signal quality along propagation distance. Other technologies combine the confinement of radiation with the optical properties of photomaterials and offer revolutionary possibilities in terms of micro- and nanotechnologies. For example, nanofibers that are silica waveguides with diameters smaller than the wavelength of transmitted light can be formed. These fibers offer a great deal of potential for microphotonic devices. The extension of these applications to the design of new sensors and other instrumentation is currently an area of intense research and development. In the areas of biophysics and life sciences, optical waveguides have been widely applied to biomedical imaging instrumentation and in designing new bio-optical fiber endoscopes.

It is of historical importance to note that in 1888, Dr. Roth and Prof. Reuss of Vienna used bent glass rods to illuminate body cavities for dentistry and surgery. A decade later, David D. Smith of Indianapolis applied for a patent for the use of a bent glass rod as a surgical lamp. In 1952, Prof. H.H. Hopkins of London's Imperial College applied for a grant to develop bundles of glass fibers as an endoscope. The Hopkins endoscope was produced and marketed by Storz, Inc. in St. Louis in 1966. Since then, biomedical applications of waveguides and optical fibers have come a long way. Applications in ophthalmology, gastroenterology, gynecology, and other medical specialties have led to treatment procedures as varied as cancer therapy, treatment of enlarged prostate, electrocautery, and laparoscopic cholecystectomy (gallbladder removal).

Other basic biophysics applications include photoreceptor optics wherein rods and cones of the retinas of vertebrate eyes act as absorbing optical waveguides. Absorbing waveguides are found in a number of other areas due to the formation of color centers in neutron-irradiated fibers and in plants. Certain deep sea sponges that grow fibers for anchorage and structural support are stronger than commercial fibers and exhibit higher light transmission. To fully portray the scope of this technology, we must mention another area closely related to optics: neutron optics.

Neutron optics and the associated neutron waveguides are not so well known to scientists and technologists. The key concept is that the wave-like behavior of slow neutron beams could lead to the development of a new area of waveguide technology: neutron waveguides.

Another nontechnical but important consideration in waveguide technologies is the issue of intellectual property rights and patent law. Fabrication expertise is often "hidden" in current text books and scientific articles. Many scientists (especially in industry) are often compelled to fabricate devices in their laboratories that are directly adapted for industrial exploitation and patent filing instead of publishing their technologies and fabrication details in the open literature.

Another factor that must be taken into account is the unpredictable fluctuation of the global economy that obviously influences the photonics industry; witness the so-called telecommunication "boom" and "bust" in the last decade of the 20th century. All these nontechnical factors affect levels of production, reliability, and consequently the commercial strength of many industries. However, photonics technology is currently enjoying great popularity. It plays an important part in everyday life and has shown resistance and resilience despite the vagaries of business and economic cycles.

In comparison with electronics, photonics is in its infancy even though the total worldwide fiber market at the end of 2003 was about 55 million km. Dense wavelength division multiplexing allows (with 40 Gbits/s available in a single mode fiber) a maximum theoretical upper limit of 2.5 billion phone calls based on bandwidth calculations over a pair of optical fibers. It should also be noted that in terms of cost performance, major decreases in long haul communication are estimated at a factor of two every year!

We can measure the growth of optical numbers in a number of ways. A study by the U.S. National Science Foundation in 2001 suggested that the bit rate distance product for WDM systems has grown steadily to approximately 10^8 GBit/km and the total capacity of DWDM systems has gone up by two orders of magnitude over the five years from 1995 to 2000.

The currently available commercial devices are mostly passive, with very low integration and with levels of functionality that clearly can be improved. For example, advancements in optical image processing arose from the inclusion of digital computational techniques. However, the techniques for optical data storage and computation are only in early stages of development. Supercomputers have just begun to use optical fibers for dense clusters of parallel processors and memory at higher bandwidths. This will significantly increase the computing power of next-generation computer systems and one should not forget the "holy grail" of all-optical (quantum) computing! We are optimistic and anticipate a future in which photonics will play a role in society similar to the one played by electronics in the 20th century.

This book is divided into nine chapters. To understand its structure we can divide it into sections. The first section is devoted to the theoretical foundations and bases of planar optical waveguides and to explaining critical optical properties such as birefringence and nonlinear optical phenomena. These areas are covered in Chapter 1 (propagation characteristics of planar optical waveguides), Chapter 2 (birefringent optical waveguides), and Chapter 3 (optically induced nonlinear waveguides). The second section is devoted to the study of current waveguiding technologies and focuses on photonic devices and telecommunication. Chapter 4 covers active optical waveguides, Chapter 5 treats wavelength dispersive planar waveguide devices, Chapter 6 deals with silicon waveguides for integrated optics, and Chapter 7 covers enabling fabrication technologies for planar waveguide devices. The third section discusses technological applications other than optical communications and photonic devices. Biomedical optical waveguides are covered in Chapter 8 and neutron waveguides are studied in Chapter 9. The chapters and sections are intended to be self-contained and flow in a linear manner. Those interested in knowing more about fundamentals may be advised to read the chapters in the first section and then pursue further reading on more applied topics. However, all sections and chapters can be read independently depending on need and interest. We hope that this book will serve as an introductory text for the novice and provide new insights for the experienced professional. It offers information on current technologies and, on the other hand, demonstrates the applicability of and improvements on the old formalisms. The authors are all well-known specialists. Readers will be surprised by the sophistication of the new generation of photonic devices that play leading roles in bringing forth a new era for the information society.

This book has been in gestation over a long period. Even though we (MLC and VL) are listed as the editors, this book is the result of an enormous team effort and the credit goes jointly to all the contributors. We believe that the whole is greater than the sum of the parts. We hope that readers will enjoy the result of this teamwork and will benefit greatly from the contents. Finally, we would like to thank our publishers, the Taylor & Francis Group, and in particular Jessica Vakili and Taisuke Soda for offering us the possibility of publishing this book and for their patience and help in bringing this book project to fruition.

María L. Calvo
Vasudevan Lakshminarayanan

About the Editors

María L. Calvo, Ph.D., is a professor of physics at the Faculty of Physics, Complutense University of Madrid (UCM, Spain) and currently Head of the Department of Optics at UCM. For more than three decades she has been working in theoretical formalisms for light scattering, optical waveguide theory, signal processing, holography, and photomaterials for optical computing. Prof. Calvo is the author or co-author of more than 120 scientific papers in the mentioned subjects. She has coordinated a new textbook on advanced optics and another one on virtual laboratory of optics, and a DVD on virtual laboratory of holography.

She is teaching undergraduate and graduate courses on optics at UCM and has delivered many seminars and been invited to conferences throughout the world. In 1993 she established the Interdisciplinary Group for Bio-Optics Research (GIBO-UCM), which later became the Interdisciplinary Group for Optical Computing (GICO-UCM). She developed current applications of optical signal theory and applications to conventional and unconventional optical processing of information, holography, and scattering theory of light. She also developed common international scientific collaborations.

Prof. Calvo is currently Secretary-General of the International Commission for Optics (ICO) and collaborates annually with the ICTP (Trieste, Italy) for the Winter College on Optics and Photonics. She also collaborates with the General Directorate for Research and Technology of the European Union in Brussels. She is a Fellow of the Optical Society of America (OSA), Fellow of the International Society for Optical Engineering (SPIE) and member of the European Optical Society (EOS) and the Spanish Optical Society (SEDOP).

Vasudevan Lakshminarayanan, Ph.D., is a professor of optometry, physics, and electrical engineering at the University of Waterloo (Canada). Prof. Lakshminarayanan conducts both experimental and theoretical studies in vision science and classical optics. Major areas of interest include visual and ophthalmic optics and mathematical modeling of visual/perceptual phenomena. In classical optics, he works on novel mathematical methods to analyze wave propagation in various media such as optical fibers, optical system design optimization using neural networks, genetic algorithms, etc. A special area of interest is in symmetry studies, which utilize group theory to study optics.

Prof. Lakshminarayanan has published more than 200 papers in areas ranging from quantum chemistry to bioengineering to applied mathematics and cognitive science in addition to clinical studies of the visual system. He is the author, editor, or co-author of eight books. He is an elected Fellow of the OSA, the Institute of Physics, the American Academy of Optometry, and SPIE – the International Society for Optical Engineering.

His various awards include most recently, the OSI Medal from the Optical Society of India, as well as awards for teaching excellence. He has been given the title of Chartered Physicist by IOP (UK). He has served as a director of the Optical Society of America and chaired the U.S. National Committee to the International Commission on Optics. He is currently a member-at-large of the U.S. IUPAP committee. He was a KITP Scholar at the Kavli Institute of Theoretical Physics, University of California at Santa Barbara, a Gulbenkian Foundation Fellow (Portugal), and more recently, a Royal Society of Edinburg Lecturer at the Glasgow Caledonian University. He has held research and teaching positions at the University of California Irvine and Berkeley campuses, the University of Missouri–St. Louis, as well as a research scientist position at Allergan Medical Optics.

Contributors

Ramón F. Alvarez-Estrada
Departamento de Fisica Teorica I
Facultad de Ciencias Fisicas
Universidad Complutense de Madrid
Madrid, Spain

Michael Cada
Department of Electrical and
Computer Engineering
Dalhousie University
Halifax, Nova Scotia, Canada

María L. Calvo
Departamento de Óptica
Facultad de Ciencias Físicas
Universidad Complutense de Madrid
Madrid, Spain

Pavel Cheben
Institute for Microstructural Sciences
National Research Council of Canada
Ottawa, Ontario, Canada

Cornelia Denz
Nichtlineare Photonik
Institut für Angewandte Physik
Westfälische Wilhelms-Universität
Müenster, Germany

Ajoy Ghatak
Indian Institute of Technology
New Delhi, India

Philip Jander
Nichtlineare Photonik
Institut für Angewandte Physik
Westfälische Wilhelms-Universität
Muenster, Germany

Paul E. Jessop
Department of Engineering Physics
McMaster University
Hamilton, Ontario, Canada

Andrew P. Knights
Department of Engineering Physics
McMaster University
Hamilton, Ontario, Canada

Vasudevan Lakshminarayanan
School of Optometry
University of Waterloo
Waterloo, Ontario, Canada

Boris Lamontagne
Institute for Microstructural Sciences
National Research Council of Canada
Ottawa, Ontario, Canada

Vadakke Matham Murukeshan
School of Mechanical and Aerospace
Engineering
Nanyang Technological University
Singapore

Contents

1 Propagation Characteristics of Planar Waveguides

Ajoy Ghatak and Vasudevan Lakshminarayanan

CONTENTS

1.1 INTRODUCTION

The simplest optical waveguide is probably the planar waveguide that consists of a thin dielectric film sandwiched between materials of slightly lower refractive indices. In such waveguides, the film is assumed to extend to infinity in one transverse direction (say y direction) and the refractive index varies along another transverse direction (i.e., along the x direction) — see Figure 1.1. A typical refractive index distribution [$n = n(x)$] is schematically shown in Figure 1.2. The refractive index profile (corresponding to Figure 1.2) can be written as

$$\begin{aligned} n(x) &= n_c & x &\le 0 \\ &= n_f(x) & 0 &\le x \le d \\ &= n_s & x &\ge d \end{aligned} \tag{1.1}$$

where n_c and n_s represent the refractive indices of the cover region ($x < 0$) and the substrate region ($x > d$), respectively; $n_f(x)$ represents the refractive index variation of the film which is assumed to be of thickness d.

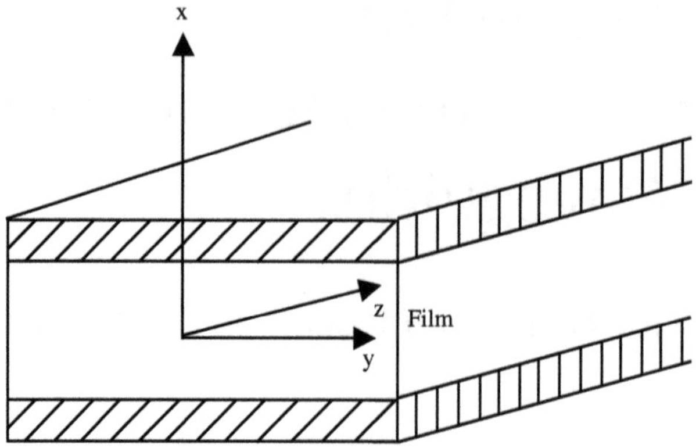

FIGURE 1.1 A planar waveguide that is assumed to be infinitely extending in the y and z directions.

1.2 TE AND TM MODES OF SLAB WAVEGUIDE

For a medium characterized by a refractive index variation that depends only on x-coordinate, i.e., for

$$n = n(x) \tag{1.2}$$

the solution of Maxwell's equation can be written in the form

$$\mathcal{E}_j(x, y, z, t) = E_j(x)e^{i(\omega t-\beta z)} \tag{1.3}$$

$$\mathcal{H}_j(x, y, z, t) = H_j(x)e^{i(\omega t-\beta z)} \tag{1.4}$$

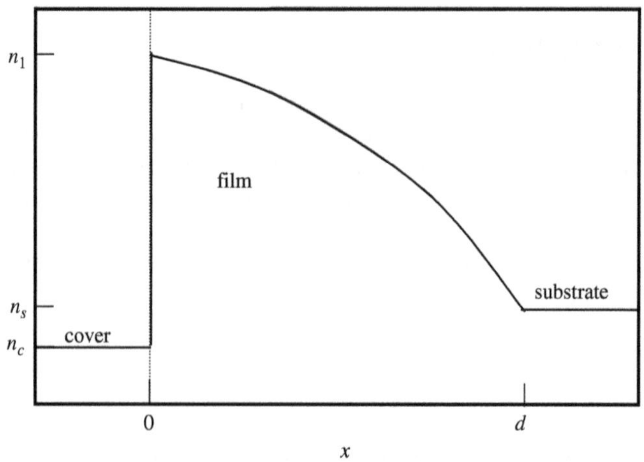

FIGURE 1.2 A typical refractive index profile for a planar waveguide.

where $j = x, y, z$, ω represents the angular frequency of the wave, and β is the propagation constant. Corresponding to a specific value of β, there is a particular field distribution described by $\mathbf{E}(x)$ and $\mathbf{H}(x)$ that remains unchanged with propagation along the waveguide; such distributions are called *modes* of the waveguide. In planar waveguides [i.e., for $n = n(x)$], if we assume the electric and magnetic field to be given by Eqs. (1.3) and (1.4) then Maxwell's equations lead to the following six equations [see, e.g., Ref. 1]:

$$i\beta E_y = -i\omega\mu_0 H_x \tag{1.5}$$

$$\frac{\partial E_y}{\partial x} = -i\omega\mu_0 H_z \tag{1.6}$$

$$-i\beta H_x - \frac{\partial H_z}{\partial x} = i\omega\varepsilon_0 n^2(x) E_y \tag{1.7}$$

$$i\beta H_y = i\omega\varepsilon_0 n^2(x) E_x \tag{1.8}$$

$$\frac{\partial H_y}{\partial x} = i\omega\varepsilon_0 n^2(x) E_z \tag{1.9}$$

$$-i\beta E_x - \frac{\partial E_z}{\partial x} = -i\omega\mu_0 H_y \tag{1.10}$$

The first three equations involve only E_y,H_x and H_z and last three equations involve only E_x,E_z, and H_y. Thus we can have two independent sets of modes; the first set corresponds to what are known as TE (transverse electric) modes where we can assume E_x,E_z, and H_y vanish — thus the electric field is transverse to the direction of propagation. Similarly, the second set corresponds to what are known as TM (transverse magnetic) modes where we can assume E_y,H_x, and H_z vanish — thus the magnetic field is transverse to the direction of propagation. Simple manipulations of Eqs. (1.5) through (1.10) would give us the following equations for TE and TM modes.

For TE modes:

$$\frac{d^2 E_y}{dx^2} + \left[k_0^2 n^2(x) - \beta^2\right] E_y(x) = 0 \tag{1.11}$$

where

$$k_0 = \frac{\omega}{c} = \omega\sqrt{\varepsilon_0\mu_0} = \frac{2\pi}{\lambda_0} \tag{1.12}$$

represents the free space wave number and λ_0 represents the free space wavelength.

For TM modes:

$$n^2(x)\frac{d}{dx}\left[\frac{1}{n^2(x)}\frac{dH_y}{dx}\right] + \left[k_0^2 n^2(x) - \beta^2\right] H_y(x) = 0 \tag{1.13}$$

Analytical solution of Eq. (1.11) exists only for limited refractive index profiles. In order to study the propagation characteristics of the mode of a waveguide with arbitrary refractive index profile $n(x)$, we can consider it as a limiting form of a layered structure as shown in Figure 1.3. Thus the first region is the cover, the Nth

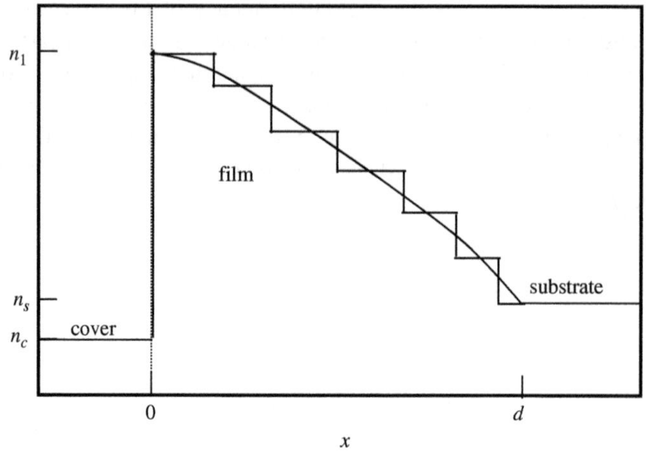

FIGURE 1.3 In the matrix method, a continuously varying refractive index is replaced by a layered structure.

region is the substrate, and the (graded index) film is represented by a layered structure consisting of $(N - 2)$ regions — each layer characterized by a constant refractive index.

It is obvious from Eq. (1.11) and Eq. (1.13) that if we assume a particular region [say, the ith region] to have a constant refractive index n_i, then both $E_y(x)$ [for TE modes] and $H_y(x)$ [for TM modes] would satisfy the scalar wave equation

$$\frac{d^2\psi_i}{dx^2} + \left[k_0^2 n_i^2 - \beta^2\right] \psi_i(x) = 0 \tag{1.14}$$

However, at any discontinuity of the refractive index we will have

$$\psi(x) \text{ and } \frac{d\psi}{dx} \text{ continuous} \tag{1.15}$$

for TE modes and

$$\psi(x) \text{ and } \frac{1}{n^2}\frac{d\psi}{dx} \text{ continuous} \tag{1.16}$$

for TM modes.

1.3 MATRIX METHOD

We rewrite Eq.(1.14) as

$$\frac{d^2\psi_i}{d\xi^2} + V^2\left[F_i - b\right] \psi_i(\xi) = 0 \tag{1.17}$$

where

$$\xi = \frac{x}{h} \tag{1.18}$$

$$V = k_0 h \left[n_m^2 - n_s^2\right]^{1/2} \tag{1.19}$$

$$b = \frac{\frac{\beta^2}{k_0^2} - n_s^2}{n_m^2 - n_s^2} \tag{1.20}$$

and

$$F_i = \frac{n_i^2 - n_s^2}{n_m^2 - n_s^2} \tag{1.21}$$

Note that n_m is the maximum value of $n_f(x)$ and $h(< d)$ is a conveniently chosen length parameter that, in some cases, may be equal to d. The parameter V is often referred to as the waveguide parameter and b is known as the normalized propagation constant of the mode. The solution of Eq. (1.14) [or of (1.17)] that satisfies the boundary conditions that $\psi(x) \to 0$ as $x \to \pm\infty$ exists only for a finite number of discrete values of b in the range

$$0 < b < 1 \tag{1.22}$$

Each discrete value of b corresponds to a *guided mode* of the waveguide. The highest value of b corresponds to the fundamental mode of the waveguide. We have assumed n_c, which represents the refractive index of the cover region which is usually air, to be less than n_s.

In order to study a multilayered structure, we have used the matrix method whose details are given in Chapter 24 of Ref. 1. The solution of Eq. (1.17) in the ith region can be written as

$$\begin{aligned}\psi_i &= A_i \cos\left[\gamma_i\left(\xi - \sigma_{i-1}\right)\right] + B_i \sin\left[\gamma_i\left(\xi - \sigma_{i-1}\right)\right] \quad \text{for } \kappa_i^2 > 0 \\ &= A_i \cosh\left[\gamma_i\left(\xi - \sigma_{i-1}\right)\right] + B_i \sinh\left[\gamma_i\left(\xi - \sigma_{i-1}\right)\right] \quad \text{for } \kappa_i^2 < 0\end{aligned} \tag{1.23}$$

where $i = 1, 2, \ldots N$ with

$$\kappa_i^2 = V^2[F_i - b] \tag{1.24}$$

$$\gamma_i = |\kappa_i| \tag{1.25}$$

$$\sigma_i = \frac{x_i}{h}; i = 1, 2, \ldots (N-1); \sigma_0 = 0 \tag{1.26}$$

and $x = x_i$ represents the right boundary of the ith layer (see Figure 1.4). In the matrix method, the coefficients A and B in two successive regions [say, i^{th} and $(i+1)^{th}$ regions] can be related through the following matrix equation:

$$\begin{pmatrix} A_{i+1} \\ B_{i+1} \end{pmatrix} = S_i \begin{pmatrix} A_i \\ B_i \end{pmatrix} \tag{1.27}$$

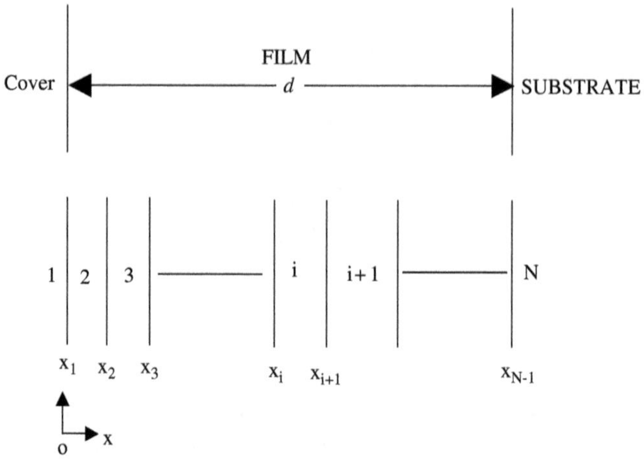

FIGURE 1.4 The layered structure for a graded index waveguide.

where S_i, which is known as the transfer matrix of the ith region, is given by

$$\begin{aligned} S_i &= \begin{pmatrix} \cos\Delta_i & \sin\Delta_i \\ -\frac{\gamma_i}{\gamma_{i+1}}\cdot\alpha_i\sin\Delta_i & \frac{\gamma_i}{\gamma_{i+1}}\cdot\alpha_i\cos\Delta_i \end{pmatrix} \quad \text{if } F_i > b \\ &= \begin{pmatrix} \cosh\Delta_i & \sinh\Delta_i \\ \frac{\gamma_i}{\gamma_{i+1}}\cdot\alpha_i\sinh\Delta_i & \frac{\gamma_i}{\gamma_{i+1}}\cdot\alpha_i\cosh\Delta_i \end{pmatrix} \quad \text{if } F_i < b \end{aligned} \tag{1.28}$$

Further,

$$\Delta_i = \gamma_i\left(\sigma_i - \sigma_{i-1}\right) \tag{1.29}$$

and

$$\begin{aligned} \alpha_i &= 1 && \text{for TE modes} \\ &= \frac{n_{i+1}^2}{n_i^2} && \text{for TM modes} \end{aligned} \tag{1.30}$$

By successive application of Eq (1.27), we can write

$$\begin{pmatrix} A_{i+1} \\ B_{i+1} \end{pmatrix} = S_i S_{i-1} \ldots S_2 S_1 \begin{pmatrix} A_1 \\ B_1 \end{pmatrix} \tag{1.31}$$

Thus

$$\begin{pmatrix} A_N \\ B_N \end{pmatrix} = G \begin{pmatrix} A_1 \\ B_1 \end{pmatrix} \tag{1.32}$$

where the G matrix is given by

$$G = S_{N-1} \ldots S_2 S_1 \tag{1.33}$$

Now in region $1(x < x_1$ — see Figure 1.4), the field must tend to zero as $x \to -\infty$, giving

$$A_1 = B_1 \tag{1.34}$$

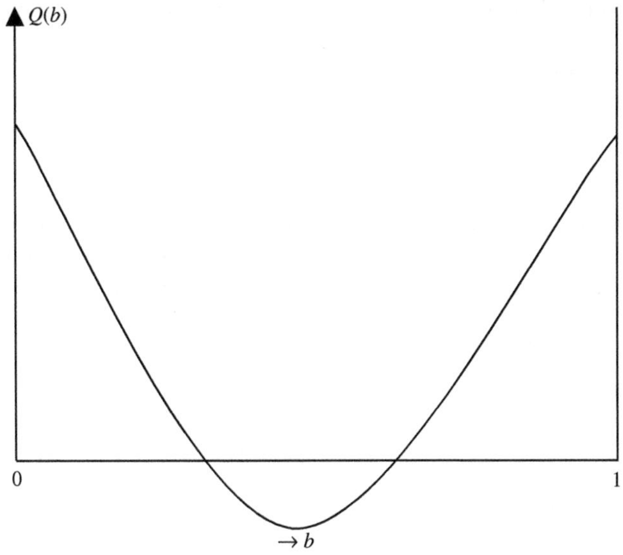

FIGURE 1.5 A typical variation of Q(b) with normalized propagation constant b.

[see Eq. (1.23)]. Without any loss of generality, we may assume $A_1 = B_1 = 1$. Thus Eq. (1.32) becomes

$$\begin{pmatrix} A_N \\ B_N \end{pmatrix} = \begin{pmatrix} G_{11} & G_{12} \\ G_{21} & G_{22} \end{pmatrix} \begin{pmatrix} 1 \\ 1 \end{pmatrix} \tag{1.35}$$

where G_{11},G_{12}, G_{21}, and G_{22} are the elements of the G matrix. Now, for a guided mode we must also have (in the Nth region) $\psi \to 0$ as $x \to \infty$. The above boundary condition would imply [see Eq. (1.23)]

$$Q(b) = A_N + B_N = 0 \tag{1.36}$$

Thus each guided mode would correspond to the zeroes of $Q(b)$ in the domain $0 < b < 1$. A typical plot of $Q(b)$ is shown in Figure 1.5. For example, for a step index (single film) waveguide with n_s = 1.47, n_c = 1.46, n_f = n_1 = 1.50, d = 5.0 μm, and λ = 1.55 μm, the function $Q(b)$ becomes zero for b > 0.84437 and 0.40458 which represent the two guided (TE) modes of the waveguides. Thus for any given refractive index variation $n(x)$, one replaces $n(x)$ by a large number of steps (as shown in Figure 1.3 and Figure 1.4) and in each region the refractive index is assumed to be a constant. Starting with an exponentially decaying solution (as $x \to -\infty$), one plots the coefficient of the exponentially amplifying solution $Q(b)$ as a function of b. The zeroes of $Q(b)$ would give the propagation constants of the bound modes.

1.4 MODIFIED AIRY FUNCTION METHOD

We will describe here another approximate method that is much more powerful than any other method and uses the Airy functions (MAF). The method was first suggested by Langer (see Ref. 2) and is applicable to both the *initial value* as well as the *eigenvalue* problem. The advantages of the method will be illustrated by means of examples.

We rewrite Eq. (1.14) as

$$\Psi''(x) + \Gamma^2(x)\Psi(x) = 0 \tag{1.37}$$

where

$$\Gamma^2(x) = k_0^2 n^2(x) - \beta^2 \tag{1.38}$$

The use of the Jeffreys-Wenzel-Kramers-Brillonin (JWKB) method is well established in problems of quantum mechanics and optical waveguides. Following the JWKB methodology, we assume a solution of Eq. (1.37) of the form (see Refs. 3 and 4)

$$\Psi(x) = F(x)Ai[\xi(x)] \text{ or } G(x)Bi[\xi(x)] \tag{1.39}$$

where $Ai(x)$ and $Bi(x)$ are solutions of the Airy equation:

$$\frac{d^2 f}{dx^2} - xf(x) = 0$$

Substitution of Eq. (1.39) in Eq.(1.37) gives

$$\begin{aligned} &F''(x)Ai(\xi) + 2F'(x)Ai'(\xi)\xi'(x) + F(x)Ai'(\xi)\xi'' \\ &+ Ai(\xi)F(x)[\xi(\xi'(x))^2 + \Gamma^2(x)] = 0 \end{aligned} \tag{1.40}$$

where the primes denote differentiation with respect to the argument. Equation (1.40) is rigorously correct. We choose $\xi(x)$ so that

$$\xi[\xi'(x)]^2 + \Gamma^2(x) = 0 \tag{1.41}$$

the solution of which gives

$$\xi(x) = \frac{3}{2}\left\{\int_x^{x_0} \sqrt{-\Gamma^2}\,dx\right\}^{2/3} \tag{1.42}$$

with x_0 being the turning point where $\Gamma^2(x)$ has a zero of order one. If we neglect the term proportional to F'' in Eq. (1.40) (that is the only approximation we will make), it becomes

$$2F'(x)Ai'(\xi)\xi'(x) + F(x)Ai'(\xi)\xi''(x) = 0 \tag{1.43}$$

the solution of which is

$$F(x) = \frac{\text{const}}{\sqrt{\xi'(x)}}. \tag{1.44}$$

The approximate solution that we will discuss, then, is

$$\Psi(x) = C_1\frac{Ai\,[\xi(x)]}{\sqrt{\xi'(x)}} + C_2\frac{Bi\,[\xi(x)]}{\sqrt{\xi'(x)}} \tag{1.45}$$

where C_1 and C_2 are constants to be determined from the initial or boundary conditions. Before illustrating the method by a few examples from optical waveguides, let us rewrite Eq. (1.37) in terms of the dimensionless variables in the following form

$$\frac{d^2\Psi}{dX^2} + V^2[N^2(x) - b]\psi(x) = 0 \tag{1.46}$$

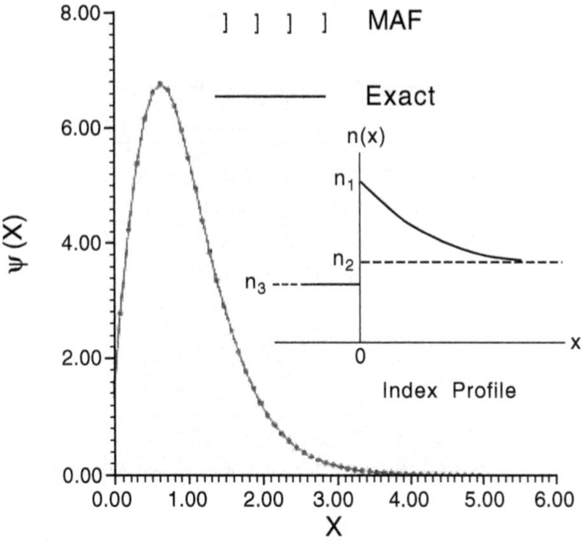

FIGURE 1.6 Variation of $\psi(X)$ with X. (Adapted from Reference 3.)

where $X = x/d$ (d is a suitably chosen length), $V = k_0 d\sqrt{n_1^2 - n_2^2}$ is the normalized waveguide parameter (n_1 is the maximum refractive index of the core and n_2 the index of the cladding), and

$$b = \frac{\frac{\beta^2}{k_0^2} - n_2^2}{n_1^2 - n_2^2} \quad \text{and} \quad N^2(x) = \frac{n^2(X) - n_2^2}{n_1^2 - n_2^2}$$

are the normalized propagation constant and the normalized index distribution, respectively.

As an example, we consider a planar optical waveguide with refractive index variation given by (see inset of Figure 1.6)

$$\begin{aligned} n^2(x) &= n_2^2 + \left(n_1^2 - n_2^2\right)\exp(-x/d) \quad \text{for } x > 0 \\ n^2(x) &= n_c^2 \quad \text{for } x < 0 \end{aligned} \tag{1.47}$$

where d is the diffusion depth of the waveguide. In this case, Eq (1.46) becomes

$$\frac{d^2\Psi}{dX^2} + V^2[\exp(-X) - b]\Psi(X) = 0 \qquad \text{for } X > 0 \tag{1.48}$$

$$\frac{d^2\Psi}{dX^2} - V^2\,[b + B]\,\Psi(X) = 0 \qquad \text{for } X < 0 \tag{1.49}$$

where $B = \left(n_2^2 - n_c^2\right)/\left(n_1^2 - n_2^2\right)$. For $X < 0$, the exact solution of Eq. (1.49) that does not diverge as $X \to -\infty$ is given by

$$\Psi(x) = \exp\left(XV\sqrt{b + B}\right) \tag{1.50}$$

For $X > 0$, the MAF [see Eq.(1.45)] solution can be written as [see Refs. 4–6]

$$\Psi(x) = \left(\frac{\xi_0'}{\xi'}\right)^{1/2} \frac{Ai(\xi)}{Ai(\xi_0)} \quad \text{for } X > 0 \tag{1.51}$$

The subscript 0 indicates the value of the function at $X = 0$. It may be noticed that in writing the solution given by Eq. (1.51) we have ignored the term proportional to $Bi(\xi)$ as it will diverge for large X; the proportionality constants in Eq. (1.50) and Eq. (1.51) have been so chosen as to satisfy the continuity of $\Psi(X)$ at $X = 0$ with $\Psi(0) = 1$. The eigenvalue equation can be written by using the continuity of $\Psi\prime(X)$ at $X = 0$:

$$\frac{V\,(B+b)^{1/2}}{\xi_0'} = \frac{Ai'(\xi_0)}{Ai(\xi_0)} - \frac{1}{2}\frac{\xi_0''}{\left(\xi_0'\right)^2}$$

We have used $n_2 = 2.177$, $n_c = 1$, and $n_1^2 - n_2^2 = 0.187$ in our calculations and compared the MAF results with the exact values and those calculated by the JWKB method (see Refs. 5–7). For $V = 1.5$, $b_{EXACT} = 0.03501$, $b_{MAF} = 0.03609$, and $b_{JWKB} = 0.03783$. The error in MAF values is only 3% (while the error in JWKB is 8%). Figure 1.6 shows Ψ_{MAF} versus X for $V = 4$ as calculated by Eq. (1.46) and Ψ_{EXACT}. The figure shows no discernable difference between the two curves even at the turning point.

1.4.1 Perturbation Correction

The fact that the MAF eigenfunction matches very well with the exact wave function can be used to improve eigenvalues with the help of the first order perturbation theory. It can be shown by simple substitution that the MAF solution given by

$$\Psi(X) = \frac{\text{const.}}{\sqrt{\xi'(X)}} Ai\left[\xi(X)\right] \tag{1.52}$$

is an exact solution of the following differential equation

$$\frac{d^2\Psi}{dX^2} + \Gamma^2(X)\Psi(X) + \left[\frac{1}{2}\frac{\xi'''}{\xi'} - \frac{3}{4}\left(\frac{\xi''}{\xi'}\right)^2\right]\Psi(X) = 0 \tag{1.53}$$

where $\Gamma^2(X) = V^2\left[N^2(X) - b\right]$ in problems of optical waveguides. Comparing Eqs. (1.53) and (1.46) and considering the last term in Eq. (1.53) as a perturbation, we get a first order correction Δb to normalized propagation constant [see Refs. 5 and 6]:

$$\Delta b \cong \frac{-\int\limits_{-\infty}^{+\infty}\left[\frac{1}{2}\frac{\xi'''}{\xi'} - \frac{3}{4}\left(\frac{\xi''}{\xi'}\right)^2\right]\Psi^2(X)dX}{V^2\int\limits_{-\infty}^{+\infty}\Psi^2(X)dX} \tag{1.54}$$

If we apply this perturbation correction to the example given above, we get $b_{MAF+Pert.} = 0.03502$ (comparison with exact shows an error of only ~0.03%). Thus, the application of the first order perturbation theory greatly improves the eigenvalues although the eigenfunctions remain the same.

1.5 NOTE ON PERTURBATION METHOD

As before, consider the eigenvalue equation rewritten in the form

$$\left[\nabla_t^2 + \kappa_u^2 n^2(x)\right] \Psi(x) = \beta^2 \Psi(x)$$

and define an operator H as

$$\mathrm{H} \equiv \nabla_t^2 + \kappa_u^2 n^2(x) \tag{1.55}$$

which results in

$$\mathrm{H}\Psi = \beta^2 \Psi \tag{1.56}$$

In the perturbation method, we estimate the eigenvalue β^2 of a mode of the waveguide of the corresponding mode (β_0^2 and $\Psi_0(x)$) of another closely related/similar waveguide is known. That is, we re-write the operator as $\mathrm{H} = \mathrm{H}_0 + \mathrm{H}'$ where H_0 is the known waveguide, and H' is a small difference between Ho and H. If $\{H'\} << \{H_0\}$, then we can show that the propagation constant β of the waveguide H is given by

$$\beta^2 = \beta_0^2 + \frac{\int \Psi_0^* \mathrm{H}' \Psi_0 \, dx}{\int |\Psi_0|^2 \, dx} \tag{1.57}$$

This can also be written as

$$\beta^2 = \beta_0^2 + \frac{k_0^2 \int \{n^2(x) - n^2(x_0)\} |\Psi_0|^2 \, dx}{\int |\Psi_0|^2 \, dx} \tag{1.58}$$

since

$$\mathrm{H}' = k_0^2 \left[n^2(x) - n^2(x_0)\right] \tag{1.59}$$

This is the first order correction.
Consider the following refractive index profile:

$$n^2(x) = n_1^2 \left[1 - 2\Delta \left(\frac{x}{a}\right)^2 + \delta \left(\frac{x}{a}\right)^4\right] \tag{1.60}$$

where $\delta << \Delta$ and is a small correction. The unperturbed profile is simply

$$n_0^2(x) = n_1^2 \left[1 - 2\Delta \left(\frac{x}{a}\right)^2\right] \tag{1.61}$$

and the fundamental mode is (see Ref. 1)

$$\Psi_0(x) = N_0 \, e^{-\frac{1}{2}\alpha^2 x^2} \tag{1.62}$$

The perturbation correction can be written as (using Eq. 59)

$$\Delta\beta^2 = \beta^2 - \beta_0^2 \; = \; \frac{k_0^2 \int_{-\infty}^{+\infty} \delta \left(\frac{x}{a}\right)^4 e^{-\alpha^2 x^2} \, dx}{\int_{-\infty}^{+\infty} e^{-\alpha^2 x^2} \, dx} \; = \; \frac{3}{8a^2} \frac{\delta}{\Delta} \tag{1.63}$$

$$\beta^2 = \beta_0^2 + \frac{3}{8a^2} \frac{\delta}{\Delta} \tag{1.64}$$

In general, we can expand β^2as a function of the perturbation correction δ

$$\beta^2(\delta) = \beta_0 + \delta \left(\frac{d\beta^2}{d\delta} |_{\delta=0} \right) + \delta^2 \left(\frac{d^2\beta^2}{d\delta^2} |_{\delta=0} \right) + \ldots \tag{1.65}$$

First order perturbation theory gives a good correction only up to $0(\delta)$, and is accurate only if the condition $0(\delta^2) << 0(\delta)$ is satisfied.

In the first order perturbation method, the corrected mode field is not obtained. The correction in β^2 is of the first order in H′. In order to estimate β, onlyβ_0^2 and Ψ_0 of that mode are required which are being corrected. It should also be noted that each mode must be corrected individually and therefore, in practice, is useful only for waveguides containing only a few modes.

The second order perturbation correction theory is much more complex to use because all modes including radiation modes of the unperturbed waveguide are needed. The first order accuracy can be estimated using the second order correction. Further discussion of the perturbation method is beyond the scope of this chapter. It should also be noted that there is a close relationship between perturbation methods and variational methods that are discussed in the next section.

1.6 VARIATIONAL METHOD

As before, we start with the scalar wave equation; however, here we write the scalar wave equation in integral form, and the crux of a variational principle is to formulate a stationary expression for the effective index n_e in terms of cross-sectional integrals involving the fields $\Psi(x, y)$. The integral term is (see Refs. 9 and 10)

$$n_e^2 = \frac{1}{k_0^2} \iint \Psi^*(x, y)\nabla\Psi(x, y)dxdy + \iint n^2(x, y)\left|\Psi(x, y)\right|^2 dxdy \tag{1.66}$$

The above equation can be transformed using the divergence theorem as

$$n_e^2 = -\frac{1}{k_0^2} \iint \left|\nabla\Psi(x, y)\right|^2 dxdy + \iint n^2(x, y)\left|\Psi(x, y)\right|^2 dxdy \tag{1.67}$$

where the limits of integration are from $-\infty$ to $+\infty$ and the modal field is normalized, i.e.,

$$\iint \left|\Psi(x, y)\right|^2 dxdy = 1 \tag{1.68}$$

It is trivial to write Eqs. (67 and 68) in terms of β (see Eq. 81). The right-hand sides of these two equations are said to be stationary expressions for the effective index n_e because it is stationary with respect to small variations in $\Psi(x, y)$ (for example, see Refs. 9 and 10).

$$\delta n_e^2(\Psi + \delta\Psi) = 0$$

Since the modal field is unknown, we use an approximation or trial field Ψ_t to get an estimate of the effective index n_{et} by evaluating Eq. (1.67) using Ψ_t. Obviously, different forms of the trial field Ψ_t will result in different values and expressions for n_{et}. What is the relationship between n_{et} and the exact value of the effective index n_e? Consider the fact that for each mode (radiation or guided), we can get an effective index and corresponding modal field. For the nth mode

$$\mathrm{H}\Psi_n(x, y) = n_{en}^2 \Psi_n(x, y)$$

or

$$n_{en}^2 = \int \Psi_n^* \mathrm{H} \Psi_n ds \tag{1.69}$$

where the integration is over the entire cross-section of the waveguide. The above equation is essentially the same as Eq. (1.66), wherein $\Psi(x, y)$ is the mode field for the nth mode. Since the modal field constitutes a complete set of orthonormal functions, we can express any function in terms of an expansion of the modal field

$$\phi(x, y) = \sum_n a_n \Psi_n(x, y) \tag{1.70}$$

or, in terms of the operator H,

$$n_{et} = \frac{\int \phi_t^* \mathrm{H} \phi_t ds}{\int \phi_t^* \phi_t ds} \tag{1.71}$$

substituting Eq. (1.70) in Eq. (1.71) and using the orthonormality condition $\int \Psi_n^* \Psi_m ds = \delta_{nm}$, it is easy to show that

$$n_{et}^2 = \frac{\sum_n a_n^* \, a_n \, n_{en}^2}{\sum_n a_n^* \, a_n} \tag{1.72}$$

If the fundamental mode index square n_{e0}^2 is subtracted from both sides, we get the fundamental result

$$n_{e0}^2 - n_{et}^2 = \frac{\sum_n |a_n|^2 \left[n_{e0}^2 - n_{en}^2 \right]}{\sum_n |a_n|^2} \tag{1.73}$$

Since the fundamental mode effective index is always highest $n_{e0}^2 \geq n_{et}^2$; or equivalently, $\beta_{exact} > \beta_t^2$. This implies that if we try a number of trial functions $\Psi_t(x, y)$ and estimate the corresponding β_t^2(or equivalently the effective index), then the largest value of β_t^2 would be closest to the exact value and the corresponding trial function $\Psi_t(x, y)$ would represent the best approximation for the model field $\Psi(x, y)$. The variational estimate of β^2 therefore represents a lower bound for the propagation constant. Appendix A gives an algorithm detailing the steps for the variational method.

The variational method, a semi-analytical tool, can be used very effectively for obtaining the propagation characteristics of guided modes in planar gradient index waveguides. In general, trial fields can be constructed in two ways. One is to model the trial field on an appropriate single function involving the variational parameters. Such fields are especially useful for the fundamental field in single moded waveguides. By using different choices for the trial function and maximizing β_t^2, we can obtain different models:

(a) Hermite-Gauss model (11, 12, 13)
(b) Secant-hyperbolic model (14)
(c) Cosine-exponential model (15)
(d) Modified Hermite-Gauss and modified secant-hyperbolic models (14, 13)

Models (a) and (b) use simple functions with only one variational parameter. Both of these models assume that the field vanishes in the cover. The Hermite-Gauss functions, which can be integrated analytically, are related to the solutions of the scalar wave equation for infinite parabolic profiles, while the secant-hyperbolic corresponds to $\mathrm{sech}(y/h)$ index profile waveguides where h is the depth defined according to the profile function $(f(y/h))$ which best models the index variation obtained during the fabrication process. The cosine-exponential model assumes a relatively complex function as the mode of an asymmetric three-layer waveguide and consists of three variational parameters; there is no assumption of vanishing field in the cover region. The modified models take into account the evanescent field in the cover region. It is obvious that the main accuracy-limiting factor is the assumption of the vanishing field. The field in the cover is taken to be an exponentially decaying function away from the interface.

The second major method, known as the Rayleigh-Ritz or Galerkin method, is an expansion on a suitable finite set of known basis functions as (in one dimension; Refs. 10 and 16)

$$\Psi(y) = \sum_{J=0}^{N-1} C_j \Phi_j(Y_j) \tag{1.74}$$

The coefficients C_j are the variational parameters. Normalized Hermite-Gauss functions are suitable choices for the basis functions, which are orthonormal solutions of the Helmholtz equation for an infinitely extended parabolic refractive index with a parameter α. A reference profile that closely resembles the true profile can lead to a rapid convergence of results. In addition, the infinite parabolic profile is selected such that it has the same peak index as the given profile and also has the same index as the given graded index profile at the point where the profile function has reduced to half its peak value (17). The normalized Hermite-Gauss functions are given by

$$\Phi_j(\alpha, y) = \sqrt{\frac{\alpha}{\sqrt{\pi}2^j j!}} \exp(-\alpha^2 \frac{y^2}{2}) H_j(\alpha, y) \tag{1.75}$$

and

$$\alpha^2 = k_0 \sqrt{n_S \Delta} / y_{0.5} \tag{1.76}$$

Substituting for the trial field by the form given by Eq. (1.74) with $\Phi_j(\alpha, y)$ defined by Eq. (1.75) into the stationary expression (Eq. 68), for effective index, we get

$$n_e^2 \sum_{m,n} c_m c_n \int \Phi_m \Phi_n \, dy = \left(\frac{1}{k_0^2}\right) \sum_{m,n} c_m c_n \int \Phi_m \frac{d^2 \Phi_n}{dy^2} dy + \sum_{m,n} c_m c_n \int n^2 \Phi_n \Phi_m \, dy \tag{1.77}$$

which can be rewritten after some algebra as

$$n_e^2 c_k = -\left[-(2k+1)\frac{\alpha^2}{k_0^2} + \int_{-\infty}^{+\infty} n^2(y)\Phi_k^2 dy + \int_{-\infty}^{+\infty} \Delta_0 y^2 \Phi_k^2 dy\right] c_k$$
$$+\sum_{n\neq k}\left[\int_{-\infty}^{+\infty} n^2(y)\Phi_k\Phi_n dy + \int_{-\infty}^{+\infty} \Delta_0 y^2 \Phi_k\Phi_n dy\right] c_n \Big/ k = 0, 1, 2 \ldots (N-1) \tag{1.78}$$

Equation (1.77) was obtained by substituting for $\frac{d^2\Phi_n}{dy^2}$ from the scalar wave equation for the infinite parabolic index profile and by using the orthonormality condition and the stationary condition $\frac{\partial n_e^2}{\partial c_k} = 0$. The above equations for an asymmetric index profile reduce to the matrix eigenvalue equation

$$A_{NxN} C_{Nx1} = n_e^2 C_{Nx1} \tag{1.79}$$

The various eigenvalues of A correspond to the effective indices and the corresponding eigenvectors are the field coefficients c_{jm} where m indicates the mode number. The matrix eigenvalue problem gives the eigenvalues, some of which may correspond to guided modes and others to a discrete representation of radiation modes. Sharma et al. (18, 19) developed a method called VOPT or the optimal variational method. This method does not require any *a priori* assumption of the form of the field and the field is generated numerically by the method itself. Consider Eq. (1.67) for the effective index. We can write equivalently a stationary expression in terms of the propagation constant β

$$\beta_t^2 = \iint k_0^2 n^2(x, y)\,|\Psi_t(x, y)|^2\, dxdy - \iint |\nabla_t|^2\, dxdy \tag{1.80}$$

As noted, Ψ_t is the trial field (*vide supra*) and the trial field is an approximation to the modal field. An important property is that all values of β_t will be smaller than the real or exact value of β. Thus higher values of β_t will be closer to the exact value.

In using this expression, we assume that the analytical trial functions assume the separability of the field in the x and y directions, i.e., $\Psi_t(x, y) = X(x)Y(y)$. Different methods differ in their ansatz for X(x) and Y(y). Under the assumption of separability of $\Psi(x, y)$, the stationary expression becomes

$$\beta_t^2 = \iint k_0^2 n^2(x, y)\ |X(x)|^2\ |Y(y)|^2\, dxdy - \int \left|\frac{dX}{dx}\right|^2 dx - \int \left|\frac{dY}{dy}\right|^2 dy \tag{1.81}$$

with the normalization conditions $\int |X|^2\, dx = \int |Y|^2\, dy = 1$. The key technique in this method is to introduce a planar index distribution $n_x^2(x)$ [or equivalently, $n_y^2(y)$] in the stationary expression to get

$$\beta_t^2 = \int k_0^2 n_x^2(x)\,|X(x)|^2\, dx - \int \left|\tfrac{dX}{dx}\right|^2 dx + \int k_0^2\,|Y(y)|^2$$
$$\cdot\left[\int \left\{n^2(x, y) - n_x^2(x)\right\} |X(x)|^2\, dx\right] dy - \int \left|\tfrac{dY}{dt}\right|^2 dy \tag{1.82}$$

The first term in the right-hand side of Eq. (1.82) is the variational expression for the planar index profile $n_x^2(x)$and is equal to the propagation constant β_x^2 of its mode. The value can be computed numerically using standard techniques and the corresponding modal field can be obtained and normalized. The second term is also in the form of the variational expression for a planar index distribution $n_y^2(y)$, which is defined as:

$$n_y^2(y) = \int \{n^2(x, y) - n^2(x)\} \, |X(x)|^2 \, dx \tag{1.83}$$

which can be evaluated using $n_x^2(x)$ and X(x) of the first term. The maximum value of the second term is β_y^2 which is the propagation constant of the waveguide defined by $n_y^2(y)$. The corresponding modal field Y(y) can be numerically obtained. We now use this value of $n_y^2(y)$to rewrite the stationary expression as

$$\begin{aligned} \beta_t = & \left[\int k_0^2 n_y^2(y) \, |Y(y)|^2 \, dy - \int \left|\frac{dY}{dy}\right|^2 dy\right] + \int k_0^2 \, |X(x)|^2 \\ & \cdot \left[\int \{n^2(x, y) - n^2(y)\} \, |Y(y)|^2 \, dy\right] - \int \left|\frac{dX}{dx}\right|^2 dx \end{aligned} \tag{1.84}$$

As before, in the above expression the maximum value of the first term is the β_y^2 of the planar waveguide mode corresponding to $n_y^2(y)$, the index distribution. The second term of the right-hand side of the equation is again a variational expression for the index profile $n_x^2(x)$, which is now defined as

$$n_x^2(x) = \int \{n^2(x, y) - n_y^2(y)\} \, |Y(y)|^2 dy \tag{1.85}$$

where $n_y^2(y)$ and $Y(y)$ are obtained from the first term. This $n_x^2(x)$ expression has a maximum value $\beta_x^2(x)$ where β_x is the propagation constant of the mode of a waveguide with profile $n_x^2(x)$ as defined by Eq. (1.85). This is one iterative cycle.

To summarize, starting from an arbitrary $n_x^2(x)$, a new $n_x^2(x)$ is generated through a variational expression for the given index profile $n^2(x, y)$. This is now the starting point for the next iterative cycle. The quantity $\beta_x^2 + \beta_y^2$ gives the propagation constant of the mode of the given waveguide $n^2(x, y)$. This approach is possible only because of the separability of the function. After each iteration, convergence should be checked. Further numerical details of this technique are beyond the scope of this chapter and the reader is referred to the literature for further details.

Early research on the variational methods for rectangular waveguides was based on cosine/sine and exponential functions (e.g., Ref. 20) as well as an iterative method (CEVAR, Ref. 21). The weighted index method (22, 23) is exactly equivalent to the CEVAR method and the CEVAR when applied to general channel waveguides is the VOPT method. A variant of the effective index method with built-in perturbation correction was developed by Chiang et al. (24, 25). This method has been shown to give much more improved accuracy and is dependent on finding an effective planar waveguide (26) [Appendix B describes this and Marcatali's method; 27] such that the perturbation correction to the given rectangular waveguide on this effective waveguide is zero. Sharma (28) has shown that this method is exactly the same as the VOPT method if the iterations are stopped after one cycle.

The relationship between the variational method and the perturbation method is discussed in Appendix C. Good reviews of various numerical and approximate methods can be found in Ref. 29 and in Koshiba's book (1.30) dealing with finite element methods.

1.7 NUMERICAL METHODS FOR SCALAR WAVE PROPAGATION

Even though the (semi-) analytical and numerical techniques described in previous sections work well for various uniform optical waveguides, these methods are not reliable when there are non-uniform structures such as bends, tapers, etc., in the propagation direction. Various techniques have been developed for the modeling of optical waveguides that treat the whole field including the guided and radiation modes. For most practical waveguide structures, the relative variation in refractive index is small enough to allow the scalar wave approximation. Under this assumption, the wave equation can be written in the form of the Helmholtz equation, the solution of which represents one of the transverse components of the electric or magnetic field. These so-called beam propagation methods (BPMs) include the fast Fourier transform (FFT-BPM) (33, 34), finite difference BPM, (35, 36, 37), finite element BPM (37, 38), collocation method (39, 40, 41), etc. In addition, beam propagation CAD software is widely available in the market.

In this section, we will briefly outline the basic ideas of BPM. We use the scalar wave approximation, namely, the relative variation of the refractive index distribution if sufficiently small. In this case, the wave equation simply becomes [substituting $E(x, y, z)$ for $\psi(x, y, z)$ as the Cartesian component of the electric field]

$$\frac{\partial^2 E}{\partial x^2} + \frac{\partial^2 E}{\partial y^2} + \frac{\partial^2 E}{\partial^2 z} + k_0^2 n^2 E(x, y, z) = 0 \tag{1.86}$$

This is a parabolic partial differential equation for the electric field E. It represents a boundary value problem for the x and y variables in that continuity and boundary conditions must be satisfied by E, $\frac{\partial E}{\partial x}$, $\frac{\partial E}{\partial y}$ for all values of Z. All propagation methods proceed from a given $E(x, y, z = 0)$ at $z = 0$ to generate $E(x, y, z)$ successively in steps of ΔZ. The accuracy of the propagation depends upon the step size ΔZ chosen. In addition, all methods use some scheme of discretization in the transverse cross-section (x, y) also to represent the continuously varying field $E(x, y, z)$ and the refractive index $n^2(x, y, z)$.

Consider the refractive index variation to be given by

$$n^2(x, y, z) = n_0^2 + \Delta n^2(x, y, z) \tag{1.87}$$

where n_0 is a uniform index and $\Delta n^2 << n_0^2$. The wave equation can be written as

$$\frac{\partial^2 E}{\partial z^2} + \frac{\partial^2 E}{\partial x^2} + \frac{\partial^2 E}{\partial y^2} + k_0^2 n_0^2 E + k_0^2 \Delta n^2 E = 0 \tag{1.88}$$

The time dependence of this field can be assumed to be $exp(i\omega t)$ and $k_0 = \omega/c$ is the free spaced wave number. As usual, writing the field as

$$E(x, y, z) = \mathrm{E}(x, y, z)e^{-ik_o n_0 z} \tag{1.89}$$

The equation for the envelope $E(x, y, z)$ becomes

$$\frac{\partial^2 E}{\partial z^2} + \frac{\partial^2 E}{\partial x^2} + \frac{\partial^2 E}{\partial y^2} - 2ik_0 n_0 \frac{\partial E}{\partial z} + k_0^2 \Delta n^2(x, y, z)E(x, y, z,) = 0 \tag{1.90}$$

In general, $\mathbf{E}(x, y, z)$ is a slowly varying function of z over the scale of one wavelength, and we can neglect the term $\frac{\partial^2 E}{\partial z^2}$ by setting it equal to 0. This is known as the paraxial or Fresnel approximation. In the paraxial approximation

$$\frac{\partial E}{\partial z} = \frac{1}{2ik_0 n_0}\left[\frac{\partial^2}{\partial x^2} + \frac{\partial^2}{\partial y^2} + k_0 \Delta n^2(x, y, z)\right] E(x, y, z) = H(x, y; z)\, E(x, y, z) \tag{1.91}$$

where

$$H(x, y, z) \equiv \frac{\left[\frac{\partial^2}{\partial x^2} + \frac{\partial^2}{\partial y^2} + k_0 \Delta n^2(x, y, z)\right]}{2ik_0 n_0} \tag{1.92}$$

is an operator defining the wave propagation and the effect due to $\Delta n^2(x, y, z)$, we have reduced the problem to one of an operator initial value problem

$$\frac{\partial E}{\partial z} = H(z)E(z) \tag{1.93}$$

In the above equation, the (x, y) dependence is implicit. Let us rewrite the operator H in terms of two different operators, H_1 and H_2

$$H(x, y, z) = H_1 + H_2$$

where

$$H_1 = \frac{1}{2ik_0 n_0}\left[\frac{\partial^2}{\partial x^2} + \frac{\partial^2}{\partial y^2}\right] \tag{1.94a}$$

and

$$H_2 = \frac{1}{2ik_0 n_0}\left[k_0 \Delta n^2(x, y, z)\right] \tag{1.94b}$$

The operator H_1 represents propagation of the beam in a uniform medium of index n_0, while H_2 represents the effect on propagation due to Δn. Appendix D describes operator algebra related to the analysis presented here.

Symmetrized splitting of operators in the above expression for $\mathbf{E}$ gives

$$E(x, y, z + \Delta z) = e^{\frac{H_1}{2}\Delta z} e^{H_2 \Delta z} e^{\frac{H_1}{2}\Delta z} E(x, y, z) \tag{1.95}$$

Here we have neglected terms involving higher powers of Δz and the expression can be rewritten in terms of new operators P and Q as

$$E(x, y, z) = PQPE(x, y, z) \tag{1.96}$$

where $P \equiv e^{H_1 \Delta z/2}$ represents propagation of the field in a medium with index n_0 over an interval $\frac{\Delta z}{2}$ and $Q \equiv e^{H_2 \Delta z}$ represents a phase change due to the variation in n over the distance Δz. This can be interpreted as nothing but a lens action. This process means that a single propagation step of length Δz represented by the above equation involves propagation over a distance from Z to $\frac{\Delta z}{2}$ in a medium of uniform index n_0, followed by a lens action of the index variation over Δz and finally another propagation in a uniform medium n_0 from $Z + \frac{\Delta z}{2}$ to $Z + \Delta z$. Such successive steps propagate the beam from $Z = 0$ to $Z = Z$ final, the desired distance.

In the fast Fourier transform-based BPM, the propagation in the uniform medium is carried out by decomposition of the field in terms of plane waves that are propagated over the propagation step and finally these are superposed to get the propagated field. The expansion and superposition are done through the fast Fourier transform. If $E(z) = e^{-ikx}$, then it is very easy to show that

$$E_1\left(z + \frac{\Delta z}{2}\right) \equiv e^{\frac{H_1}{2}\Delta z}\varepsilon(z) \equiv e^{-i\frac{\Delta}{2}\frac{k^2}{k_0 n_0 + (k_0^2 n_0^2 - k^2)^{\frac{1}{2}}}} \tag{1.97a}$$

and the effect of the phase change is a simple multiplication

$$E_2 \equiv e^{H_2 \Delta z} E(z) = e^{\frac{-i\Delta z k_0 \Delta n^2}{2n_0}} E(z) \tag{1.97b}$$

Even though FFT-BPM has been widely applied, it suffers from a number of disadvantages, all related to the nature of the FFT:

1. The discretizations must be uniform, but cannot be very small.
2. They are inadequate for large index differences.
3. The propagation steps must be small.
4. The number of sampling points must be a power of 2.
5. A long computational time is required.
6. Polarization cannot be treated.

The previous BPM technique used the Fresnel or paraxial approximation. More recently, Sharma et al. (42, 43) developed a methodology wherein this is not involved and the second order wave equation is solved directly. Briefly, we look for solutions of the two-dimensional Helmholtz equation $E(x, z)$ as

$$E(x, z) = \sum_{n=1}^{N} c_n(z)\Phi_n(x) \tag{1.98}$$

where $c_n(z)$ are expansion coefficients and $\Phi_n(x)$ are orthogonal functions, n is the order of the basis functions and N is the number of basis functions used in the expansion. The choice of $\Phi_n(x)$ depends on boundary conditions and the symmetry of the guide. The literature has reported both Hermite-Gauss functions and sinusoidal functions. The coefficients of expansion are unknown and physically represent the z variation of the field. The collocation method requires that the Helmholtz equation be satisfied *exactly* by the series expansion at N collocation points X_j, $j = 1, 2, 3 \ldots N$, which are chosen such that these are the zeros of $\Phi_{n+1}(x)$. Using these conditions, we arrive at the matrix ordinary differential equation (Appendix E) known as the collocation equation

$$\frac{d^2E}{dz^2} + SE(z) = 0 \tag{1.99}$$

where $\mathbf{S} = \mathbf{BA}^{-1} + \mathbf{R}(z) = \mathbf{S}_0 + \mathbf{R}(z)$ and the matrices $\mathbf{E}(z)$ and $\mathbf{R}(z)$ are given by

$$\mathbf{E}(z) = \begin{pmatrix} E(x_1, z) \\ E(x_2, z) \\ E(x_3, z) \\ \cdot \\ \cdot \\ \cdot \\ E(x_N, z) \end{pmatrix}$$

$$\mathbf{R}(z) = k_0^2\, diag\left[n^2(x_1, z), n^2(x_2, z) \ldots n^2(x_N, z)\right]$$

Matrices A and B are constant matrices, depending only on the choice of orthogonal functions and the value of N.

It should be noted that the collocation equation is exactly equivalent to the Helmholtz equation as $N \to \infty$. The collocation equation can be solved directly or the paraxial equation can be used to get the equation for the envelope directly, in which case a simple first order matrix differential equation is obtained and can be solved directly. The reader is referred to the literature for details on numerical analysis, stability, etc. The accuracy of the collocation method increases as N is increased. It is straightforward to obtain propagation constants and modal fields, and it is possible to take advantage of any symmetries present in the waveguiding structure. For example, a three-dimensional fiber exhibiting circular symmetry can be treated as a two-dimensional problem.

We have discussed in this section beam propagation methods based on the paraxial approximation. Hadley (44; see also 45) introduced the wide-angle beam propagation method based on Padé approximant operators. In addition, a technique called the multistep method, also developed by Hadley (46), has proven useful for treating non-Fresnel regime propagation. The splitting of the propagation operator has been successfully applied to wide-angle problems, and has been implemented in both the collocation and FD-BPM.

1.8 SUMMARY

Considerable progress has been made in modeling and simulation, both analytical and numerical, of various waveguides and photonic devices. In this chapter, particular emphasis has been placed on planar waveguides. This is because dielectric optical waveguides in planar geometry are the building blocks of integrated optical devices. The realization of planar components such as gratings, etc. requires a useful description of the modal fields. Additionally, since planar waveguides are relatively easy to analyze, their analysis gives us insight into the physics of the basic guidance phenomena.

It should be noted that all approximate methods for analyzing channel waveguides are based on analyzing suitable planar waveguides. Keeping this in mind, we have briefly described the matrix method, the modified Airy function method, variational and perturbation methods, as well as the Beam propagation methods. Newer techniques such as the dynamic programming technique (47–49) and the decomposition method (50) have not been included due to space constraints. The reader is referred to the literature for further details. With the ever-increasing computer power and consequent reduction in computational time, these semi-analytical or numerical techniques lend themselves to ever more realistic applications to be exploited by the optical scientist. To paraphrase the physicist Richard Feynmann, there is plenty of room at the bottom, where the modeling and simulation are done before the fabrication stage of optoelectronic devices such as waveguides.

Appendix A

Algorithm for Variational Method

1. Set up a trial field $\Psi_t(x, y; P_1, P_2, P_3 \ldots P_n)$ where P_i are parameters which are adjustable. Ψ_t is suitably normalized.
2. The functional dependence of Ψ_t on (x, y) is chosen such that it resembles the actual modal field as much as possible.
3. This field is then used in the stationary expression (Eq. 67) and n_{et}^2 (or β_t^2) is *maximized* with respect to the parameters P_i.
4. The maximum value of n_{et}^2 (or β_t^2) is the estimate for the effective index (propagation constant) and the function $\Psi_t(x, y; P_1, P_2, P_3 \ldots P_n)$ with optimized values for P_i is the approximation for the modal field.
5. In general, increasing the number of parameters increases the accuracy of the trial field. However, due to computational complexity constraints as well as for modeling simplicity a judicious choice of a trial field with fewer numbers of parameters is always the better choice.

Appendix B

Effective Index Method

The effective index method (26) is a simple analytic method. For real life optical waveguides, the analytical methods are less accurate than numerical methods, but they are easier to use and clarify the physics. Another analytic technique, Marcatili's method (27), is also briefly described in this appendix.

In general, consider the scalar wave equation

$$\frac{\partial^2 \Psi(x,y)}{\partial x^2} + \frac{\partial^2 \Psi(x,y)}{\partial y^2} + k_0^2(\varepsilon_r(x,y) - n_{eff}^2)\Psi(x,y) = 0 \tag{B.1}$$

The symbols have the usual meaning; ε_r is the relative permittivity of the material, k_0 is the wave number in vacuum, and n_{eff} is the effective index to be found. The main assumption made is $\Psi(x,y) = f(x)g(y)$. We assume that the variables x and y are independent. Substituting for $\Psi(x,y)$ in Eq. B.1 and dividing the results by the wave function $\Psi(x,y)$, we get

$$\frac{1}{f(x)}\frac{d^2 f(x)}{dx^2} + \frac{1}{g(y)}\frac{d^2 g(y)}{dy^2} + k_0^2(\varepsilon_r(x,y) - n_{eff}^2) = 0 \tag{B.2}$$

Setting the sum of the third and second terms in the above equation equal to $k_0^2 N^2(x)$,

$$\frac{1}{g(y)}\frac{d^2 g(y)}{dy^2} + k_0^2\varepsilon_r(x,y) = k_0^2 N^2(x)$$

This implies that the sum of the fourth and first terms is equal to $-k_0^2 N^2(x)$:

$$\frac{1}{f(x)}\frac{d^2 f(x)}{dx^2} - k_0^2 n_{eff}^2 = -k_0^2 N^2(x)$$

Therefore, we get two independent equations:

$$\frac{d^2 g(y)}{dy^2} + k_0^2[\varepsilon_r(x,y) - N^2(x)]g(y) = 0 \tag{B.3}$$

and

$$\frac{d^2 f(x)}{dx^2} + k_0^2\left[N^2(x) - n_{eff}^2\right] f(x) = 0 \tag{B.4}$$

The effective index calculation procedure can be summed up as:

1. Replace the two-dimensional optical waveguide with a combination of two one-dimensional optical waveguides.
2. For each one-dimensional waveguide, calculate the effective index along the y axis.
3. Model the waveguide by using the effective index calculated in Step (2) along the x axis.
4. Obtain the effective index by solving the model in Step 3 along the x axis.

In the Marcatili method also, the main assumption is that the wave function can be separated into x and y. The functions $f(x)$ and $g(y)$ are separated into x and y. The functions $f(x)$ and $g(y)$ are the symmetric modes of the slab/rectangular waveguide, which are independent. If the refractive index profile is given as (horizontal length $2a$, vertical height $2b$)

$$\begin{aligned} n^2(x, y) &= n_1^2 \quad |x| < a, |y| < b \\ & n_2^2 \quad \text{otherwise} \end{aligned}$$

then the field $\Psi(x, y) = f(x)g(y)$ is the modal field of the profile

$$\tilde{n}^2(x, y) = n_x^2(x) + n_y^2(y) - n_1^2 \tag{B.5}$$

where

$$\begin{aligned} n_x^2(x) &= n_1^2 \; |x| < a \\ & n_2^2 \; |x| > a \\ n_y^2(y) &= n_1^2 \; |y| < b \\ & n_2^2 \; |y| > b \end{aligned}$$

If β_x^2 and β_y^2 are the propagation constants obtained for the x and y slabs, then the propagation constant for the $\tilde{n}^2(x, y)$ index profile is

$$\tilde{\beta}^2 = \beta_x^2 + \beta_y^2 - k_0^2 n_1^2 \tag{B.6}$$

In the corner regions, the index is given by $2n_2^2 - n_1^2$; if $n_1 - n_2 << n_1$, then the corner regions have index approximately equal to n_2^2. The reader is referred to the literature for further details (27).

Appendix C
Relationship of Variational and Perturbation Methods

The formalism for analyzing the problem of scalar modes in a waveguide is similar to the problem of quantum mechanical states for a given potential well and there is a close connection between optics and mechanics (30, 31). Using the Bra and Ket notation, we can write the scalar wave equation as

$$H\,|\Phi\rangle = E\,|\Phi\rangle \tag{C.1}$$

where H is a Hamiltonian operator $H = \nabla^2 + k_0^2 n^2(x, y)$, E is the eigenvalue β^2, the propagation constant and $|\Phi\rangle$ represents the eigenfunction corresponding to the modal field and is normalized $\langle\Phi \mid \Phi\rangle = 1$. In the variational method, the Rayleigh quotient $\langle\Phi\,|H|\,\Phi\rangle$ is used to get an estimate for the eigenvalue. Consider a trial function $|\Phi_t(p)\rangle$ where p are some parameters that can be adjusted.

$$E_t(p) = \langle\Phi_t(p)\,|H|\,\Phi_t(p)\rangle \tag{C.2}$$

The "best" estimate of E_t is given by

$$\frac{\partial E_t}{\partial p} = 0 \tag{C.3}$$

In the perturbation method, a trial Hamiltonian H_0 is chosen, which is close to H and is exactly solvable. If this is so, then the eigenvalues E_0 and $|\Phi_0\rangle$, the fields corresponding to this are known exactly

$$E_0 = \langle\Phi_0\,|H_0|\,\Phi_0\rangle$$

In the first order perturbation method (see, for example, any standard book on quantum mechanics), the approximate eigenvalue of the Hamiltonian H is given by

$$\begin{aligned} E_t &= E_0 + \langle\Phi_0\,|H - H_0|\,\Phi_0\rangle \\ &= \langle\Phi_0\,|H_0|\,\Phi_0\rangle + \langle\Phi_0\,|H - H_0|\,\Phi_0\rangle \\ &= \langle\Phi_0\,|H|\,\Phi_0\rangle \end{aligned} \tag{C.4}$$

This shows that the first order perturbed eigenvalue is the same as the Rayleigh quotient with the unperturbed field as trial field.

In the perturbation method with built-in correction, we seek an unperturbed eigenvalue problem in which the first order correction is zero. Let $H_t(p)$ be a Hamiltonian with p being a set of adjustable parameters. Let the corresponding unperturbed field be $|\Phi_t(p)\rangle$. The perturbed eigenvalue for the Hamiltonian is

$$E_t(p) = \langle \Phi_t(p)\,|H|\,\Phi_t(p)\rangle$$

Now expand $E_t(p)$ into a Taylor series around $p = p_0$ (where the first order correction to the eigenvalue is zero)

$$E_t(p = p_0 + \Delta p) = E(p_0) + \Delta p \left.\frac{\partial E_t}{\partial p}\right|_{p=p_0} \tag{C.5}$$

It is recognized that the second term on the right-hand side of Eq. (C.5) is nothing but the first order perturbation correction to the eigenvalue and is a function of p_0. The condition that the first order correction must vanish is

$$\left.\frac{\partial E_t}{\partial p}\right|_{p=p_0} = 0 \tag{C.6}$$

which gives the required p_0 and the eigenvalue estimate $E_t(p_0)$. Equation C.6 is the same as Eq. C.3.

Appendix D

BPM and Operator Algebra

Consider an operator equation of the form:

$$\frac{dx}{dz} = H(z)X(z) \tag{D.1}$$

If $H(z)$ is constant, the solution of Eq. D.1 is straightforward:

$$X(z) = e^{Hz}x(0)$$

However, if $H(z)$ is z-dependent and is a slowly varying function of z such that it can be taken as a constant over a small length Δz, then the solution can be written as

$$X(z + \Delta z) = e^{H(z)\Delta z}x(z) \tag{D.2}$$

If the $H(z)$ value in the exponential is taken to be at $z + \frac{\Delta z}{z}$, then the error is of the order of $(\Delta z)^3$. In general, the operator $H(z)$ can be written as the sum of two operators $H(z) = H_1(z) + H_2(z)$ where H_1 and H_2 do not commute. For example, using H_1 and H_2 given in BPM (Eq. 1.94), it is convenient and/or necessary to split the exponentials

$$e^{(H_1+H_2)\Delta z} \neq e^{H_1\Delta z} + e^{H_2\Delta z} \tag{D.3}$$

since $[H_1, H_2] = 0$. However, we can write

$$\begin{aligned} H &= \tfrac{H_1}{2} + H_2 + \tfrac{H_1}{2} \equiv H_1 + H_2 \\ e^{(H_1+H_2)\Delta z} &= e^{(\frac{H_1}{2}+H_2+\frac{H_1}{2})\Delta z} \\ &= e^{\frac{H_1}{2}\Delta z}\, e^{H_2\Delta z}\, e^{\frac{H_1}{2}\Delta z} + \mathrm{O}((\Delta z)^3) \end{aligned} \tag{D.4}$$

This is known as symmetrized splitting of operators. It is possible to split into several more operators, but this will increase the computational complexity.

Appendix E

Collocation Equation

The derivation of the collocation equation is achieved by working the Helmholtz equation at each of the collocation points $X_j,\ j = 1, 2, 3 \ldots N$.

$$\left.\frac{\partial^2\Psi}{\partial x^2}\right|_{x=x_j} + \frac{\partial^2\Psi_j}{\partial z^2} + k_0 n^2(x, z)\Psi_j(z) = 0$$

$$\Psi_j(z) = \Psi(x = x_j, z) \tag{E.1}$$

or in matrix form

$$\frac{d^2\Psi}{dz^2} + D + R\Psi = 0 \tag{E.2}$$

where

$$\begin{aligned}
\Psi(z) &= \text{col.}\,[\Psi(x_1, z), \Psi(x_2, z) \ldots \Psi(x_n, z)] \\
R(z) &= k_0^2 \,\text{diag}\left[n^2(x_1, z), n^2(x_2, z) \ldots n^2(x_n, z)\right] \\
D(z) &= \text{col.}\left[\left.\frac{\partial^2\Psi}{\partial x^2}\right|_{x=x_1} \left.\frac{\partial^2\Psi}{\partial x^2}\right|_{x=x_2} \cdots \left.\frac{\partial^2\Psi}{\partial x^2}\right|_{x=x_N}\right]
\end{aligned}$$

Now, write the expansion at the collocation points as

$$\begin{aligned}
\Psi(x_j, z) &= \sum_{n-1}^{N} C_n(z)\Phi_n(x_j) \\
\Psi &= AC(z) \\
C(z) &= \text{col}(C_1(z), C_2(z) \ldots C_N(z)) \\
A &= \begin{pmatrix} \Phi_1(x_1) & \Phi_2(x_1) & \cdots & \Phi_N(x_1) \\ \Phi_1(x_2) & \Phi_2(x_2) & & \Phi_N(x_2) \\ \vdots & & & \\ \Phi_1(x_N) & \Phi_2(x_N) & & \Phi_N(x_N) \end{pmatrix}
\end{aligned} \tag{E.3}$$

We differentiate the series expansion twice with respect to x and write the resulting expansion at each of the collection points; we get

$$D(z) = BC(z)$$

where

$$B = \begin{pmatrix} \left.\frac{\partial^2 \Phi_1}{\partial x^2}\right|_{x=x_1} & \left.\frac{\partial^2 \Phi_2}{\partial x^2}\right|_{x=x_1} & \cdots & \left.\frac{\partial^2 \Phi_N}{\partial x^2}\right|_{x=x_1} \\ \left.\frac{\partial^2 \Phi_1}{\partial x^2}\right|_{x=x_2} & \left.\frac{\partial^2 \Phi_2}{\partial x^2}\right|_{x=x_2} & & \left.\frac{\partial^2 \Phi_N}{\partial x^2}\right|_{x=x_2} \\ \vdots & & & \vdots \\ \left.\frac{\partial^2 \Phi_1}{\partial x^2}\right|_{x=x_N} & \left.\frac{\partial^2 \Phi_2}{\partial x^2}\right|_{x=x_2} & & \left.\frac{\partial^2 \Phi_N}{\partial x^2}\right|_{x=x_N} \end{pmatrix} \tag{E.4}$$

We have from Eq. (E.2)

$$C = A^{-1}\Psi(z)$$

or

$$D = BA^{-1}\Psi(z) \tag{E.5}$$

and Eq. (E.2) becomes

$$\frac{\partial^2 \Psi}{\partial z^2} + S\Psi(z) = 0$$

where

$$S = BA^{-1} + R(z) = S_0 + R(z) \tag{E.6}$$

which is the collocation equation.

If we assume that the field E varies very rapidly with z and the refractive index changes slowly with z, we can write

$$E(z) = E(z)\exp(-ikz) \tag{E.7}$$

where $E(z)$ is the envelope and $k = k_0 n_0$. The envelope of the field satisfies

$$\frac{\partial^2 E}{\partial z^2} - 2ik\frac{dE}{dz} + (S - k^2 I)E(z) = 0 \tag{E.8}$$

Since $E(z)$ is a slowly varying function of z and the second derivative can be neglected, we get in the Fresnel approximation

$$\left(\frac{d}{dz}\right) E = (S - k^2 I)E(z)/2ik \tag{E.9}$$

REFERENCES

1. A.K. Ghatak and K. Thyagarajan, *Introduction to Fiber Optics*, Cambridge University Press, Cambridge, U.K. (1998).
2. R.E. Langer, On the Asymptotic Solutions of Ordinary Differential Equations, with an Application to the Bessel Functions of Large Order, *Trans. Am. Math. Soc. 33*, 23 (1931).
3. A.K. Ghatak, R.L. Gallawa, and I.C. Goyal, Modified Airy Function and WKB Solutions to the Wave Equation, National Institute of Standards & Technology Monograph 176, U.S. Government Printing Office, Washington, D.C. (1991).
4. I.C. Goyal, R. L. Gallawa, and A.K. Ghatak, Approximate Solution to the Wave Equation, Revised, *J. Electromagnetic Waves Appl. Opt. 5*, 623 (1991).
5. I.C. Goyal, R.L. Gallawa, and A.K. Ghatak, Approximate Solution to the Scalar Wave Equation for Planar Optical Waveguides, *Appl. Opt. 30*, 2985 (1991).
6. I.C. Goyal, R.L. Gallawa, and A.K. Ghatak, Method of Analyzing Planar Optical Waveguides, *Opt. Lett. 16*, 30 (1991).
7. E.M. Conwell, Modes in Optical Waveguides Formed by Diffusion, *Appl. Phys. Lett. 23*,328 (1973).
8. A. Snyder and J. Love, *Optical Waveguide Theory,* Chapman & Hall, London (1987).
9. M.J. Adams, *An Introduction to Optical Waveguides*, Wiley, New York (1981).
10. K. Kawano and T. Kitoh, *An Introduction to Optical Waveguide Analysis: Solving Maxwell's Equations and the Schroedinger Equation*, Wiley, New York (2001).
11. S.K. Korotoky et al., Mode Size and Method for Estimating the Propagation Constant for Single Mode Ti:LiNBO$_3$ Strip Wave Guides, *IEEE J. Quantum Electron. QE-18*, 1196 (1982).
12. A. Sharma and P. Bindal, Analysis of Diffused Planar and Channel Waveguides, *IEEE J. Quantum Electron. QE-29*,150 (1993).
13. A.K. Taneja, S. Srivatsava, and E.K. Sharma, Closed Form Expressions for Propagation Characteristics of Diffused Planar Optical Waveguides, *Microwave Optical Technol. Lett. 5*, 305 (1997).
14. A. Sharma and P. Bindal, Variational Analysis of Diffused Planar and Channel Waveguides and Directional Couplers, *J. Opt. Soc. Am. A11*, 2244 (1994).
15. M. Saini and E.K. Sharma, Equivalent Refractive Index of Multiple Quantum Well Waveguides by Variational Analysis, *Opt. Lett. 20*,2081 (1995).
16. J.P. Meunier, J. Pigeon, and J.N. Massott, A Numerical Technique for the Determination of Propagation Characteristics of Inhomogeneous Planar Optical Waveguides, *Opt. Quantum Electron 15*,77 (1983).
17. E.K. Sharma, S. Srivatsava, and J.P. Meunier, Mode Coupling Analysis for the Design of Refractive Ion Exchange Integrated Optical Components, *IEEE J. Sel. Topics Quantum Electron. 2*, 165 (1996).
18. A. Sharma and P. Bindal, An Accurate Variational Analysis of Single Mode Diffused Channel Waveguides, *Opt. Quantum Electron. 24*, 1359 (1992).
19. A. Sharma and P. Bindal, Solutions of the 2-D Helmholtz Equation for Optical Waveguides: Semi-Analytical and Numerical Variational Approaches, *LAMP Series Report, LAMP/92/2,* International Center for Theoretical Physics, Trieste, Italy (1992).
20. A. Sharma, P.K. Mishra, and A.K. Ghatak, Single Mode Optical Waveguides and Directional Couplers with Rectangular Cross-Section: A Simple and Accurate Method of Analysis, *J. Lightwave Tech. 6*, 1119 (1988).
21. A. Sharma, On Approximate Theories of Single Mode Rectangular Waveguides, *Opt. Quantum Electron. 21*, 517 (1989).

22. F.P. Payne, A New Theory of Rectangular Optical Waveguides, *Opt. Quantum Electron. 14*, 525 (1982).
23. T.M. Benson and P.C. Kendall, Variational Techniques Including Effective and Weighted Index Methods, *Prog. Electromag. Res., 10*, 1 (1995).
24. K.S. Chiang, Analysis of Rectangular Dielectric Waveguides: Effective Index Method with Built-in Perturbation Correction, *Electron. Lett., 28*, 388 (1992).
25. K.S. Chiang, K.M. Lo, and K.S. Kwok, Effective Index Method with Built-in Perturbation Correction for Optical Waveguides, *J. Lightwave Tech. 14*, 223 (1997).
26. G.B. Hocher and W.K. Burns, Mode Dispersion in Diffused Channel Waveguide by the Effective Index Method, *Applied Opt. 16,* 113 (1997).
27. E.A. Marcatili, Dielectric Rectangular Waveguide and Directional Coupler for Integrated Optics, *Bell System Tech. J. 48*, 2071 (1969).
28. A. Sharma, Analysis of Integrated Optical Waveguides: Variational Method and Effective Index Method with Built-in Perturbation Correction, *J. Opt. Soc. Am. A18*, 1383 (2001).
29. K.S. Chiang, Review of Numerical and Approximate Methods for the Modal Analysis of General Optical Dielectric Waveguides, *Opt. Quantum Electron. 26*, S113 (1994).
30. M. Koshiba, *Optical Waveguide Theory by the Finite Element Method*, Kluwer, Dordrecht, Netherlands (1992).
31. R.J. Black and A. Ankiewicz, Fiber Optic Analogies with Mechanics, *Am. J. Phys. 53*, 554 (1985).
32. V. Lakshminarayanan, A.K. Ghatak, and K. Thyagarajan, *Lagrangian Optics,* Kluwer, Boston (2001).
33. L. Thyelen, The Beam Propagation Method: An Analysis of its Applicability, *Opt. Quantum Electron. 15*, 433 (1983).
34. M.D. Feit and J.A. Fleck, Jr., Computation of Mode Eigenfunction in Gradient Index Optical Fiber Waveguides by a Propagating Beam Method, *Appl. Opt., 19*, 1154 (1980).
35. D. Yevick and B. Hermansson, Efficient Beam Propagation Techniques, *IEEE J. Quantum Electron. 26*,109 (1990).
36. Y. Chung and N. Dagle, Assessment of Finite Difference Beam Propagation, *IEEE J. Quantum Electron. 26*,1335 (1990).
37. T. Koch, J. Davies, and D. Wickramasinghe, Finite Element/Finite Difference Propagation Algorithm for Integrated Optical Devices, *Electron. Lett. 25*, 514 (1989).
38. M. Koshiba and Y. Tsuji, Design and Modeling of Microwave Photonic Devices, *Opt. Quant. Electron. 30*, 995 (1998).
39. A. Sharma, Collocation Method for Wave Propagation through Optical Waveguiding Structures, in *Methods for Modeling and Simulation of Guided Wave OptoElectric Devices*, W. P. Huang, Ed., EMW Publishing, Cambridge, MA (1995) pp. 143–198.
40. A. Taneja and A. Sharma, Propagation of Beams through Optical Waveguiding Structures: Comparison of the Beam Propagation Method and the Collocation Method, *J. Opt. Soc. Am. A 10*, 1739 (1993).
41. A. Sharma and A. Taneja, Variable Transformed Collocation Method for Field Propagation through Optical Waveguide Structures, *Opt. Lett. 17*, 804 (1992).
42. A. Sharma and A. Agarwal, Wide Angle and Bidirectional Beam Propagation Using the Collocation Method for the Non-Paraxial Wave Equation, *Opt. Commun. 216*, 41 (2003).
43. A. Sharma, Collocation Method for Numerical Scalar Wave Propagation through Optical Waveguiding Structure, *Asian J. Phys. 12*, 143 (2003).
44. G.R. Hadley, Wide Angle Beam Propagation using Padé Approximant Operators, *Opt. Lett., 17*, 1426 (1992).

45. J. Yamauchi, J. Shibayama, and H. Nakano, Beam Propagation Method Using Padé Approximant Operators, *Trans. IEICE Jpn., J77-C-1*, 490 (1994).
46. G.R. Hadley, A Multi-Step Method for Wide Angle Beam Propagation, *Integrated Photon. Res., 1Tu 15-1*, 387 (1993).
47. M. L. Calvo and V. Lakshminarayanan, Spatial Pulse Characterization in Periodically Segmented Waveguides by Using Dynamic Programming, *Opt. Commun. 169*, 223 (1999).
48. M. L. Calvo and V. Lakshminarayanan, Optimal Design Using Dynamic Programming: Application to Gainguided Segmented Planar Waveguides, in *Optics and OptoElecronics: Theory, Devices and Applications,* Vol. 2, O.P. Nijhawan et al., Eds., Springer-Narosa, New Delhi (1999) pp. 1206–1214.
49. M. L. Calvo and V. Lakshminarayanan, Light Propagation in Optical Waveguides: A Dynamic Programming Approach, *J. Opt. Soc. Am. A14*, 872 (1997).
50. V. Lakshminarayanan and S. Varadharajan, Approximate Solutions to the Wave Equation: The Decomposition Method, *J. Opt. Soc. Am. A15*, 1394 (1998).

2 Birefringent Optical Waveguides

María L. Calvo and Ramón F. Alvarez-Estrada

CONTENTS

2.1 INTRODUCTION

At the macroscopic scale (the one considered throughout this article), a dielectric material is optically isotropic if, at any given spatial location in it, its optical properties are the same for any direction.[1–3] Then, at a given spatial location in that medium, there is only one dielectric permittivity (for a given frequency of light) and, hence, only one refractive index of light. Gases, liquids and amorphous solids constitute relevant optically isotropic dielectric materials (in particular, glass is a remarkably important example of such an amorphous solid). Various general and specific aspects of the propagation and scattering of the electromagnetic field in optically isotropic materials are, by now, well understood and are well documented.[2,4–7]

Several decades ago, the confined propagation of light was achieved experimentally in optically isotropic dielectric fibers (optical fibers), namely, thin solid waveguides with transverse dimensions about tens of wavelengths. Optical fibers, with transverse sizes in the micron range, have had great technological impact. Very interesting physics in that confined propagation lies in the fact that the electromagnetic wave aspects of light have to be taken into account (say, a geometrical–optics formulation, although very useful, does not describe various important features). Thus, light is transmitted along the fiber only in the form of some specific distributions of the electromagnetic field (solving Maxwell's equations), named propagation modes.

Confined propagation of light in optically isotropic fibers has been studied in detail in a number of articles and books.[8–15] Similar and very important phenomena occur in the eye, namely the propagation of light along the retinal photoreceptors, regarded as biological waveguides (the transverse dimensions of which are only a few wavelengths).[16,17] Various aspects of confined propagation of light are also well documented, in an updated way, in different chapters of this book.

Back to optical fibers, it is a fact that the preservation of the polarization state of light, propagating confined along them, is indispensable for their efficient use in coherent optical transmission, for polarization sensitive sensors and for connections with integrated optical waveguide devices.[18] The last fact has stimulated many developments by a large number of researchers, some of which will be summarized succinctly in the present chapter.

On the other hand, an optically anisotropic dielectric material is, by definition, one in which, for a given macroscopically small volume element, the optical properties depend on the chosen direction (also, for a given frequency of light).[1–3] Then, the dielectric permittivity becomes a 3×3 symmetric tensor and there are more than one refractive indices of light (birefringence). This fact implies different and important optical phenomena. The description of the electromagnetic field of light propagating through an optically anisotropic medium is more complicated than that for an isotropic one. Crystals and liquid crystals provide important different classes of optically anisotropic media. Besides crystals and liquid crystals, at least other two important classes of optically anisotropic dielectric materials exist, namely, those due to form birefringence[1] and to the photo-elastic effect (or stress birefringence).[1–3]

The confined propagation of light in optically anisotropic dielectric waveguides is another interesting phenomenon (which, as we shall see, has also become important in the last two decades). It had already attracted attention at some earlier stage.[10,18,19] One motivation for that interest has been the technological importance of preserving, as much as possible, the polarization of the propagation modes subject to various perturbations in optical fibers, as commented above. Thus, it appeared that optically anisotropic dielectric waveguides, due (in particular, but not exclusively, to the photo-elastic effect) could offer some interesting possibilities aimed towards retaining a high degree of polarization, thereby improving optical devices.[18]

On the other hand, by analyzing the scattering of light by a system displaying the photo-elastic effect, one could obtain useful information about that system (say, the stresses in it). Waveguides with induced anisotropy may have additional interest, as they could play a role in biological waveguides (mostly under conditions of pathological functioning), in polarimetric fiber-optic sensors, and in mechanical stress sensors.[20]

We shall comment on those different subjects throughout this chapter. Our emphasis will be essentially methodological. We shall provide some basic descriptions of various systems, in which optical anisotropy arises or could arise. On the other hand, we shall treat certain theoretical foundations of electromagnetic wave scattering by and confined propagation in optically anisotropic systems. Detailed applications of the theoretical formulations to all systems described lie outside our scope. In spite of that, we shall outline some limited applications. We shall refer to the bibliography for specific and additional details.

The contents of the present chapter are the following. Section 2.2 will summarize classical electromagnetic wave propagation in an anisotropic dielectric material. As an introductory example, the propagation of a plane electromagnetic wave in an extended homogeneous anisotropic dielectric medium is discussed briefly in Section 2.2. We overview physically relevant cases in which optical anisotropy occurs in Sections 2.3, 2.4 and 2.5. Section 2.3 outlines it in crystals and liquid crystals. Section 2.4 deals with form birefringence. An account of the photo-elastic effect is given in Section 2.5. Section 2.6 outlines a few aspects of high birefringence in polarimetric optical fibers and sensors. Section 2.7 discusses, rather qualitatively, some aspects of biological waveguides (in the eye) as well as form birefringence in and light scattering by them. Section 2.8 presents a rather general treatment for the scattering of classical electromagnetic radiation by arbitrary three-dimensional optically anisotropic objects, and a multiple-scattering reformulation thereof is outlined in Section 2.9. Section 2.10 deals with scattering by parallel cylindrical dielectric waveguides: one and several waveguides as well as long wavelength approximations are treated. Section 2.11 summarizes recent work on the scattering of light by parallel dielectric waveguides, in which the anisotropy has been generated by the photo-elastic effect, and discusses shortly possible applications to stress sensing. Section 2.12 deals with confined propagation of classical electromagnetic radiation along cylindrical anisotropic dielectric waveguides: in particular, the long wavelength approximation for one waveguide (monomode confined propagation) is analyzed. Section 2.13 contains the conclusions. Various derivations, computations and formulas are collected in several appendixes.

2.2 CLASSICAL ELECTROMAGNETIC WAVES IN ANISOTROPIC DIELECTRIC MEDIA

2.2.1 MAXWELL'S EQUATIONS FOR MEDIUM

We shall follow the same conventions as Born and Wolf.[1] Let us consider an optically anisotropic dielectric material. The latter has zero electrical conductivity, real scalar magnetic permeability (μ_0) equal to that of vacuum, and a dielectric permittivity tensor given as the product: $\epsilon_0\tilde{\epsilon}$. ϵ_0 is the real constant (scalar) dielectric permittivity of some optically isotropic and homogeneous dielectric reference medium (also with magnetic permeability μ_0), to be discussed later. Then, the constant refractive index of that isotropic reference material is $n_0 = (\epsilon_0\mu_0)^{1/2}$. On the other hand, $\tilde{\epsilon}$ is the dielectric permittivity tensor of the anisotropic material relative to (the permittivity of) the isotropic reference one, mentioned above. $\tilde{\epsilon}$ is a symmetric 3×3 matrix with components ϵ_{ij}, $i, j = 1, 2, 3$ ($\epsilon_{ij} = \epsilon_{ji}$). Each ϵ_{ij} could depend on the three-dimensional spatial position ($\bar{x} = (x_1, x_2, x_3)$) and on the frequency ($\omega$) of a propagating electromagnetic field, and could be either real or complex, in principle. In the general study, we shall allow for ϵ_{ij} to be complex and we shall restrict them to be real, at the appropriate places. If the optically anisotropic medium reduces to an isotropic one, then $\tilde{\epsilon}$ becomes proportional to the 3×3 unit matrix I_3.

Let classical monochromatic (complex) electric ($\mathbf{E}' \exp(-i\omega t)$) and magnetic ($\mathbf{H}' \exp(-i\omega t)$) fields, with real frequency ω, propagate in the anisotropic dielectric medium. As ω will be held fixed, we shall not specify any dependence of ϵ_{ij} on it. The time-dependent factor ($\exp(-i\omega t)$) will always be factored out. The t-independent fields $\mathbf{E}' = \mathbf{E}'(\bar{x})$ and $\mathbf{H}' = \mathbf{H}'(\bar{x})$ fulfill Maxwell's equations for that medium, in the absence of charges:[1]

$$\nabla \times \mathbf{E}' = \frac{i\omega}{v}\left(\frac{\mu_0}{\epsilon_0}\right)^{1/2}\mathbf{H}', \nabla \cdot (\tilde{\epsilon}\mathbf{E}') = 0 \tag{2.1}$$

$$\nabla \times \mathbf{H}' = -\frac{i\omega}{v}\left(\frac{\epsilon_0}{\mu_0}\right)^{1/2}\tilde{\epsilon}\mathbf{E}', \nabla \cdot \mathbf{H}' = 0 \tag{2.2}$$

Here, v denotes the velocity of light in the isotropic reference medium: $v = c/n_0$. c will always denote the velocity of light in vacuum (where $n_0 = 1$).

Let $(\tilde{\epsilon})^{-1}$ denote the (3×3) inverse matrix of $\tilde{\epsilon}$ ($(\tilde{\epsilon})^{-1} \cdot \tilde{\epsilon} = I_3$, I_3 being the unit 3×3 matrix). Using $(\tilde{\epsilon})^{-1}$, it is easy to reduce exactly the above system in Eqs. (2.1) and (2.2) involving both $\mathbf{E}'$ and $\mathbf{H}'$ to one system depending only on $\mathbf{H}'$, in which $\mathbf{E}'$ will be eliminated. For that purpose, we perform, successively, the following operations: i) we apply $(\tilde{\epsilon})^{-1}$ to the first equation in Eq. (2.2), ii) we apply $\nabla\times$ to the result of i), iii) in the result of ii), we replace $\nabla \times \mathbf{E}'$ by $(i\omega/v)(\mu_0/\epsilon_0)^{1/2}\mathbf{H}'$. The result is the announced system involving only $\mathbf{H}'$:

$$\nabla \times ((\tilde{\epsilon})^{-1}\nabla \times \mathbf{H}') = \frac{\omega^2}{v^2}\mathbf{H}', \nabla \cdot \mathbf{H}' = 0 \tag{2.3}$$

Clearly, the divergence $(\nabla \cdot)$ of the first equation in Eq. (2.3) is consistent with the second equation in Eq. (2.3). It is also easy to get another system depending only on $\mathbf{E}'$,[10] in which $\mathbf{H}'$ has been eliminated, but we shall not consider it. Our interest in and preference for the first equation in Eq. (2.3) rely upon the fact that $\mathbf{H}'$ is divergenceless, as expressed by the second equation in Eq. (2.3).

Let us now consider two different dielectric media, separated only by a surface Σ (one goes from one medium to the other just by traversing Σ). The boundary conditions to be fulfilled by the electromagnetic field are the following.[1] Across Σ and at any point of it, the following six quantities have to be continuous: i) $\mathbf{H}'$ (its three components), ii) the projection of $\tilde{\epsilon}\mathbf{E}'$ orthogonal to Σ (one quantity), iii) the two tangential components of $\mathbf{E}'$ (namely, the components of the projection of $\mathbf{E}'$ on Σ). In order to simplify the analysis, we shall always suppose that all components of the tensor $\tilde{\epsilon}$ and all their spatial derivatives of first order are continuous inside each dielectric media. At this stage, the behaviors of the components of the tensor $\tilde{\epsilon}$ (and of the derivatives thereof) across the interface Σ separating two different dielectric media have not yet been specified, but we shall have to make suitable assumptions also about them as we proceed. Of course, Eqs. (2.1), (2.2) and (2.3) hold independently on assumptions about $\tilde{\epsilon}$ and their first spatial derivatives across any such interface.

2.2.2 Extended and Homogeneous Anisotropic Medium

As a simple illustration of electromagnetic wave propagation in an anisotropic dielectric medium, we shall suppose that the latter is very extended and has no inhomogeneities, that is, $\tilde{\epsilon}$ is constant ($\bar{x}$-independent) and it is not proportional to I_3. In such a case, we shall choose the three axis x_1, x_2, x_3 of Cartesian coordinates so that they coincide with the principal axis of the symmetric tensor $\tilde{\epsilon}$. That is, with respect to such a coordinate system, $\tilde{\epsilon}$ is represented by a diagonal matrix ($\epsilon_{ij} = 0$, for $i \neq j$). The anisotropic medium is called uniaxial, if two out of the three non-vanishing diagonal elements (eigenvalues) ϵ_{ii} are equal to each other, the third one being different from them. The medium is named biaxial, if all three non-vanishing diagonal values ϵ_{ii} are different from one another.

In such a coordinate system where $\tilde{\epsilon}$ is diagonal, we shall consider the propagation of an electromagnetic plane wave with frequency ω and wavevector $\bar{k} = (k_1, k_2, k_3)$. Such an electromagnetic wave has to be a solution of Eqs. (2.1) and (2.2) (or, equivalently, of Eq. (2.3)) for any $\bar{x}$. Those equations imply an important relationship (the dispersion relation) between $\bar{k}$ and ω. The dispersion relation for an anisotropic dielectric medium is more complicated than (and contains new physical information compared to) that for an isotropic one. The anisotropic dispersion relation is known as the Fresnel equation. We use the known structure of the latter[1] and recast it in terms of $\bar{k}$ and ω. Then, the Fresnel equation reads:

$$\frac{(k_1/\mid\bar{k}\mid)^2}{[v\mid\bar{k}\mid/\omega]^2-\mu_0\epsilon_{11}}+\frac{(k_2/\mid\bar{k}\mid)^2}{[v\mid\bar{k}\mid/\omega]^2-\mu_0\epsilon_{22}}+\frac{(k_3/\mid\bar{k}\mid)^2}{[v\mid\bar{k}\mid/\omega]^2-\mu_0\epsilon_{33}} = \frac{1}{[v\mid\bar{k}\mid/\omega]^2} \quad (2.4)$$

Various derivations of the Fresnel equation[1–3] employ all Eqs. (2.1) and (2.2). An important physical feature is that, by virtue of anisotropy, the dispersion relation (2.4) has two (but not more) different solutions for ω^2, for a given $\bar{k}$ (birefringence). There is an important physical reason for that (namely, that there are only two independent states of polarization for an electromagnetic plane wave): such a motivation has been succinctly stated,[2] and also explained in more algebraic detail[3] in terms of $\tilde{\epsilon}\mathbf{E}'$ (the displacement vector). In Appendix A, we shall outline that physical reason, based upon $\mathbf{H}'$ and Eq. (2.3) (thereby allowing us to put them to work). Appendix A will employ the standard unit polarization vectors, which will be also useful when dealing with scattering by optically anisotropic objects in three dimensions. We shall not stop to use $\mathbf{H}'$ and Eq. (2.3) in order to provide another algebraic derivation of the Fresnel equation. $\mathbf{H}'$ and (2.3) will turn out to be quite useful for general studies about scattering and confined propagation, later in this chapter.

Below, we shall allow, in principle, for the possibility that $\tilde{\epsilon}$ is not only ω-dependent but $\bar{x}$-dependent as well. We shall study phenomena determined by one or several essentially parallel cylindrical anisotropic dielectric waveguides: wherever possible in those cases, it will certainly be advantageous to choose the x_3-axis so that it is parallel to the axis of the cylindrical waveguides. That x_3-axis will, in principle, be different from any of the principal axes of the symmetric tensor $\tilde{\epsilon}$. Then, $\tilde{\epsilon}$ will not be represented by a diagonal matrix for any $\bar{x}$, in general (although it may be, in some special cases).

2.3 OPTICAL ANISOTROPY(I): CRYSTALS AND LIQUID CRYSTALS

Let us consider a crystal, the size of which is not too small. It is formed by a suitably large number of atoms or molecules, the locations of which form a regular pattern (essentially periodic in the spatial region occupied by the crystal): positional order. The atoms and molecules have essentially fixed and well-defined positions and orientations in the crystal. Moreover, the electrical properties of those atoms and molecules are typically anisotropic. Altogether, those facts imply that, in principle, the properties of the crystal are not the same in all directions, namely, anisotropy.[1]

The propagation of electromagnetic waves (and, in particular, of light) in crystals depends, typically but not necessarily, on the direction. That is, crystals may display, but not necessarily, optical anisotropy. In fact, there are crystals in which the relative dielectric permittivity tensor reduces to a scalar (that is, $\tilde{\epsilon}$ becomes proportional, but not equal, to the 3 × 3 unit matrix I_3). Such crystals are those belonging to the so-called cubic system (in which three mutually orthogonal and crystallographically equivalent directions may be chosen). Crystals belonging not to the cubic system, but to either the trigonal or the tetragonal or the hexagonal systems (in all of which, two or more crystallographically equivalent directions may be chosen in one plane) are uniaxial and, so, optically anisotropic. Crystals belonging not to any of the above four systems but to either the orthorhombic or the monoclinic or the triclinic systems (in all of which, no two crystallographically equivalent directions may be chosen) are biaxial and, so, optically anisotropic as well. Quartz and calcite are examples of uniaxial crystals. Aragonite and Brazil topaz provide examples of biaxial crystals.

A crystalline grain can be regarded as a crystal, the size of which is quite small. As the number of atoms forming it is not small, and since their spatial locations resemble those for a proper crystal, as outlined above, a crystalline grain may also display, in general, a similar type of optical anisotropy. On the other hand, what one usually meets and handles easily is a polycrystalline material, which typically has a size much larger than that of a crystalline grain. A polycrystalline material is an aggregate formed by a large number of crystalline grains (disjoint but closely packed among themselves), which are randomly oriented relative to one another. It follows that the optical anisotropy properties of one crystalline grain in the aggregate tend to be compensated for by those of the surrounding ones. Then, the average (macroscopic) optical properties of the polycrystalline material may appear to be approximately isotropic.[1]

A liquid crystal has no positional order, but it displays an orientational one. That medium is formed by molecules, for each of which the structure and properties are not the same for different directions in principle, as commented above. The positions of the different molecules in the liquid are fully random. However, the orientations of those (anisotropic) molecules may turn out to be not completely random. It follows that, by virtue of that orientational order, the medium is anisotropic and, in particular, the liquid crystal turns out to be optically anisotropic.[3]

2.4 OPTICAL ANISOTROPY(II): FORM BIREFRINGENCE

A different and quite interesting case of optical anisotropy is form birefringence,[1] which can be described as follows. The electrical properties of atoms and molecules are typically anisotropic. Let s_{is} be some bound system formed by a sufficiently large number of those (individually anisotropic) atoms and molecules, which has a size naturally large compared to typical molecular dimensions and, on the other hand, small compared to the wavelength of light. We also suppose that s_{is} turns out to be optically isotropic (so that the anisotropies of the molecules constituting s_{is} cancel out with one another, on the average).

Let S_{anis} be another system of larger size, constituted by an ordered arrangement of a certain number of similar bound subsystems, each of the them identical to s_{is}. To fix the ideas, we shall also suppose that the separations among neighboring subsystems of type s_{is} in the assembly S_{anis} are smaller than typical wavelengths of light. Then, it follows that the ordered arrangement S_{anis} appears to display optical anisotropy (in some effective sense, as explained below).

Two explicit idealized models, displaying that anisotropy phenomenon, have been analyzed in certain detail. Several important, and by now classical, results have been obtained by Lord Rayleigh and O. Wiener, and we refer to Born and Wolf[1] for an account of them. Another clear presentation has been made by Harosi.[21] In one model (model p), s_{is} is chosen to be a thin plate, while S_{anis} is a regular assembly of parallel plates, all identical to s_{is}. A key role is played by the axis perpendicular to the planes of the plates (optic axis). In the second model (model r), s_{is} is a thin cylindrical dielectric rod and S_{anis} is an assembly of rods with parallel axes, all similar to s_{is}. Then, an axis parallel to those of the rods is important (optic axis). For either of those two models, let us suppose that a plane electromagnetic wave is incident on S_{anis}. For both models, illumination (the wave vector of the incoming light) is perpendicular to the corresponding optic axis. The wavelength (in the typical optical range) is supposed to be larger than: i) either the thickness of the plates (model p) or the transverse diameter of the rods (model r), and ii) typical separations between two neighboring plates or rods, as commented above.

For both models, let the electric field of the electromagnetic wave be perpendicular to the corresponding optic axis: ordinary (o) wave case. Then, let ϵ_o be the effective dielectric permittivity of the assembly S_{anis} (after having performed suitable averages), for both models. Next, and also for both models, we suppose that the electric field of the electromagnetic wave is parallel to the corresponding optic axis: extraordinary (e) wave case. For both models, let ϵ_e be the (effective) dielectric permittivity of the assembly S_{anis}. We remark, in order to facilitate the comparison with Born and Wolf,[1] that the dielectric permittivities considered in this section are not the relative ones, that is, they equal ϵ_0 multiplied by the corresponding relative dielectric permittivities. It follows that:[1] i) $\epsilon_e - \epsilon_o \neq 0$, ii) then, S_{anis} behaves as a uniaxial optically anisotropic system. Such an anisotropy (on a scale much larger than the molecular one) is named form birefringence. Quantitatively, the latter is characterized by the coefficient $\epsilon_e - \epsilon_o$ or by the equivalent expression that results if dielectric permittivities are replaced by refractive indices (so that those dielectric permittivitties times μ_0 equal the corresponding squared refractive indices).

For model p, let us consider a medium consisting of plates of refractive index n_1 and fractional volume f_1 embedded in a surrounding substance of index n_2 and fractional volume f_2. f_1 and $f_2(= 1 - f_1)$ are the fractions of the total volume occupied by the plates and by the surrounding medium, respectively. Then:

$$n_e^2 - n_o^2 = -\frac{f_1 f_2(n_1^2 - n_2^2)^2}{f_1 n_2^2 + f_2 n_1^2} \tag{2.5}$$

For model r, let us consider a medium consisting of small, parallel, well-separated rods of refractive index n_1 and fractional volume f_1 embedded in a surrounding substance of index n_2 and fractional volume f_2. f_1 and f_2 are, as in the above case of model p, the fractions of the total volume occupied by the rods and by the surrounding medium, respectively. In the actual case (model r), it is assumed that $f_1 \ll 1$. Then, the form birefringence coefficient is approximately:

$$n_e^2 - n_o^2 = \frac{f_1 f_2(n_1^2 - n_2^2)^2}{(1 + f_1)n_2^2 + f_2 n_1^2} \tag{2.6}$$

In this formula, there is no restriction on the refractive index difference between the rods and the surrounding substance.

We stress that in both Eqs. (2.5) and (2.6), n_e is the (extraordinary) refractive index for light linearly polarized parallel to the optic axis. n_o is the (ordinary) refractive index for light linearly polarized perpendicular to the optic axis. See Born and Wolf[1] and references therein. Notice that the right-hand side of Eq. (2.5) is negative ("negative uniaxial crystal") while that of (2.6) is positive ("positive uniaxial crystal").

The phenomenon of form birefringence has some important applications in biological microscopy. For instance, the experimental measurement of that optical anisotropy (that is, of those effective dielectric permittivities) may yield information regarding the form of s_{is}, the fraction of volume occupied by all s_{is} (over that of the surrounding medium in which they are located).

The form birefringence phenomenon has continued to attract further interesting researches. We shall limit ourselves to quote three of them, to which we refer for a wider perspective and, in turn, for further references. The form birefringence of lamellar systems containing three or more components has been studied by Thornburg.[22] The form birefringence of an array of parallel cylinders of arbitrary sizes and separations, under the assumption of small variation among the refractive indices of its components, has been analyzed further by Hemenger:[23] a new expression for the form birefringence coefficient of that assembly has been obtained. The latter expression depends on the correlation function of the fluctuations in refractive index through the medium (those fluctuations being small, by assumption). Such a formula,[23] which applies for cylinders with dimensions comparable to or greater than a typical wavelength of light, generalizes Eq. (2.6). Further interesting investigations of form birefringence for thin fibers and membranes have been carried out by Zhou and Knighton.[24]

2.5 OPTICAL ANISOTROPY(III): PHOTO-ELASTIC EFFECT

Another important class of optically anisotropic dielectric media results when originally (unstressed) isotropic materials are subject to mechanical stresses. The latter could be due to externally applied forces. Internal mechanical deformations are produced in those materials which, in turn, give rise to optical anisotropies. This is frequently referred to as the photo-elastic effect.[3] It is also known as stress birefringence[1] and dynamo-optical effect.[2]

Brewster provided evidence of the photo-elastic effect through a fundamental experiment.[25] See the quite interesting discussion in Bruhat.[26] The photo-elastic effect is documented in some already classical monographs.[27,28]

We shall discuss some general features of stress birefringence.[2] Let us consider a solid medium at rest and at equilibrium, which appears to behave as isotropic (both from the mechanical and optical standpoints), as a zeroth-order approximation. Then, the dielectric permittivity tensor $\tilde{\epsilon}$ of the material relative to that of another (optically isotropic) reference material is: $\epsilon_{ij} = \epsilon_e \delta_{ij}$, $i, j = 1, 2, 3$. ϵ_e is the relative (dimensionless) dielectric permittivity of the solid medium and δ_{ij} denotes the Kronecker delta symbol ($\delta_{ii} = 1, \delta_{ij} = 0$ if $i \neq j$). In the latter description, one neglects possible small local deformations of the material due to internal stresses. In the application of specific interest in this article, the reference material will have dielectric permittivity ϵ_0 and be chosen as the surrounding medium in which the above solid one (in the form of one or several waveguides) will turn out to be located.

A more accurate description of that solid medium requires us to take into account small elastic deformations in it (arising from small elastic stresses, in the domain of reliability of Hooke's law.[29]) Those small deformations are potentially important for the optical behavior of the solid medium since, in principle, they give rise to some (small) optical anisotropy. This is the photo-elastic effect.

In order to outline a quantitative description of such an effect, a few notions of elasticity theory are required,[29] to which we now turn. The deformed solid medium is also supposed to be at rest and at equilibrium, for the small elastic deformations to be considered here. Let us consider, in the absence of deformations in the solid medium, a generic small volume element of it $(dx_1dx_2dx_3)$, located at the generic position $\bar{x} = (x_1, x_2, x_3)$. When the solid medium is subject to small deformations, that small volume element becomes a new small one $(dx_1'dx_2'dx_3')$, located at the new position $\bar{x}' = (x_1', x_2', x_3') = (x_1 + u_1, x_2 + u_2, x_3 + u_3)$. $u_i = u_i(x_1, x_2, x_3)$ is the displacement, which will always be assumed to be small. Let $u_{ik} = u_{ik}(x_1, x_2, x_3)$, $i, k = 1, 2, 3$, be the (small) components of the deformation tensor in the deformed solid medium. They are given in terms of the displacements u_i, since the latter are small, as follows: $u_{ik} \simeq 2^{-1}(\partial u_i/\partial x_k + \partial u_k/\partial x_i)$. To grasp the physical interest of u_{ik}, we shall limit ourselves to recall that the size of the new (deformed) small volume element is given, in terms of the one before deformation, as follows: $dx_1'dx_2'dx_3' \simeq dx_1dx_2dx_3(1 + \sum_{i=1}^{3} u_{ii})$. Notice that $u_{ik} = u_{ki}$ and that all u_{ik} are dimensionless.

The small deformations are caused by small internal elastic stresses. Specifically, small internal elastic forces act upon any given volume element of the deformed solid medium, due to the rest of the material. The physical origin of all those internal elastic forces could be some force applied externally to the solid material under consideration. A reliable physical idea is that the net result of all those internal forces on that volume element acts across its boundary (and also that the external force applied to the solid material acts across its external surface).[29] This idea implies that the net elastic force per unit volume (F_1, F_2, F_3), on a small domain (having volume $dx_1dx_2dx_3$) of the deformed solid medium is given by $F_i = \sum_{k=1}^{3} \partial\sigma_{ik}/\partial x_k$. $\sigma_{ik} = \sigma_{ik}(x_1, x_2, x_3)$ is the stress tensor, and it has dimension *force/(surface area)*. It is symmetric ($\sigma_{ik} = \sigma_{ki}$), under suitable conditions,[29] which are supposed to hold here. A basic approximation that holds for small deformations is a suitable linear relationship between the components of the stress tensor and those of the deformation one and conversely (Hooke's law). Hooke's law, expressed as that linear relationship and its inverse, reads:[29]

$$\sigma_{ik} \simeq \left(\lambda_{La} + \frac{2\mu_{La}}{3}\right)\delta_{ik} + 2\mu_{La}\left[-\frac{1}{3}\delta_{ik}\sum_{l=1}^{3} u_{ll} + u_{ik}\right] \tag{2.7}$$

$$u_{ik} \simeq \frac{1}{9(\lambda_{La} + (2/3)\mu_{La})}\delta_{ik}\sum_{l=1}^{3}\sigma_{ll} + \frac{1}{2\mu_{La}}\left[-\frac{1}{3}\delta_{ik}\sum_{l=1}^{3}\sigma_{ll} + \sigma_{ik}\right] \tag{2.8}$$

λ_{La} and μ_{La} (both positive and with dimension *force*/(*surface area*)) are constants, characteristic of the solid medium. They are known as the Lame coefficients.

We shall now come back to our main subject, namely optics. When the effects of small elastic deformations of the solid medium are taken into account as discussed above, one accepts that the relative dielectric permittivity tensor of the latter medium is:[2]

$$\epsilon_{ij} \simeq \epsilon_e \delta_{ij} + a_1' u_{ij} + a_2' \left(\sum_{l=1}^{3} u_{ll} \right) \delta_{ij} \tag{2.9}$$

a_1' and a_2' are (dimensionless) photo-elastic constants, characteristic of the deformed medium, and u_{ij} is its deformation tensor. It is convenient to recast the above ϵ_{ik} in terms of the stress tensor σ_{ik}, by using Hooke's law (Eq. (2.8)). One finds easily:

$$\epsilon_{ik} \simeq \epsilon_e \delta_{ik} + a_1 \sigma_{ik} + a_2 \left(\sum_{l=1}^{3} \sigma_{ll} \right) \delta_{ik} \tag{2.10}$$

where a_1 and a_2 are elastic-optical constants (with dimension (*surface area*)/(*force*)). They are given, in terms of the above a_1' and a_2', by:

$$a_1 = \frac{a_1'}{2\mu_L}, a_2 = -\frac{\lambda_L a_1'}{2\mu_L(3\lambda_L + 2\mu_L)} + \frac{a_2'}{3\lambda_L + 2\mu_L} \tag{2.11}$$

As an example, we quote the following values of a_1 and a for silica: $a_1 = 0.38 \times 10^{-12}$ $m^2/Newton$ and $a_2 = 2.68 \times 10^{-12}$ $m^2/Newton$. The photo-elastic effect has certain interesting consequences regarding optical wave propagation in waveguides formed by such materials.[20,30,31]

2.6 HIGHLY BIREFRINGENT WAVEGUIDES

We shall review a few interesting aspects of propagation modes in an ideal optically isotropic homogeneous fiber with circular cross-section of radius R and refractive index n_{co}. That waveguide is placed in another homogeneous optically isotropic medium (cladding) with refractive index $n_{cl}(< n_{co})$. Let light, with wavelength λ, propagate confined along the fiber. One introduces the useful dimensionless parameter

$$V = \frac{2\pi R}{\lambda} \left(n_{co}^2 - n_{cl}^2 \right)^{1/2} \tag{2.12}$$

The fundamental (or lowest order) mode is degenerate, that is, there are two lowest order polarization modes (denoted as HE_{11}^x and HE_{11}^y), orthogonal to each other and with the same spatial intensity distribution: they both correspond to $0 < V < 2.40$. The next four modes of first order correspond to $2.40 \leq V < 3.83$: they are denoted as TE_{01}, TM_{01}, HE_{21}^{even} and HE_{21}^{odd}. Suitable linear combinations of these are also currently employed, in order to describe the set of four first order modes, just above the two ones of zeroth order (namely, the fundamental ones). The set of all propagation modes of second order correspond to $3.83 \leq V < 5.13$. For single-mode or monomode optical fibers, only the (degenerate) fundamental mode propagates.

Perturbations (induced by pressure, strain, bend, twist, temperature, . . .) in a real or non-ideal waveguide eventually upset the properties of the propagation modes in it. Let us consider a real (non-ideal) homogeneous optically isotropic homogeneous fiber with circular cross-section, and let us suppose that it corresponds to a monomode waveguide (say, $V < 2.40$). Let light penetrate into it and start to propagate confined along it, in a definite polarization state corresponding to one of the degenerate fundamental modes. Estimates indicate that, due to those perturbations in the real fiber and after propagation through a length about 1 m, the polarization of the initially launched fundamental mode will be modified in an unpredictable way. The control of that polarization state launched into the monomode waveguide turns out to be essential for efficient performance. This has led to an added recognition of the importance of polarization and of its preservation in those single-mode fibers. Single-mode optical waveguides, in which the polarization state of its propagating modes is preserved to a high degree, have acquired an increasing technological importance in the last two decades, in particular through their applications for optical communications, as high-performance polarimetric optical fibers and sensors.

Research on confined propagation in optically anisotropic dielectric waveguides has arisen from attempts to preserve (linear) polarization of the propagation modes in single-mode optical fibers. That preservation has been attempted in two ways. One has been the minimization of the stress-induced birefringence.[32] The other has been to maximize birefringence, which breaks the degeneracy of the fundamental modes. The latter can be accomplished: i) by introducing asymmetric profiles for the refractive index distribution (for instance, through non-circular shapes of the transverse cross-sections of the waveguides),[33,34] and ii) through anisotropic internal stresses.[34,35] Just as an example, we note that a proposal has been reported[18] about confined propagation modes in twisted fibers. The twist gives rise to shearing stresses which, in turn, introduce optical anisotropy: the dielectric permittivity becomes a tensor (photo-elastic effect). For a sufficiently large twist, the propagation modes in the anisotropic fiber retain a high degree of polarization, as analyzed theoretically and demonstrated experimentally. There is a very active and well documented research[36] on birefringent dielectric waveguides aimed at improving optical devices.

Over the last two decades, there have been an increasing number of proposals for producing sensor optical waveguides and characterizing significant thermal and mechanical parameters like temperature tolerance and elasticity and optical properties, such as refractive index distribution.[37]

The basic idea of stress sensing by means of optical fibers is the following. Let us consider a dielectric fiber that had been optically isotropic first and, at a later stage became optically anisotropic. The latter behavior has been due to some externally applied force, giving rise to the photo-elastic effect in that fiber. Let light beams propagate confined along the waveguide, first when the latter was optically isotropic and, later, when it became optically anisotropic. Upon analyzing comparatively the features of both confined beams, physically relevant information can be obtained about the state of stress of the anisotropic waveguide, and, eventually, about the external force that has given rise to it. Alternatively, one could scatter light by the fiber, first when it was optically isotropic and, later, when it had been perturbed, by developing the optical anisotropy. Suitable differences between the light scattering distributions in both cases can now yield information about the additional internal stress of the perturbed waveguide and its cause (the external force on it).

In many applications in civil engineering (mining, construction projects, concrete structures), it is important to evaluate the response of a structure for safety purposes and damage assessment. Sensing pressure in those environments can be acomplished very efficiently by using technology based upon optical fiber sensors, specifically those employing highly birefringent waveguides (polarimetric sensors). Similar statements appear to apply to evaluating machinery in engineering mechanics and aerospace industries (strains in advanced aircraft and space vehicles). One very interesting possibility consists of embedding sensors based upon highly birefringent fibers into the structures of composite materials.

Different kinds of optical fiber sensors based upon the confined propagation of light for stress sensing have been investigated, for example, interferometric and grating-based systems.[36]

In previous work on light scattering applied to stress sensing, experimental determination of backward-scattering data yielded information on the refractive index distribution. Therefore, changes in the latter due to stresses and strains could be detected by analyzing the distribution of the scattered radiation at 180°.[38]

2.7 BIOLOGICAL WAVEGUIDES IN THE EYE

The interaction of light with biological tissues as dielectric matter in the eye has enormous relevance, at least: i) in order to understand the behavior of some very important biological phenomena, and ii) for its application as an optical method to investigate certain, also very important, elements in the eye, under both normal and pathological functioning conditions. We shall limit ourselves to discussing a few general aspects.

The retina is a soft biological tissue with thickness smaller than about 0.5 mm and transparent to light.[16] Its outermost part is the pigment epithelium (in which the optical image is formed), and its innermost part (the inner limiting membrane) is connected to another crucial part of the eye system, namely the optic nerve. The connection between those inner parts of the retina and the optic nerve is named the optic disk.

Leaving aside other elements in the retina, we shall focus on the retinal photoreceptors (that start below the pigment epithelium and are roughly perpendicular to it) and on the retinal nerve fiber layer (which is just above the inner limiting membrane). Input light that penetrates into the eye through the cornea and traverses the pigment epithelium enters into the retinal photoreceptors and propagates confined along them. At some later stage, and through some involved processes (not to be discussed in this chapter), light is absorbed by certain organic molecules (rhodopsin) and the resulting output biological signal is transmitted to the optic nerve. There are two kinds of photoreceptors: the rods (operating maximally at low light intensities, say, at night) and the cones (operating maximally at high light intensities and in color vision). Some features of the retinal nerve fiber layer and some further ones of photoreceptors will be commented on later in this section.

The retinal photoreceptors and the retinal nerve fiber layer are very complex structures. Following previous researchers[16] and as a zeroth-order approximation, one models the retinal photoreceptors as arrays of approximately parallel cylindrical dielectric waveguides (each having a length larger than its transverse diameter). Qualitatively similar models (but unrelated to those for photoreceptors) will be assumed for the elements of which the retinal nerve fiber layer is composed.[24]

In order to motivate our further discussions in this section, we shall simply notice that the macula (a central part of the retina) has significant involvement in ocular pathology and can display considerable birefringence. The measurement of birefringence in the macula may indicate the existence of a pathology.[23]

In the following subsections of this section, we shall offer a qualitative partial overview of various arrays of (more or less) approximately parallel cylindrical dielectric waveguides located inside the eye (in the cornea and in the retina). The propagation of light inside the eye gives rise to certain important phenomena in those arrays, which will also be discussed in those subsections. The average refractive index of the retina (regarded as a surrounding biological medium in which the various retinal waveguides are embedded) is about 1.34 (very close to that for water).[39] Other authors take 1.36 an as average value for the refractive index of the retina.[24] See also Subsection 2.7.1. The various retinal waveguides are regarded as optically isotropic, at least as a zeroth-order approximation in normal functioning conditions. In other situations (namely, beyond the zeroth-order approximation in normal functioning conditions or in pathological functioning ones), optical anisotropy may be considered or does arise. Form birefringence and scattering occur. Multiple scattering effects can be approximately neglected in some situations, but not in others.

We note that the presence of some birefringence in retinal photoreceptors and in the nerve fiber layer of the human retina appears to be known from various experimental data and procedures. That birefringence is associated, at least to a certain degree with dichroism[1,21] in the material structure. We employ the term *dichroism* to refer to different amounts of absorption by the material for various light beams propagating through it and polarized in different directions. For our purposes (and, at least, in a qualitative or general sense), we may understand such dichroism by recalling that the components ϵ_{ij} of the relative dielectric permittivity tensor were allowed to be complex. Dichroism is anisotropy in absorption. Let a composite body, formed by light- absorbing particles, be similar to those described in Section 2.4 (the wavelength of light being also adequately larger than typical dimensions). Then, the composite body may exhibit anisotropic absorption of light (form dichroism).[21]

2.7.1 Scattering by Cylindrical Waveguides in Cornea

The normal cornea is nearly transparent to light, except for some (weak) scattering phenomena which occur in it. The corneal stroma (the tissue forming its ground substance) is formed by stacked sheets (named lamellae). Each sheet contains long thin parallel collagen fibers (fibrils), embedded in a ground substance (composed of glycosaminoglycans in a salt solution). The ground substance can be regarded to be optically homogeneous. The fibril axes are parallel to the surfaces of the cornea. Fibrils in normal or swollen corneas seem to have essentially the same diameter (about a few tens of nanometers, nm), while those in scarred corneas have a rather wide range of diameters. The distribution of fibrils in a normal cornea is different from those for swollen corneas. Light propagating through the cornea is scattered by the individual fibrils. Scattering phenomena in normal corneas will be different, eventually, from those taking place in swollen or scarred corneas. The scattering of light by the fibrils in a normal cornea, in the so-called Debye-Born-Rayleigh-Gans approximation (to be discussed in Sections 2.8 and 2.10) for each individual fibril, has been studied.[40] The fact that the fibrils in a normal cornea do not behave as independent scatterers gives rise to important cancellations in the differential cross-section. The latter, in turn, account for the near transparency of a normal cornea. Smaller cancellations would take place for scatterings in swollen or scarred corneas, which would imply less transparency for them.[40] Multiple scattering in the cornea has been investigated,[41] and birefringence in the cornea has been studied.[42] See also next subsection.

2.7.2 Retinal Nerve Fiber Layer: Form Birefringence and Scattering

The retinal nerve fiber layer (that part of the retina that is connected to the optic nerve) is a complex structure which, regarding light propagation, is nearly transparent and behaves as weakly reflecting. Consequently, the retinal nerve fiber layer scatters some light (to a certain limited extent). It seems that such a scattering is due to structures (cellular organelles), approximately cylindrical and parallel to one another that are distributed throughout the thickness of the retinal nerve fiber layer. Cellular organelles at the retinal nerve fiber layer (RNFL) provide another important example of approximately parallel cylindrical biological waveguides as microtubules.[24]

As revealed by electron microscopy, the RNFL contains bundles of nerve cells (nerve-cell axons) that run just under and across the surface of the retina. A nerve fiber bundle contains four different kinds of approximately cylindrical structures: a) axonal cell membranes around nerve axons (axonal membranes), b) microtubules, c) neurofilaments, and d) mitochondria. Those structures lie in a rather complex cellular environment (some aqueous solution of proteins and other cellular constituents). Within the scope of this chapter, it is interesting to describe briefly those structures and their sizes.[24]

Axonal membranes are thin phospholipid bilayers (with transverse sizes about 6 to 10 nm) that form cylindrical shells enclosing the axonal cytoplasm. Microtubules are long (10 to 25 microns) tubular cylinders of the protein tubulin; they have inner and outer diameters about 15 and 25 nm, respectively. Neurofilaments are stable proteins, having a diameter about 10 nm. Mitochondria are ellipsoidal organelles that contain membranes of lipid and protein: their length and thickness are about 1 to 2 microns and about 100 to 200 nm, respectively. It follows that the transverse dimensions of axonal membranes, microtubules and neurofilaments are appreciably smaller than typical wavelengths of light (thin fibers) while those of mitochondria are not. The typical separations between two neighboring cylindrical structures of those kinds appear to be smaller than the wavelength of light.

The refractive indices of the various organelles in the RNFL range from about 1.36 up to about 1.64.[24]

The resulting arrays of approximately parallel cylinders appear to display some form birefringence: compare with Section 2.4 (model r). Then, cellular organelles at the retinal nerve fiber layer (RNFL) provide an example of optical anisotropy.[24]

To appreciate the potential interest of researchers about light scattering by the latter cylindrical structures, we shall limit ourselves to point out the following. The retinal nerve fiber layer is damaged in glaucoma and other diseases affecting the optic nerve. Detectable modifications of the retinal nerve fiber layer (say, its thinning) may occur before losses of vision become manifest and measurable. Then, optical methods that enable us to assess the state (thickness) of the retinal nerve fiber layer may provide aids for diagnosis. The optical methods involved include the analysis of the reflected light (arising from its scattering by the retinal nerve fiber layer into the backward direction).[24]

Further and interesting discussions of form birefringence for the RNFL arose,[24] by means of a relatively recent formula for an assembly of parallel cylinders. Some detailed analysis of the scattering of light by the organelles in the RNFL, each of them treated as an optically isotropic waveguide, have also been carried out in the same investigation.[24]

Various experiments implying birefringence in the RNFL have been reported in recent decades.[43,44] Other experiments,[45] using a laser ellipsometry technique demonstrated that the RNFL of the human retina exhibits substantial uniaxial birefringence. Additional studies have demonstrated that the RNFL possesses dichroism (interpreted possibly, as form dichroism, due to microtubules and neurofilaments).[46] Those studies[45,46] also deal with birefringence and dichroism in the cornea.

2.7.3 Retinal Photoreceptors

Retinal photoreceptors in the eye can be regarded at least as a zeroth order of approximation, as arrays of optically isotropic cylindrical dielectric cylinders with circular cross-sections of radius R.[16,17] The transverse diameters of the various photoreceptors can be taken to vary between about 0.6 μ m and about 2 μ m. The average separations between two neighboring photoreceptors can be regarded be to about 0.5 μ m (or, at most, a bit larger, up to about 2 μ m).

Let n_1 and n_2 be the refractive indices of the photoreceptor and of the surrounding medium, respectively.

The refractive indices n_1 of the various photoreceptors have been estimated to range from about 1.36 up to about 1.41.[16] One may also estimate the value for n_1 as 1.35, since the fibers of the RNFL (see Subsection 2.7.2) are not coated with white matter or myelin. The considered value corresponds more to "grey matter" or axonal material.

The refractive index n_2 of the surrounding medium can be estimated as follows. As established earlier by Sidman,[39] the interstitial organic material surrounding a biological waveguide has a refractive index very close to that for water (H_2O): $n_2 = n_{H_2O} + \alpha C$. α is a numerical coefficient and C is the percentage of concentration of solids (proteins, lipoproteins or lipids). We point out that α is rather stable for bio-materials, and $|\alpha C| << 1$. Since: $n_{H_2O} = 1.3334$, we assume $n_2 \simeq 1.34$ (for a wavelength of light in free space corresponding to an average over the visible spectrum, namely, for $\lambda = 0.5\ \mu$m). Notice that 1.34 is also used as a value for the refractive index of the cytoplasm (see, for example, Sidman[39] and references therein). Thus, the refractive indices of the various photoreceptors minus that of the surrounding medium (say, the average refractive index of the retina) in which the former are embedded may be about 0.01.

As was established by Enoch,[16] optical photoreceptors are guiding light under conditions of total internal reflection and behave as waveguides under ordinary conditions (normal functioning). In a geometrical optics description, light that has penetrated into one photoreceptor is guided along it subject to successive total internal reflections. In the electromagnetic wave description, light propagates confined along the photoreceptor in the form of propagation modes. For photoreceptors with circular cross-section of radius R, each set of modes is characterized by the values of the so-called V parameter given in Eq. (2.12), where: $n_1(= n_{co})$ and $n_2(= n_{cl})$ are the refractive indices of the photoreceptor (core) and of the surrounding media (cladding), discussed above, and $2R$ is the diameter of the photoreceptor. In more technical terms, the V parameter determines the specific cut-off frequency for the corresponding set of modes.[10] We consider $2R \simeq 2.0\ \mu$m.

It is accepted that human optical photoreceptors (regarded, in a zeroth order approximation, as homogeneous optically isotropic dielectric fibers of circular cross-section) support about six propagation modes. The latter include the set of two fundamental modes (zeroth order) and other higher order modes (first and second order ones).[16] Recall the short summary about propagation modes in Section 2.6. Modes of order two or higher transport a much smaller percentage of energy than the fundamental ones.[17] For a waveguide supporting low-order modal patterns, as is the case with human receptors, one may take $0 < V < 10.9$, where the upper limit has been taken as rather high, so as to include a set of low order modes.

The absorption of light by the organic material in photoreceptors during the confined propagation appears to be small and it constitutes an interesting phenomenon.

Birefringence and dichroism in retinal photoreceptors have been established and analyzed in detail.[21]

Mitochondria may introduce (small) anisotropies in photoreceptors.[47]

The birefringence of biological waveguides and, in particular, of retinal optical photoreceptors has been studied and measured by several authors during recent decades by using various techniques in mammalians and also for the human eye.[39,48] The birefringence in biological waveguides is due to the particular assembly of anisotropic organic molecules constituting the material of each one of the segments forming the whole waveguide structure and to the photochemical properties of the components. Recall that for the RNFL, Dreher et al.[45] measured the degree of preserved polarization and amount of dichroism as functions of angular position of measurement locations in eight postmortem human eyes. The authors obtained two maxima that correlate with the areas of the thickest RNFL.

To summarize, one can make a quantitative assessment of the amount of birefringence by means of the so-called net birefringence (Δn), which is a suitable sum of differences of refractive indices (each difference being due to a specific physical effect). A non-vanishing value for the net birefringence ($\Delta n \neq 0$) is a manifestation of overall optical anisotropy. In principle, the net birefringence is the sum of the intrinsic, form and chromatic birefringences: $\Delta n = \Delta n_I + \Delta n_F + \Delta n_C$. We anticipate that the contribution of the chromatic birefringence (Δn_C) will be neglected.

Liebman[48] succeeded in measuring the intrinsic (Δn_I), form (Δn_F) and net birefringence (Δn) in unbleached and bleached rod outer segments (ROS) in *R. Pipiens*. He assumed that the ROS behaves as a uniaxial crystal. By measuring the percentage of retardation of light, he found for the net birefringence: $\Delta n \simeq +0.0015$ (for a wavelength of light in free space $\lambda = 0.5\ \mu$m). Since birefringence depends on wavelength, we can estimate $\Delta n \simeq +0.0010$ for $\lambda = 0.68\ \mu$m , by following the mentioned results. The measured value of the net birefringence is affected by contributions with negative signs. It turns out that Δn_F takes on negative values.

For the purposes of characterizing the birefringence, one needs particular experimental values of $\Delta n_F = n_e - n_o = c/v_e - c/v_o$, where v_e and v_o are the velocities associated with the extraordinary and ordinary rays, respectively, in relation with the direction of the optic axis. We notice that the measured experimental values used to be very much influenced by the experimental procedure and by photochemical reactions induced in the structure of the organic components of the waveguide at the time of carrying through the measurements.

At this point, let us recall the form birefringence coefficient given in Eq. (2.5), which turns out to be negative. In the case of interest for us here, this behavior could correspond, for example, to the real lamellar structure originated by the disposition of fibers (n_1) and the interstitial matrix (n_2) in the array of the RNFL of the retina (polarization properties) as theoretically considered in the present study ($n_1 > n_2$). In so doing, we are identifying (or approximating) these n_1 and n_2, respectively, with the values for the refractive indices of the photoreceptor (n_1) and of the surounding medium (n_2), discussed previously in this subsection. Upon proceeding with the form birefringence coefficient given in Eq. (2.5), let $f_1 = f_2 = 1/2$ (namely, the case of two components with equal fractional volumes) and we consider small birefringence, so that $n_e + n_o \simeq n_1 + n_2$. Then, the form birefringence coefficient given in Eq. (2.5) (actually, a simplification of Wiener's formula[1]) yields the following approximate formula for the form birefringence:

$$\Delta n_F = n_e - n_o \simeq = -\frac{(n_1 - n_2)^2 (n_1 + n_2)}{2\left(n_1^2 + n_2^2\right)} \tag{2.13}$$

This behavior is characterizing a particular form birefringence.[45] See Figure 2.1. We observe that Δn_F has negative small values (negative uniaxial crystal).

Thus, from a theoretical point of view, this degree of birefringence of photoreceptors (for one single receptor) is related to their waveguiding properties.

The degree of birefringence is also a function of the V parameter of the waveguide: $n_1 = n_1(V)$. In order to understand this dependence, we have analyzed the behavior of $\Delta n_F = \Delta n_F(V)$, for two values of the wavelength: $\lambda = 0.6328\ \mu$m and $\lambda = 0.5\ \mu$m. For a fiber behaving as a single mode waveguide (or for $0 < V \leq 4$), the difference of the values of Δn_F for those two close values of λ appears to be very small (almost zero). On the other hand, as we approach the behavior of the fiber as multimode supporting two sets of modes or as V increases (say, for $6 < V \leq 10$), the

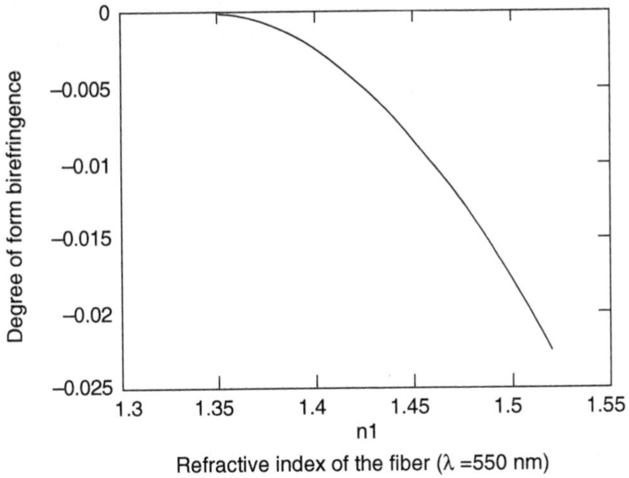

FIGURE 2.1 Behavior of Δn_F as a function of the refractive index n_1, for a wavelength $\lambda = 0.55\ \mu$ m of light.

difference between the two values of Δn_F for those two wavelengths dramatically increases as V does.

Assuming that $n_2 \simeq 1.34$ and $\Delta n_F = -0.009$ (as theoretically calculated by Harosi,[21]) we can observe that this value is obtained for $V = 8.4\ (< 10.9)$ for $\lambda = 0.5$ μ m and $V = 6.6$ ($\lambda = 0.6328\ \mu$ m). From this result, we can interpret that the value $\Delta n_F = -0.009$ is acceptable, and the fiber is far from being considered as a single mode one. Nevertheless, it is not warranted, *a priori*, that all allowed high-order propagation modes will be effectively excited. We can still keep the single mode approximation for the sake of simplicity of the present model.

Notice that in order to obtain positive values for Δn, we still need to include the additional contribution corresponding to the intrinsic birefringence properties of the concentration of solids along the axial direction of the fiber (in particular, lipid molecules), namely, Δn_I. Upon neglecting the chromatic birefringence, we have $\Delta n \simeq \Delta n_F + \Delta n_I$. Specifically, in order to obtain the positive experimental values reported in the literature ($\Delta n \simeq 0.0015$), and since $\Delta n_F = -0.009$, one has $\Delta n_I = 0.0105$, corresponding to a higher birefringence value. The latter, in turn, may be originated by a permanent organization of lipid molecules.

Let us regard each photoreceptor as a uniaxial birefringent medium in the form of a cylindrical waveguide and assume the x_3 axis lies along the axis of the waveguide. The values of the non-diagonal elements of the relative dielectric permittivity tensor are associated with optical activity (related, in turn, to the presence of non-symmetric molecules). At this point, we can mention that rhodopsin contains non-symmetric molecules like α-helical structures. A measurement of the optical activity can be performed by determining circular birefringence, but the importance of the induced phenomenon would be very weak.[21] Then, one may estimate approximately that, in the actual case, the corresponding relative dielectric permittivity tensor has $\epsilon_{12} \simeq 0$.

There are some important effects regarding the capture of light by retinal photoreceptors and visual response. One of them is the Stiles-Crawford effect of first kind (*SCE-I*) that evidences the behavior of retinal photoreceptors as antennas and optical dielectric waveguides.[16,17] In the *SCE-I*, the profile of the experimental intensity distribution of the luminous energy scattered by the retinal photoreceptors exhibits a maximum in the direction of the center of the eye pupil. For the particular case of the *SCE-I*, and due to the limited range of the measured angles (limited eye pupil), the experimental results turn out to correspond to a selective angular range that is close to the forward scattering direction. The numerical representation of the angular distribution of the luminous energy evidences a maximum for the forward scattering direction. This result indicates the orientational properties of groups of retinal photoreceptors.

Recall that the receptor orientation is the result of several forces: a) phototropic (orienting the photoreceptors toward the center of the pupil), b) a force perpendicular to the retinal epithelium (for alignment of the photoreceptors toward the center of the retinal sphere), c) retinal tractions, and d) receptor packing. Results obtained earlier for the *SCE-I* in observers having ectasia (Fuch's coloboma) show receptor alignment abnormality or disturbance.[49]

Further experimental results in relation to the *SCE-I*, with dramatic changes of the data have revealed systematic strains causing orderly alterations in photoreceptor orientations in observers with high myopia.[50] One possible qualitative explanation of the latter experimental results could be that, under such anomalous conditions (high myopia), some optical anisotropy (implied by the photo-elastic effect) could have arisen in those biological cylindrical waveguides, namely in the retinal photoreceptors and in the retinal nerve fiber layer.

2.8 CLASSICAL ELECTROMAGNETIC SCATTERING BY ANISOTROPIC OBJECTS IN ISOTROPIC MEDIUM

Let us consider in three-dimensional space an infinite isotropic and homogeneous dielectric reference medium, that has real constant dielectric permittivity ϵ_0, zero electrical conductivity, magnetic permeability equal to that of vacuum (μ_0, real) and, hence, constant refractive index n_0. We also consider N separate (non-overlapping) optically anisotropic dielectric objects (with arbitrary shapes and relative orientations with respect to one another) located at fixed positions in the reference medium. Each object has zero electrical conductivity, magnetic permeability equal to μ_0 and a (symmetric) dielectric permittivity tensor $\epsilon_0 \cdot \tilde{\epsilon}$, where $\tilde{\epsilon}$ is the dielectric permittivity tensor of the object relative to that of the surrounding reference medium. ϵ_{ij} is, in general, complex.

We shall suppose that each ϵ_{ij} and its first spatial derivatives are continuous everywhere, that is, also across the interface separating any two different dielectric media. Strictly speaking, each ϵ_{ij} and the first spatial derivatives thereof, being continuous across the limiting surface of any of those objects, must vary in a certain three-dimensional region containing those boundary surfaces. We suppose that the sizes (or transverse widths) of those regions in the directions orthogonal to the corresponding surfaces are adequately small. Except for the last assumptions regarding ϵ_{ij}, the formulation in this section is rather general, and it applies regardless of the fact that the objects are or are not identical cylindrical waveguides. The generalization of that three-dimensional formulation when ϵ_{ij} or the first spatial derivatives thereof are allowed to develop discontinuities at the boundaries of the dielectric objects, is rather complicated and it will not be attempted here. That generalization for electromagnetic scattering by optically anisotropic cylindrical waveguides, which becomes a two-dimensional situation, has already been carried out and reference to it will be given in Appendix E.

Let $\bar{\epsilon}(\bar{k},\sigma)$ be the standard linear (real) or circular (complex) unit polarization vectors; they are given by the same expressions as for electromagnetic wave propagation in vacuum.[51] The (linear or circular) polarization index σ takes on only two values. Their basic properties are summarized in Appendix A. Let incoming classical monochromatic (complex) electric ($\mathbf{E}'_{in}\exp(-i\omega t)$) and magnetic ($\mathbf{H}'_{in}\exp(-i\omega t)$) fields, with real frequency ω and wavevector $\bar{k}$, propagate freely in the surrounding reference medium at very large distances from the N objects (say, for distances much larger than their characteristic sizes and larger than the wavelength). Certain shapes of the objects require specific discussions and treatments of what is meant by "distances much larger than their characteristic sizes." The case of cylindrical objects that have (relatively) large sizes along certain directions will be dealt with in Section 2.10. Both $\mathbf{E}'_{in}$ and $\mathbf{H}'_{in}$ fulfill Maxwell's equations (that is, Eqs. (2.1) and (2.2) with $\tilde{\epsilon} = I_3$). Then, both equations in (2.3) for $\mathbf{H}'_{in} = \mathbf{H}'_0 \exp i\bar{k}\bar{x}$, with constant $\mathbf{H}'_0$, read for any $\bar{x}$:

$$\nabla \times (\nabla \times \mathbf{H}'_{in}) = \frac{\omega^2}{v^2}\mathbf{H}'_{in},\ \nabla \cdot \mathbf{H}'_{in} = 0 \tag{2.14}$$

The v now denotes the velocity of the electromagnetic wave in the surrounding reference medium. In general, $\mathbf{H}'_0$ is a linear combination of the two vectors $\bar{\epsilon}(\bar{k},\sigma)$, for both values of σ. In order to simplify the analysis without an essential loss of generality, we choose $\mathbf{H}'_0 = \bar{\epsilon}(\bar{k},\sigma_{in})$, with some given initial (linear or circular) polarization index σ_{in}. The first equation in (2.14) can be re-expressed using the second one as:

$$[\Delta + K_{cl}^2]\mathbf{H}'_{in} = 0,\ K_{cl}^2 = \frac{\omega^2}{v^2} \tag{2.15}$$

where $\Delta = \partial^2/\partial x_1^2 + \partial^2/\partial x_2^2 + \partial^2/\partial x_3^2$.

Let those incoming fields approach the anisotropic objects and be scattered by them. The total electric field ($\mathbf{E}'\exp(-i\omega t)$) equals the incoming plus the scattered field and so on for the total magnetic field ($\mathbf{H}'\exp(-i\omega t)$). Both $\mathbf{E}'$ and $\mathbf{H}'$ fulfill Maxwell's equations (2.1) and (2.2) for any $\bar{x}$ (in the surrounding reference medium and inside the objects). Notice that $\tilde{\epsilon} \neq I_3$ inside the objects. We shall concentrate on the total magnetic field that satisfies Eq. (2.3). For convenience, we shall recast Eq. (2.3) for $\mathbf{H}'$ at any $\bar{x}$ as:

$$[\Delta + K_{cl}^2]\mathbf{H}' = \mathbf{j}',\ \nabla \cdot \mathbf{H}' = 0 \tag{2.16}$$

$$\mathbf{j}' = \nabla \times ([(\tilde{\epsilon})^{-1} - I_3]\nabla \times \mathbf{H}') \tag{2.17}$$

It is easy to check that $\nabla \cdot \mathbf{j}' = 0$, so that Eq. (2.17) is consistent with the second Eq. (2.16). The structures of Eqs. (2.16) and (2.17) can be easily justified as follows. Let there be no anisotropic objects (ϵ_{ij} thereby becoming equal to δ_{ij}), so that the region of space occupied by them becomes similar to the surrounding reference medium. Then, there would be no scattered waves and, hence, $\mathbf{H}'$ would become $\mathbf{H}'_{in}$, $\mathbf{j}' = 0$ (the right-hand side of Eq. (2.17) vanishing identically) and the first Eq. (2.16) would become the first Eq. (2.15).

A Green's function ($G_{Tr}(\bar{x})$) adequate for dealing with both equations in (2.16) (and, specifically, for solving the difficulties arising from the second equation) has been proposed and adequately studied.[52,53] $G_{Tr}(\bar{x})$ is named the transverse Green's function and it is a 3×3 tensor, characterized in Appendix B. Upon applying standard Green's function techniques with $G_{Tr}(\bar{x} - \bar{x}')$, the partial differential equations in (2.16) can be recast as the following linear integral equation valid for any $\bar{x}$ (that is, both in the surrounding reference medium and inside any anisotropic object):

$$\mathbf{H}'(\bar{x}) = \mathbf{H}'_{in}(\bar{x}) + \int d^3\bar{x}' G_{Tr}(\bar{x} - \bar{x}')\mathbf{j}'(\bar{x}') \tag{2.18}$$

The integration over $\bar{x}'$ is extended to the whole regions occupied by the N separate optically anisotropic dielectric objects. Using the properties of $G_{Tr}(\bar{x})$, as outlined in Appendix B, one sees easily that $\mathbf{H}'(\bar{x})$, as given by the right-hand side of (2.18) fulfills, throughout all space, both the first (2.16) (for $\bar{x}$ inside the anisotropic objects and outside them) and the second (2.16). Eq. (2.18) is the basic integral equation describing the scattering of the classical electromagnetic wave by the separate optically anisotropic objects.

For $\mid \bar{x} \mid \rightarrow +\infty$, along a fixed direction (say, fixed $x_l / \mid \bar{x} \mid, l = 1, 2, 3$), by using Eq. (B.7), with $\bar{k}' = K_{cl}(\bar{x} / \mid \bar{x} \mid)$, Eq. (2.18) becomes:

$$\mathbf{H}'(\bar{x}) \rightarrow \mathbf{H}'_{in}(\bar{x}) + \frac{\exp i K_{cl} \mid \bar{x} \mid}{\mid \bar{x} \mid} \sum_{\sigma} \bar{\epsilon}(\bar{k}', \sigma) f(\bar{k}', \sigma; \bar{k}, \sigma_{in}) \tag{2.19}$$

$$f(\bar{k}', \sigma; \bar{k}, \sigma_{in}) = -\frac{1}{4\pi} \int d^3\bar{x}' \exp[-i\bar{k}'\bar{x}'] \bar{\epsilon}(\bar{k}', \sigma)^{+} \mathbf{j}'(\bar{x}') \tag{2.20}$$

The two $\bar{\epsilon}(\bar{k}', \sigma)$'s are also standard linear (real) or circular (complex) polarization vectors in three dimensions (see Appendix A). Recalling that $\mathbf{H}'_{in} = \bar{\epsilon}(\bar{k}, \sigma_{in}) \exp i\bar{k}\bar{x}$, we choose, for consistency, the $\bar{\epsilon}(\bar{k}', \sigma)$'s in (2.19) to represent the same type of polarization (linear or circular) as $\bar{\epsilon}(\bar{k}, \sigma_{in})$. The row vector $\bar{\epsilon}(\bar{k}', \sigma)^{+}$ is the adjoint (involving complex conjugation and matrix transposition) of the column vector $\bar{\epsilon}(\bar{k}', \sigma)$. Then, $\epsilon(\bar{k}, \sigma)^{+}\mathbf{j}'(\bar{x}')$ denotes the scalar product of the vector $\bar{\epsilon}(\bar{k}', \sigma)$ (involving the complex conjugate of its components) times $\mathbf{j}'(\bar{x}')$. The sum over the polarization index σ is carried out only over its two allowed values. $f(\bar{k}', \sigma; \bar{k}, \sigma_{in})$ is the three-dimensional electromagnetic scattering amplitude by the assembly of all anisotropic objects. See Figure 2.2.

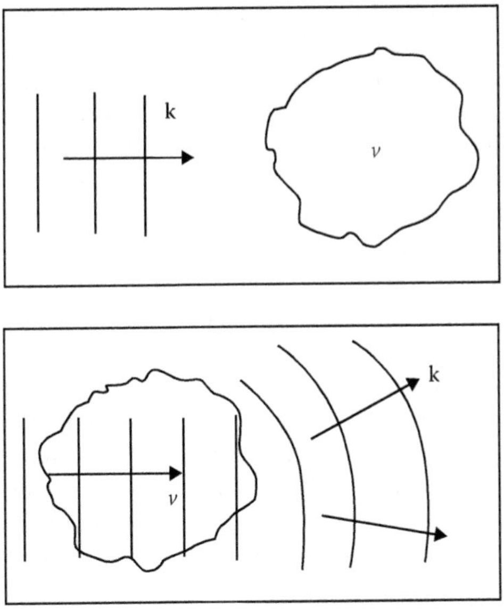

FIGURE 2.2 Top: Incoming electromagnetic radiation with wavevector $\bar{k}$ approaches an optically anisotropic three-dimensional object (V). Bottom: Scattered radiation by the optically anisotropic object in the direction corresponding to the outgoing wavevectors $\bar{k}'$.

Let us consider the flux of the Poynting vector determined by the incoming fields $\mathbf{E}'_{in}\exp(-i\omega t)$ and $\mathbf{H}'_{in}\exp(-i\omega t)$ (across a unit planar surface orthogonal to $\bar{k}$) and that corresponding to (2.19) and to its associated electric field (across a very small surface element orthogonal to $\bar{k}'$, located at $\mid \bar{x} \mid \rightarrow +\infty$). The ratio of the second flux over the first one yields the differential cross-section for the scattering by that assembly of anisotropic objects. That differential cross-section is proportional to $\mid f(\bar{k}', \sigma; \bar{k}, \sigma_{in}) \mid^2$. We shall omit further details, as one may see similar computations elsewhere.[54]

Eq. (2.18) is a linear inhomogeneous integral equation that can be solved formally by successive iterations. The latter generate an infinite series for $\mathbf{H}'(\bar{x})$, which solves Eq. (2.18). For weak scattering processes, it suffices to approximate $\mathbf{H}'(\bar{x})$ by $\mathbf{H}'_{in}(\bar{x})$ plus the result of the first iteration. This yields the following approximate explicit representation for $\mathbf{H}'(\bar{x})$:

$$\mathbf{H}'(\bar{x}) \simeq \mathbf{H}'_{in}(\bar{x}) + \int d^3\bar{x}' G_{Tr}(\bar{x} - \bar{x}')\mathbf{j}'^{(1)}(\bar{x}') \tag{2.21}$$

$$\mathbf{j}'^{(1)}(\bar{x}') = \nabla \times ([(\tilde{\epsilon})^{-1} - I_3]\nabla \times \mathbf{H}'_{in}(\bar{x}')) \tag{2.22}$$

In this approximation (named the Debye-Born-Rayleigh-Gans or DBRG), $f(\bar{k}', \sigma; \bar{k}, \sigma_{in})$ is obtained by replacing $\mathbf{j}'(\bar{x}')$ in (2.20) by $\mathbf{j}'^{(1)}(\bar{x}')$:

$$f(\bar{k}', \sigma; \bar{k}, \sigma_{in}) \simeq f(\bar{k}', \sigma; \bar{k}, \sigma_{in})_{DBRG}$$
$$= -\frac{1}{4\pi}\int d^3\bar{x}' \exp[-i\bar{k}'\bar{x}']\bar{\epsilon}(\bar{k}', \sigma)^{+}\mathbf{j}'^{(1)}(\bar{x}') \tag{2.23}$$

Some necessary conditions for this approximation to hold are: i) $\tilde{\epsilon}$ differs little from I_3, that is, all $\mid \epsilon_{ii} - 1 \mid$ and all $\mid \epsilon_{ij} \mid$, $i \neq j$, are small compared to unity, and ii) all ϵ_{ij} vary slowly inside the objects. We shall leave the approximate explicit integral representation for $\mathbf{H}'(\bar{x})$ given by Eqs. (2.21) and (2.22) as it stands. In principle, it could be applied directly to some physically interesting scattering situations, say, through numerical integration. However, as we shall see, one may easily understand that its practical reliability would become reduced, at least, as the number of scattering objects increases; say, as the possibility of multiple scattering effects becomes non-negligible. That is, the above necessary conditions i) and ii) will not be sufficient for the reliability of (2.21) and (2.22) if the number of scattering objects is not adequately small.

Intuitively, one could imagine that the incoming field $\mathbf{H}'_{in}(\bar{x})$ is scattered once by each of the N anisotropic objects. Then, let the magnetic field arising from the scattering by the i-th anisotropic object alone be, in turn and subsequently, scattered by the j-th object ($j \neq i$) and so on. This gives rise to imagining a whole series of successive multiple scatterings of the electromagnetic wave by the different objects. At this stage, one could argue, quite naturally, that the exact total magnetic field $\mathbf{H}'(\bar{x})$, given by Eq. (2.18), should also be equal to the total magnetic field resulting from all those multiple scatterings. Such a multiple scattering point of view is physically correct and mathematically consistent, and it has given rise to further insight and useful approximations. Of course, the approximate (2.21) and (2.22) amount to neglecting all multiple scatterings (double, triple,. . .). So, *a posteriori*, we see that the approximate reliability of (2.21) and (2.22) would require that multiple scattering effects be adequately small, intuitively. The latter condition appears not to be fulfilled in many physically interesting cases (eventually not, if the number of scattering objects is adequately large).

Then, it makes sense to develop further alternative and general procedures that deal with multiple scattering specifically. A rather general formulation will be outlined in the next section. Its applicability (for N parallel cylindrical long waveguides) will be explained later in this chapter.

2.9 MULTIPLE SCATTERING BY OPTICALLY ANISOTROPIC OBJECTS IN THREE DIMENSIONS: AN OUTLINE

Let there be only one optically anisotropic object, namely, the i-th one (as if all the remaining $N - 1$ ones had been removed). Let $\mathbf{H}'(\bar{x})_i^{(a)}$ be the total magnetic field describing the scattering of the incoming field $\mathbf{H}'_{in}(\bar{x})$ by the i-th anisotropic object alone (superscript a). $\mathbf{H}'(\bar{x})_i^{(a)}$ fulfills Eq. (2.18) with only the i-th scattering object, which we cast as:

$$\mathbf{H}'(\bar{x})_i^{(a)} = \mathbf{H}'_{in}(\bar{x}) + \int_i d^3\bar{x}' G_{Tr}(\bar{x} - \bar{x}')\mathbf{j}'(\bar{x}')_i \tag{2.24}$$

$\mathbf{j}'(\bar{x}')_i$ is obtained from $\mathbf{j}'(\bar{x}')$ (Eq. (2.17)) upon replacing in it $\mathbf{H}'(\bar{x})$ by $\mathbf{H}'(\bar{x})_i^{(a)}$. The subscript i in the integral in Eq. (2.24) indicates that the integration over $\bar{x}'$ is extended only over the region occupied by the i-th object. The series for $\mathbf{H}'(\bar{x})_i^{(a)}$ formed by the all successive iterations of Eq. (2.24) can be recast symbolically as:

$$\mathbf{H}'_i(a) = \mathbf{H}'_{in} + G_{Tr}\tau_i\mathbf{H}'_{in} \tag{2.25}$$

Eq. (2.25) defines the scattering integral operator τ_i (which is also a 3×3 matrix) for the i-th object. More specifically, τ_i is given by a series, so that the n-th term in it is obtained from the n-th iteration of Eq. (2.24). For instance, the first iteration of Eq. (2.24) gives the first contribution, $\tau_i^{(1)}$, to τ_i ($\tau_i \simeq \tau_i^{(1)}$):

$$G_{Tr}\tau_i^{(1)}\mathbf{H}'_{in} = \int_i d^3\bar{x}' G_{Tr}(\bar{x} - \bar{x}')\mathbf{j}'^{(1)}(\bar{x}')_i \tag{2.26}$$

$\mathbf{j}'^{(1)}(\bar{x}')_i$ is given in (2.22) only for $\bar{x}'$ in the region occupied by the i-th scattering object. One can identify the explicit form of $\tau_i^{(1)}$ from (2.26) but, for brevity, we shall not write it (a similar identification will be carried out later in the simpler case of cylindrical waveguides). Similarly, the second iteration of Eq. (2.24) gives the second contribution, $\tau_i^{(2)}$, to τ_i ($\tau_i \simeq \tau_i^{(1)} + \tau_i^{(2)}$). In order to illustrate the potential usefulness of these studies, we note that the second iteration of (a far simpler version of) Eq. (2.24) was one key element in deriving the form birefringence coefficient for an array of parallel cylinders of arbitrary sizes and separations.[23] And so on for higher order contributions to τ_i. In the remainder of this section, we shall work with τ_i but no longer with $\mathbf{H}'(\bar{x})_i^{(a)}$.

Next, we again consider the case with all (N) optically anisotropic objects present and we treat all possible scatterings of the magnetic field by them. That is, we come back to the total magnetic field $\mathbf{H}'(\bar{x})$ describing the scattering of the incoming field $\mathbf{H}'_{in}(\bar{x})$ and to Eq. (2.18). The key features of the alternative multiple scattering formulation of Eq. (2.18) are the following. The same total magnetic field $\mathbf{H}'(\bar{x}) \equiv \mathbf{H}'$, which solves Eq. (2.18), is also given by the representation:

$$\mathbf{H}' = \mathbf{H}'_{in} + \sum_{i=1}^{N} G_{Tr}\tau_i\mathbf{H}'_i \tag{2.27}$$

In turn, the new "partial" magnetic fields $\mathbf{H}'_i = \mathbf{H}'(\bar{x})_i, i = 1, \ldots, N$, are the solution of the system:

$$\mathbf{H}'_i = \mathbf{H}'_{in} + \sum_{j=1, j\neq i}^{N} G_{Tr}\tau_j\mathbf{H}'_j \tag{2.28}$$

τ_i in both Eqs. (2.27) and (2.28) is the same as in Eq. (2.25), namely, the scattering integral operator for the i-th object. Eq. (2.28) is an inhomogeneous linear system of N integral equations for all $\mathbf{H}'_i$, $i = 1, \ldots, N$. Notice that the i-th Eq. (2.28) does not contain $\mathbf{H}'_i$ in its right-hand side. Physically, this can be interpreted through the statement that the magnetic field that reaches the i-th object is the sum of the incoming one plus the superposition of all those generated by the remaining $N - 1$ objects. We emphasize that $\mathbf{H}'(\bar{x})_i$ is physically different from $\mathbf{H}'(\bar{x})_i^{(a)}$ and it would be erroneous to confuse them; only the i-th object was present when $\mathbf{H}'(\bar{x})_i^{(a)}$ was defined (in Eq. (2.24), while all (N) objects were considered in Eq. (2.28) for $\mathbf{H}'(\bar{x})_i$. A key property is that the solution for $\mathbf{H}'$ obtained by all iterations of Eq. (2.28) together with Eq. (2.27) is equivalent to a rearrangement of the series obtained by iterating Eq. (2.18). It can be proved upon comparing the series formed by all successive iterations for Eq. (2.18) with the one provided by Eq. (2.27) and all successive iterations of Eq. (2.28).

The scattering amplitude $f(\bar{k}', \sigma; \bar{k}, \sigma_{in})$ ((2.17)) can be obtained, for instance, through the substitution of $\mathbf{H}'$ in (2.17) by the right-hand side of (2.27).

The multiple scattering formulation in quantum mechanics is well documented.[54–56] Its extension for the multiple scattering formulation of electromagnetic waves had to deal with the additional difficulties arising from the second equation in Eq. (2.16). The transverse Green's function G_{Tr} allowed us to solve those difficulties, as shown through further computations and estimates.[57]

In practice, the above multiple scattering formulation, although physically quite appealing, is not easy to apply, with the above generality, in three dimensions as the number N of scattering objects increases. Fortunately, the application of such formulation for parallel cylindrical waveguides (which boil down to treating two-dimensional situations) under additional simplifying assumptions will be more tractable, as we shall see. The rather general formulations in this and the preceding section could also provide some starting point for the study of electromagnetic wave scattering by objects, when those simplifying assumptions are not met with sufficient approximation (or are not fulfilled at all): for instance, for cylindrical waveguides with misalignments.

2.10 SCATTERING BY ANISOTROPIC CYLINDRICAL WAVEGUIDES

2.10.1 Two-Dimensional Maxwell's Equations for Anisotropic Cylindrical Waveguides

We shall now assume that each of the N objects is an optically anisotropic cylindrical waveguide with arbitrary transverse cross-section Ω and suitably large length L. Let R be some length scale so that the area of Ω is of order R^2. L is supposed to be much larger than R and than typical wavelengths of light, so, for practical purposes, the waveguide will be regarded to have infinite length ($L \simeq \infty$). See Figure 2.3. We suppose that all N waveguides have their axes parallel to one another. See Figure 2.4. We shall choose the x_3-axis to be parallel to the axis of the waveguides (for both $N = 1$ and $N > 1$) and write: $\bar{x} = (\boldsymbol{x}, x_3)$, so that $\boldsymbol{x} = (x_1, x_2)$. We shall suppose that $\epsilon_{13} = \epsilon_{23} = 0$, and that the non-vanishing ϵ_{ij} are real for typical frequencies of light, at least, approximately (absence of absorption). If, in addition, $\epsilon_{12} = \epsilon_{21} = 0$, the material of which the waveguide is made is, by definition, not optically active. Let a material be not optically active, then, the case $\epsilon_{11} = \epsilon_{22} \neq \epsilon_{33}$ corresponds to uniaxial anisotropy, while if all ϵ_{ii} differ from one another, one meets the biaxial one. In the general study that follows, and unless explicitly stated, we shall allow for optically active materials ($\epsilon_{12} = \epsilon_{21} \neq 0$).

Under the above additional simplifications, the general three-dimensional scattering problem treated in Section 2.8 can be reduced to a somewhat simpler one, in two spatial dimensions. In fact, with $L \simeq +\infty$, it is consistent to search for

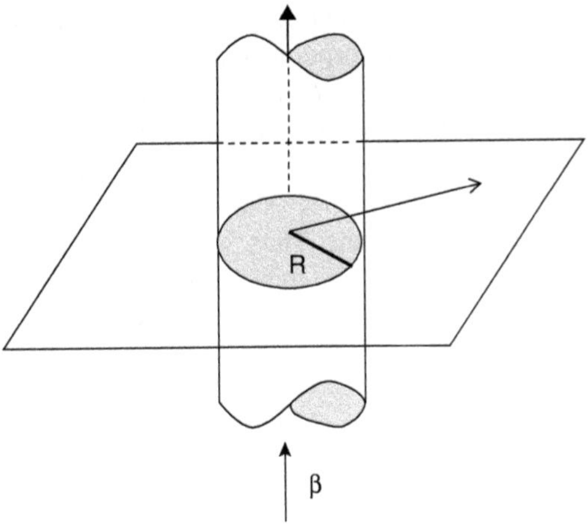

FIGURE 2.3 One single anisotropic object having the form of a cylindrical waveguide of radius R. The constant β appearing in Eq. (2.30) is displayed. β also gives the component of the wavevector of the incoming radiation parallel to the axis of the waveguide.

physically consistent solutions of Maxwell's equations, in which the x_3-dependences are factorized in the total electric and magnetic fields:

$$\mathbf{E}' = \exp(i\beta x_3)\mathbf{E} \tag{2.29}$$

$$\mathbf{H}' = \exp(i\beta x_3)\mathbf{H} \tag{2.30}$$

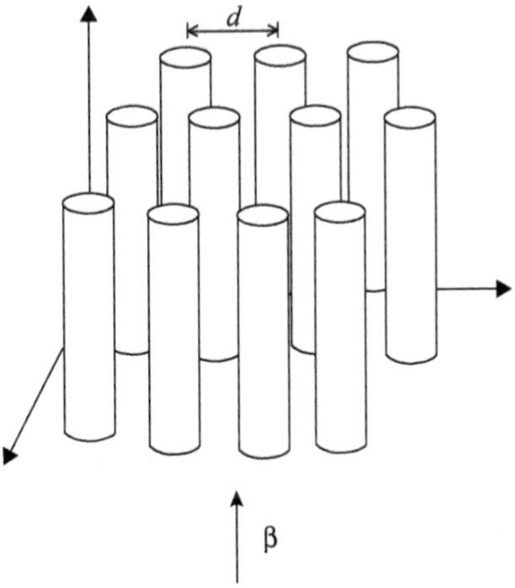

FIGURE 2.4 N anisotropic objects, each having the form of a cylindrical waveguide. The axes of the N waveguides are parallel to one another. The average distance between any pair of neighboring waveguides is d. The constant β appearing in Eq. (2.30) is displayed (see also caption of Figure 2.3). For comparison with Figure 2.6, the optic axis of each waveguide is supposed to be parallel to its geometric axis.

Both **E** and **H** are independent on x_3, but they do depend on $\boldsymbol{x} = (x_1, x_2)$. On the other hand, β is a real constant (as absorption effects are regarded as negligible).

Let $\mathbf{E} = (E_1, E_2, E_3)$ and $\mathbf{H} = (H_1, H_2, H_3)$. In this case, we shall give a simpler (two-dimensional) form of Maxwell's equations, involving only H_1 and H_2. Such a form will turn out to be equivalent to the two equations in (2.3) plus Eq. (2.30). For that purpose, one could use Eq. (2.3), as we shall comment in Appendix C, but we shall proceed through an alternative route. Thus, we consider the first components of the first (vector) equation in (2.2) and the third component in the first (also vector) equation in (2.1) and take Eq. (2.29) into account. All those yield:

$$\frac{\partial H_2}{\partial x_1} - \frac{\partial H_1}{\partial x_2} = -\frac{i\omega}{v}\left(\frac{\epsilon_0}{\mu_0}\right)^{1/2} \epsilon_{33} E_3 \tag{2.31}$$

$$\frac{\partial E_3}{\partial x_2} - i\beta E_2 = \frac{i\omega}{v}\left(\frac{\mu_0}{\epsilon_0}\right)^{1/2} H_1, \quad -\frac{\partial E_3}{\partial x_1} + i\beta E_1 = \frac{i\omega}{v}(\frac{\mu_0}{\epsilon_0})^{1/2} H_2 \tag{2.32}$$

Eqs. (2.31) and (2.32) allow us to obtain, successively, E_3 and, then, E_2 and E_1 in terms of H_1 and H_2. Then, we consider $\nabla \times (\nabla \times \mathbf{H}') = -i(\omega/v)(\epsilon_0/\mu_0)^{1/2} \nabla \times (\epsilon \mathbf{E}')$, take its first two components, use the second equation in (2.2) and Eq. (2.30) and employ the above expressions for E_3, E_2 and E_1 in terms of H_1 and H_2 and spatial derivatives thereof. Thus, one finds the announced system depending only on H_1 and H_2 (with $\Delta_T \equiv \partial^2/\partial x_1^2 + \partial^2/\partial x_2^2$ and $K^2 = (\omega/v)^2 - \beta^2$):

$$[\Delta_T + K^2] H_i = j_i, \quad i = 1, 2 \tag{2.33}$$

$$j_1 = -(\omega/v)^2(\epsilon_{22} - 1)H_1 + (\omega/v)^2 \epsilon_{21} H_2 - \frac{\partial \ln \epsilon_{33}}{\partial x_2}\left[\frac{\partial H_2}{\partial x_1} - \frac{\partial H_1}{\partial x_2}\right]$$
$$+ \left(\epsilon_{21}\frac{\partial}{\partial x_1} - (\epsilon_{33} - \epsilon_{22})\frac{\partial}{\partial x_2}\right)\left[\frac{1}{\epsilon_{33}}\left(\frac{\partial H_2}{\partial x_1} - \frac{\partial H_1}{\partial x_2}\right)\right] \tag{2.34}$$

$$j_2 = -(\omega/v)^2(\epsilon_{11} - 1)H_2 + (\omega/v)^2 \epsilon_{12} H_1 + \frac{\partial \ln \epsilon_{33}}{\partial x_1}\left[\frac{\partial H_2}{\partial x_1} - \frac{\partial H_1}{\partial x_2}\right]$$
$$- \left(\epsilon_{12}\frac{\partial}{\partial x_2} - (\epsilon_{33} - \epsilon_{11})\frac{\partial}{\partial x_1}\right)\left[\frac{1}{\epsilon_{33}}\left(\frac{\partial H_2}{\partial x_1} - \frac{\partial H_1}{\partial x_2}\right)\right] \tag{2.35}$$

for any $\boldsymbol{x}$. Both j_1 and j_2 vanish outside the transverse cross-section of any anisotropic waveguide. A distinguishing feature of (2.34) and (2.35) is that the components of $\tilde{\epsilon}^{-1}$ do not appear explicitly in them. Notice that the above Eqs. (2.34) and (2.35) correct for a sign missprint in a previous work[19] (specifically, in the second term on the right-hand side of Eq. (2.1.8) in that work.[19]) Eqs. (2.33), (2.34) and (2.35) hold independently on any assumption about the behavior of the components of $\tilde{\epsilon}$ and their first spatial derivatives across any interface between the cylindrical waveguides and the surrounding medium (cladding).

In turn, Eqs. (2.33), (2.34) and (2.35) can be recast as

$$[\Delta_T + K^2] h(\boldsymbol{x}) = j(\boldsymbol{x}) \tag{2.36}$$

where $h = h(\boldsymbol{x})$ and $j = j(\boldsymbol{x})$ are the two column vectors formed, respectively, by $H_i, i = 1, 2$ and by $j_i, i = 1, 2$.

In what follows, and in order to simplify the analysis, we shall suppose that $\epsilon_{ij}, i, j = 1, 2, \epsilon_{33}$ and their first spatial derivatives are continuous throughout all space (that is, and in particular, also across any interface separating any anisotropic waveguide and the surrounding cladding). This is the counterpart for the actual cylindrical waveguides of similar assumptions for the three-dimensional case made in Section 2.8.

2.10.2 Scattering by N Anisotropic Cylindrical Waveguides: General Analysis

Let $H_0^{(1)}(K \mid \boldsymbol{x} \mid)$ be Hankel's outgoing function of zeroth order.[58] It fulfills:

$$[\Delta_T + K^2]\frac{H_0^{(1)}(K \mid \boldsymbol{x} - \boldsymbol{x}' \mid)}{4i} = \delta^{(2)}(\boldsymbol{x} - \boldsymbol{x}') \tag{2.37}$$

$\delta^{(2)}$ being the two-dimensional Dirac delta function. That is, $H_0^{(1)}(K \mid \boldsymbol{x} - \boldsymbol{x}' \mid)/4i$ is a Green's function for $\Delta_T + K^2$, in two spatial dimensions. Some useful properties of $H_0^{(1)}(K \mid \boldsymbol{x} \mid)$ are collected in Appendix D. For the general case of N cylindrical waveguides, and upon applying standard Green's function techniques, the partial differential equation (2.36) can be recast as the following linear integral equation

$$h(\boldsymbol{x}) = h^{(0)}(\boldsymbol{x}) + \int d^2\boldsymbol{x}' \frac{H_0^{(1)}(K \mid \boldsymbol{x} - \boldsymbol{x}' \mid)}{4i} j(\boldsymbol{x}') \tag{2.38}$$

The integral in Eq. (2.38) is extended over the transverse cross-sections of the N anisotropic waveguides. The inhomogeneous term $h^{(0)}(\boldsymbol{x})$ fulfills $[\Delta_T + K^2]h^{(0)}(\boldsymbol{x}) = 0$ for any $\boldsymbol{x}$. For the scattering situation, there certainly is one physically propagating wave very far from the waveguide, namely the incoming plane wave: in this case, $h^{(0)}(\boldsymbol{x}) = h^{(0)} \exp i\boldsymbol{x}\boldsymbol{k}$, with $\boldsymbol{k}^2 = K^2 > 0$, that is, K is real. On the other hand, $h^{(0)}$ is $\boldsymbol{x}$-independent. Without an essential loss of generality, we can choose $h^{(0)}$ to equal a two-component linear (real) or circular (complex) polarization vector $\epsilon(\boldsymbol{k}, \sigma_{in})$ in two dimensions (see Appendix D). σ_{in} denotes the initial (linear or circular) polarization index. The three-dimensional Eq. (2.18) has simplified to the two-dimensional Eq. (2.38). $h^{(0)}(\boldsymbol{x})$ in (2.38) is the present two-dimensional counterpart of $\mathbf{H}'_{in}$ for (2.18).

For $\mid \boldsymbol{x} \mid \to +\infty$, along a fixed direction (say, fixed $x_l / \mid \boldsymbol{x} \mid, l = 1, 2$), and with $\boldsymbol{k}' \equiv K(\boldsymbol{x}/ \mid \boldsymbol{x} \mid)$, Eqs. (2.38) and (D.2) lead to:

$$h(\boldsymbol{x}) \to h^{(0)}(\boldsymbol{x}) + \frac{\exp iK \mid \boldsymbol{x} \mid}{\mid \boldsymbol{x} \mid^{1/2}} \sum_{\sigma} f_w(\boldsymbol{k}', \sigma; \boldsymbol{k}, \sigma_{in})\epsilon(\boldsymbol{k}', \sigma) \tag{2.39}$$

$$f_w(\boldsymbol{k}', \sigma; \boldsymbol{k}, \sigma_{in}) = \frac{1}{4i}\left[\frac{2}{\pi K}\right]^{1/2} \exp[-i\pi/4]\int d^2\boldsymbol{x}' \exp[-i\boldsymbol{k}'\boldsymbol{x}']\epsilon(\boldsymbol{k}', \sigma)^+ j(\boldsymbol{x}') \tag{2.40}$$

$\epsilon(\boldsymbol{k}', \sigma)$ is a possible two-component linear (real) or circular (complex) polarization vector in two dimensions (Appendix D). The sum over the polarization index σ is carried over only two values. $f_w(\boldsymbol{k}', \sigma; \boldsymbol{k}, \sigma_{in})$ is the two-dimensional electromagnetic scattering amplitude by the N anisotropic cylindrical waveguides.

Eq. (2.38) also applies for confined propagation (with $h^{(0)}(\boldsymbol{x}) = 0$ and K pure imaginary); Section 2.12 will be devoted to studying that physical situation.

2.10.3 One Anisotropic Cylindrical Waveguide: Weak Scattering

We now restrict the study in the precedent subsection to the case of only one cylindrical waveguide (all the remaining $N - 1$ ones being removed by assumption). We concentrate on Eq. (2.38), when the integral in it refers only to that unique waveguide.

Notice that $h(\boldsymbol{x}')$ and derivatives thereof appear inside $j(\boldsymbol{x}')$ (as implied by Eqs. (2.34) and (2.35)). It is possible to recast Eq. (2.38) into an equivalent form in which only $h(\boldsymbol{x}')$ (but not its derivatives) appears inside the integral over $\boldsymbol{x}'$. This is at the expense of having derivatives of $H_0^{(1)}(K \mid \boldsymbol{x} - \boldsymbol{x}' \mid)$ inside that integral. The alternative form of Eq. (2.38) is commented succinctly in Appendix E and has been treated in detail previously.[19]

If the anisotropic waveguide is a weak scatterer (say, if all $\mid \epsilon_{ii} - 1 \mid$ and $\mid \epsilon_{12} \mid$ are small compared to unity), it suffices to approximate $h(\boldsymbol{x})$ by $h^{(0)}(\boldsymbol{x})$ plus the result of the first iteration of Eq. (2.38). One gets:

$$h(\boldsymbol{x}) \simeq h^{(0)}(\boldsymbol{x}) + \int d^2\boldsymbol{x}' \frac{H_0^{(1)}(K \mid \boldsymbol{x} - \boldsymbol{x}' \mid)}{4i} j^{(1)}(\boldsymbol{x}') \tag{2.41}$$

where $j^{(1)}(\boldsymbol{x}')$ is obtained upon replacing $h(\boldsymbol{x}')$ by $h^{(0)}(\boldsymbol{x}')$ in $j(\boldsymbol{x}')$ (notice that a similar procedure was employed in Eqs. (2.21) and (2.22)). The integral in the right-hand side of Eq. (2.41) is extended only over the transverse cross-section of the unique waveguide considered here. In turn, the replacement of $h(\boldsymbol{x}')$ by $h^{(0)}(\boldsymbol{x}')$ in Eq. (2.40) provides the Debye-Born-Rayleigh-Gans approximation $f_w(\boldsymbol{k}', \sigma; \boldsymbol{k}, \sigma_{in})_{DBRG}$ for $f_w(\boldsymbol{k}', \sigma; \boldsymbol{k}, \sigma_{in})$:

$$f_w(\boldsymbol{k}', \sigma; \boldsymbol{k}, \sigma_{in})_{DBRG} = \frac{1}{4i}\left[\frac{2}{\pi K}\right]^{1/2} \exp[-i\pi/4]$$
$$\cdot \int d^2\boldsymbol{x}' \exp[-i\boldsymbol{k}'\boldsymbol{x}')]\epsilon(\boldsymbol{k}', \sigma)^{+} j^{(1)}(\boldsymbol{x}') \tag{2.42}$$

In previous works,[20,31] the scattering by one anisotropic waveguide has been studied using Eq. (2.41) and a detailed numerical analysis. In those works, the following useful and practical conclusions have been reached for weak scattering. Let: i) ϵ_{12} and $\epsilon_{11} - \epsilon_{22}$ be suitably small (say, small optical activity and anisotropic waveguide relatively close to uniaxial behavior), ii) ϵ_{11} and ϵ_{33} be weakly dependent on $\boldsymbol{x}$, and iii) let $\boldsymbol{k}$ be suitably small, so that $h^{(0)}(\boldsymbol{x}')$ varies little as $\boldsymbol{x}'$ does inside the transverse section of the waveguide. Then, the contribution of $j^{(1)}(\boldsymbol{x}')$ to the integral in (2.41) is dominated by the terms independent on spatial derivatives of $H_0^{(1)}(K \mid \boldsymbol{x} - \boldsymbol{x}' \mid)$.[20,31]

2.10.4 Scattering by One Anisotropic Cylindrical Waveguide: Long Wavelength Approximation

We shall now outline a long-wavelength approximation for the scattering by one waveguide. Besides the above weak scatterer assumptions in the previous subsection, let us suppose that K (real) is such that the dimensionless product $\mid K \mid R$ is smaller than unity (the area of the transverse cross-section Ω of the cylindrical waveguide being of order R^2). Consistently with the last assumption, we neglect the $\boldsymbol{x}'$ dependence in $h(\boldsymbol{x}')$ inside the integral, as we integrate over $\boldsymbol{x}'$ with $\mid \boldsymbol{x}' \mid$ less than about R. Also, we shall restrict to $\boldsymbol{x}$ such that $\mid K \mid\mid \boldsymbol{x} \mid \leq 1$. Then $h^{(0)}(\boldsymbol{y}) = h^{(0)} \exp i\boldsymbol{y}\boldsymbol{k} \simeq h^{(0)}$ for $\boldsymbol{y} = \boldsymbol{x}$, and $\boldsymbol{x}'$. Also, $h = h(\boldsymbol{x})$ is essentially independent on $\boldsymbol{x}$, and we also denote its two constant components by $H_i, i = 1, 2$. One can apply the approximation (D.5) to the integral in the second term of the right-hand side of the exact Eq. (2.38) (but, in principle, not to that of the first-order correction to Eq. (2.38), given in Eq. (2.41)). Then:

$$h \simeq h^{(0)} + \frac{1}{2\pi} \ln[-2^{-1} i K R \exp \gamma] \int d^2\boldsymbol{x}'\, j(\boldsymbol{x}') \tag{2.43}$$

γ being Euler's constant.[58] In turn, as we have neglected the $\boldsymbol{x}'$ dependence of h inside the integral, one has $\partial H_i/\partial x_j = 0$ for $i, j = 1, 2$, and the two components of the actual $j(\boldsymbol{x}')$ in (2.43) reduce to:

$$j(\boldsymbol{x}')_1 = \left(\frac{\omega}{v}\right)^2 [-(\epsilon_{22}(\boldsymbol{x}') - 1)H_1 + \epsilon_{21}(\boldsymbol{x}')H_2] \equiv \sum_{j=1}^{2} (A_2(\boldsymbol{x}'))_{1j} H_j \tag{2.44}$$

$$j(\boldsymbol{x}')_2 = \left(\frac{\omega}{v}\right)^2 [-(\epsilon_{11}(\boldsymbol{x}') - 1)H_2 + \epsilon_{12}(\boldsymbol{x}')H_1] \equiv \sum_{j=1}^{2} (A_2(\boldsymbol{x}'))_{2j} H_j \tag{2.45}$$

The right-hand sides of (2.44) and (2.45) define the four elements $(A_2(\boldsymbol{x}'))_{ij}$ of the 2×2 matrix $A_2(\boldsymbol{x}')$. The notation indicates the fact that ϵ_{ij} may still depend on $\boldsymbol{x}'$. Eq. (2.43) should not be confused with Eq. (2.41). Thus, Eq. (2.41) requires weak scattering and includes only the first iteration, but it is not restricted to long wavelength. On the other hand, Eq. (2.43) requires weak scattering as well, but it amounts to treating approximately all iterations of Eq. (2.38), although it is restricted to long wavelength. Thus, even if, for physical consistency, the weak scatterer assumptions are also involved in Eq. (2.43), all higher order iterations of Eq. (2.38)) are to be taken into account in the approximation. Eq. (2.43) is an inhomogeneous linear system of two equations for $H_i, i = 1, 2$ that can be trivially solved and so provides the basis for an explicit approximate solution valid for long wavelength scattering. One gets:

$$h \simeq [I_2 - (2\pi)^{-1} \ln(-2^{-1} i K R \exp \gamma) A_2]^{-1} h^{(0)} \tag{2.46}$$

$$A_2 \equiv \int d^2\boldsymbol{x}'\, A_2(\boldsymbol{x}') \tag{2.47}$$

I_2 is the 2×2 unit matrix. Notice that $[I_2 - (2\pi)^{-1} \ln(-2^{-1} i K R \exp \gamma) A_2]^{-1}$ is a 2×2 matrix.

Using Eqs. (2.40) and (2.46), the scattering amplitude in the long wavelength approximation reads:

$$f_w(\mathbf{k}', \sigma; \mathbf{k}, \sigma_{in}) \simeq \frac{1}{4i} \exp[-i\pi/4] \left[\frac{2}{\pi K}\right]^{1/2}$$
$$\times \epsilon(\mathbf{k}', \sigma)^{+} A_2[I_2 - (2\pi)^{-1} \ln(-2^{-1} i K R \exp\gamma) A_2]^{-1} h^{(0)} \quad (2.48)$$

The well known optical theorem,[54,56] is a general consistency requirement that must be satisfied by the scattering amplitude. It is easy to see that the long wavelength approximation given in Eq. (2.48) satisfies approximately the optical theorem. For brevity, we shall omit details.

2.10.5 Multiple Scattering by N Parallel Anisotropic Waveguides: Approximations

With a view toward the analysis of the interaction of light with N parallel optically anisotropic cylindrical waveguides and a multiple scattering formulation thereof (restricting the general three-dimensional approach outlined in Section 2.9 to the actual two-dimensional case), we shall start with the case of only one cylindrical waveguide (all the remaining $N-1$ ones being removed for the time being). Then, we consider Eq. (2.38) for only one cylindrical waveguide. We notice that the series for $h(\mathbf{x})$ formed by the all successive iterations of Eq. (2.38) (providing its solution for one waveguide) can be rewritten formally as follows. By omitting explicit $\mathbf{x}$ dependences, such a series solution of Eq. (2.38) can be cast as:

$$h = h^{(0)} + \frac{H_0^{(1)}}{4i} \tau_w h^{(0)} \quad (2.49)$$

which defines the waveguide scattering integral operator τ_w (which is also a 2×2 matrix). In detail, Eq. (2.49) reads:

$$h(\mathbf{x}) = h^{(0)}(\mathbf{x}) + \int d^2\mathbf{x}' \int d^2\mathbf{x}'' \frac{H_0^{(1)}(K \mid \mathbf{x} - \mathbf{x}' \mid)}{4i} \tau_w(\mathbf{x}', \mathbf{x}'') h^{(0)}(\mathbf{x}'') \quad (2.50)$$

The integrations over both $\mathbf{x}'$ and $\mathbf{x}''$ in (2.50) are carried out over the transverse cross-section of the waveguide. The comparison between the exact Eq. (2.50) and the approximate Eq. (2.41) would allow us to characterize the first approximation $\tau_w^{(1)}$ for τ_w. One has:

$$\int d^2\mathbf{x}'' \tau_w^{(1)}(\mathbf{x}', \mathbf{x}'') h^{(0)}(\mathbf{x}'') = j^{(1)}(\mathbf{x}') \quad (2.51)$$

$\tau_w^{(1)}(\mathbf{x}', \mathbf{x}'')$ could be inferred from the validity of Eq. (2.51) for any $h^{(0)}$. We shall not pursue this and we shall now give an explicit long wavelength approximation for τ_w, based upon the previous subsection. Thus, under the long wavelength approximation (so that both h and $h^{(0)}$ are $\mathbf{x}$-independent), the comparison between (2.50) and (2.46) yields an explicit approximation, $\int d^2\boldsymbol{x}' \int d^2\boldsymbol{x}'' \tau_{w,lw}$, for the double integral $\int d^2\boldsymbol{x}' \int d^2\boldsymbol{x}'' \tau_w$:

$$\int d^2\boldsymbol{x}' \int d^2\boldsymbol{x}'' \tau_{w,lw}(\boldsymbol{x}', \boldsymbol{x}'') \\ = A_2[I_2 - (2\pi)^{-1} \ln(-2^{-1} i K R \exp\gamma) A_2]^{-1} \tag{2.52}$$

We now come back to the scattering of the incoming field $h^{(0)}(\boldsymbol{x})$ by all (N) parallel anisotropic cylindrical waveguides, by restricting to two dimensions the three-dimensional multiple scattering approach outlined in Section 2.9. Neither weak scatterer nor long wavelength approximations will be made, unless otherwise stated. In such an application, an important additional simplification occurs: one no longer works with the more complicated transverse Green's function G_{Tr} but with the simpler Green's function $H_0^{(1)}/4i$.

We shall also have in mind Eqs. (2.38) and (2.49) for each single waveguide: they characterize the waveguide scattering (2×2 matrix) integral operator $\tau_{w,i}$ (= $\tau_{w,i}(\boldsymbol{x}', \boldsymbol{x}'')$) for the i-th anisotropic waveguide. Let $h(\boldsymbol{x})$ be the total magnetic field describing the scattering of the incoming field $h^{(0)}(\boldsymbol{x})$ by the N waveguides: $h(\boldsymbol{x}) \equiv h$ (which is a column vector with two components) fulfills Eq. (2.38), the integration over $\boldsymbol{x}'$ being extended over the N transverse cross-sections of the waveguides. In the multiple scattering formulation, the same total magnetic field h, is also given by:

$$h = h^{(0)} + \sum_{i=1}^{N} \frac{H_0^{(1)}}{4i} \tau_{w,i} h_i \tag{2.53}$$

In turn, the new "partial" magnetic fields $h_i = h_i(\boldsymbol{x}), i = 1, \ldots, N$, are the solution of:

$$h_i = h^{(0)} + \sum_{j=1, j \neq i}^{N} \frac{H_0^{(1)}}{4i} \tau_{w,j} h_j \tag{2.54}$$

Eq. (2.54) is an inhomogeneous linear system of N integral equations for all h_j, $j = 1, \ldots, N$. Recall that each h_j is a column vector with two components. At this stage, we take the long distance limit in Eq. (2.53) ($\mid \boldsymbol{x} \mid \to +\infty$, along a fixed direction, with $\boldsymbol{k}' \equiv K(\boldsymbol{x}/\mid \boldsymbol{x} \mid)$) and recall Eq. (D.2). One gets again Eq. (2.39), with an alternative representation for the scattering amplitude $f_w(\boldsymbol{k}', \sigma; \boldsymbol{k}, \sigma_{in})$ in terms of all h_i, namely:

$$f_w(\boldsymbol{k}', \sigma; \boldsymbol{k}, \sigma_{in}) = \frac{1}{4i} \left[\frac{2}{\pi K}\right]^{1/2} \exp[-i\pi/4] \sum_{i=1}^{N} \int d^2\boldsymbol{x}' d^2\boldsymbol{x}'' \exp[-i\boldsymbol{k}'\boldsymbol{x}'] \\ \times \epsilon(\boldsymbol{k}', \sigma)^{+} \tau_{w,i}(\boldsymbol{x}', \boldsymbol{x}'') h_i(\boldsymbol{x}') \tag{2.55}$$

We emphasize that no approximations are involved in either Eq. (2.53) or Eq. (2.54). We shall turn to perform approximations on them in the remainder of this subsection and in the following one.

Let us choose some fixed axis inside the i-th cylindrical waveguide and another one in the j-th one ($i \neq j$), both of them being parallel to the x_3 axis. Let $D_{ij}, i \neq j$, be the distance between those two axes (measured, of course, in a plane orthogonal to the x_3 axis). D_{ij} gives information about the average separation between those waveguides. In various cases of physical interest, like several that apply to biological waveguides in the eye (Section 2.7), D_{ij}, R (the typical transverse dimension of any waveguide) and the typical wavelengths of light have about similar orders of magnitude. Then, Eq. (2.54) is not easy to analyze and one may have to turn to numerical studies directly.

Certain cases, in which some length parameters are larger than the others, become more tractable from the theoretical point of view, as we shall see. Let us suppose that all D_{ij} are adequately larger than R (the typical transverse dimension of any waveguide) and than $2\pi K^{-1}$ (the wavelength of light). Let $\boldsymbol{x}$ lie inside the transverse cross-section of any of the N waveguides. Then, we replace, at least formally, all $H_0^{(1)}/4i$ in Eq. (2.54) by their large separation approximation, as given in (D.6). Thus, Eq. (2.54) becomes:

$$h_i(\boldsymbol{x}) \simeq h^{(0)}(\boldsymbol{x}) + \sum_{j=1,j\neq i}^{N} \frac{1}{4i}\left[\frac{2}{\pi K D_{ij}}\right]^{1/2} \exp i(K D_{ij} - \pi/4) \times \exp[iK\boldsymbol{u}_{ij}(\boldsymbol{x} - \boldsymbol{z}_i)]Y_j(\boldsymbol{u}_{ij}) \tag{2.56}$$

$$Y_j(\boldsymbol{u}_{ij}) \equiv \int d^2\boldsymbol{x}' \int d^2\boldsymbol{x}'' \exp[-iK\boldsymbol{u}_{ij}(\boldsymbol{x}' - \boldsymbol{z}_j)]\tau_{w,j}(\boldsymbol{x}', \boldsymbol{x}'')h_j(\boldsymbol{x}'') \tag{2.57}$$

The integrations over both $\boldsymbol{x}'$ and $\boldsymbol{x}''$ in (2.57) are carried out over the transverse cross-section of the j-th waveguide. Notice the key feature that all $Y_j(\boldsymbol{u}_{ij})$ are column vectors with two ($\boldsymbol{x}$-independent but $\boldsymbol{u}_{ij}$-dependent) components. Next, we consider (2.56) for $i = 1$ and $\boldsymbol{x}$ lying inside the transverse cross-section of that first waveguide and we multiply, successively, that equation by $\exp[-iK\boldsymbol{u}_{s1}(\boldsymbol{x}''' - \boldsymbol{z}_1)]\tau_{w,1}(\boldsymbol{x}''', \boldsymbol{x})$, with $s = 2, \ldots N$. This gives rise to $N-1$ separate equations, for the different values of s. In each of the $N-1$ equations so resulting, we integrate the above products over both $\boldsymbol{x}'''$ and $\boldsymbol{x}$ inside the transverse cross-section of the first cylindrical waveguide. Then, the left-hand sides of the resulting $N-1$ equations become $Y_1(\boldsymbol{u}_{s1})$. We perform similar operations successively with (2.56) for $i = 2, \ldots N$ and $\boldsymbol{x}$ lying, in each case, inside the transverse cross-section of that i-th waveguide. Then, (2.56) becomes an inhomogeneous linear and purely algebraic system for all $Y_i(\boldsymbol{u}_{si})$, $s \neq i$:

$$Y_i(\boldsymbol{u}_{si}) = Y_i^{(0)}(\boldsymbol{u}_{si}) + \sum_{j=1,j\neq i}^{N} \Lambda_{i,j}(\boldsymbol{u}_{si}, \boldsymbol{u}_{ij})Y_j(\boldsymbol{u}_{ij}) \tag{2.58}$$

where:

$$Y_i^{(0)}(\boldsymbol{u}_{si}) = \int d^2\boldsymbol{x}''' \int d^2\boldsymbol{x}_i \exp[-iK\boldsymbol{u}_{si}(\boldsymbol{x}''' - \boldsymbol{z}_i)]\tau_{w,i}(\boldsymbol{x}''', \boldsymbol{x})h^{(0)}(\boldsymbol{x}) \tag{2.59}$$

$$\Lambda_{i,j}(\boldsymbol{u}_{si}, \boldsymbol{u}_{ij}) = \frac{1}{4i}\left[\frac{2}{\pi K D_{ij}}\right]^{1/2} \exp i(K D_{ij} - \pi/4) \int d^2\boldsymbol{x}''' \int d^2\boldsymbol{x} \times \exp[-iK\boldsymbol{u}_{si}(\boldsymbol{x}''' - \boldsymbol{z}_i)]\tau_{w,i}(\boldsymbol{x}''', \boldsymbol{x}) \exp[iK\boldsymbol{u}_{ij}(\boldsymbol{x} - \boldsymbol{z}_i)] \tag{2.60}$$

All $Y_j^{(0)}(\boldsymbol{u}_{si})$ are column vectors with two $\boldsymbol{x}$-independent components, while each $\Lambda_{ij}(\boldsymbol{u}_{si}, \boldsymbol{u}_{ij})$ is a 2×2 matrix with four $\boldsymbol{x}$-independent components.

To summarize: all $Y_i(\boldsymbol{u}_{si})$, $s \neq i$, are given by solving the inhomogeneous algebraic system (2.58). Then, Eq. (2.56) provides $h_i(\boldsymbol{x})$ for $\boldsymbol{x}$ inside the i-th waveguide. Finally, Eq. (2.55) gives the scattering amplitude in terms of all $h_i(\boldsymbol{x})$, as $\boldsymbol{x}$ varies inside the i-th waveguide. Then, when all D_{ij} are suitably large, the solution of the multiple scattering of an electromagnetic wave by N parallel anisotropic cylindrical waveguides is, in essence, reduced to the explicit knowledge of suitable integrals of all $\tau_{w,i}(\boldsymbol{x}''', \boldsymbol{x})$ and to the solution of an inhomogeneous system of algebraic system (for all $Y_i(\boldsymbol{u}_{si})$, $s \neq i$). $\tau_{w,i}(\boldsymbol{x}''', \boldsymbol{x})$ are discussed further below.

In general, there still remain the difficulties involved in the computations of integrals of all $\tau_{w,i}(\boldsymbol{x}''', \boldsymbol{x})$, which, in particular, are required in order to know the constant column vectors $Y_i^{(0)}(\boldsymbol{u}_{si})$ and the constant 2×2 matrices $\Lambda_{i,j}(\boldsymbol{u}_{si}, \boldsymbol{u}_{ij})$. One could proceed further, upon replacing each $\tau_{w,i}$ by its first approximation $\tau_w^{(1)}$, characterized previously in this subsection. Another possibility consists in applying the long wavelength approximation, to which we now turn. As not only all D_{ij} are large but the wavelength is also large, we have to be more specific about their relative magnitudes. For the validity of the approximate equations below, we shall suppose that all separations are larger than the wavelength which, in turn, is larger than the typical transverse sizes of all waveguides: $D_{ij} > 2\pi K^{-1} > R$. Then, upon integrating inside the transverse cross-section of each cylindrical waveguide in Eqs. (2.59), (2.60), $h^{(0)}(\boldsymbol{x})$ can be regarded as approximately constant and all K-dependent exponentials can be approximated by unity. The new simplifying feature is that Eq. (2.52) can now be applied for each waveguide. Then, Eqs. (2.59) and (2.60) become:

$$Y_i^{(0)}(\boldsymbol{u}_{si}) \simeq \int d^2\boldsymbol{x}''' \int d^2\boldsymbol{x}\,\tau_{w,i}(\boldsymbol{x}''', \boldsymbol{x})h^{(0)} \tag{2.61}$$

$$\Lambda_{i,j}(\boldsymbol{u}_{si}, \boldsymbol{u}_{ij}) \simeq \frac{1}{4i}\left[\frac{2}{\pi K D_{ij}}\right]^{1/2} \exp i(K D_{ij} - \pi/4)$$
$$\times \int d^2\boldsymbol{x}''' \int d^2\boldsymbol{x}\,\tau_{w,i}(\boldsymbol{x}''', \boldsymbol{x}) \tag{2.62}$$

The double integral $\int d^2\boldsymbol{x}' \int d^2\boldsymbol{x}''\tau_{w,i}(\boldsymbol{x}', \boldsymbol{x}'')$ (a constant 2×2 matrix) is given by the right-hand side of (2.52) for each waveguide. In this long wavelength approximation, all inhomogeneous terms and all coefficients in the linear system of equations for all Y_i in (2.58) are fully explicit. Then, the solution of the multiple scattering of an electromagnetic wave by N parallel anisotropic cylindrical waveguides boils down to solve such an algebraic system for all Y_i. It may have some interest to have treated the above cases in which all D_{ij} are adequately larger than the other length parameters. In fact, such cases could provide some clue or some starting approximation for other situations in which D_{ij}'s be only slightly or barely larger than the transverse dimensions of the waveguides and the typical wavelengths of light.

2.10.6 Multiple Scattering by N Parallel Anisotropic Waveguides: Approximate Calculation Using Sampling Theorem

The computation of the scattered field for an array of parallel cylindrical anisotropic waveguides is not an easy task. We shall now outline the results of an approximate calculation in which the separation between two neighboring waveguides is not appreciably larger than their transverse size (R), based on the application of sampling methods.[59] The latter methods have already been employed in previous works in optical diffraction theory[60] and biological waveguides.[61]

A summary of the sampling theorem is given in Appendix F. For convenience, we shall change some notations. Thus, (x, y) and $1/(2B)$, employed in Appendix F, are replaced here by (x_1, x_2) and $d/2$, respectively. A point $(nd/2, md/2)$ for each (n, m) is associated to one waveguide. The distance $d/2$ gives a scale characterizing the average separation between any two adjacent waveguides (say, between two axes suitably chosen in them). Roughly speaking, the transverse size of each waveguide (say, twice its radius, R) should not exceed lengths that are about that average separation $(d/2)$; this is the so-called Nyquist limit. In some cases, $2R$ may even be allowed to exceed $d/2$ slightly, but not much; in such a case, numerical studies are required in order to check the reliability of those values. Let us recall the form of the sampling theorem as given in Eq. (F.4) with the function g replaced by $h = h(\boldsymbol{x}) = h(x_1, x_2)$, which is a column vector with two components. Each of the latter are, in turn, supposed to be band-limited functions (see Appendix F). One has the following identity, provided by the sampling theorem:

$$h(x_1, x_2) = \sum_{n=-\infty}^{+\infty} \sum_{m=-\infty}^{+\infty} h\left(\frac{nd}{2}, \frac{md}{2}\right) \frac{\pi}{4} \times \frac{2J_1(2\pi((x_1 - nd/2)^2 + (x_2 - md/2)^2)^{1/2}/d)}{2\pi((x_1 - nd/2)^2 + (x_2 - md/2)^2)^{1/2}/d} \tag{2.63}$$

$J_1(x)$ is the ordinary Bessel's function of order 1.[56]

In particular, Eq. (2.63) becomes an identity if $h(\boldsymbol{x})$ and $h(\frac{nd}{2}, \frac{md}{2})$ are replaced by $h^{(0)}(\boldsymbol{x}) = h^{(0)}(x_1, x_2)$ and $h^{(0)}(\frac{nd}{2}, \frac{md}{2})$, respectively.

Next, we perform the following approximation on Eq. (2.63): we replace the double infinite series in it by two finite sums, so that the total number of terms in the latter coincides with the total number of parallel cylindrical anisotropic waveguides. Then, the identity (2.63) becomes the approximate formula:

$$h(x_1, x_2) \simeq \sum_{n,m} h\left(\frac{nd}{2}, \frac{md}{2}\right) \frac{\pi}{4} \times \frac{2J_1((2\pi/d)((x_1 - nd/2)^2 + (x_2 - md/2)^2)^{1/2})}{(2\pi/d)((x_1 - nd/2)^2 + (x_2 - md/2)^2)^{1/2}} \tag{2.64}$$

After that approximation, notice the following. Let $h(\boldsymbol{x})$ and $h(\frac{nd}{2}, \frac{md}{2})$ be replaced by $h^{(0)}(\boldsymbol{x}) = h^{(0)}(x_1, x_2)$ and $h^{(0)}(\frac{nd}{2}, \frac{md}{2})$, respectively, in Eq. (2.64). Then, one no longer has an identity, that is, the left-hand side of Eq. (2.64) and its right-hand side are not equal to each other. Clearly, their discrepancy becomes the smaller; the larger is the total number of terms in the two finite sums in Eq. (2.64). Of course, these remarks (the discrepancy and its reduction, as the total number of terms retained increases) also apply if, in the application of Eq. (2.64) for an array of parallel cylindrical anisotropic waveguides, both $h(x_1, x_2)$ and $h(\frac{nd}{2}, \frac{md}{2})$ stand for the total magnetic field (that given by Eq. (2.53)). However, the latter application would appear to be useless. In fact, one would not know, *a priori*, the total field ($h(\frac{nd}{2}, \frac{md}{2})$) at the finite number of points $(nd/2, md/2)$ associated to the terms retained in the two finite sums in Eq. (2.64).

Fortunately, at this stage, we can still employ Eq. (2.64) for an approximate computation of the total magnetic field (given by (2.53)), provided that we make the following additional assumption. We approximate $h(\frac{nd}{2}, \frac{md}{2})$ in the right-hand side of Eq. (2.64) by the total magnetic field corresponding to the scattering by just one waveguide, namely the one corresponding to $(x_1 = nd/2, x_2 = md/2)$. In this way, one profits from the knowledge of the total field corresponding to the scattering by single waveguides which is simpler to obtain. In other words, such a $h(\frac{nd}{2}, \frac{md}{2})$ is approximated by the solution of Eq. (2.49) for the single waveguide corresponding to $(nd/2, md/2)$. Equivalently, which may add a bit more insight, $h(\frac{nd}{2}, \frac{md}{2})$ is replaced by the right-hand side of (2.53), when, in the latter, we approximate the full $h^{(0)} + \sum_{i=1}^{N}(H_0^{(1)}/4i)\tau_{w,i}h_i$ just by $h^{(0)} + (H_0^{(1)}/4i)\tau_{w,i}h^{(0)}$, the retained i corresponding to $(nd/2, md/2)$. In turn, $h^{(0)} + (H_0^{(1)}/4i)\tau_{w,i}h^{(0)}$ is approximated by the right-hand side of Eq. (2.41), with the adequate $j^{(1)}(\boldsymbol{x}')$ (obtained upon replacing $h(\boldsymbol{x}')$ by $h^{(0)}(\boldsymbol{x}')$ in $j(\boldsymbol{x}')$). That is, $h^{(0)} + (H_0^{(1)}/4i)\tau_{w,i}h^{(0)}$ is approximated by the Debye-Born-Rayleigh-Gans approximation. Of course, if there were no scattering ($\tau_{w,i} = 0$ for all i, say, for all $(nd/2, md/2)$), then the resulting (2.64) would reduce to the approximate form of the sampling theorem corresponding to $h^{(0)}(x_1, x_2)$ discussed above.

The spirit of the above approximations is that the resulting (2.64) could still yield an acceptable approximation for the total magnetic field when multiple scattering effects are included, that is, that provided by the multiple scattering approach. Alternatively, one expects that (2.64) would provide a reasonable approximation for (2.53).

Depending on the symmetry of the problem, different sampling lattices may be used. Eqs. (2.63) and (2.64) correspond to one of such choices. See Figure 2.5.

Applications of (2.64) for transverse sizes of the waveguides and average separations between adjacent ones about the micron range have been carried out.[20,31] In the numerical calculations carried out in the latter references, the values $d = 2\ \mu m$ and $R = 0.75\ \mu m$ have been employed. Therefore, $2R = 1.5\ \mu m$ is, *a priori*, slightly beyond the $2R \leq d/2$ bound (Nyquist limit). However, no artifacts have been observed upon performing the numerical computations, so that the results of the different analysis[20,31] appear to be reasonable.

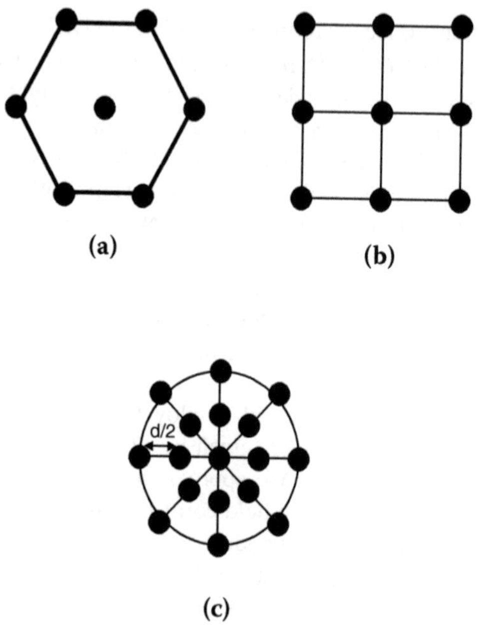

FIGURE 2.5 Three different geometries of the sampling lattice. Eqs. (2.63), (2.64) and (F.4) correspond to c), while Eq. (F.2) corresponds to b).

2.11 ANISOTROPIC DIELECTRIC WAVEGUIDES WITH PHOTO-ELASTIC EFFECT

We continue to assume that the non-vanishing ϵ_{ij} are real for typical frequencies of light, at least, approximately (absence of absorption).

2.11.1 SCATTERING

Let us consider an extended isotropic and homogeneous reference medium, with real constant dielectric permittivity ϵ_0. Let N (≥ 1) parallel cylindrical dielectric waveguides be located inside that medium. The latter are supposed to be isotropic (and optically isotropic) to some zeroth-order (or leading order) approximation or in the absence of externally applied forces or actions. Those waveguides have dielectric permittivity tensor $\epsilon_{ik} = \epsilon_e \delta_{ik}$, $i, k = 1, 2, 3$, relative to that of the surrounding medium. In a next-to-leading order approximation or under externally applied forces, small additional internal stresses σ_{ik} inside those waveguides should be taken into account, so that the latter become optically anisotropic by virtue of the photo-elastic effect (as explained in Section 2.5). Then, the induced relative dielectric permittivity tensor ϵ_{ik} of the waveguides is given by Eq. (2.10).

In a strict sense, when the photo-elastic effect is taken into account, ϵ_{ik} as given by Eq. (2.10) lies outside the domain of validity of the simplifications considered in Subsection 2.10.1, which enabled reduction of the three-dimensional problem to a two-dimensional one. In fact, one may expect that, by virtue of the photo-elastic effect, one has $\epsilon_{13} \neq 0$ and $\epsilon_{23} \neq 0$, in general and, so, the simpler (two-dimensional) scattering theory of Section 2.10 would not apply. Instead, the more general (three-dimensional) one as outlined in Section 2.8 should be employed, but its use involves a larger amount of computational work. In order to avoid three-dimensional complications wherever possible, we shall suppose, at least tentatively, that ϵ_{13} and ϵ_{23} can still be regarded as negligible, say, at the present next-to-leading order approximation. With this simplifying assumption, strained cylindrical parallel waveguides, in which the induced anisotropy comes from the photo-elastic effect, can still be treated through the simpler two-dimensional procedures cited in Section 2.10, as we shall now outline. We shall summarize the analysis carried out,[20,31] in which further simplifying assumptions were also made. Suitable components of the stresses σ_{ik} were taken as proportional to the transverse force per unit length applied to each waveguide. Thus, for a force (with absolute magnitude F) applied parallel to the x_2 axis, the mechanical effects will be produced in the (x_1, x_2) plane. Then, σ_{11} is estimated to be $F/(\pi R^2)$, and $\sigma_{22} = -3\sigma_{11}, \sigma_{33} = 0$ with all the remaining $\sigma_{ik} \simeq 0, i \neq k$. See Figure 2.6. With the induced anisotropy due to the photo-elastic effect, $\epsilon_{12} = \epsilon_{21} \simeq 0$ were regarded also as negligible. Then, the non-vanishing elements of ϵ_{ik} are:

$$
\begin{aligned}
\epsilon_{11} &\simeq \epsilon_e + a_1\sigma_{11} + a_2\sigma_{22} + a_2\sigma_{33} \\
\epsilon_{22} &\simeq \epsilon_e + a_2\sigma_{11} + a_1\sigma_{22} + a_2\sigma_{33} \\
\epsilon_{33} &\simeq \epsilon_e + a_1\sigma_{11} + a_2\sigma_{22} + a_1\sigma_{33}
\end{aligned}
\tag{2.65}
$$

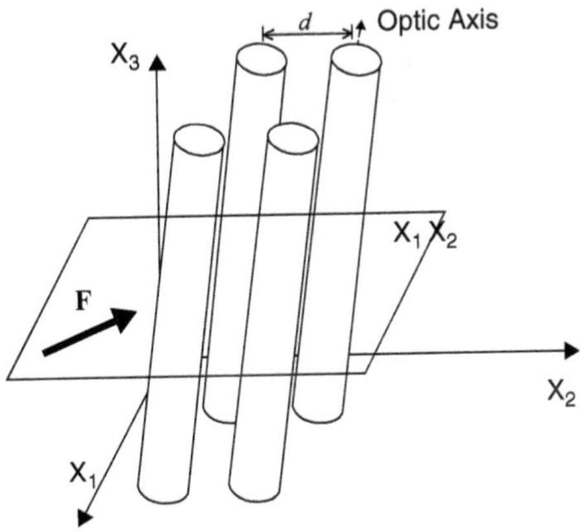

FIGURE 2.6 Model for the photo-elastic effect. A weak external force is applied in the (x_1, x_2) plane. The optic axis of each waveguide, when the external force is applied, is displayed (the optic axis of each waveguide is parallel to the x_3 axis, in the absence of external force). The angle between the optic axis of each waveguide (when the external force is applied) and the x_3 axis is assumed to be small.

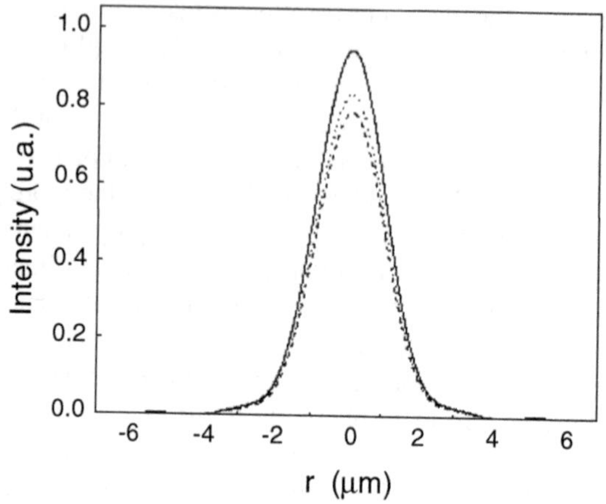

FIGURE 2.7 Spatial dependence of the forward-scattering light intensity in an interval about the forward direction (a.u.= arbitrary units). Dotted curve: zero applied force ($F = 0$). Dashed curve: $F = 0.01$ *Newton*, light polarized along x_1. Solid curve: $F = 0.01$ *Newton*, light polarized light along x_2. The external force F is applied along the x_2 direction.

a_1 and a_2 being the elastic-optical constants. Then, under the action of the externally applied force, each waveguide behaves as a uniaxial crystal. At this stage, the analysis of the total field for one single waveguide (Subsections 2.10.3 and 2.10.5) and for N parallel waveguides (using the approximations in Subsection 2.10.6 and the sampling theorem) can be directly employed. This has led to detailed numerical computations.[20,31]

The results of those computations for the intensity of light scattered in an interval about the forward direction are displayed in Figure 2.7. This figure shows the intensities for scattered light without external force ($F = 0$) and with a given value of the applied force ($F \neq 0$). In the latter case, results are also displayed for light polarized along different directions in the (x_1, x_2) plane. Figure 2.8 shows the intensity for light scattered in the forward direction against the strength of the applied force F, also for different polarizations of light in the (x_1, x_2) plane.

2.11.2 Possible Applications to Stress Sensing

The multiple scattering of light by an array of birefringent optical waveguides (arising from mechanical stresses), as described in the previous subsection, could provide another alternative for stress sensors.[31] The question of physical interest is to estimate, upon combining experimental data regarding light scattering by the strained waveguides (due to the photo-elastic effect) and theory, what the stresses σ_{ik} are; this is what stress sensing is about. One would study experimentally the scattering of

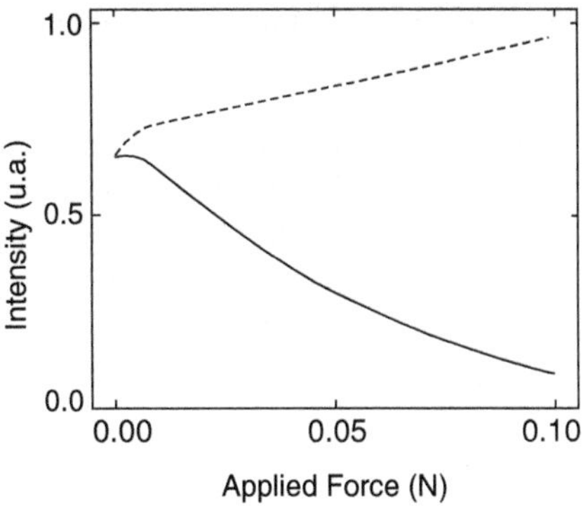

FIGURE 2.8 Forward-scattering intensity of light versus the strength F of the external force (applied along the x_2 direction) in *Newton* (N) (a.u. = arbitrary units). The solid curve corresponds to X-polarized light and the dotted curve to Y-polarized light.

light by the anisotropic waveguides (with $\sigma_{ik} \neq 0$). We suppose that we know, with adequate accuracy, ϵ_e and the elastic-optical constants a_1 and a_2 for the waveguides. One would compare the resulting experimental intensity distribution of scattered light with the one for the case in which those waveguides were isotropic (not subject to stresses, with $\sigma_{ik} = 0$). The comparison of those distributions would provide information about σ_{ik}, that is, of the state of internal stress of the waveguides and of the externally applied force (which gave rise to σ_{ik}).

Detailed numerical studies[20,31] show that the major impact of stress occurs at the central peak of the scattered light profile, that corresponds to forward scattering. This fact is what the numerical results given in Figure 2.7 display in a very neat way. When the distribution of the scattered light profile is plotted in an interval about the forward direction, it shows a central peak corresponding precisely to forward scattering. This result may be very convenient for practical purposes, since the experimental detection of light scattered in the forward direction may be performed easily in various situations. Figure 2.8 shows the forward scattering intensity for x_1- and x_2-polarized light as a function of the strength F of the externally applied force along the x_2 direction.

Even for moderate values as $F = 0.05$ *Newton*, a difference around 60 per cent between the intensities of light corresponding to both polarizations is observed. Thus, since the amount of change of the intensities for both light polarizations is directly related to the strength of the applied force, Figure 2.8 could provide a calibration curve for a stress-sensing system.[20] One main consequence of stress is the loss of symmetry due to the appearance of two privileged directions in the (x_1, x_2) plane, orthogonal to the axis of the waveguides (the x_3-axis): the axis parallel to the external force (say, the x_2-axis) and the perpendicular axis (the x_1-axis). The asymmetry induced by stress between x_1- and x_2-polarized light provides a method to assess the direction of application of the external force.[62]

2.12 CONFINED PROPAGATION OF LIGHT ALONG ANISOTROPIC WAVEGUIDES

As in Section 2.10, we shall also assume that $\epsilon_{13} = \epsilon_{23} = 0$ and that the remaining non-vanishing ϵ_{ij} are real for typical frequencies of light at least approximately (negligible absorption). When confined propagation of the classical electromagnetic wave along the anisotropic cylindrical waveguides occurs, β (also real) is named the propagation constant. For confined propagation, there are neither an incoming electromagnetic field nor physically propagating waves very far from the waveguides in the $\boldsymbol{x}$-plane.

The confined propagation along the anisotropic cylindrical waveguides is described by a set of propagation modes $h(\boldsymbol{x})_\alpha$. α denotes a suitable set of labels, so that one choice of values for those labels characterizes and distinguishes one propagation mode. Each $h(\boldsymbol{x})_\alpha$ fulfills the homogeneous version of Eq. (2.38) corresponding to $h^{(0)}(\boldsymbol{x}) \equiv 0$ (absence of propagating waves very far from the waveguides) and pure imaginary K ($K^2 < 0$). Then:

$$h(\boldsymbol{x})_\alpha = \int d^2\boldsymbol{x}' \frac{H_0^{(1)}(K \mid \boldsymbol{x} - \boldsymbol{x}' \mid)}{4i} j(\boldsymbol{x}')_\alpha \tag{2.66}$$

where $j(\boldsymbol{x}')_\alpha$ is given by the right-hand side of Eqs. (2.34) and (2.35), with $h(\boldsymbol{x})$ replaced by $h(\boldsymbol{x})_\alpha$. The integration is extended over the whole cross-sections of all waveguides. In the case of $N > 1$, one can interpret physically the confined propagation by stating that an exchange of energy of the electromagnetic field among the N waveguides occurs. There is, then, an effective coupling among those waveguides ("cross-talk").

2.12.1 One Cylindrical Waveguide

Equation (2.66) describes the confined propagation along one anisotropic cylindrical waveguide. The integration in (2.66) is now performed over the cross-sections of that unique waveguide. Let us suppose that all ϵ_{ii} are slightly larger than unity and that ϵ_{12} is small. Let us also assume that K (pure imaginary) is small in the sense that the dimensionless product $\mid K \mid R$ is smaller than unity. Physically, this will amount to restricting the analysis to the case where the anisotropic cylindrical waveguide can support only one pair of non-degenerate propagation modes, eventually those of lowest order (in the long wavelength and weakly guiding approximations). With the understanding that such a pair is regarded as one propagation mode which, in turn, could exist with two different values for K, we regard and denote such an anisotropic waveguide as a monomode. Such a mode is the fundamental one and it is non-degenerate. Then, for such values of K, one can employ the approximation (D.5) in the second term of the right-hand side of Eq. (2.66). For consistency, we neglect the $\boldsymbol{x}'$ dependence in $h(\boldsymbol{x})_\alpha$ both inside and outside the integral, so that $h(\boldsymbol{x})_\alpha \simeq h_\alpha$, with constant components $H_1\alpha$ and $H_2\alpha$. Hence:

$$h_\alpha \simeq \frac{1}{2\pi} \ln[-2^{-1} i K R \exp\gamma] \int d^2\boldsymbol{x}' j(\boldsymbol{x}')_\alpha \tag{2.67}$$

Notice that ϵ_{ij} may still depend (smoothly) on $\boldsymbol{x}'$. Then, due to the latter, there could still be some $\boldsymbol{x}'$-dependence in $j(\boldsymbol{x}')_\alpha$, which now has, approximately, the following components:

$$(j(\boldsymbol{x}')_\alpha)_1 = -(\omega/v)^2(\epsilon_{22} - 1)H_{1\alpha} + (\omega/v)^2\epsilon_{21}H_{2\alpha} \tag{2.68}$$

$$(j(\boldsymbol{x}')_\alpha)_2 = -(\omega/v)^2(\epsilon_{11} - 1)H_{2\alpha} + (\omega/v)^2\epsilon_{12}H_{1\alpha} \tag{2.69}$$

Compare the approximate Eqs. (2.68) and (2.69) with the exact Eqs. (2.34) and (2.35). Eq. (2.67) with Eqs. (2.68) and (2.69) constitutes a linear homogeneous system of two equations for the two constant magnetic field components $H_{1\alpha}$ and $H_{2\alpha}$ of h_α. The necessary and sufficient condition for the existence of a non-trivial solution of that system (namely, that $H_{1\alpha}$ and $H_{2\alpha}$ do not vanish identically) is that the determinant of the 2×2 matrix for that system vanishes. Physically, the vanishing of that determinant yields the dispersion relation for the propagation modes of the monomode anisotropic waveguide. Recall that $K^2 = (\omega/v)^2 - \beta^2$ and that K is pure imaginary, so that we write $K = iK_\pm$, where both K_+ and K_- are real. Through some algebra, Eq. (2.67) with Eqs. (2.68) and (2.69) yields:

$$\ln[2^{-1}K_\pm R \exp\gamma] \simeq -\pi \frac{[N_1 - (\pm)N_2]}{D} \tag{2.70}$$

$$N_1 = \int d^2\boldsymbol{x}(\epsilon_{11} + \epsilon_{22} - 2) \tag{2.71}$$

$$N_2 = \left[\left(\int d^2\boldsymbol{x}(\epsilon_{11} - \epsilon_{22})\right)^2 + 4\left(\int d^2\boldsymbol{x}\epsilon_{12}\right)^2\right]^{1/2} \tag{2.72}$$

$$D = (\omega/c)^2\left[\left(\int d^2\boldsymbol{x}(\epsilon_{11} - 1)\right)\left(\int d^2\boldsymbol{x}(\epsilon_{22} - 1)\right) - \left(\int d^2\boldsymbol{x}\epsilon_{12}\right)^2\right] \tag{2.73}$$

which confirm that the anisotropic waveguide under the long wavelength and weakly guiding approximations, does support only one pair of non-degenerate propagation modes.

If the waveguide is optically isotropic, then $\epsilon_{11} = \epsilon_{22}$ and $\epsilon_{12} = 0$, so that $N_2 = 0$ and $K_+ = K_-$ (degeneracy). Birefringence disappears and Eq. (2.70) becomes an approximate dispersion relation for a monomode isotropic waveguide, also in the long wavelength and weakly guiding approximations. This agrees with the properties of the degenerate fundamental (or lowest order) mode of a monomode isotropic waveguide outlined in Section 2.6.

The existence of two solutions K_+ and K_- is the actual counterpart of the fact that the Fresnel equation had two solutions (recall Subsection 2.2.2).

The book by Snyder and Love[10] summarizes some detailed studies about the confined propagation of light in one homogeneous planar waveguide with step profile for the relative dielectric permittivity tensor. As the components of the latter are constant, except for finite discontinuities at the surfaces of the waveguide, the planar geometry allows for the partial differential equations to be separable in Cartesian coordinates and to be exactly solvable. Those authors[10] also discussed, quite succinctly, some aspects of the approximate analysis for one anisotropic waveguide with other geometries in the weakly guiding approximation. Our analysis in this subsection, based upon a previous work,[19] provides a non-trivial generalization of their studies.[10]

2.12.2 N Parallel Cylindrical Waveguides: Approximations

We again consider the case with all (N) parallel anisotropic waveguides and we now treat the confined propagation of electromagnetic radiation along those N waveguides (so that K is also pure imaginary).[63] We shall concentrate in modifying succinctly the multiple scattering approach for N parallel anisotropic waveguides discussed in Subsection 2.10.5. The total magnetic field is $h_\alpha(\boldsymbol{x})$. The basic modification consists in setting $h^{(0)}(\boldsymbol{x}) \equiv 0$ in the exact Eqs. (2.53) and (2.54) (as there are neither incoming electromagnetic field nor physically propagating waves very far from any of the parallel anisotropic waveguides), h and h_i being replaced by h_α and $h_{\alpha,i}$, respectively. Then:

$$h_\alpha = \sum_{i=1}^{N} \frac{H_0^{(1)}}{4i} \tau_{w,i} h_{\alpha,i} \tag{2.74}$$

$$h_{\alpha,i} = \sum_{j=1,j\neq i}^{N} \frac{H_0^{(1)}}{4i} \tau_{w,j} h_{\alpha,j} \tag{2.75}$$

Eq. (2.75) now becomes a homogeneous system of linear equations for all $h_{\alpha,i}$, $i = 1, \ldots, N$.

Let all separations among waveguides be adequately larger than their transverse dimensions (R). Then, one has new constant column vectors $Y_{\alpha,i}(\boldsymbol{u}_{si})$, $s \neq i$, also given by (2.57). The latter fulfills a homogeneous linear system implied by Eq. (2.75) and similar to (2.58), with two crucial modifications: K is pure imaginary and, of course, $Y_j^{(0)}(\boldsymbol{u}_{si}) \equiv 0$. Each $\Lambda_{ij}(\boldsymbol{u}_{si}, \boldsymbol{u}_{ij})$ continues to be given by (2.60), also with pure imaginary K. The homogeneous linear and purely algebraic system for all $Y_i(\boldsymbol{u}_{si})$, $s \neq i$, reads:

$$Y_i(\boldsymbol{u}_{si}) = \sum_{j=1,j\neq i}^{N} \Lambda_{i,j}(\boldsymbol{u}_{si}, \boldsymbol{u}_{ij}) Y_j(\boldsymbol{u}_{ij}) \tag{2.76}$$

Moreover, let us suppose that $D_{ij} > 2\pi/ \mid K \mid > R$, that is, each waveguide is monomode and the separation between any pair of waveguides is larger than the transverse effective width of each individual propagation mode ($2\pi/ \mid K \mid$). Then, $\Lambda_{ij}(\boldsymbol{u}_{si}, \boldsymbol{u}_{ij})$ is given by (2.62), with pure imaginary K. In turn, the double integral $\int d^2\mathbf{x}' \int d^2\mathbf{x}'' \tau_{w,i}$ is given approximately by (2.52), also with pure imaginary K. In this case, the homogeneous system for $Y_{\alpha,i}(\boldsymbol{u}_{si})$ can be solved explicitly. Its solution gives the dispersion relation for the propagation modes of the N-coupled monomode anisotropic waveguides (say, the counterpart of (2.70) which includes "cross-talk"). Some numerical estimates and applications for $N = 2$ have also been carried out.[63]

2.13 CONCLUSION AND FINAL COMMENTS

We have summarized two closely interrelated subjects: a) several physically relevant cases in which optical anisotropy occurs, and b) some general features of classical electromagnetic wave scattering by and confined propagation along optically anisotropic materials. Detailed applications of b) to all systems included in a) lie outside our scope here, but we tried to emphasize their relevance and reciprocal relationship and we have concentrated on certain applications. More specifically:

1. We have summarized a rather general formulation of the scattering (and multiple scattering) of a classical electromagnetic wave by N optically anisotropic objects of arbitrary shapes, relative orientations and separations. The interest in presenting this formalism lies in its generality and flexibility. It includes scattering by optically isotropic objects (in which the dielectric permittivity tensor reduces to a scalar). As a particular but important case, it also provides the basis for analyzing the scattering by parallel cylindrical optically anisotropic waveguides. It could also provide a starting point for studying more general configurations, for instance, the inclusion of misalignments of cylindrical waveguides.
2. We have concentrated on the electromagnetic scattering by $N(\geq 1)$ parallel cylindrical optically anisotropic dielectric waveguides including long wavelength approximations and the case in which the anisotropy is due to stress birefringence.
3. We have summarized recent research about the effect of external stress on scattering by an array of parallel cylindrical waveguides (with anisotropies due to the photo-elastic effect). By analyzing the light scattered (specifically in the forward direction) by dielectric fibers that were optically isotropic first and at a later stage became optically anisotropic due to stress, physically relevant information can be obtained about the externally applied force giving rise to the photo-elastic effect. That could be the basis of another kind of optical stress sensor.
4. We have treated the confined propagation along cylindrical anisotropic dielectric waveguides; in particular, the long wavelength approximation for one waveguide (monomode confined propagation) has been analyzed.

The dominant effects in the biological waveguides in the eye (those in the retinal nerve fiber layer and in the retinal photoreceptors) seem to correspond to optical isotropy, at least under normal functioning conditions. Optical anisotropies in those waveguides may arise under pathological functioning conditions (say, due to the abnormal stresses giving rise to the photo-elastic effect). Then, (weak) scattering by anisotropic waveguides could play a role. In this connection, we mention some experimental results on the Stiles-Crawford effect of first kind (*SCE-I*).[50] Such results seem to describe changes appearing in the retinal nerve fiber layer as it is exposed to some tractions and external mechanical stresses. The latter have motivated further modeling and approximate numerical estimates for the retinal nerve fiber layer, regarded as a bundle of biological waveguides in which the photo-elastic effect is included.[64] In the latter work, the intensity distribution of the light scattered by the array of waveguides is connected with the *SCE-I*, since it would reflect the waveguiding properties of the retina and it would be a psychophysical consequence of the directional sensitivity of the retinal photoreceptors.

One open important question is whether optical fibers of the nerve disk require any specific birefringence properties along with the photo-elastic effect (and dichroism) or, opposite to it, the behavior they exhibit (in relation to the optical mechanism for waveguiding and scattering of the incoming luminous radiation) is due to accumulative stress forces acting permanently throughout the whole life of each individual (infancy and adult stages would not exhibit the same photo-elastic properties).

ACKNOWLEDGMENTS

M. L. Calvo wishes to acknowledge the support of Ministerio de Ciencia y Tecnología (Project TEC2005-02186) of Spain. We are also grateful to Dr. G. F. Calvo for having provided valuable help with the LaTex files. The kind help provided by Profs. A. Munoz Sudupe and L. A. Fernandez regarding the figures is also acknowledged.

Appendix A

Fresnel Equation, Birefringence and Polarization Vectors

We consider Eqs. (2.3) for an extended homogeneous anisotropic dielectric medium and the propagation in it of an electromagnetic plane wave with frequency ω and wavevector $\bar{k} = (k_1, k_2, k_3)$. That wave is described by $\mathbf{H}' = \mathbf{H}'_0 \exp i\bar{k}\bar{x}$, $\mathbf{H}'_0$ being constant. Moreover, we shall impose that $\mathbf{H}'_0$ fulfills $\bar{k}\mathbf{H}'_0 = 0$, so that the second relation in (2.3) be automatically fulfilled. Then, the first relation in (2.3) becomes:

$$\mathbf{H}''_0 + (\omega/v)^2 \mathbf{H}'_0 = 0 \tag{A.1}$$

$$\mathbf{H}''_0 = \bar{k} \times [(\tilde{\epsilon})^{-1} \bar{k} \times \mathbf{H}'_0] \tag{A.2}$$

$\times$ denoting vector product. Eq. (A.1) is also consistent with the second relation in (2.3). Let σ be an index which takes on only two values (say, σ_1 and σ_2). Also, let $\bar{\epsilon}(\bar{k}, \sigma)$ be two real or complex three-dimensional (column) vectors fulfilling $\bar{k}\bar{\epsilon}(\bar{k}, \sigma) = 0$ for $\sigma = \sigma_1, \sigma_2$, as well as $\bar{\epsilon}(\bar{k}, \sigma)^+ \bar{\epsilon}(\bar{k}, \sigma') = \delta_{\sigma,\sigma'}$ and $\sum_\sigma (\bar{\epsilon}(\bar{k}, \sigma))_i (\bar{\epsilon}(\bar{k}, \sigma)^+)_j = \delta_{i,j} - (k_i k_j / \bar{k}^2)$, $i, j = 1, 2, 3$. $\delta_{i,j}$ denotes here the Kronecker delta symbol, namely, $\delta_{i,i} = 1$ and $\delta_{i,j} = 0$ for $i \neq j$. The superscript $+$ denotes adjoint (that is, complex conjugate and matrix transposition), so that $\bar{\epsilon}(\bar{k}, \sigma)^+$ is a row vector. The $\bar{\epsilon}(\bar{k}, \sigma)$'s are nothing but the standard linear (real) or circular (complex) unit polarization vectors.[51] They are given by the same expressions as for electromagnetic wave propagation in vacuum.

It follows that $\mathbf{H}'_0 = \sum_\sigma (\mathbf{H}'_0)_\sigma \bar{\epsilon}(\bar{k}, \sigma)$, $\mathbf{H}''_0 = \sum_\sigma (\mathbf{H}''_0)_\sigma \bar{\epsilon}(\bar{k}, \sigma)$, where $(\mathbf{H}'_0)_\sigma$ and $(\mathbf{H}''_0)_\sigma$ are, respectively, the two components of $\mathbf{H}'_0$ and $\mathbf{H}''_0$ with respect to the basis formed by the two vectors $\bar{\epsilon}(\bar{k}, \sigma)$. Eq. (A.2) implies necessarily $(\mathbf{H}''_0)_\sigma = \sum_{\sigma'} (\tilde{\Lambda}_2)_{\sigma,\sigma'} (\mathbf{H}'_0)_{\sigma'}$. The elements $(\tilde{\Lambda}_2)_{\sigma,\sigma'}$ of the 2×2 matrix $\tilde{\Lambda}_2$ (which depend quadratically on $\bar{k}$, bilinearly on $\bar{\epsilon}(\bar{k}, \sigma)$ and $\bar{\epsilon}(\bar{k}, \sigma)^+$ and linearly on $((\tilde{\epsilon})^{-1})_{ii}$) are somewhat lengthy. Neither will they be given here nor will their explicit expressions be necessary for the conclusion that follows. The latter equation for $(\mathbf{H}''_0)_\sigma$ and Eq. (A.1) yield:

$$\sum_{\sigma'} (\tilde{\Lambda}_2)_{\sigma,\sigma'} (\mathbf{H}'_0)_{\sigma'} + (\omega/v)^2 (\mathbf{H}'_0)_{\sigma'} = 0 \tag{A.3}$$

Eq. (A.3) is a linear homogeneous system of two equations for the two amplitudes $(\mathbf{H}_0')_{\sigma'}$. The compatibility condition for that system to have a nonzero solution for those $(\mathbf{H}_0')_{\sigma'}$ (namely the vanishing of the determinant) is a quadratic equation in ω^2, for given $\bar{k}$. The latter property, then, follows from the physical fact that only two independent states of polarization are possible for an electromagnetic plane wave. That compatibility condition is nothing but the Fresnel equation.

Appendix B

Three-Dimensional Transverse Green's Function

We write $G_{Tr}(\bar{x}) = (G_{Tr}(\bar{x})_{l,l'})$, $l, l' = 1, 2, 3$ where $G_{Tr}(\bar{x})_{l,l'}$ is the (l, l')-th component of the transverse tensor Green's function $G_{Tr}(\bar{x})$. By definition, $G_{Tr}(\bar{x})$ is constructed so that for any $\bar{x}$ in three-dimensional space:

$$\left[\Delta + K_{cl}^2\right] G_{Tr}(\bar{x})_{l,l'} = \delta_{Tr}^{(3)}(\bar{x})_{l,l'}, \quad \sum_{l=1^3} \frac{\partial G_{Tr}(\bar{x})_{l,l'}}{\partial x_l} = 0 \tag{B.1}$$

$\delta_{Tr}^{(3)}(\bar{x})_{l,l'}$ stands for the (l, l')-th component of the three-dimensional transverse Dirac delta function. It is given by:

$$\delta_{Tr}^{(3)}(\bar{x})_{l,l'} = \frac{1}{(2\pi)^3} \int d^3\bar{q} \left[\delta_{l,l'} - \frac{q_l q_{l'}}{\bar{q}^2}\right] \exp i\bar{q}\bar{x} \tag{B.2}$$

In turn, $\delta_{l,l'}$ is the Kronecker delta symbol and $\bar{q} = (q_1, q_2, q_3)$. One finds:[53]

$$G_{Tr}(\bar{x})_{l,l'} = \frac{1}{(2\pi)^3 K_{cl}^2} \int d^3\bar{q} \left[\frac{q_l q_{l'} - K_{cl}^2 \delta_{l,l'}}{\bar{q}^2 - (K_{cl}^2 + i\eta)} - \frac{q_l q_{l'}}{\bar{q}^2}\right] \exp i\bar{q}\bar{x} \tag{B.3}$$

where it is assumed that the limit $\eta \to 0$, with $\eta > 0$, is taken after having carried out the integration over $\bar{q}$.

Upon performing the integration over $\bar{q}$ (in particular, the one over $|\bar{q}|$ by means of Cauchy's residue theorem), one arrives at the explicit expression for $G_{Tr}(\bar{x})_{l,l'}$:

$$G_{Tr}(\bar{x})_{l,l'} = -\frac{\exp iK_{cl} |\bar{x}|}{4\pi |\bar{x}|} \delta_{l,l'} + \frac{1}{4\pi K_{cl}^2} \frac{\partial^2}{\partial x_l \partial x_{l'}} \left[\frac{1 - \exp iK_{cl} |\bar{x}|}{|\bar{x}|}\right] \tag{B.4}$$

Two interesting properties of $G_{Tr}(\bar{x})_{l,l'}$ are its short and long distance behaviors. For $|\bar{x}| \to 0$

$$G_{Tr}(\bar{x})_{l,l'} \simeq -\frac{1}{8\pi |\bar{x}|} \left[\delta_{l,l'} + \frac{x_l x_{l'}}{|\bar{x}|^2}\right] \tag{B.5}$$

For fixed $\bar{x}'$ and $|\bar{x}| \rightarrow +\infty$, along a fixed direction (say, fixed $x_l/|\bar{x}|, l = 1, 2, 3$):

$$G_{Tr}(\bar{x} - \bar{x}')_{l,l'} \simeq -\frac{1}{4\pi |\bar{x}|}\left[\delta_{l,l'} - \frac{x_l x_{l'}}{|\bar{x}|^2}\right] \cdot \exp iK_{cl} |\bar{x}| . \exp[-iK_{cl}(\bar{x}\bar{x}')/|\bar{x}|] \tag{B.6}$$

We introduce the outgoing wavevector $\bar{k}' \equiv K_{cl}(\bar{x}/|\bar{x}|)$ and, in terms of it, we recall the two standard (real or complex) three-dimensional polarization vectors $\bar{\epsilon}(\bar{k}', \sigma)$ discussed in Appendix A. Then, the long-distance Eq. (B.6) can be recast as:

$$G_{Tr}(\bar{x} - \bar{x}')_{l,l'} \simeq -\frac{1}{4\pi |\bar{x}|}\sum_{\sigma}(\bar{\epsilon}(\bar{k}', \sigma))_l(\bar{\epsilon}(\bar{k}', \sigma)^+)_{l'} \cdot \exp iK_{cl} |\bar{x}| . \exp[-i\bar{k}'\bar{x}'] \tag{B.7}$$

Appendix C

Alternative Two-Dimensional Formulations

We shall discuss equivalent alternatives for the two-dimensional Maxwell's equations (2.33), (2.34) and (2.35), for anisotropic cylindrical waveguides. One can also start from the system of four coupled equations for the three components H_1', H_2', H_3' in Eqs. (2.16) and (2.17), under the simplifying assumptions embodied in $\tilde{\epsilon}$, as formulated in Subsection 2.10.1. The subsystem for the first two components of the first Eq. (2.16) reads:

$$[\Delta + K_{cl}^2]H_i' = j_i', i = 1, 2 \tag{C.1}$$

$$\begin{aligned} j_1' = \frac{\partial}{\partial x_2}\left[\left(\frac{1}{\epsilon_{33}} - 1\right)\left(\frac{\partial H_2'}{\partial x_1} - \frac{\partial H_1'}{\partial x_2}\right)\right] - \frac{\partial}{\partial x_3}\left[-\frac{\epsilon_{21}}{det\tilde{\epsilon}_2}\left(\frac{\partial H_3'}{\partial x_2} - \frac{\partial H_2'}{\partial x_3}\right)\right. \\ \left. + \left(\frac{\epsilon_{11}}{det\tilde{\epsilon}_2} - 1\right)\left(\frac{\partial H_1'}{\partial x_3} - \frac{\partial H_3'}{\partial x_1}\right)\right] \end{aligned} \tag{C.2}$$

$$\begin{aligned} j_2' = \frac{\partial}{\partial x_3}\left[\left(\frac{\epsilon_{22}}{det\tilde{\epsilon}_2} - 1\right)\left(\frac{\partial H_3'}{\partial x_2} - \frac{\partial H_2'}{\partial x_3}\right) - \frac{\epsilon_{12}}{det\tilde{\epsilon}_2}\left(\frac{\partial H_1'}{\partial x_3} - \frac{\partial H_3'}{\partial x_1}\right)\right] \\ - \frac{\partial}{\partial x_1}\left[\left(\frac{1}{\epsilon_{33}} - 1\right)\left(\frac{\partial H_2'}{\partial x_1} - \frac{\partial H_1'}{\partial x_2}\right)\right] \end{aligned} \tag{C.3}$$

for any $\boldsymbol{x}$. j_1' and j_1' are the first two components of $\mathbf{j}'$. Notice that Eq. (2.30) has not been used so that H_3' has not been eliminated. One has $det\tilde{\epsilon}_2 = \epsilon_{11}\epsilon_{22} - \epsilon_{12}^2$ (which is a reminder of the fact that one is dealing with $(\tilde{\epsilon}^{-1})$.

Next, one employs the second equation in (2.3) together with (2.30) in order to express H_3 in terms of H_1, H_2. One plugs the resulting H_3 into Eqs. (C.1), (C.2) and (C.3). Then, Eqs. (C.1), (C.2) and (C.3) become a system of two equations depending only on H_1, H_2, in which $det\tilde{\epsilon}_2$ (which comes from $(\tilde{\epsilon}^{-1})$ does appear explicitly. Consequently, such a new system for H_1, H_2 does not seem to coincide in an explicitly obvious way with Eqs. (2.33), (2.34) and (2.35). The solution to the apparent paradox comes from Eq. (2.1). Thus, using the assumptions on $\tilde{\epsilon}$ and (2.29), one expresses E_1, E_2, E_3 in terms of H_1, H_2. Consequently, $\nabla \cdot (\tilde{\epsilon}\mathbf{E}') = 0$ becomes a relationship among components of $\tilde{\epsilon}$, H_1, H_2 and spatial derivatives of them. Such a constraint equation indicates that various, apparently different, systems for H_1, H_2 have to be related (not to say equivalent) to one another. The system for H_1, H_2 arising from Eqs. (C.1), (C.2) and (C.3), as derived in this appendix, and Eqs. (2.33), (2.34) and (2.35) are certainly and strictly equivalent to each other, as they both come from Maxwell's equations. We have indeed checked that fact, under the assumption that the elements of $\tilde{\epsilon} - I_3$, K^2, derivatives of the elements of $\tilde{\epsilon}$ and derivatives of H_1, H_2 are small. Then, Eqs. (2.34) and (2.35) reduce approximately to (2.33) with:

$$j_1 \simeq -(\omega/v)^2(\epsilon_{22} - 1)H_1 + (\omega/v)^2\epsilon_{21}H_2 \equiv j_{1,app} \tag{C.4}$$

$$j_2 \simeq -(\omega/v)^2(\epsilon_{11} - 1)H_2 + (\omega/v)^2\epsilon_{12}H_1 \equiv j_{2,app} \tag{C.5}$$

which are to be combined with (2.33). It is easy to see that, under similar approximations, the system for H_1, H_2 implied by Eqs. (C.1), (C.2) and (C.3) as derived in this appendix does coincide with Eqs. (2.33), (C.4) and (C.5).

Another equivalent alternative is the following. After having used Eqs. (2.29) and (2.30) in order to factor out the x_3-dependence, one continues with $\mathbf{H} = (H_1, H_2, H_3)$, without eliminating H_3, and with Eqs. (2.16) and (2.17). Then, one transforms those x_3-independent Eqs. (2.16) and (2.17) into another inhomogeneous scattering integral equation by means of a cylindrical transverse tensor Green's function.[65] The analysis in the latter reference for an isotropic cylindrical waveguide indicates the essentials of the alternative formulation, and it can be extended to the actual case of an anisotropic cylindrical waveguide. We shall omit details.

Appendix D

Outgoing Hankel's Function of Zeroth Order and Useful Approximations for It

A very important property of $H_0^{(1)}(K \mid \boldsymbol{x}-\boldsymbol{x}' \mid)/4i$ is its large distance behavior. For fixed $\boldsymbol{x}'$ and $\mid \boldsymbol{x} \mid \to +\infty$, along a fixed direction (say, fixed $x_l / \mid \boldsymbol{x} \mid, l = 1, 2$):

$$\frac{H_0^{(1)}(K|\boldsymbol{x}-\boldsymbol{x}'|)}{4i} \to \frac{1}{4i}\left[\frac{2}{\pi K|\boldsymbol{x}|}\right]^{1/2} \exp i(K|\boldsymbol{x}| - \pi/4). \exp[-iK(\boldsymbol{x}\boldsymbol{x}')/|\boldsymbol{x}|] \tag{D.1}$$

Let $\epsilon(\boldsymbol{k}, \sigma)$, $\sigma = \sigma_1, \sigma_2$, be two real or complex two-dimensional vectors fulfilling $\epsilon(\boldsymbol{k}, \sigma)^+\epsilon(\boldsymbol{k}, \sigma') = \delta_{\sigma,\sigma'}$ and $\sum_\sigma(\epsilon(\boldsymbol{k}, \sigma))_i(\epsilon(\boldsymbol{k}, \sigma)^+)_j = \delta_{i,j}$, $i, j = 1, 2$. The superscript + also denotes adjoint. Then, with $\boldsymbol{k}' \equiv K(\boldsymbol{x}/ \mid \boldsymbol{x} \mid)$, Eq. (D.1) can be reformulated as:

$$\frac{H_0^{(1)}(K \mid \boldsymbol{x}-\boldsymbol{x}' \mid)}{4i}\delta_{i,j} \to \frac{1}{4i}\left[\frac{2}{\pi K \mid \boldsymbol{x} \mid}\right]^{1/2} \exp i(K \mid \boldsymbol{x} \mid -\pi/4)$$
$$\times \exp[-i\boldsymbol{k}'\boldsymbol{x}'] \sum_\sigma (\epsilon(\boldsymbol{k}', \sigma))_i(\epsilon(\boldsymbol{k}', \sigma)^+)_j \tag{D.2}$$

Let K be either real (scattering) or pure imaginary (confined propagation). One has the following identity, for any $\mid \boldsymbol{x}-\boldsymbol{x}' \mid$:

$$\frac{H_0^{(1)}(K \mid \boldsymbol{x}-\boldsymbol{x}' \mid)}{4i} = \frac{1}{2\pi}\ln[-2^{-1}iK \mid \boldsymbol{x}-\boldsymbol{x}' \mid \exp\gamma] + G_1(K \mid \boldsymbol{x}-\boldsymbol{x}' \mid) \tag{D.3}$$

which defines the function $G_1(K \mid \boldsymbol{x}-\boldsymbol{x}' \mid)$, trivially. Here, γ denotes Euler's constant. Explicit forms of $G_1(K \mid \boldsymbol{x}-\boldsymbol{x}' \mid)$ can be obtained upon comparing Eq. (D.3) with various known representations for $H_0^{(1)}(K \mid \boldsymbol{x}-\boldsymbol{x}' \mid)$,[58] but they will not be needed here. For fixed $\mid \boldsymbol{x}-\boldsymbol{x}' \mid \neq 0$, let $K \to 0$. Then, in such a limit, the dominant term on the right-hand side of Eq. (D.3) is the first one:[58]

$$\frac{H_0^{(1)}(K \mid \boldsymbol{x}-\boldsymbol{x}' \mid)}{4i} \to \frac{1}{2\pi}\ln[-2^{-1}iK \mid \boldsymbol{x}-\boldsymbol{x}' \mid \exp\gamma] \tag{D.4}$$

The following remark will display the physical interest of Eq. (D.4), having in mind applications for scattering and confined propagation with long wavelength. Let the area of the transverse cross-section Ω of the anisotropic waveguide be of order R^2. Let us assume that K is small in the sense that the dimensionless product $|\ K\ |\ R$ is small compared to unity. Let $\boldsymbol{x}$ vary arbitrarily inside the transverse cross-section Ω of the anisotropic waveguide. Then, for such values of K and for any suitably smooth function $f(\boldsymbol{x}')$ that vanishes outside the transverse cross-section Ω:

$$\int d^2\boldsymbol{x}' \frac{H_0^{(1)}(K\mid \boldsymbol{x}-\boldsymbol{x}'\mid)}{4i} f(\boldsymbol{x}') \simeq \frac{1}{2\pi} \ln[-2^{-1} i K R \exp\gamma] \int d^2\boldsymbol{x}' f(\boldsymbol{x}') \quad \text{(D.5)}$$

Notice that $\boldsymbol{x}'$ also varies inside Ω. Then, as $\boldsymbol{x}$ also varies inside the latter, $|\ \boldsymbol{x}-\boldsymbol{x}'\ |$ is, on the average, of order R. Eq. (D.5) constitutes the physical approximation based upon Eq. (D.4).

Let us now consider two separate cylindrical waveguides (say, the i-th and the j-th ones, $i \neq j$), both of them parallel to the x_3 axis. We now consider $H_0^{(1)}(K|\boldsymbol{x}-\boldsymbol{x}'|)/4i$, for $\boldsymbol{x}$ on the i-th waveguide and $\boldsymbol{x}'$ on the j-th one. In a plane orthogonal to the x_3 axis, and with respect to an arbitrary origin located in that plane, let z_i and z_j be the two-dimensional position vectors of two suitably chosen fixed points, one inside the i-th waveguide and the other inside the j-th one. Let $D_{ij} =|\ z_i - z_j\ |$ and we suppose that D_{ij} is larger than the sum of the transverse sizes of both waveguides ($D_{ij} > R_i + R_j$). We introduce, for later convenience $\boldsymbol{u}_{ij} \equiv (z_i - z_j)/D_{ij}$. We write the above $\boldsymbol{x}$ and $\boldsymbol{x}'$ as: $\boldsymbol{x} = z_i + \boldsymbol{x} - z_i$ and $\boldsymbol{x}' = z_j + \boldsymbol{x}' - z_j$. We are interested in the large distance behavior of $H_0^{(1)}(K\ |\ \boldsymbol{x}-\boldsymbol{x}'\ |)/4i$ for suitably large D_{ij}, for fixed $\boldsymbol{u}_{ij}$, when $|\ \boldsymbol{x}-z_i\ |$ and $|\ \boldsymbol{x}'-z_j\ |$ are smaller than D_{ij}. A computation which generalizes directly (D.1) yields:

$$\frac{H_0^{(1)}(K\mid \boldsymbol{x}-\boldsymbol{x}'\mid)}{4i} \simeq \frac{1}{4i}\left[\frac{2}{\pi K D_{ij}}\right]^{1/2} \exp i(K D_{ij} - \pi/4) \times \exp[i K \boldsymbol{u}_{ij}((\boldsymbol{x}-z_i)-(\boldsymbol{x}'-z_j))] \quad \text{(D.6)}$$

Appendix E

Equivalent Form of Integral Equations and Generalizations for Discontinuous Permittivities

Notice that in Eq. (2.38) both $h(\boldsymbol{x}')$ and spatial derivatives thereof appear inside the integral. It is possible to recast Eq. (2.38) into an equivalent form without spatial derivations of $h(\boldsymbol{x}')$. Still with the assumption that all components of $\tilde{\epsilon}$ and their first spatial derivatives are continuous throughout all space (and also across the boundaries of all waveguides), Eq. (2.38) can be re-expressed as:

$$h(\boldsymbol{x}) = h^{(0)}(\boldsymbol{x}) + \int d^2\boldsymbol{x}' \left[\frac{H_0^{(1)}(K \mid \boldsymbol{x} - \boldsymbol{x}' \mid)}{4i} P_1(\boldsymbol{x}') + P_2(\boldsymbol{x}, \boldsymbol{x}') \right] h(\boldsymbol{x}') \quad \text{(E.1)}$$

The 2×2 matrices P_1 and P_2 have been obtained explicitly.[19] The derivation of Eq. (E.1) from Eq. (2.38) proceeds by performing in the latter two integrations by parts, successively.[19]

We now comment shortly about the generalization for discontinuous $\tilde{\epsilon}$. Thus, let the components of $\tilde{\epsilon}$ and their first spatial derivatives be continuous everywhere, except across the interfaces (Σ) separating the anisotropic cylindrical waveguides from the surrounding isotropic medium. Then, generalizations of the integral Eq. (2.38) and of (E.1) (including, in both cases, boundary terms given as integrals over Σ) have also been obtained.[19] We shall omit details of those generalizations, and we limit ourselves to remark that the relevant equations are Eqs. (3.2) and Eqs. (A.11) through (A.19) in Reference 19. We also remark that the following misprints should be corrected in the formulas given in a previous work.[19] In both Eqs. (A.13) and (A.14) of that previous work,[19] a_{t2} should be replaced by a_{t1} while a_{t1} should be substituted by $-a_{t2}$.

Appendix F
Sampling Theorem

The so-called sampling theorem provides a specific procedure for representing a class of functions with interest in information and communication theories and optics. There is a very interesting account of the sampling theorem with a view toward its applications in optics.[59] For our purposes, the summary of such an account[59] to be given below will suffice.

We consider the plane (x, y) (x and y being two independent real variables in $-\infty < x < +\infty$ and $-\infty < y < +\infty$) and real or complex functions $g = g(x, y)$ (or column vectors, formed by the latter functions). To fix the ideas, we shall continue referring to functions g in this appendix, although the basics of the sampling theorem are applied to column vectors in Subsection 2.10.6 and Section 2.11. Qualitatively, the basic sampling idea is the following. We consider a suitable discrete (finite or denumerably infinite) set of points (x_j, y_j), $j = 1, \ldots$, and the set of all the resulting values $g_j = g(x_j, y_j)$ ("the array of all sampled values or data"). We would like such an array of sampled data to provide an adequate representation of the complete function g. That is, we expect that it is possible to reconstruct with sufficient accuracy, $g(x, y)$ for any (x, y) from that array of data g_j ($j = 1, \ldots$) by interpolation or by other means. Clearly, in order for such a "sufficient accuracy" to be possible, g and the chosen discrete set of points (x_j, y_j), $j = 1, \ldots$, have to fulfill certain restrictions. The sampling theorem given below characterizes certain suitable g's and discrete sets of points (x_j, y_j) for which such a representation is exact.

It will be crucial to introduce the Fourier transform $F[g] = G = G(f_x, f_y)$ of g:

$$F[g] = G(f_x, f_y) = \int_{-\infty}^{+\infty} dx \int_{-\infty}^{+\infty} g(x, y) \exp[-i2\pi(xf_x + yf_y)]dy \quad \text{(F.1)}$$

f_x and f_y (the "spatial frequencies") are real. In the two-dimensional spatial frequency space, a "point" is formed by one pair (f_x, f_y). The function g is, by definition, a band-limited one, if $G(f_x, f_y)$ is non-vanishing over only a finite region of the spatial frequency space (f_x, f_y).

Let g be a band-limited function. Let B_X, B_Y denote certain fixed values of the spatial frequencies f_x, f_y, respectively. One has, for any (x, y), the following identity[59]

$$g(x, y) = \sum_{n=-\infty}^{+\infty} \sum_{m=-\infty}^{+\infty} g\left(\frac{n}{2B_X}, \frac{m}{2B_Y}\right) \sin c\left(2B_X\left(x - \frac{n}{2B_X}\right)\right) \times \sin c\left(2B_Y\left(y - \frac{m}{2B_Y}\right)\right) \tag{F.2}$$

We recall that the function $\sin c(x)$ is defined as:

$$\sin c(x) \equiv \frac{\sin(\pi x)}{\pi x} \tag{F.3}$$

In this case, the discrete set of points (x_j, y_j), $j = 1, \ldots$ referred to above is denumerably infinite, and it is formed by all $(n/2B_X, m/2B_Y)$, $n = -\infty, \ldots, +\infty$, $m = -\infty, \ldots, +\infty$. The actual array of all sampled data is now formed by all $g(n/2B_X, m/2B_Y)$, $n = -\infty, \ldots, +\infty$, $m = -\infty, \ldots, +\infty$. One sees immediately that Eq. (F.2) enables us to reconstruct exactly the band-limited function g for any (x, y) in terms of the actual array of all sampled data. It can be justified[59] that the maximum spacings in that discrete set of all points (x_j, y_j), that allow for exact recovery of the original function $g(x, y)$, are $1/2B_X$ and $1/2B_Y$. The set of all products $\sin c(2B_X(x - n/2B_X)) \sin c(2B_Y(y - m/2B_Y))$ gives rise in an effective way to the interpolations necessary to yield g, for any (x, y) from all $g(n/2B_X, m/2B_Y)$'s.

Eq. (F.2) is one form of the so-called Whittaker-Shannon sampling theorem, namely, the one associated with the chosen discrete set of points (x_j, y_j). There are other forms of that sampling theorem for other choices of the discrete sets of points (x_j, y_j). We shall consider one of those alternative forms (more adequate for our purposes), namely that for band-limited functions g such that $G(f_x, f_y)$ vanishes outside a circle of radius R and centered about the origin in the spatial frequency space (f_x, f_y). One now has the following alternative form of the sampling theorem:[59]

$$g(x, y) = \sum_{n=-\infty}^{+\infty} \sum_{m=-\infty}^{+\infty} g\left(\frac{n}{2B}, \frac{m}{2B}\right) \frac{\pi}{4} \times \frac{2J_1(2\pi B((x - n/2B)^2 + (y - m/2B)^2)^{1/2})}{2\pi B((x - n/2B)^2 + (y - m/2B)^2)^{1/2}} \tag{F.4}$$

where $J_1(x)$ denotes the ordinary Bessel's function of order 1.[58]

REFERENCES

1. M. Born and E. Wolf, *Principles of Optics*, 7th Ed. (Cambridge University Press, Cambridge, 1999).
2. L. D. Landau and E. M.Lifchitz, *Electrodynamics of Continuous Media* (Pergamon Press, Oxford, 1960).
3. B. E. A. Saleh and M. C. Teich, *Fundamentals of Photonics*, Wiley Series in Pure and Applied Optics (John Wiley & Sons, New York, 1991).
4. H. C. van de Hulst, *Light Scattering by Small Particles* (Dover, New York, 1981).
5. M. Kerker, *The Scattering of Light and Other Electromagnetic Radiation*, (Academic, Orlando, Florida, 1969).
6. C. F. Bohren and D. R. Huffman, *Absorption and Scattering of Light by Small Particles* (John Wiley & Sons, New York, 1983).
7. S. K. Sharma and D. J. Somerford, Scattering of Light in the Eikonal Approximation, in *Progress in Optics*, Vol. XXXIX, E. Wolf, Ed. (Elsevier Science, Amsterdam, 1997).
8. D. Marcuse, *Light Transmission Optics* (Van Nostrand Reinhold, New York, 1972).
9. D. Marcuse, *Theory of Dielectric Optical Waveguides* (Academic Press, New York, 1974).
10. A. W. Snyder and J. D. Love, *Optical Waveguide Theory* (Chapman & Hall, London, 1983).
11. E. Snitzer, *J. Opt. Soc. Am.* **51**, 491 (1961).
12. P. J. B. Clarricoats, *Optical Fibre Waveguides: A Review* in *Progress in Optics*, Vol. XIV, E. Wolf, Ed. (Elsevier, Amsterdam, 1976).
13. M. J. Adams, *An Introduction to Optical Waveguide* (John Wiley & Sons, New York, 1981).
14. M. S. Sodha and A. Ghatak, *Inhomogeneous Optical Waveguide* (Plenum Press, London, 1977).
15. A. Ghatak and K. Thyagarajan, *Introduction to Fibre Optics* (Cambridge University Press, Cambridge, 1998).
16. J. M. Enoch and F. L. Tobey, Eds., *Vertebrate Photoreceptor Optics* (Springer-Verlag, Berlin, 1981).
17. V. Lakshminarayanan and J. M. Enoch, Biological Waveguides Optics, in *Handbook of Optics*, Volume III, 2nd Ed., M. Bass, et al., Eds. (McGraw-Hill, New York, 2001).
18. J. Sakai, S. Machida and T. Kimura, *Opt. Lett.* **6**, 496 (1981).
19. R. F. Alvarez-Estrada and M. L. Calvo, *Optica Acta* **30**, 481 (1983).
20. J. Limeres, M. L. Calvo, V. Lakshminarayanan and J. M. Enoch, *J. Opt. A: Pure. Appl. Opt.* **5**, S370 (2003).
21. F. I. Harosi, Microspectrophotometry and Optical Phenomena: Birefringence, Dichroism and Anomalous Dispersion, in *Vertebrate Photoreceptor Optics*, J. M. Enoch and F. L. Tobey, Eds. (Springer-Verlag, Berlin, 1981).
22. W. Thornburg, *J. Biophys. Biochem. Cytol.* **3**, 413 (1957).
23. R. P. Hemenger, *Appl. Optics* **28**, 4030 (1989).
24. Q. Zhou and R. W. Knighton, *Appl. Opt.* **36**, 2273 (1997).
25. D. Brewster, *Phil. Trans.* 60 (1815), 156 (1816), and *Trans. Royal Soc. Ed.* **8**, 369 (1818).
26. G. Bruhat, *Optique*, Chapter XXV (Masson & Cie., Paris, 1965).
27. E. G. Coker and L. N. G. Filon, *A Treatise on Photo-Elasticity* (Cambridge University Press, Cambridge, 1931).
28. M. M. Frocht, *Photoelasticity*, Vol. I and II (John Wiley & Sons, New York, 1948).
29. L. D. Landau and E. M.Lifchitz, *Theory of Elasticity*, revised and enlarged by E. M. Lifchitz, A. M. Kosevich and L. P. Pitaevskii (Butterworth-Heinemann, Oxford, 1997).

30. R. Dandliker, Rotational Effects of Polarization in Optical Fibres, in *Anisotropic and Non-linear Optical Waveguides*, C. G. Someda and G. Stegman, Eds. (Elsevier, Amsterdam, 1992).
31. J. Limeres, M. L. Calvo, V. Lakshminarayanan and J. M. Enoch, *J. Opt. Soc. Am.* B **20**, 1542 (2003).
32. S. R. Norman, D. N. Payne, M. J. Adams and A. M. Smith, *Electron. Lett.* **15**, 309 (1979).
33. V. Ramaswamy, W. G. French and R. D. Stanley, *Appl. Opt.* **17**, 3014 (1978).
34. I. P. Kaminow and V. Ramaswamy, *Appl. Phys. Lett.* **34**, 268 (1979).
35. R. H. Stolen, V. Ramaswamy, P. Kaiser and W. Pleibel, *Appl. Phys. Lett.* **33**, 699 (1978).
36. T. R. Wolinski, Polarimetric Optical Fibres and Sensors in *Progress in Optics*, Vol. XL, E. Wolf, Ed. (Elsevier, Amsterdam, 2000).
37. K. T. V. Grattan and T. Sun, *Sensors Actuators* **82**, 40 (200).
38. R. F. Cordero Ianarella, *J. Opt. Soc. Am.* **70**, 799 (1980).
39. R. L. Sidman, *J. Biophys. Biochem. Cytol.* **3**, 15 (1957).
40. D. E. Freund, R. L. McCally and R. A. Farrell, *Appl. Opt.* **25**, 2739 (1986).
41. T. B. Smith, *J. Mod. Opt.* **35**, 93 (1988).
42. G. J. Van Brokland and S. C. Verhelst, *J. Opt. Soc. Am.* A **4**, 82 (1987).
43. A. Sommer, H. A. Kues, S. A. D'Anna, S. Arkell, A. Robin and H. A. Quigley, *Arch. Ophthalmol.* **102**, 864 (1984).
44. A. Plesch, V. Klingbell and J. Bille, *Appl. Opt.* **26**, 1480 (1987).
45. A. W. Dreher, K. Reiter and R. N. Weinreb, *Appl. Opt.* **31**, 3730 (1992).
46. A. M. Benoit, K. Naoun, V. Louis-Dorr, L. Mala and A. Raspiller, *Appl. Opt.* **40**, 565 (2001).
47. B. R. Horowitz, Theoretical Considerations of the Retinal Photoreceptor as a Waveguide, in *Vertebrate Photoreceptor Optics*, J.M. Enoch and F.L. Tobey, Eds. (Springer-Verlag, Berlin, 1981).
48. P. A. Liebman, Birefringence, Dichroism and Rod Outer Segment Structure, in *Photoreceptor Optics*, A. W. Snyder and R. Menzel, Eds. (Springer-Verlag, Berlin, 1975).
49. V. Lakshminarayanan, J. E. Bailey and J. M. Enoch, *Optom. Vision Sci.* **74**, 1011 (1997).
50. S. Choi, M. Kono and J. M. Enoch, Evidence for Transient Strains at the Optic Disk and Nerve in Myopia: I Stiles-Crawford Effect Studies Performed over Time, presented at the Annual Meeting of Association for Research in Vision and Ophthalmology, Fort Lauderdale, Florida, May, 5–10 (2002).
51. L. Mandel and E. Wolf, *Optical Coherence and Quantum Optics* (Cambridge University Press, Cambridge, 1995).
52. P. M. Morse and H. Feshbach, *Methods of Theoretical Physics*, Vol. II, Ch. 13 (McGraw-Hill, New York, 1953).
53. M. L. Calvo and A. Duran, *Il Nuovo Cimento* **29B**, 277 (1975).
54. R. G. Newton, *Scattering Theory of Waves and Particles*, 2nd. Ed. (Dover, New York, 2002).
55. M. Lax, *Rev. Mod. Phys.* **23**, 287 (1951).
56. M. L. Goldberger and K. M. Watson, *Collision Theory* (Dover, New York, 2004).
57. M. L. Calvo and A. Duran, *Il Nuovo Cimento* **45B**, 68 (1978).
58. M. Abramowitz and I. A. Stegun, Eds. *Handbook of Mathematical Functions* (Dover, New York, 1955).
59. J. W. Goodman, *Introduction to Fourier Optics*, 2nd Ed. (McGraw-Hill, New York, 1996).

60. R. Barakat, *J. Opt. Soc. Am.* **54**, 921 (1964).
61. D. R. Williams and N. J. Coletta, *J. Opt. Soc. Am.* A **4**, 1514 (1987).
62. J. Limeres, M. L. Calvo, V. Lakshminarayanan and J. M. Enoch, Stress Sensor Based on Light Scattering by an Array of Birefringent Optical Waveguides in *International Commission for Optics XIX: Optics for the Quality of Life*, A. Consortini and G. C. Righini, Eds. *Proc. SPIE*, **4829**, 881 (2002).
63. M. L. Calvo and R. F. Alvarez-Estrada, Coupling of Dielectric Waveguide in *Max Born Centenary Conference*, M. J. Colles and D. Swift, Eds. *Proc. SPIE*, **369**, 401 (1982).
64. J. M. Enoch, S. S. Choi, M. Kono, V. Lakshminarayanan and M. L. Calvo, Receptor Alignments and Visual Fields in *High and Low Myopia*, Proceedings of Halifax Conference, September (2000).
65. R. F. Alvarez-Estrada and M. L. Calvo, *J. Math. Phys.* **21**, 389 (1980).

3 Optically Induced Nonlinear Waveguides

Philip Jander and Cornelia Denz

CONTENTS

3.1 INTRODUCTION

Compensating the diffraction of an optical signal is crucial to efficiently transmitting and switching data or performing logical operations with optical beams and pulses. Linear optics provides various methods to keep optical waves localized as they propagate, most prominently optical fibers that exist in numerous different designs. Common to all such waveguides is a prefabricated transverse variation of the refractive index. This effectively limits their application to static waveguiding. Thus, static waveguides need to be combined with external devices in order to realize switching or logical operations.

In the light of increasing interest in all-optical information processing, the ability to dynamically generate and reconfigure waveguides by employing light itself as a means of control is very desirable. Nonlinear effects in optical materials provide the opportunity to dynamically generate refractive index profiles in initially unstructured media, thus allowing optical guiding to be combined with switching operations.

Among various schemes one can consider for utilizing optical nonlinearities, the generation of self-induced adaptive waveguides, commonly called spatial optical solitons,[1] is distinguished by a variety of special properties stemming from the particle-like behavior of these entities and their ability for controlled interaction. In general, a soliton[2] is a localized solution of a nonlinear wave equation where the nonlinearity exactly balances spreading of the wavepacket in the temporal dimension through dispersion or in the transverse spatial dimension through diffraction or both. The first type is already being employed in optical high-speed communications.[3] Here, dispersion-managed temporal solitons[4,5] play an increasing role in keeping the information capacity of long distance waveguides up to market demands. The second kind of soliton is relevant to the problem outlined above. While spatial solitons are supported by many different kinds of optical nonlinearities, we can identify three conditions relevant for applications.

If we aim to utilize solitons as waveguides for probe or data waves, the soliton should exist due to a change of the refractive index (Figure 3.1) which also affects the probe beams. Additionally, interaction between individual solitons should be inelastic in order to provide a broader spectrum of interaction scenarios that can be employed in waveguide coupling. Finally, only low beam power should be required for the generation of induced waveguides. Ideally, such waveguides will be able to guide data beams of higher power at other frequencies.

Photorefractive materials[6] fulfill those conditions and consequently have been the primary nonlinearity in experiments concerning the generation of nonlinear waveguides. Therefore, most of this chapter will present considerations based on experiments performed in these kinds of media. While the formation of a single self-induced

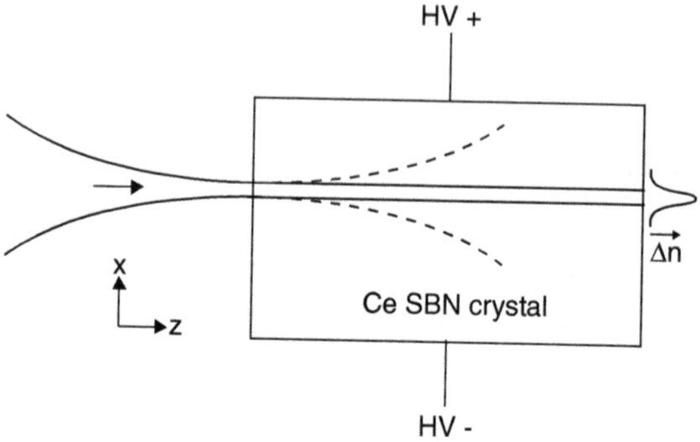

FIGURE 3.1 Principle of soliton propagation by self-focussing. A focussed beam is incident onto a nonlinear medium. Although it would diffract in a linear medium, the focussing nonlinearity creates a refractive index change acting as a lens that compensates diffraction. The beam becomes a guided mode of its own induced waveguide.

waveguide is a rather straightforward process, we will note fascinating additional features once one considers many such waveguides propagating in parallel. Rich interaction of individual solitons ranging from phase modulation via energy transfer to beam fusion and fission provides promising opportunities for applications.

Of special interest is the interaction of solitons propagating in opposite directions. In principle, such counterpropagating beams could be utilized to create coupling devices providing quite complex functionality with well defined input and output interfaces. However, instabilities intrinsic to the counterpropagating geometry lead to highly nonlinear behavior. While those instabilities pose a hindrance for simpler devices, the self-sustained dynamics connected with the instabilities are very desirable for any kinds of active devices.

Finally, one of the primary advantages of optical data processing is inherent parallelity in two transverse dimensions. Naturally, one strives to transfer this to nonlinear optical devices as well. Since spatial solitons are laterally confined by definition, the transverse dimensions are available for multiplexing of channels as well as functional interaction of multiple channels.

This chapter is organized as follows. In the second section, we introduce optical spatial solitons as the central concept of optically induced nonlinear waveguides. Following a brief look into the mathematical descriptions of these entities, we present realizations of optical spatial solitons in various media. Generic properties and exemplary applications of such nonlinear waves are discussed. In the third section we introduce the photorefractive nonlinearity as a model system in which many experiments are performed due to a wide range of practical advantages. The generation of self-trapped optical beams in photorefractive media is presented along with their interaction properties. Naturally, we look at the usability of the generated refractive index profiles as waveguides with a special emphasis on waveguide interaction. The fourth section presents dynamic effects that can be exploited to provide dynamically rearrangeable waveguides and discusses the realization of optical couplers and switching devices in photorefractive materials. Additionally, we look into intrinsic dynamics encountered in the interaction of counterpropagating self-trapped beams. Finally, we consider the generation of periodic materials consisting of many parallel waveguides by solitons in order to take advantage of the parallel information processing capabilities of optical systems.

3.2 SPATIAL OPTICAL SOLITONS AS WAVEGUIDES

3.2.1 CONCEPT OF SOLITON

In the most general definition, a soliton is a localized solution of a nonlinear wave equation whose transverse or longitudinal profile remains constant under propagation. This implies a nonlinear self-action compensating either diffraction or dispersion. If the wave equation is integrable, each solution maintains its identity throughout propagation as well as collisions and displays particle-like characteristics, hence the term *soliton*.[7]

The first type of optical soliton that has received much attention in the past is the *temporal soliton*. In a dispersive medium, linear propagation leads to spreading of the wave packet due to different propagation velocities for different wave lengths. An appropriate nonlinear term of the wave equation can balance the dispersive spreading, resulting in a nonlinearily localized wave packet.

As we are interested in the nonlinear generation of waveguides, we restrict ourselves to the second type. *Spatial solitons* denote the diffractive spreading of a single frequency continous wave in various nonlinear media. In the strictest definition, the term *soliton* only applies to constant solutions of integrable nonlinear wave equations. However, the definition has long since been extended to nonintegrable systems and the experimental observation of compensation of diffraction by nonlinearity in general. To avoid the term *soliton*, such waves are often also called solitary waves or localized waves. Throughout this chapter, we use the term *soliton* in the broadest definition. In the following, we will consider electromagnetic problems with soliton solutions.

3.2.2 Theory of Spatial Solitons

Starting from Maxwell's equations, one obtains the general electromagnetic wave equation for a non-magnetic material

$$\Delta\, \mathbf{E}(\mathbf{r}, t) - \frac{\mathrm{n}(\mathbf{r}, t)^2}{c^2} \partial_t^2\, \mathbf{E}(\mathbf{r}, t) = 0 \tag{3.1}$$

where $\mathbf{E}$ is the electromagnetic wave and $\mathrm{n}(\mathbf{r}, t)$ is the refractive index which is a static tensor for linear materials but will depend on $\mathbf{E}$ for any nonlinear medium.

We can now separate the optical frequency and consider only the case of a wave linearly polarized in the x-direction:

$$\mathbf{E}(\mathbf{r}, t) = A(\mathbf{r}, t) \exp\left[\mathrm{i}(k_0 n_0 z - \omega t)\right]\, \mathbf{e}_x + \mathrm{c.c} \tag{3.2}$$

where k_0 is the optical wave number in the vacuum, n_0 is the linear refractive index for x-polarized light and ω is the optical frequency of the electromagnetic wave. Assuming only slow variation of the envelope amplitudes A (SVEA approximation) and small angles between all wave vectors and the propagation direction along the z-axis of our coordinate system we arrive at the nonlinear paraxial wave equation for the envelope amplitudes

$$2in_0 k_0 \partial_z A + \nabla_\perp^2 A + (n^2 - n_0^2) k_0^2 A = 0 \tag{3.3}$$

with the transverse Laplacian $\nabla_\perp^2 = \partial_x^2 + \partial_y^2$, and A and n being variables of all spatial and the temporal dimension.

The nonlinear term $(n^2 - n_0^2) \approx 2n_0 \Delta n$ defines the characteristics of the medium. For self-focussing, we look for nonlinearities where the refractive index locally depends on the intensity: $\Delta n = \Delta n(|A|^2)$.

Optical Kerr nonlinearity — The simplest form of nonlinearity providing self-focussing is the Kerr nonlinearity or weak symmetric anharmonic response. It can be expressed as an intensity-dependent refractive index modification $\Delta n = n_2 |A|^2$ where n_2 is the second coefficient of the polynomial expansion of the nonlinear refractive index. Self-trapping in Kerr media was considered by Chiao et al.[8] who gave an initial estimate for self-focussing. Rescaling the spatial coordinates and the field A to absorb the constants and considering only one transverse dimension x, one obtains the following equation which describes the simplest case of self-focussing

$$\partial_z A - \frac{\mathrm{i}}{2}\partial_x^2 A - \mathrm{i}|A|^2 A = 0 \tag{3.4}$$

The Kerr or quadratic electrooptic effect represents an instantaneous linear dependence of the refractive index on the local intensity. It can be observed for sufficiently high intensities whenever the linear electrooptic effect is inhibited due to a centrosymmetry of the medium. Due to the similarity to the well-known Schrödinger equation, (3.4) is called the nonlinear Schrödinger equation (NLSE). In analogy to the quantum mechanical case, the nonlinear refractive index, i.e., the intensity pattern, provides a potential in which the optical wave can be trapped if $n_2 > 0$, as the potential induces self-focussing which counteracts spreading due to the diffraction term. Self-trapping, i.e., propagation with neither spreading nor collapse, can only be achieved if both terms exactly cancel. The problem is integrable and an infinite number of solutions can be found. In particular, one is interested in those solutions which have a constant profile along the propagation direction:

$$A(x, z) = \Phi_\beta(x) \cdot \mathrm{e}^{\mathrm{i}\beta z} \tag{3.5}$$

we will call all such solutions solitons. Zakharov and Shabat[9] solved (3.4) analytically. One possible solution is

$$A(x, z) = \operatorname{sech}(x)\mathrm{e}^{\mathrm{i}z/2} \tag{3.6}$$

from which families of more complicated solutions can be found using transformations which exploit the infinite number of conserved quantities of (3.4).[2] A general procedure to find solutions of this problem is the inverse scattering method.[9,10] As a result of the integrability, soliton solutions are nonlinearily superponable. Hence, collisions between solitons conserve all properties of a soliton (except for a phase shift) and are therefore called elastic. This behavior is reminiscent of particle solutions in quantum mechanics which is the reason for the term *soliton*. This property will change drastically if we consider different kinds of nonlinearities. Additionally, the Kerr nonlinearity does not have any stable soliton solutions when considered in two transverse dimensions. Only spreading solutions or solutions collapsing into a singularity can be found.

Experimentally, this of course corresponds to an inadequate description of the nonlinearity by the Kerr term; however, permanent material damage due to self-focussing is in fact often observed in Kerr media.[11] This behavior stems from the number of transverse dimensions represented in the nonlinear term. While self-focussing can still compensate diffraction in two transverse dimensions, the corresponding solutions are unstable with respect to small variations and hence not observable. In order to stabilize solutions with constant profile, one requires additional terms in the nonlinear expansion saturating the Kerr term. This is available for instance with the cubic-quintic ($\Delta n = n_2|A|^2 + n_4|A|^4$) nonlinearity, where $n_4 < 0$ avoids singularities. The cubic-quintic nonlinearity is often treated in analytical work due to its simplicity.

Real materials, however, are often better modeled by a saturable Kerr nonlinearity, retaining an infinite number of expansion terms at the cost of non-integrability.

Saturable nonlinearity — If we consider (3.3) with a nonlinearity of the form

$$\Delta n(|A|^2) = \Delta n_{\text{sat}} \frac{|A|^2}{1 + |A|^2/I_{\text{sat}}} \tag{3.7}$$

we obtain a saturable Kerr model, again absorbing all constants in the scaling of the variables:

$$\partial_z A - \frac{\mathrm{i}}{2}\nabla_\perp^2 A - \mathrm{i}\frac{|A|^2}{1 + \sigma|A|^2} A = 0 \tag{3.8}$$

where the free saturation parameter $\sigma = n_0/(k_0 n_2 I_{\text{sat}})$ contains the saturation intensity and the strength of the nonlinearity. This wave equation asymptotically follows Kerr behavior for small intensities but saturates and finally yields a constant refractive index modification as the intensity is further increased. Saturation of the nonlinearity avoids singularities and solitary solutions become possible in two transverse dimensions too. Because a change to the refractive index is always somehow limited in real media, this nonlinearity describes many experimental systems better than a pure Kerr model. The wave equation with a saturable nonlinearity (or saturable Kerr nonlinearity) is no longer integrable. As a result, the number of conserved quantities is not infinite as in the Kerr case. Hence, the saturation in such systems dramatically changes the properties of the soliton solutions, especially their interaction scenarios.

Again looking for solutions with constant profiles, Gatz and Herrmann[12,13] arrived at an expression for two families of solitary solutions that cannot be found analytically. The families are discriminated by different peak amplitudes, such that for a given set of parameters, multiple solitons may exist. The profile of a soliton differs slightly from a Gaussian beam. Figure 3.2 gives the soliton existence curves for the two families and exemplary transverse soliton profiles for one saturation parameter value.

Properties of spatial solitons in saturable media — The major difference between solutions of the Kerr and saturable Kerr wave equations is the fact that a nonlinear superposition of two solutions is no longer necessarily a solution of the wave equation. As a result, two colliding solitary beams may strongly affect each other and are characterized as inelastic. Both energy and shape of a single soliton are no longer conserved but can be exchanged between them. Even the number of solitons is no longer conserved. Consequently, new solitons can result from collisions and individual solitons can be annihilated in collisions. Finally, the solitons are coupled to the radiation field. Still, apart from the interactions, solitary solutions have much in common with soliton solutions of the Kerr medium and hence share the name *soliton* under the broader definition.

Interesting from an experimentalist point of view is the evolution of input conditions that do not exactly match a soliton solution. In reality, this is always the case as one usually inputs beams with a Gaussian profile and the soliton profile is different. Gatz and Herrmann[13] numerically investigated different input conditions and found that the stability of the soliton solutions leads to oscillation of a beam if

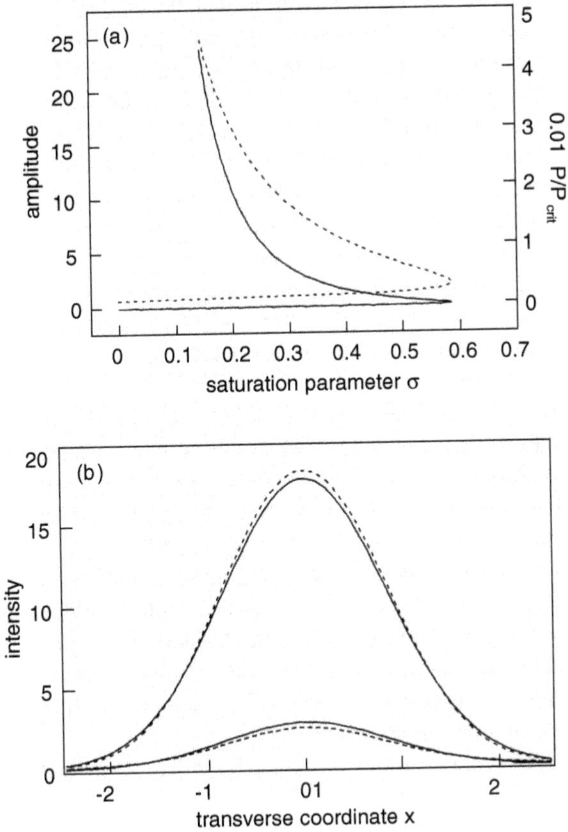

FIGURE 3.2 **(a)** Dependence of the amplitude (dashed curve) and the normalized soliton power (solid curve) on the saturation parameter σ. **(b)** Soliton shapes (dashed curves) for $\sigma = 0.5$ in comparison with corresponding Gaussian beams with the same power (solid lines) (after Gatz and Herrmann[13]).

the input condition is slightly different from the solitary profile. This is significant for experiments as it shows that the stable self-focussing does not critically depend on matching the exact soliton profile. Instead a whole range of input conditions still propagate localized, albeit displaying an oscillating amplitude.

3.2.3 Spatial Solitons in Experiments

Spatial solitons have been observed in numerous media with as many different mechanisms for self-focussing. Kerr-type solitons have been observed in different media. As was to be expected from the analytical treatment, only one transverse dimension allows for the generation of stable self-trapped propagation of optical beams. Single stripe solitons were reported for the first time in liquid CS_2 by Barthelemey et al.[14] but required stabilization to counteract transverse instabilities. Shortly thereafter,

Aitchison et al.[15] observed the first Kerr solitons in a planar glass waveguide induced by a femtosecond laser source. Subsequently, both groups investigated the properties of elastic collisions of multiple Kerr solitons.

The observation of self-focussing in saturable media actually predates the observation of the Kerr case. In 1974, Bjorkholm and Ashkin[16] reported stable soliton propagation in sodium vapor with a laser source detuned slightly off a resonance, investigated the conditions for soliton existence, and attributed their stability to saturation. However, due to the presence of strong absorption close to the resonance, no further experiments were conducted. Also in the mid 1970s, Monot et al.[17] investigated self-channeling of extremely high laser pulses in dense plasmas. Since the 1990s, several new types of spatial solitons in saturable media have been at the focus of research, most of them in noncentrosymmetric materials.

In photorefractive media, the index of refraction is slowly changed by an interplay of charge carrier transport and linear electrooptic effect. We will look at photorefractive solitons in detail in the next section. Media with linear electrooptic effect also support so-called quadratic solitons that mainly rely on second harmonic generation as nonlinearity. Here, the medium is not modified as is the case for photorefractive solitons. Instead, the energy exchange between one or multiple pumps and harmonic beams in parametric processes leads to a mutual trapping of all beams on the fly. The parametric wave mixing requires certain geometries to attain phase matching, so the possibilities for soliton generation are somewhat limited. Nevertheless, soliton generation, soliton interaction and also elementary operations have been demonstrated using quadratic solitons.[18] Liquid crystals also feature a photorefractive-like nonlinearity that provides solitons called nematicons.[19]

So far, we considered only solitons supported by self-focussing. However, stationary beam propagation is also possible in media with a defocussing nonlinearity. In that case, the localized structure is not a bright beam but rather a dark line (one transverse dimension) or a dark spot (in two transverse dimensions) instead, which is sustained by the nonlinearity. The principle is quite analogous to the bright soliton case, with a small twist to it. The defocussing nonlinearity leads to an increased refractive index at the dark spot or line. This induced lens refracts intensity from the adjoint bright regions into the dark area. Intuitively, this should result in immediate disappearance of any dark area. The twist is that the adjoint beam regions on both sides of a dark stripe soliton must be out of phase by π, so that destructive interference in combination with the induced lens sustains the soliton. In case of a two dimensional soliton, i.e., a dark spot, the same principle applies if any beam regions on opposite sides of the dark soliton are out of phase by π. This is the case for an optical vortex, for which the phase increases by integer multiples of 2π around the dark center. Such dark and vortex solitons have been predicted[20,21] and experimentally verified in various media which have in common that the material is altered by the nonlinearity and the resulting refractive index change can be experienced by guided probe beams too.[1]

Finally, the last and still very new soliton class we should mention are discrete solitons.[22,23] Here, the nonlinearity in a material with a photonic structure compensates the discrete diffraction present by virtue of the same photonic structure. As for the continous case, a host of different media have been investigated, among them Kerr media, quadratic media, photorefractive materials, and media exploiting the nonlinearity of the band gap induced by the photonoic structure, many supporting both focussing and defocussing nonlinearities. Due to the possibility of altering the photonic structure supporting the discrete solitons, they are very interesting candidates for the design of all-optical integrated "circuits." For a well-written overview on this rapidly developing field, we refer to Christodoulides et al.[23]

As we are looking for solitons that can be utilized as waveguides for probe beams carrying data, only a few of the experimentally demonstrated types of spatial solitons are of interest. The primary condition is that the refractive index change induced by the nonlinearity is not only acting on the self-trapped beam but also on any additional optical waves. Ideally, the change to the material is persistent even if the soliton itself is removed from the medium. Additionally, the generation of the soliton should require as modest beam powers as possible. These are exactly the properties of spatial solitons in photorefractive media, which we shall explore in more detail in the following sections.

3.3 SOLITONS IN PHOTOREFRACTIVE MEDIA

3.3.1 Photorefractive Solitons as Experimental Models

Photorefractive materials[6] offer an intensity-dependent saturable nonlinearity that has extensively been used for generating spatial solitons.[24] The photorefractive effect describes the nonlinear generation of a refractive index change depending on incident light. In its most common definition, it encompasses a local photoexcitation of charge carriers, several charge transport mechanisms giving rise to local static fields and finally a change of the refractive index via the linear electrooptic effect as a result of the generated fields. Photorefractivity has been known for more than three decades but was initially only considered a hindrance to the observation of nonlinear optical effects and hence called optical damage.[25]

The photorefractive effect has been observed in various materials that support the linear electrooptic effect, usually dielectric or semiconducting crystals such as lithium barium niobate ($LiNbO_3$), strontium barium niobate ($Sr_{1-x}Ba_xNb_2O_6$, SBN), barium titanate ($BaTiO_3$) or gallium arsenide (GaAs). Crystals typically include metal dopants that create energy levels within the bandgap that provide donors and traps for the photoexcitation and trapping of charge carriers (Figure 3.3). Today, the physics of photorefractive media is well understood[26,27] and a wide range of applications of photorefractive materials have been invented, from beam amplification[28] and phase conjugate mirrors[29] via optical data storage schemes[30] up to all-optical image processing devices.[31,32]

Under a number of approximations, the photorefractive effect can be reduced to a saturable Kerr nonlinearity, which is exactly what we identified above as a useful model in which stable spatial solitons can be found. Consequently, self-focussing and stable propagation of localized beams are observed in experiment, although the reduction to the saturable Kerr nonlinearity is an oversimplification that does not cover experimental reality very well. Nevertheless, the ease with which spatial solitons can be generated in photorefractives leads to such materials being highly advantageous in experimental investigations. Photorefractive materials are commonly highly nonlinear

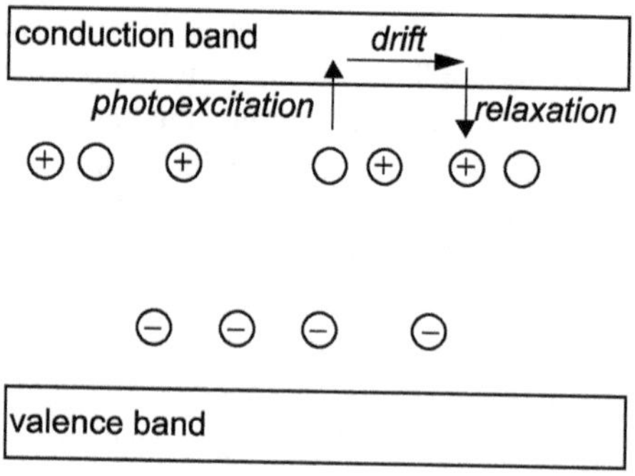

FIGURE 3.3 Band transport model. The photorefractive medium is insulating, with dopants creating additional energy levels in the band gap. Photoexcitation and drift in the externally applied field lead to charge redistribution and eventually to the screening of the external field.

and only require beam powers below one microwatt to observe stable self-focussing. The time constants involved can be influenced by the beam power and the choice of materials and vary from the order of microseconds for semiconductors to seconds for ferroelectric crystals.

Due to the photoexcitation, materials are sensitive over a wide range of frequencies but can also be designed to be sensitive to short optical wavelengths only while remaining quite linear for the infrared telecom regime. This is exploited to utilize solitons in the visible light to generate photorefractive waveguides for infrared radiation. The waveguides remain after the visible light is removed and are not distorted by the infrared signal beams guided by them. Again using visible wavelengths, the induced waveguides can be erased or otherwise changed, which allows for dynamically reconfigurable waveguiding. The latter has well been demonstrated, along with the implementation of basic switching and coupling devices that we will examine below.

3.3.2 Self-Focussing in Photorefractive Media

Photorefractive media support very different optical nonlinearities. Among them is one which is closely connected to the saturable Kerr nonlinearity and therefore supports not only self-focussing but also stable solitary solutions. Basically, the microscopic principle of the photorefractive effect for the case of self-focussing can be outlined as follows (Figure 3.4): a beam that is incident on the material locally excites charge carriers from either the valence band or dopant energy levels into the conduction band. Depending on the material, the charge carriers can be described either as electrons or holes. As they gain mobility, different charge transport mechanisms lead to a redistribution of the charges. Relevant for self-focussing is the charge carrier drift in an external dc field. This field can in principle be induced by means of the photovoltaic effect.

More common for soliton experiments is the application of an external field through electrodes which, for the example of strontium barium niobate, is on the order of 1 kV/cm. The drift of charge carriers creates a local space charge field

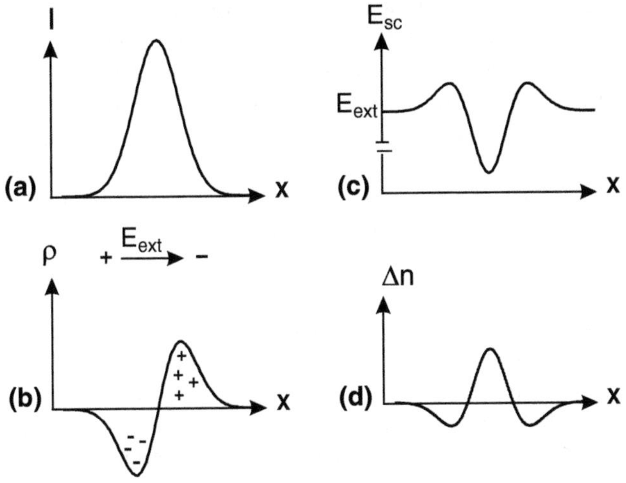

FIGURE 3.4 Photorefractive effect for the generation of screening spatial solitons. **(a)** Intensity distribution of the beam. **(b)** Charge carrier redistribution through photoexcitation, drift in the external electric field and trapping. **(c)** Modification of the static external field through the redistribution of charge carriers. **(d)** Refractive index change resulting from the local static field through the linear electrooptic effect.

that eventually screens the external field, stopping the charge migration. For this reason, solitons in photorefractives are often called screening photorefractive solitons. The total electric field roughly follows the intensity modulation although it is not proportional to it due to the saturation and a medium anisotropy discussed below. The saturation in turn results from the continuous and homogeneous photoexcitation provided by background illumination, thus generating a steady supply of mobile charge carriers.

In the case of self-focussing in photorefractive SBN, the nonlinearity provided for by the linear electrooptic effect is of the form

$$n^2 - n_0^2 = n_0^4 r_{\text{eff}} E_{\text{SC}} = -n_0^4 r_{\text{eff}} \partial_x \phi \tag{3.9}$$

where r_{eff} is the effective electrooptic coefficient for the selected geometry. The latter is chosen to maximize r_{eff} while allowing for a focussing nonlinearity; beam polarization along the crystal's c-axis and external electric field along the same direction which is denoted x in our coordinate system. ϕ is the electrostatic potential whose spatial derivative with respect to the x-axis yields the total static electric space charge field E_{SC} present (external field plus local space charge fields).

The potential ϕ[33,34] is induced by the total intensity and follows the following time-dependent relaxation equation

$$\begin{aligned} \tau \partial_t (\nabla^2 \phi) + \nabla^2 \phi + \nabla \phi \cdot \nabla \ln(|A|^2 + 1) &= E_e \partial_x \ln(|A|^2 + 1) \\ &+ \frac{k_B T}{e} [\nabla^2 \ln(|A|^2 + 1) + (\nabla \ln(|A|^2 + 1))^2] \end{aligned} \tag{3.10}$$

Here, τ is the material relaxation time (that is also a function of the intensity — more light, faster dynamics) and E_e is the external dc field. The right-hand side of (3.10) describes the charge carrier drift in the external field and charge carrier diffusion.

Combined, Eqs. (3.3), (3.9) and (3.10) describe the temporal evolution of photorefractive screening solitons with good accurancy. They can be solved numerically using a technique developed by Petviashvili,[35,36] or be simplified with three possible approximations. First, if one considers only beams propagating in parallel in a single direction, a stationary steady state can always be found which depends only on the input conditions. Therefore, the temporal evolution can be disregarded in favor of calculating only steady states. This approximation breaks down once one considers counterpropagating solitons. Second, diffusion can be eliminated by setting the temperature $T = 0$. This specifically removes the beam bending effect from the description which we will introduce below. Finally, (3.10) contains an anisotropy in $E_e\partial_x \ln(|A|^2 + 1)$ that stems from the screening of the externally applied field (distinguishes the x-axis over all other transverse axes). An isotropic approximation[37]

$$\tau\partial_t E_{\mathrm{SC}} + E_{\mathrm{SC}} = -E_e\frac{|A|^2}{|A|^2+1} \tag{3.11}$$

significantly simplifies the nonlinearity. Thereby however, one loses significant properties of photorefractive solitons, foremost the anomalous interaction also discussed below.

3.3.3 Spatial Solitons in Photorefractive Media

The experimental observation of spatial solitons in photorefractive is quite straightforward. A typical arrangement is depicted in Figure 3.5. Nearly all groups actively engaged in photorefractive soliton research today use cerium-doped strontium

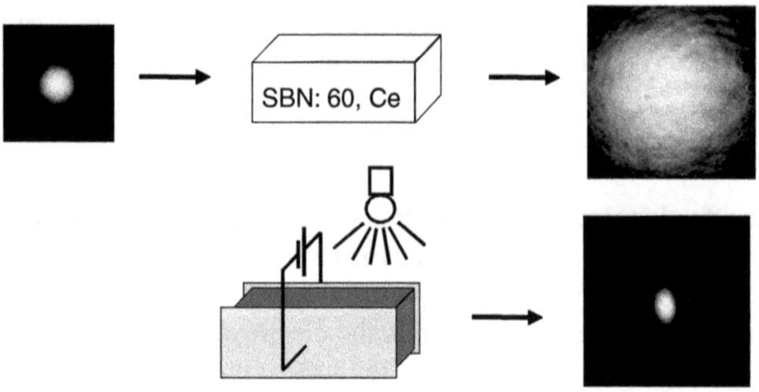

FIGURE 3.5 Schematic experiment for the observation of solitons in SBN. The SBN crystal is illuminated by a focussed laser beam (left image). The exit face of the crystal is imaged onto a CCD camera. The crystal itself is biased by an external high voltage field applied in parallel to the crystallographic c-axis. Incoherent background illumination is applied to saturate the nonlinearity. Without the field external, linear propagation is observed (right top), with the nonlinearity active, a spatial soliton is generated (right bottom).

FIGURE 3.6 Top-view photograph of a 10 μm wide beam propagating in a 5 mm long photorefractive crystal. (**a**) Linear diffraction with the nonlinearity deactivated. (**b**) Nonlinear self-induced waveguide when the nonlinearity is applied. (reprinted from Shih et al.[70]).

barium niobate single crystals. Typical crystal dimensions are 5 mm by 5 mm for the transverse axes (crystallographic b and c axes) and between 5 and 25 mm along the propagation direction. As the soliton propagation within the crystal is usually extremely hard to observe (Figure 3.6), especially if the interaction of multiple solitons is considered, identical crystals with different lengths serve to study the propagation. The crystal is biased with an external dc field of 1–2 kV/cm, applied along the c-axis. A circular Gaussian cw beam typically derived from a frequency doubled Nd:YAG laser with a diameter of a few tens of microns in the range of a few microwatts is focussed onto the crystal front face. The polarization is chosen along the c-axis, taking advantage of the large electrooptic r_{33} coefficient of SBN. The saturation parameter, i.e., the ratio between the soliton peak intensity and the background illumination is usually between 1 and 3.

The process of self-focussing is depicted in Figure 3.7. Due to the large time constants associated with photorefractive self-focussing in SBN, the beam initially diffracts linearly and self-focusses slowly into a soliton. After the beam diameter is reduced, an additional effect leads to bending of the soliton (discussed below). Finally a steady state is reached which in principle lasts indefinitely.

3.3.4 Properties of Experimental Solitary Beams

Real input conditions — In experimental situations, it would be highly challenging to generate an incident beam which corresponds to the soliton solution in the input plane. Instead, a Gaussian beam with a diameter closely resembling the soliton solution is used. As the beam adjusts to the profile required for stable propagation in a self-induced waveguide, it undergoes several oscillations while radiating off any excess energy.[38]

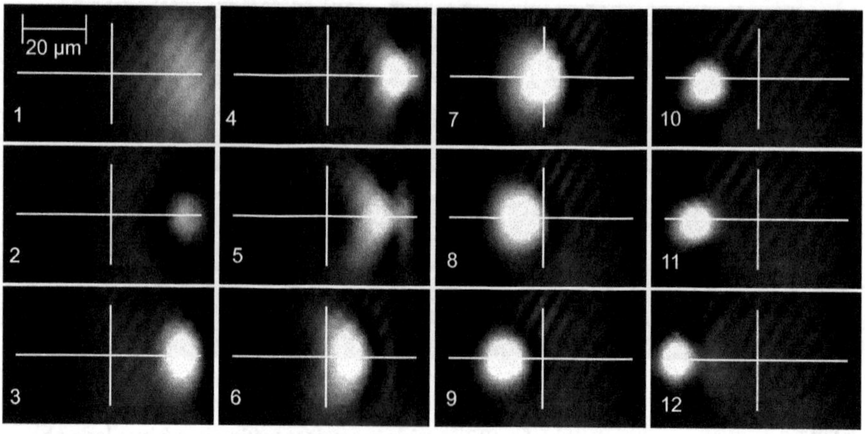

FIGURE 3.7 Temporal formation of a photorefractive screening soliton. The exit face of the crystal is imaged. **(1)** Linear regime, the beam diffracts. **(2–4)** Self-focussing of the beam towards a soliton. **(5–12)** The beam bending effect laterally displaces the beam output.

Consequences of anisotropy — Due to the anisotropy with respect to the transverse coordinates, spatial solitons in strontium barium niobate are always elliptical. The ellipticity is related to the saturation and hence can be somewhat controlled by changing the background illumination, beam power and beam diameter. There are two immediate consequences. First, any waveguide obtained by generating a photorefractive soliton will also have an elliptical cross-section. Second, if the soliton is generated using an incident Gaussian beam, the refractive index carries an azimuthal modulation as depicted in Figure 3.8. This influences the interaction of two experimental solitons as we will see in the next section.

Beam bending — A dominant property of solitons in experiment is the bending of a beam in the direction of the crystallographic *c*-axis. This effect results from the diffusive charge carrier transport that is always present in any real crystal. The refractive index profile is shifted along the *c*-axis with respect to the intensity distribution. Therefore the beam is laterally displaced as it propagates, leading to the formation of a bent soliton.[39,40] Beam bending or self-deflection increases as the beam diameter decreases because diffusion plays an increasing role in the charge redistribution process. It becomes less dominant if the externally applied field is increased. Additionally, beam bending can be controlled by adjusting the beam and background illumination intensities.

Filamentation — Related to solitons is the propagation of a broader beam. If its diameter is well above the width of the soliton solution, transverse modulational instability leads to filamentation of the beam (Figure 3.9).[41] Thin, near one-dimensional stripe beams, as well as broad two-dimensional beams, locally self-focus into equidistant spots of comparable size. As the filaments arising from local self-focussing are of the same size as solitons, beam filamentation is generally seen as a precursor to soliton

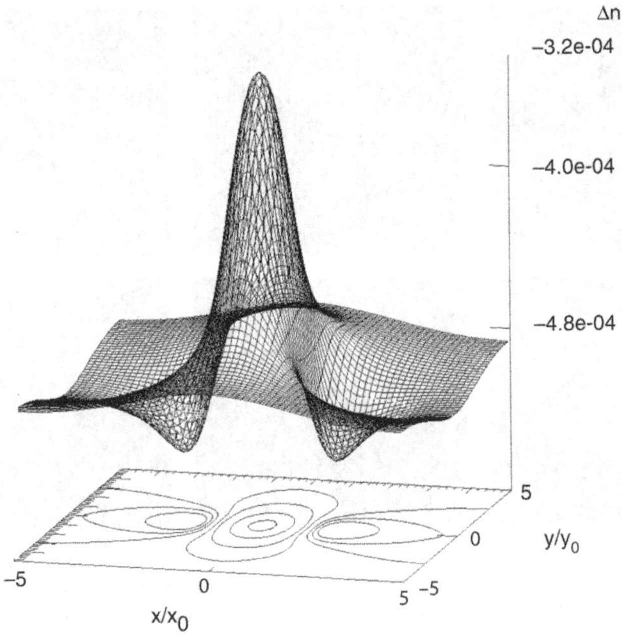

FIGURE 3.8 Anisotropic refractive index modulation in photorefractive SBN. The refractive index is increased where the beam displays high intensity. The medium anisotropy is manifested as two side lobes in the x-direction where the refractive index is actually lower than for dark regions.

formation. Due to the anisotropy of photorefractive media, a broad two-dimensional beam first breaks into an array of stripe beams. However, since the stripes are also unstable with respect to filamentation, they again break up into individual spots after short propagation.

The modulation developing in the beams arises from initial noise, given a sufficiently long propagating distance. The further evolution of the filaments is dynamic and does not reach a symmetric state, as patterns in systems with feedback do.

Recently, the filamentation of incoherent beams in nonlinear media has attracted increased interest, following the observation of partially incoherent solitons in photorefractive materials. Soljacic et al. predicted a threshold for the onset of modulational instability in noninstantaneous saturable media.[42] Increasing transverse incoherence as well as saturation were seen to eventually take the system below that threshold. Subsequent experiments using SBN as a focussing nonlinear medium confirmed filamenting beam propagation of incoherent beams as predicted.[43–45]

The properties of single photorefractive solitons outlined in this section become highly interesting, once one considers the interaction of two or more solitons. The anisotropy and the general inelastic nature of soliton collisions in saturable media lead toward a fascinating range of interaction scenarios that we will present in the following section.

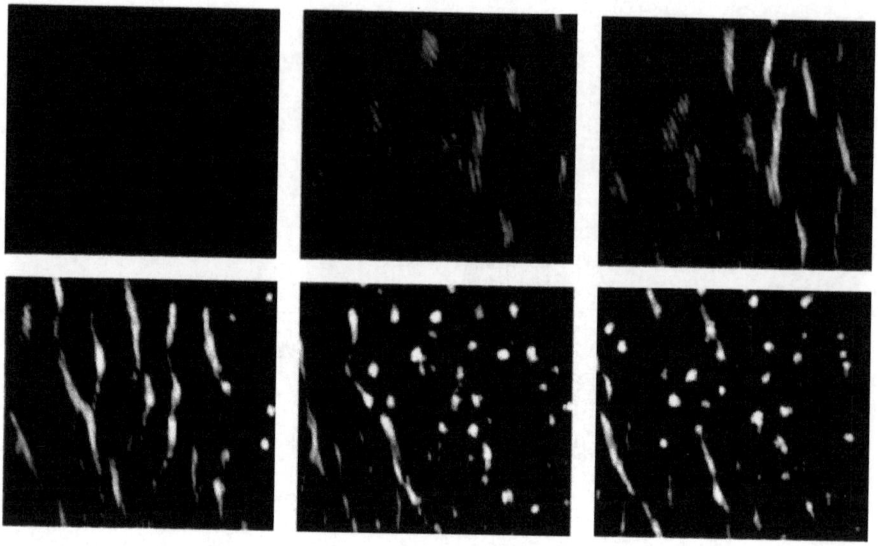

FIGURE 3.9 Filamentation of a broad beam into filaments of soliton scale. The beam is too large in diameter to self-trap completely. Instead, regions with appropriate power collapse into solitons. Under an increasing external field, the anisotropy first leads to stripe filaments which then filament into spots.

3.4 GUIDING LIGHT BY LIGHT

3.4.1 USING PHOTOREFRACTIVE SOLITONS AS WAVEGUIDES

A conventional waveguide is a path of high refractive index within a lower index transparent medium. Total internal reflection at the index boundary or continuous focussing caused by either step index or gradient index profiles compensates diffraction and traps the electromagnetic wave within the waveguide. As a spatial soliton propagates as a fundamental mode within its own induced refractive index profile, the latter can also be used to guide additional beams.[46–54] Comparable to a gradient index fiber, a spatial soliton provides a high index core with radially decreasing refractive index. The index change along the radius is typically (for strontium barium niobate) on the order of 10^{-5}. Although this is very small compared with manufactured fibers, it is nevertheless still sufficient to achieve waveguiding. With different materials, it can be as high as 10^{-2} for certain photorefractive polymers.[55] The low index difference requires a very small angle for efficient coupling into the waveguide, with the limiting angle for total internal reflection being approximately

$$\theta_c = \frac{n_{\max} - n_{\min}}{n_{\max}} \tag{3.12}$$

where $n_{\max}$ and $n_{\max}$ are the core and base refractive indices.

Waveguiding in photorefractive screening solitons was first demonstrated by Shih et al. in 1996.[56] A red probe beam of 15 microwatts derived from a HeNe laser was guided by a soliton formed by a 488 nm beam at only 1.5 microwatt. The soliton provided a single mode waveguide for the probe which had a diameter of about 12 microns. The coupling efficiency was reported to be 85%, with corrections for material absorption and Fresnel reflections already accounted for. The difference in laser power is possible due to the different frequencies used: while the generation of a soliton requires only a modest laser power, the energy of the single photons must be high enough to bridge the gap to the conduction band in the crystal. It is then possible to guide stronger beams of a longer wavelength that cannot excite charge carriers and hence only linearly propagate in the refractive index profile of the soliton. In fact, it is even possible to turn off the writing beam and still make use of the previously written refractive index. Probe beams with different frequencies compared to the writing beam also have the advantage that separation of both beams is feasible by simple spectroscopic methods.

Figure 3.10 shows the build-up process of a soliton which is simultaneously utilized as a waveguide for a probe beam of longer wavelength. While the beam bending effect complicates the transient build-up, the final position of the soliton output is well defined. The probe beam is guided during most of the transient and in the stationary steady state.

As we already noted in the preceding section, the material anisotropy of strontium barium niobate permits only elliptical solitons resulting in the refractive index profile depicted in Figure 3.8. While using the waveguide as a single mode guide, the guided wave also acquires a slightly elliptical profile. For a high order mode guided wave, only TEM_{0y} modes are possible as the waveguide is strongly confining in one transverse dimension.[36,57,58] Additionally, the anisotropy provides for fascinating interactions between multiple solitons.

3.4.2 Interactions of Spatial Solitons: Coupling and Switching

If one aims to conceive all-optical devices capable of guiding and switching light by light itself, induced waveguiding is not sufficient. It needs to be complemented by the ability to manipulate such waveguides in dependence on the presence of other solitons. Therefore, we will now look into the interaction of spatial solitons. Collisions

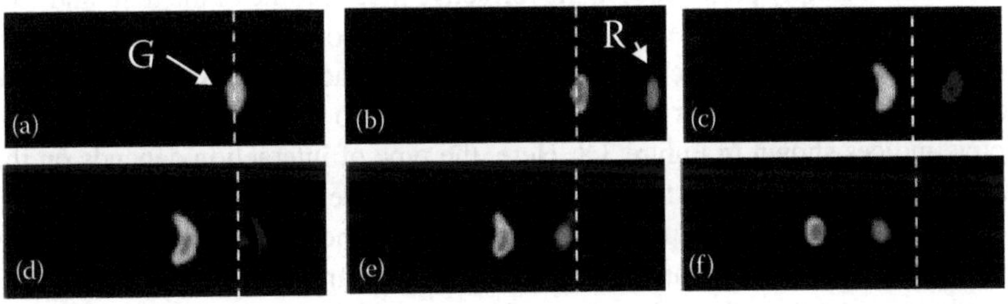

FIGURE 3.10 Dynamic generation of a waveguide. Example of waveguiding of a 632 nm probe beam (R) in a 532 nm soliton (G) in SBN. The outputs are spectroscopically separated by means of a prism in order to display both guiding and guided beam independently. Initially (**a**) the guiding and guided beams propagate linearly. Then, self-focussing and beam bending set in. The probe beam is visibly guided by solitons during the whole process. The timescale of the complete build-up is on the order of a few seconds.

of solitons in the strict mathematical sense are not very interesting. They are fully elastic as a result of the integrability of the nonlinear Schrödinger equation. Here, the soliton solutions do not couple to the radiation, no losses are possible and each soliton maintains all its properties — identity, shape, energy and velocity — through a collision. Spatial Kerr solitons display somewhat richer interaction scenarios where coherent solitons can attract or repel each other, depending on their relative phase. However, solitons in Kerr media are limited to stripe beams in order to avoid collapse and hence still not very interesting for applications.

Fortunately, spatial optical solitons such as available in photorefractive media are of a quite different nature. As they couple to the linear radiation field, collisions result in the transfer of energy from one soliton to the other or in dissipation and are therefore inelastic. This feature gives rise to new phenomena such as fusion,[56,59] fission[60] and annihilation[61] or bound states[62] of stable solitons. Along with the availability of both transverse dimensions, the range of possible interactions is very attractive for guiding, steering and manipulating light only with light — without any intervening prefabricated structures. This behavior is typical for a broad range of non-Kerr nonlinearities, starting from saturable Kerr media, which may still be quite close to the Kerr case, to quite different nonlinearities. Connecting to our examples cited in the previous sections, we again restrict ourselves to the photorefractive nonlinearity, which has been extensively investigated in the past and hence today serves as the experimental model for a more general class of spatial optical solitons.

There are two possible realizations for the interaction of two solitons through a common refractive index change, depending on the mutual coherence of the solitons: coherent and incoherent interaction. Coherent interaction occurs whenever the relative phase between the interacting beams is constant for a sufficiently long time so that the nonlinear medium can respond to intensity variation due to interference. Here the local field is the sum of both fields and the solitons are described by the same equations as before.

If the relative phase changes too fast, the time constants of the nonlinear medium render any interference effectively invisible to the nonlinearity. In that case the beams are considered mutually incoherent and the nonlinearity only depends on the sum of their intensities as the interference term averages to zero. Under incoherent interaction, a second soliton simply propagates in the refractive index profile of the first (and vice versa). In an isotropic saturable nonlinearity, incoherent solitons would always attract as they would always be diffracted toward the higher refractive index. For anisotropic photorefractive solitons, this is somewhat different due to the sidelobes with reduced refractive indices shown in Figure 3.8. Here, the type of interaction depends on the initial position and distance of the beams: if the beams are arranged along the y-axis, they will always experience a positive gradient toward the second and hence always be attracted. If the beams are arranged along the x-axis and close enough to not feel the low index sidelobes, they also attract. However, if the initial distance is larger, the sidelobes result in a mutual repulsion of the two beams, which is exceptional for mutually incoherent solitons and a result of the photorefractive anisotropy. In between is a fixed point with zero forces which is unstable.

This anomalous interaction of photorefractive solitons can in principle be exploited to discriminate an input encoded in the initial soliton distance with the

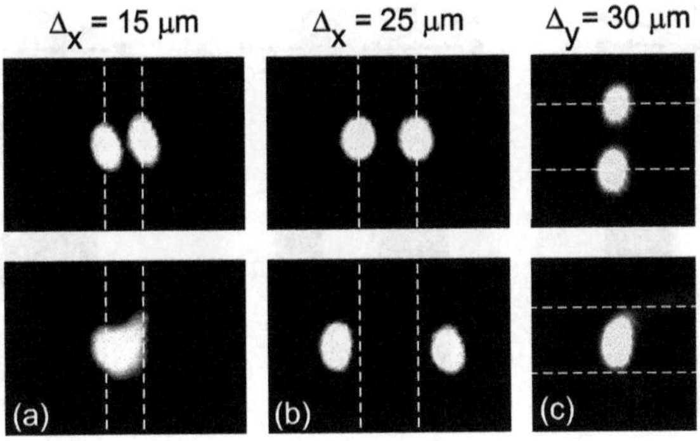

FIGURE 3.11 Anomalous interaction in experiment. Top row: solitons without interaction (images taken separately and combined for illustrative purposes). Bottom row: result of the soliton interaction for different distances (a,b), where the change from attraction to repulsion with increasing initial distance is clearly visible. For alignment along the y-axis (c), only attractive interaction exists.

output being either repulsion and subsequent separation or attraction and subsequent fusion which we will consider below. The anomalous interaction is easily demonstrated in experiment as shown in Figure 3.11.

For mutually coherent solitons, interference directly influences the interaction. Attraction and repulsion can both be observed, depending on the relative phase: in-phase beams constructively interfere and thereby cause an additional increase of the refractive index encompassing both beams. Therefore they are attracted toward their common center. This is comparable to the attraction of incoherent solitons but stronger in effect. For very small interaction angles, the beams fuse into a single brighter soliton, radiating away excess energy. Solitons out of phase by π destructively interfere, creating a dark region between them in which the refractive index is as low as far away from the beams. Consequently, both solitons are deflected from their common center and a repulsive force is observed. For these effects to be observable, both solitons must be equal in power and diameter. If they vary in intensity, the phase of one soliton will rotate faster than that of the other beam due to the different refractive index at the beam core. As the relative phase changes, the nature of the interaction also varies. For relative phases that are not integer multiples of π, the soliton interaction becomes asymmetric under exchange of the solitons; energy is transferred from one beam into the other, leading to the eventual disappearance of one beam.

These interactions depend strongly on the angle under which the beams collide. For very large angles, the refractive index change becomes too minute to have a strong effect and the beams pass through each other essentially unaltered — the collision becomes elastic. For very small angles, well below one degree for the SBN experiments with parameters as outlined in the previous section, the effects described above are observed. In between, for collision angles about one degree, two special cases have been reported: coherent in phase beams may result in three strong interference fringes, where the central one is of the right diameter and power to create its own solitary

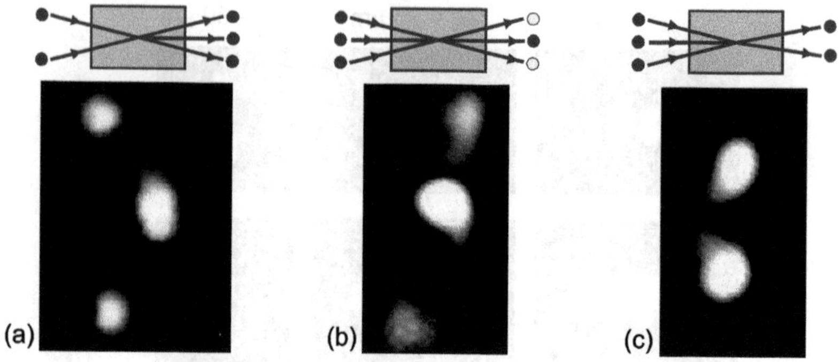

FIGURE 3.12 Coherent interaction of in-phase beams with a control beam: creation, fusion and annihilation. **(a)** No control beam present: a new beam is created from the constructive interference pattern. Beam bending leads to a lateral separation of the created and the original beams. **(b), (c)** A control beam can lead to fusion of all three beams or to its own annihilation depending on the relative phase.

waveguide. In that case two colliding solitons create a third soliton in their center which takes on part of their energy. Alternatively, two beams can be manipulated, depending on the presence and phase of a third beam. Depicted in Figure 3.12 are two in-phase beams colliding at an angle just above one degree. Their collision is close to elastic and both beams continue to exist throughout the collision. The presence of a third beam with a phase advanced by $\pi/2$ leads to fusion of all three beams. Instead, if the central beam is retarded in phase, its energy is completely transferred into the two outer beams that propagate unaffected, aside from the increase in beam power and a displacement as a result of their shared collision with the central beam. Thus, different kind of gates or optical switches can be realized by exploiting the properties of coherent soliton collisions.

Again, as was the case for a single soliton, the refractive index profiles resulting from the interacting solitons can serve as waveguides for additional probe beams. Where a probe beam is inserted into only one input of a soliton collision, its energy is distributed among the outputs, serving as an X- or Y-coupling device. All the considerations mentioned earlier regarding the guiding of probe beams, efficiency and persistence of the refractive index profiles after the writing beams are switched off apply. Soliton collisions can therefore be used to dynamically create optical junctions that change their switching behavior for probe beams, depending on the presence of writing beams.

3.4.3 Properties of Counterpropagating Waveguides

All interactions between solitons considered so far have been in copropagating geometry. If one imagines complex devices consisting of self-induced waveguides, it is easy to see that counterpropagating beams will be required for many schemes. As an example, it has been proposed to use counterpropagating solitons[63] that attract each other to build a device for automatically aligning fiber waveguides.

Two major differences distinguish the counterpropagating geometry from the ordinary case. First, the coherent interaction between the two beams that depends on the phase relation is more complex. As the beams propagate in opposite directions, no constant phase difference exists but it rather oscillates with twice the optical frequency, giving rise to an interference pattern between the two beams. Depending on the exact geometry in the experiment, the interference can create an additional refractive index change or lead to energy exchange between the two counterpropagating solitons through photorefractive two-wave mixing.[26,27] Second, counterpropagation of optical beams in nonlinear media generally opens the possibility for dynamical instabilities and requires the explicit accounting of time in the model.[64] In contrast, copropagating models permit us to give up explicit time dependence and consider a stationary steady state for the input plane which in turn establishes a steady state in every transverse plane along the propagation direction. Under counterpropagation, stationary states may be unstable and yield to either periodic or chaotic dynamic states. While the first complication may be avoided by using mutually incoherent beams, the second is a general property of such systems.

Analytically, the interaction of two counterpropagating solitons is readily described by the following set of equations:

$$i\partial_z A_1 + \nabla_\perp^2 A_1 = E_{SC} A_1 \tag{3.13}$$

$$-i\partial_z A_2 + \nabla_\perp^2 A_2 = E_{SC} A_2 \tag{3.14}$$

$$\tau\partial_t E_{SC} + E_{SC} = -E_e \frac{|A_1|^2 + |A_2|^2}{1 + |A_1|^2 + |A_2|^2} \tag{3.15}$$

which describe the propagation of two beams with envelope amplitudes A_1 and A_2 in the $+z$ and $-z$ directions in a self-induced refractive index distribution, considering only mutually incoherent beams. Thus, the propagation equations are not directly coupled, but only via the slow change of the refractive index which is described by static electric field given by the third equation. The material evolution contains a temporal derivative that cannot be neglected as the majority of solutions are dynamic. In the propagation equations on the other hand, all temporal derivatives have again been dropped which signifies that the optical propagation is many orders of magnitude faster than the photorefractive medium and hence is slaved to its dynamics.

Numerical simulations and experimental investigations[65] both demonstrate that three qualitatively distinct parameter regimes can be distinguished. For small values of the nonlinearity, the counterpropagating beams attract each other and as a result overlap as far as the initial lateral separation permits. In principle, this is comparable to the attraction of copropagating solitons. If one increases the medium length or the strength of the nonlinearity, the system undergoes a bifurcation at a certain threshold. Beyond, the beams tend to separate although they interact only through attractive forces mediated by a common refractive index profile (Figure 3.13). While this seems counterintuitive at first glance, the mechanism leading to the separation is straightforward and has been demonstrated in a pseudo-mechanical model of the optical system.[66] Finally, by further increasing the strength of the nonlinearity, a second bifurcation renders the stationary states unstable and only states with perpetual dynamics exist (Figure 3.14).

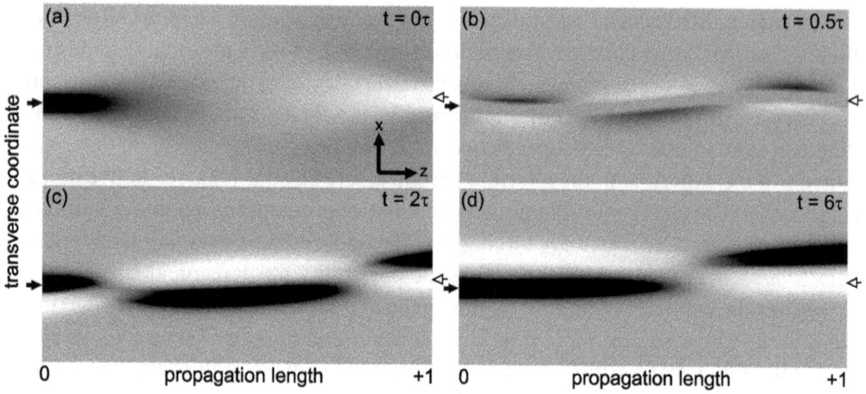

FIGURE 3.13 Numerical time series of the unstable separation of counterpropagating mutually incoherent solitary beams. The beams are color-coded black and white for easy visual separation. **(a)** Linear propagation, both beams diffract. **(b)** Initial self-focussing into a joint waveguide. A small initial lateral offset serves to seed the developing instability. **(c)** Metastable separated states with two intersections of the beams. **(d)** Final steady state with separate beams. Obtained by numerical integration of (3.13)–(3.15).

The concept of counterpropagating self-induced waveguides with attractive forces initially led to some ideas for applications. However, the existence of instabilities associated with counterpropagation will pose a significant hindrance to many such schemes. At the time of writing, the interaction of counterpropagating solitons is still a very new field under active research. Waveguiding has not yet been demonstrated; neither have there been reports concerning the stabilization of unstable states, which is a prerequisite for their use in all-optical devices. Beyond suppression of the instability, the self-sustained dynamics connected with the instability of counterpropagation solitary beams may become interesting for active optical devices in the long run.

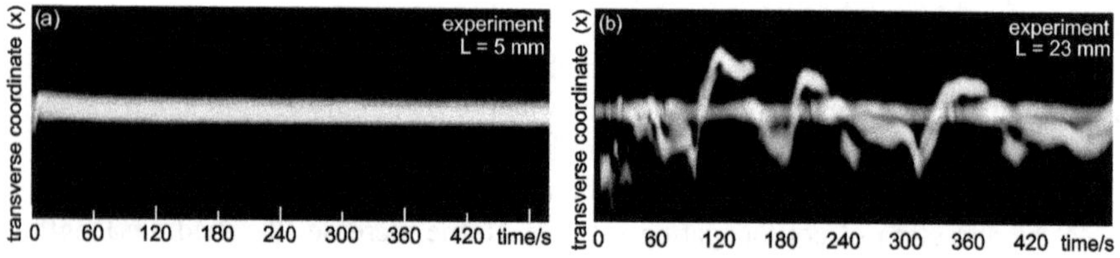

FIGURE 3.14 Experimental time series of the dynamic instability of two counterpropagating solitons. The two time series are obtained by dropping one transverse direction of camera frames and plotting the x coordinate over time. Each series shows the soliton output (bright beam) superimposed on the input (reflection, faint beam) of its counterpropagating partner. **(a)** Time series of below-threshold dynamics. After a brief transient, both beams overlap and form a common induced waveguide. **(b)** Above-threshold time series. The beams move irregularly about each other, driven by mutual attraction but separated by the instability.

3.5 WAVEGUIDE ARRAYS AND PHOTONIC LATTICES

A singular advantage of optical data processing is the inherent parallelity provided by the two transverse dimensions. While a single self-induced waveguide is a point-like structure by comparison with linear images, transverse space can be utilized by arranging a large number of soliton pixels on a grid of rectangular or other symmetry. Considering that at any time each soliton represents one bit, one easily arrives at very large "words" available for data processing purposes.

We distinguish two main topics connected with parallel soliton waveguides. First, if the individual solitons are propagating at a sufficient distance to be noninteracting, data may be transferred independent of neighboring channels. Within such an *array of solitons*, single channels may be manipulated by introducing additional control beams between channels, providing the full range of interaction scenarios presented in the previous section of this chapter. Second, if the distance of the solitons is reduced, they start to interact with their neighbors. The resulting *soliton lattices* feature a smoother refractive index profile which can in principle form a photonic crystal for longer wavelength beams propagating nonparallel to the constituent solitons. However, the interaction renders most lattices highly unstable. We will look into possibilities for their stabilization.

3.5.1 SOLITON WAVEGUIDE ARRAYS

To create large two-dimensional structures of photorefractive solitons, conditions for non-interacting propagation need to be considered. Crucial to the parallel propagation of photorefractive solitons are the anisotropic properties of the induced waveguides, which lead to different interactions along the two transverse dimensions. Experimentally, one finds a minimum distance above which stable parallel propagation is still obtainable. A typical experiment for the generation of soliton arrays is depicted in Figure 3.15.[67]

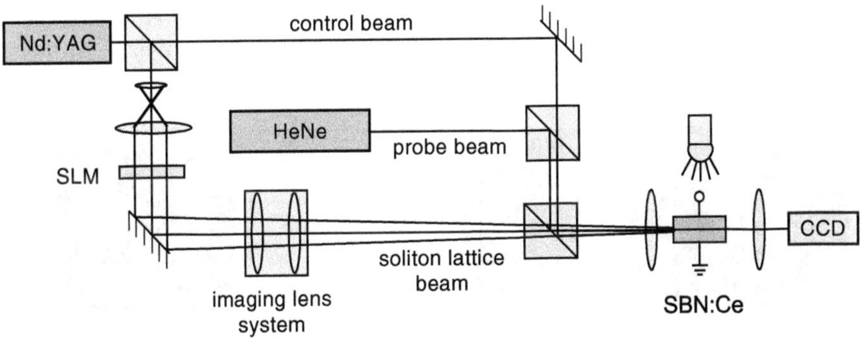

FIGURE 3.15 Experimental setup for the generation and investigation of soliton arrays. The array is seeded by a spatial light modulator (SLM) that is imaged onto the photorefractive crystals. Additional beams serve to investigate controlled interactions and wave guiding in the array.

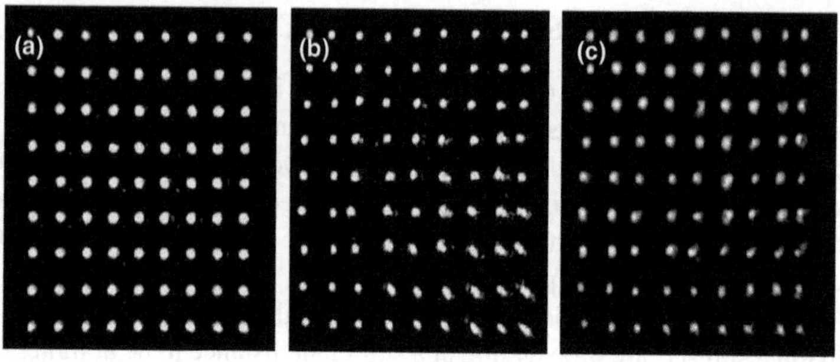

FIGURE 3.16 Waveguide array with 9 × 9 channels induced by photorefractive solitons. (**a**) Crystal input face with the incident spot array (distance between spots $\Delta x = 60$ μm, $\Delta y = 80$ μm; intensity in each channel 110 nW; diameter of each spot 15 μm). (**b**) Array of 81 focused solitons on the output face, (**c**) Waveguiding in separate channels of the array, using a probe beam at $\lambda = 633$ nm. Reprinted from Trager et al.[71]

The beam of a frequency-doubled Nd:YAG laser is extended to illuminate a spatial light modulator, which can be a fixed mask of liquid crystal display. The modulator imprints the array onto the widened beam which is then imaged onto the front face of the photorefractive crystal. For the 25×25 pixel arrays shown in Figure 3.16, the power of a single soliton is 110 nW, with the remaining parameters comparable to above descriptions for single solitons. The minimum distance for which no interaction was observed was 100 μm along the *c*-axis and 125 μm along the *b*-axis. Separations below the critical limit lead to fusion of channels, primarily along the *b*-axis which is easily explained by the anisotropic refractive index profile associated with each channel.

Combining the possible soliton interactions and the propagation of several solitons in an array naturally leads to the concept of controlled interaction of channels with an array. Figure 3.17 depicts a Y-coupler created in a 3 × 3 waveguide array by introduction of an additional control beam.

The input to the soliton array consists of nine coherent in-phase beams obtained from a frequency-doubled Nd:YAG laser. The initial beam diameter is 12 μm, the lateral separation is 70 μm along the *b*-axis and 50 μm along the *c*-axis. At this

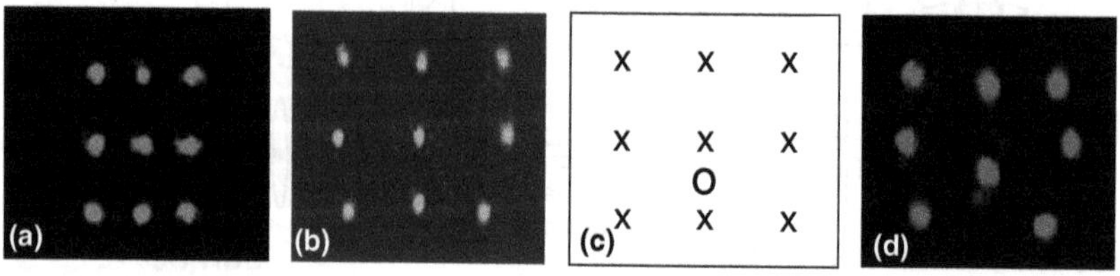

FIGURE 3.17 Modification of a soliton array with an additional control beam. (**a**) Writing beams creating the soliton array at the crystal input face. (**b**) Soliton array on the output face. (**c**) Sketch of the control beam insertation. (**d**) Guiding of longer wavelength probe beams in the soliton array in which the control beam has created a Y-junction of two channels. Reprinted from Petter et al.[72]

configuration, the individual solitons propagate independently and probe beams coupled into any of the nine input channels are guided to the corresponding output. If an additional beam derived from the same laser source is inserted between two of the waveguides arranged along the *b*-axis, they can be fused, resulting in a Y-coupler with two inputs and a single output. Figure 3.17 shows 633 nm probe beams coupled into all input channels. Probe beams guided in any of the two fused inputs are invariably guided to the common output. This example shows that all-optical manipulation of individual channels in soliton arrays can be achieved with controlled soliton interaction. The induction of interaction via control beams relies on the individual solitons having a sufficient distance to be non-interacting without the control. If their distance is reduced such that they start to interact strongly with their nearest neighbors, the soliton structure becomes unstable.

3.5.2 Stabilizing Soliton Lattices by Phase Engineering

If we reduce the distance between solitons in a waveguide array well below the noninteraction limit, we arrive at a nonlinear photonic lattice. In general, a photonic lattice is an optical material with a periodic modulation of the refractive index, also called a photonic crystal. The study of nonlinear effects in periodic photonic structures recently attracted strong interest because of many novel possibilities to control light propagation, steering and trapping. Periodic modulation of the refractive index modifies the linear spectrum and wave diffraction and consequently strongly affects the nonlinear propagation and localization of light, including the formation of novel types of self-trapped optical beams, spatial solitons. Photonic lattices can be induced optically by linear diffraction-free light patterns created by interfering several plane waves.[68] This is achieved by selecting a polarization of which the effective electrooptic coefficient is very small (for SBN; ordinary polarized light). While still modulating the refractive index for extraordinary polarized light, the writing beams only encounter a homogeneous refractive index. However, this procedure is essentially limited to completely periodic media with only trivial defects.

If one endeavors to create optically induced lattices with richer internal structures, a prerequisite to their effective application as reconfigurable waveguides and functional devices, nonlinear intensity patterns offer increased flexibility. Nonlinear diffraction-free light patterns in the form of stable self-trapped nonlinear periodic waves can propagate without change in their profile, becoming the eigenmodes of the self-induced periodic potentials. This behavior is generic, since nonlinear periodic waves can exist in many types of nonlinear systems, and they provide a simple realization of nonlinear photonic crystals. Such structures can be considered as flexible because the lattice is modified and shaped by the nonlinear medium. Such flexible nonlinear photonic lattices extend the concept of optically induced gratings beyond the limits of weak material nonlinearity. Moreover, nonlinear lattices offer many novel possibilities for the study of nonlinear effects in periodic systems because they can interact with localized signal beams via cross-phase modulation and can form composite bound states. Nonlinear photonic lattices created by two-dimensional arrays of pixel-like spatial solitons have recently been demonstrated experimentally in parametric processes and in photorefractive crystals with both coherent and partially incoherent light.

Such soliton lattices consisting of closely interacting solitons, in essence photonic crystals created by their own guided modes, allow us to locally modify the structures imprinted into the nonlinear medium. For the case of two-dimensional arrays of in-phase spatial solitons created by the amplitude modulation, every pixel of the lattice induces a waveguide that can be manipulated by an external steering beam.

However, the strong interaction of the no longer separated waveguides poses a significant problem to their stability. Due to the attractive forces between incoherent as well as coherent in-phase self-induced waveguides, soliton lattices tend to fuse. The fusion is primarily observed in one transverse dimension as the lowered refractive index due to the photorefractive medium anisotropy separates individual columns stronger than individual rows of beamlets. This observation gives a hint at a mechanism that can be utilized to control the stability of nonlinear photonic lattices. If next neighbor beamlets are coherent and consistently out of phase by π, the intensity will drop to zero between waveguides due to the destructive interference and result in repulsion of the beamlets instead of attraction. Consequently, phase-engineered soliton lattices have been implemented are are seen to be very stable in comparison to their in-phase counterparts.[69]

One can obtain very closely packed lattices of self-guided waves that are stable in bulk (Figure 3.18). However, the repulsive forces lead to the evaporation of beamlets on the boundaries which eventually destroys the structure if the medium length is too long. Apart from this limitation, nonlinear soliton lattices offer the possibility to dynamically create photonic materials with the ability to adaptively change the local structure by appropriately modifying the input condition.

3.5.3 Discrete Solitons in Photorefractives

The propagation of optical beams within these periodic lattices is fundamentally different from that occurring in a homogeneous medium. For example, when the light is focussed into one waveguide, linear propagation along the waveguides results in tunnelling to adjacent sites, exhibiting a characteristic diffraction pattern with the

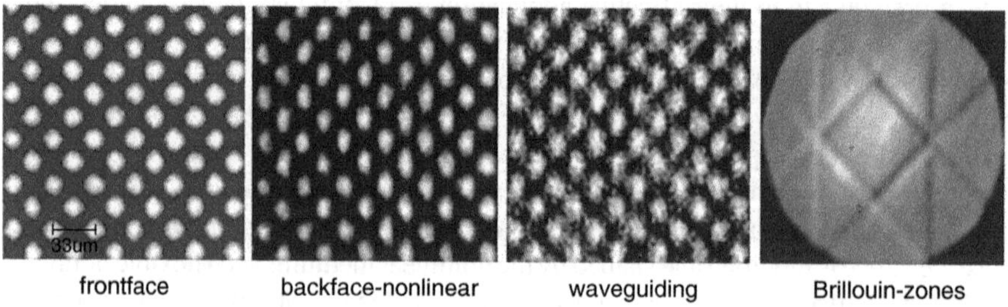

FIGURE 3.18 Photonic lattice composed of photorefractive screening solitons. Nearest neighbors are out-of-phase by π to stabilize the lattice. Additionally, the lattice is rotated by 45 degrees to lessen the impact of the anisotropy. Displayed are input conditions (left), nonlinear output (center left), a strongly divergent, spatially incoherent guided wave (center right) and the Fourier transform of the guided wave showing the band gaps of the photonic lattice as dark lines. Images kindly provided by B. Terhalle, WWU Münster.

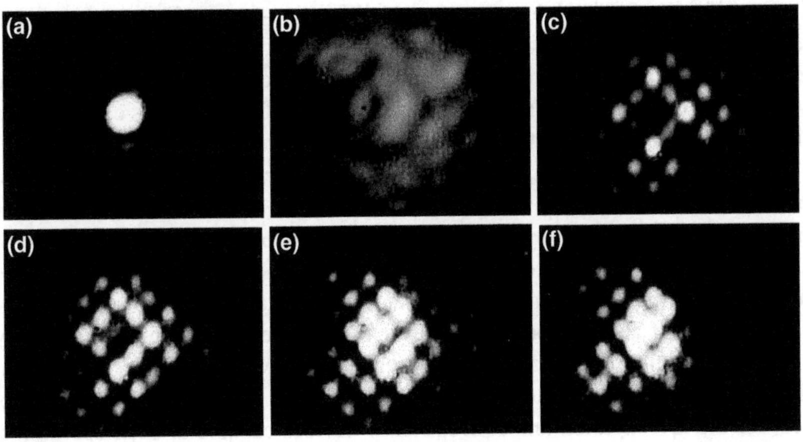

FIGURE 3.19 Discrete soliton formed on a photorefractive lattice. The lattice is generated by interference of four ordinary polarized beams. The extraordinary soliton seed is shown without the lattice on the front face (a) and back face (b). With the lattice present, it displays discrete diffraction for low beam power (c, 25 nW) and increasing localization to a few central lattice sites with an increase in beam power (d–f, up to 250 nW).

intensity mainly concentrated in the outer lobes. For a sufficiently high nonlinearity, self-focussing can balance this effect, leading to a lattice (discrete) soliton. It is this nonlinear effect that can be exploited in optically induced photonic lattices due to the photorefractive effect in order to realize waveguiding using the formation of discrete solitons.

For light propagating at a certain angle with respect to the array, the periodicity of the lattice becomes important, as the corresponding Bloch momentum can satisfy Bragg reflection conditions within the Brillouin zone. Near the edge of this zone, diffraction becomes anomalous (negative), leading to such effects as diffraction management and staggered (out-of-phase) solitons. In the center of the zone, in contrast, nonlinear lattice interaction leads to the formation of a lattice soliton. Figure 3.19 demonstrates a two-dimensional lattice soliton that can be exploited for waveguiding at the base of the first Brilloiun zone. These solitons have been demonstrated by Fleischer et al.[68] using a grating induced by interfering beams.

The utilization of optically induced lattices opens the possibility of tailoring the lattice — and thereby the nonlinearity responsible for the generation of discrete solitons — in real time and hence allows for the generation of adaptive nonlinearities in which discrete diffraction can be balanced to achieve this still very novel kind of soliton.

3.6 CONCLUSION AND OUTLOOK

Adaptive self-induced optical waveguides are promising milestones on the road to all-optical information processing. They can be realized in nonlinear media that change the refractive indices as a result of intensity. The balance of linear diffraction with nonlinear self-focussing leads to the concept of diffraction-free nonlinear propagation, to the concept of a spatial optical soliton. Such optical beams that remain constant under propagation exist in numerous nonlinear media and feature properties that are very attractive for application in guiding and switching of probe beams carrying data and in all-optical logical operations.

Among the different media, photorefractive nonlinearity has stood out as a prime experimental model for the investigation first of single spatial solitons and subsequently interactions of a few or many such beams. Some of the properties of photorefractive systems that lead to their strong presence in this field are the extremely low beam power required for soliton generation, the time constants that are experimentally very accessible but also can be adjusted over a few orders of magnitude, and very interesting interaction scenarios. While extremely fast switching is not the domain of photorefractive media due to the long time constants, they still provide complex models for the investigation and conception of devices capable of such operations. However, the ability to guide probe beams in the soliton's refractive index profile, especially in the longer telecom wavelength range that does not affect the nonlinear medium, marks photorefractive media as a candidate for dynamically reconfigurable waveguides.

The interaction properties of photorefractive solitons range from simple, particle-like behavior to complex scenarios in which the identities of solitons are no longer preserved. The particle-like interactions include attractive and repulsive forces that non-trivially depend on the soliton alignment and the distance between them due to the anisotropy of the photorefractive effect. The interaction can additionally be controlled by appropriately selecting the mutual coherence and the optical phase differences among individual solitons. Even bound and spiralling states consisting of two solitons can be generated, underlining the particle-like properties of such solitons even under the broadest definition.

As one focusses on the wave-like properties of solitons, the interactions become even more promising. The creation of new beams out of a soliton collision, and the deterministic fusion and annihilation controllable by additional optical steering beams provide the basic set of tools for successful implementation of logical operations. Almost any interaction scenario can be controlled by adjusting a variety of parameters, such as phase, coherence, polarization, beam position, orientation and distance, collision angles, strength of the nonlinearity, presence or absence of control beams and of additional background intensity.

Striving to utilize the transverse dimensions, one can access the inherent parallelity of optical propagation by creating arrays of multiple solitons arranged in the transverse plane. Initially only seen as a method of multiplexing channels, the application of soliton interactions to such soliton arrays opens fascinating opportunities for complex optical devices. As the ability to control close-range soliton interactions was developed, the generation of photonic lattices through very closed-packed solitons became feasible. Today, it is possible to generate adaptive photonic crystals by means of non-diffracting nonlinear beams that may be influenced locally to generate defects. In such photonic lattices, probe beams experience band gaps for certain wave vectors, opening the field of possibilities connected with the concept of photonics. Among these possibilities is the utilization of nonlinear dispersion at the band gap to again observe nonlinear self-focussing, this time balancing discrete diffraction and this novel kind of nonlinearity.

As one sums up the features of today's soliton generation, the rich features of single solitons, the complex interaction properties and the range of media in which experiments have been conducted, including and beyond the photorefractive materials, and the concept of self-induced, hence adaptive, nonlinear optical waveguides hold an immense potential for applications in all-optical data processing devices.

ACKNOWLEDGMENTS

The achievements of our group, some of which are presented in this chapter, owe much to the work of present and past collegues at the Institutes of Applied Physics at the University of Münster and Darmstadt Technical University, Denis Träger, Anton Desyatnikov, Nina Sagemerten, Jochen Schröder, Bernd Terhalle, Jürgen Petter and Carsten Weilnau, and to fruitful collaborations with the groups of Yuri Kivshar and Wieslav Krolikowski at ANU Canberra and Milivoj Belic at the Institute of Physics in Belgrade. The results presented throughout this chapter have been supported by the Deutsche Forschungsgemeinschaft, among others in the frame of the graduate research school programme GK 217, Nonlinear Continuous Systems and Their Analysis with Analytical, Numerical and Experimental Techniques.

REFERENCES

1. M. Segev. Optical spatial solitons. *Opt. Quant. Electron.*, 30:503, 1998.
2. N.N. Akhmediev and A. Ankiewicz, Editors. *Solitons: Nonlinear pulses and beams.* Chapman & Hall, London, 1997.
3. A. Hasegawa. Soliton-based ultra-high speed optical communications. *Pramana*, 57(5):1097, 2001.
4. A. Hasegawa and F. Tappert. Transmission of stationary nonlinear optical pulses in dispersive dielectric fibers. *Appl. Phys. Lett.*, 23:142, 1973.
5. L.F. Mollenhauer, R.H. Stolen, and J.P. Gordon. Experimental observation of picosecond pulse narrowing and solitons in optical fibers. *Phys. Rev. Lett.*, 45:1095, 1980.
6. P. Günter and J.P. Huignard, Editors. *Photorefractive materials and their applications.* Springer, New York, 2006.
7. N.J. Zabusky and M.D. Kruskal. Interaction of 'solitons' in a collisionless plasma and the recurrence of initial states. *Phys. Rev. Lett.*, 15(6):240, 1965.
8. R.Y. Chiao, E. Garmire, and C.H. Townes. Self-trapping of optical beams. *Phys. Rev. Lett.*, 13(15):479, 1964.
9. V.E. Zakharov and A.B. Shabat. Exact theory of two-dimensional self-focusing and one-dimensional self-modulation of waves in nonliner media. *Sov. Phys. Journ. Exp. Theo. Phys.*, 34:62, 1972.
10. C.S. Gardner, J.M. Greene, M.D. Kruskal, and R.M. Miura. Method of solving the Korteweg–de Vries equation. *Phys. Rev. Lett.*, 19:1095, 1967.
11. M. Hercher. Laser-induced damage in transparent media. *J. Opt. Soc. Am.*, 54:563, 1964.

12. S. Gatz and J. Herrmann. Soliton propagation in materials with saturable nonlinearity. *J. Opt. Soc. Am. B*, 8:2296, 1991.
13. S. Gatz and J. Herrmann. Propagation of optical beams and the properties of two-dimensional spatial solitons in media with a local saturable nonlinear refractive index. *J. Opt. Soc. Am. B*, 14(7):1795, 1997.
14. A. Barthelemy, S. Maneuf, and C. Froehly. Soliton propagation and self-trapping of laser beams by a Kerr optical nonlinearity. *Opt. Commun.*, 55:201, 1985.
15. J.S. Aitchinson, A.M. Weiner, Y. Silberberg, M.K. Oliver, J.L. Jackel, D.E. Leaird, E.M. Vogel, and P.W.E. Smith. Observation of spatial optical solitons in a nonlinear glass waveguide. *Opt. Lett.*, 15(9):471, 1990.
16. J.W. Bjorkholm and A. Ashkin. Cw self-modulation and self-focusing of light in sodium vapour. *Phys. Rev. Lett.*, 32:129, 1974.
17. P. Monot, T. Auguste, P. Gibbon, F. Jokober, G. Mainfray, A. Duliev, M. Louis-Jacquet, G. Malka, and I.L. Miguel. Experimental demonstration of relativistic self-channeling of a multiterawatt laser pulse in an underdense plasma. *Phys. Rev. Lett.*, 74:2953, 1975.
18. G. Assanto and G.I. Stegeman. Simple physics of quadratic spatial solitons. *Opt. Expr.*, 10(9):388, 2002.
19. G. Assanto, M. Peccianti, K.A. Brzdakiewicz, A. De Luca, and C. Umeton. Nonlinear wave propagation and spatial solitons in nematic liquid crystals. *J. Nonl. Opt. Phys. Mat.*, 12(2):123, 2003.
20. V.E. Zakharov and A.B. Shabat. Interaction between solitons in a stable medium. *Sov. Phys. Journ. Exp. Theo. Phys.*, 37:823, 1973.
21. A.W. Snyder, L. Polodian, and D.J. Mitchell. Stable dark self-guided beams of circular symmetry in a bulk Kerr medium. *Opt. Lett.*, 17:789, 1992.
22. H.S. Eisenberg, Y. Silberberg, R. Morandotti, A.R. Boyd, and J.S. Aitchison. Discrete spatial optical solitons in waveguide arrays. *Phys. Rev. Lett.*, 81(16):3383, 1998.
23. D.N. Christodoulides, F. Lederer, and Y. Silberberg. Discretizing light behaviour in linear and nonlinear waveguide lattices. *Nature*, 424:817, 2003.
24. G.C. Duree, J.L. Shultz, G.J. Salamo, M. Segev, A. Yariv, B. Crosignani, P. Di Porto, E.J. Sharp, and R.R. Neurgaonkar. Observation of self-trapping of an optical beam due to the photorefractive effect. *Phys. Rev. Lett.*, 71(4):533, 1993.
25. A. Ashkin, G.D. Boyd, J.M. Dziedzic, R.G. Smith, A.A. Ballmann, J.J. Levinstein, and K. Nassau. Optically-induced refractive index inhomogeneities in $LiNbo_3$ and $LiTao_3$. *Appl. Phys. Lett.*, 9:72, 1966.
26. P. Yeh, Editor. *Introduction to photorefractive nonlinear optics*. John Wiley & Sons, New York, 1993.
27. L. Solymar, D.J. Webb, and A. Grunnert-Jepsen, Editors. *The physics and applications of photorefractive materials*. Clarendon Press, Oxford, 1996.
28. J.P. Huignard and A. Marrakchi. Two-wave mixing and energy transfer in bso crystal: application to image amplification and vibration analysis. *Opt. Lett.*, 6:622, 1981.
29. J. Feinberg. Self-pumped, continuous-wave phase conjugator using internal reflection. *Opt. Lett.*, 7:486, 1982.
30. L. Hesselink, S.S. Orlov, and M.C. Bashaw. Holographic data storage systems. *Proc. IEEE*, 92(8):1231, 2004.
31. H. Rajbenbach, S. Bann, P. Refregier, P. Joffre, J.P. Huignard, H.-S. Buchkremer, A.S. Jensen, E. Rasmussen, K.-H. Brenner, and G. Lohmann. Compact photorefractive correlator for robotic applications. *Appl. Opt.*, 31:5666, 1992.
32. D.Z. Anderson and J. Feinberg. Optical novelty filters. *IEEE J. Quant. Elec.*, 25:635, 1989.

33. A.A. Zozulya and D.Z. Anderson. Propagation of an optical beam in a photorefractive medium in the presence of a photogalvanic nonlinearity or an externally applied electric field. *Phys. Rev. A*, 51(2):1520, 1995.
34. A. Stepken, F. Kaiser, M.R. Belic, and W. Krolikowski. Interaction of incoherent two-dimensional photorefractive solitons. *Phys. Rev. E*, 58:R4112, 1998.
35. I.V. Petviashvili. On the equation of a nonuniform soliton. *Sov. Journ. Plasma Phys.*, 2:257, 1976.
36. A.A. Zozulya, A.D. Anderson, A.V. Mamaev, and M. Saffman. Solitary attractors and low-order filamentation in anisotropic self-focusing media. *Phys. Rev. A*, 57(1):522, 1997.
37. D.N. Christodoulides and M.I. Carvalho. Bright, dark, and gray spatial soliton states in photorefractive media. *J. Opt. Soc. Am. B*, 12(9):1628, 1995.
38. A.A. Zozulya, D.Z. Anderson, A.V. Mamaev, and M. Saffman. Self-focusing and soliton formation in media with anisotropic nonlocal material response. *Europhys. Lett.*, 36(6):419, 1996.
39. M.I. Carvalho, S.R. Singh, and D.N. Christodoulides. Self-deflection of steady-state bright spatial solitons in biased photorefractive crystals. *Opt. Commun.*, 120:311, 1995.
40. M. Shih, M. Segev, G.C. Valley, G. Salomono, B. Crosignani, and P. DiPorto. Observation of two-dimensional steady-state photorefractive screening solitons. *Electron. Lett.*, 31:826, 1995.
41. A.V. Mamaev, M. Saffman, D.Z. Anderson, and A.A. Zozulya. Propagation of light beams in anisotropic nonlinear media: from symmetry breaking to spatial turbulence. *Phys. Rev. A*, 54(1):870, 1996.
42. M. Soljacic, M. Segev, T. Coskun, D.N. Christodoulides, and A. Vishwanath. Modulation instability of incoherent beams in noninstantaneous nonlinear media. *Phys. Rev. Lett.*, 84:467, 2000.
43. D. Kip, M. Soljacic, M. Segev, E. Eugenieva, and D.N. Christodoulides. Modulation instability and pattern formation in spatially incoherent light beams. *Science*, 290:495, 2000.
44. D. Kip, M. Soljacic, M. Segev, S.M. Sears, and D.N. Christodoulides. (1+1)-dimensional modulation instability of spatially incoherent light. *J. Opt. Soc. Am. B*, 19(3):502, 2002.
45. Z. Chen, J. Klinger, and D.N. Christodoulides. Induced modulation instability of partially spatially incoherent light with varying perturbation periods. *Phys. Rev. E*, 66:066601, 2002.
46. T.T. Shi and S. Chi. Nonlinear photonic switching by using the spatial soliton collision. *Opt. Lett.*, 15:1123, 1990.
47. R. McLeod, K. Wagner, and S. Blair. (3+1)-dimensional optical soliton dragging logic. *Phys. Rev. A*, 52(4):3254, 1995.
48. P.D. Miller and N.N. Akhmediev. Transfer matrices for multiport devices made from solitons. *Phys. Rev. E*, 53:4098, 1996.
49. B. Luther-Davies and X. Yang. Steerable optical waveguides formed in self-defocusing media by using dark spatial solitons. *Opt. Lett.*, 17:1775, 1992.
50. B. Luther-Davies, X. Yang, and W. Krolikowski. On the properties of waveguide x-junctions written by spatial solitons. *Int. J. Nonlinear Opt. Phys.*, 2:339, 1993.
51. W.J. Firth and A.J. Scroggie. Optical bullet holes: robust controllable localized states of a nonlinear cavity. *Phys. Rev. Lett.*, 76:1623, 1996.
52. M. Brambilla, L.A. Lugiato, and M. Stefani. Interaction and control of optical localized structures. *Europhys. Lett.*, 34:109, 1996.

53. R. de la Fuente, A. Barthelemy, and C. Froehly. Spatial-soliton-induced guided waves in a homogeneous nonlinear Kerr medium. *Opt. Lett.*, 16:793, 1991.
54. J. Petter and C. Denz. Guiding and dividing waves with photorefractive solitons. *Opt. Commun.*, 188:55, 2001.
55. F. Sheu and M. Shih. Photorefractive polymeric solitons supported by orientational enhanced birefringent and electro-optic effects. *J. Opt. Soc. Am. B*, 6:785, 2001.
56. M. Shih, M. Segev, and G. Salamo. Incoherent collisions between one-dimensional steady-state photorefractive screening solitons. *Appl. Phys. Lett.*, 69:4151, 1996.
57. S. Gatz and J. Hermann. Anisotropy, nonlocality and space-charge field displacement in $(2 + 1)$-dimensional self-trapping in biased photorefractive crystals. *Opt. Lett.*, 23:1176, 1998.
58. M. Saffman and A.A. Zozulya. Circular solitons do not exist in photorefractive media. *Opt. Lett.*, 23:1579, 1998.
59. M. Shih and M. Segev. Incoherent collisions between two-dimensional bright steady-state photorefractive spatial screening solitons. *Opt. Lett.*, 21:1538, 1996.
60. W. Krolikowski and S.A. Holmstrom. Fusion and birth of spatial solitons upon collision. *Opt. Lett.*, 22:369, 1997.
61. W. Krolikowski, B. Luther-Davies, C. Denz, and T. Tschudi. Annihilation of photorefractive solitons. *Opt. Lett.*, 23:97, 1998.
62. M. Shih, M. Segev, and G. Salamo. Three-dimensional spiraling of interacting spatial solitons. *Phys. Rev. Lett.*, 78:2551, 1997.
63. O. Cohen, R. Uzdin, T. Carmon, J.W. Fleischer, M. Segev, and S. Odoulov. Collisions between optical spatial solitons propagating in opposite directions. *Phys. Rev. Lett.*, 89(13):133901, 2002.
64. Y. Silberberg and I. Bar Joseph. Instabilities, self-oscillation, and chaos in a simple nonlinear optical interaction. *Phys. Rev. Lett.*, 48:1541, 1982.
65. Ph. Jander, J. Schröder, C. Denz, M. Petrovic, and M.R. Belic. Dynamic instability of self-induced bidirectional waveguides in photorefractive media. *Opt. Lett.*, 30(7):750, 2005.
66. K. Motzek, Ph. Jander, A. Desyatnikov, M. Belic, C. Denz, and F. Kaiser. Dynamic counterpropagating vector solitons in saturable self-focusing media. *Phys. Rev. E*, 68:066611, 2003.
67. C. Denz, A. Desyatnikov, Ph. Jander, J. Schröder, D. Träger, M. Belic, M. Petrovic, A. Strinic, and J. Petter. Photonic applications of spatial photorefractive solitons: soliton lattices, bidirectional waveguides and waveguide couplers. *OSA TOPS*, 87:382, 2003.
68. J.W. Fleischer, M. Segev, N.K. Efremidis, and D.N. Christodoulides. Observation of two-dimensional discrete solitons in optically induced nonlinear photonic lattices. *Nature*, 422:147, 2003.
69. A.S. Desyatnikov, D.N. Neshev, Y.S. Kivshar, N. Sagemerten, D. Träger, J. Jägers, C. Denz, and Y. V. Kartashov. Nonlinear photonic lattices in anisotropic nonlocal self-focusing media. *Opt. Lett.*, 30(8):869, 2005.
70. M. Shih, M. Segev, G.C. Valley, G. Salamo, B. Crosignani, and P. DiPorto. Two-dimensional steady-state photorefractive screening solitons. *Opt. Lett.*, 21:324, 1996.
71. D. Träger, A. Strinic, J. Schröder, C. Denz, M. Belic, M. Petrovic, S. Matern and H.G. Purwins, Interaction in large arrays of solitons in photorefractive crystals, *J. Opt. A*, 5(6):518, 2003.
72. J. Petter, J. Schröder, D. Träger and C. Denz, Optical control of arrays of photorefractive screening solitons, *Opt. Lett.*, 28(6):438, 2003.

4 Active Optical Waveguides

Michael Cada

CONTENTS

4.1 INTRODUCTION

The spectacular progress in laser technology since the early 1960s has led to the development of solid state, gas, and semiconductor lasers used in communications, medicine, material processing, metrology, military, sensing and many other fields. The compact disc player, a marvel of engineering ingenuity, truly revolutionized consumer electronics by marrying physics, communications, materials, and signal processing in an unprecedented way that brought the tiny laser diode chip to a "dollar-store" level. The successful development of low-loss low-dispersion optical fibers at acceptable prices fundamentally changed the whole area of telecommunications. That revolution is still ongoing, with Internet serving as the main driving force for the development of novel photonic devices with new functionalities and/or better performance properties.

Although photonics is a pervasive technology that has entered many facets of our everyday life and has shown a resilient nature with respect to business and economic cycles, it is still in its infancy if a comparison with electronics is warranted. Present devices are mostly passive, with very low integration, at primitive levels of functionality, and with no built-in intelligence in the form of programming capability and flexibility. No one has yet developed software for photonic chips.

Besides the classical use of light in imaging, it is still best exploited in the straight transmission of information rather than in its processing and storage; those are the exclusive domains of electronics. Some skeptics even suggest that the use of light for transmission may be its only and ultimate practical application because electronic devices have always accommodated new high performance processing and storage requirements better. The material side of photonics appears even more dismal if one considers functional and technological compatibility issues. Much effort has been expended in attempts to develop integrated optic and/or optoelectronic circuits analogous to the powerful electronic integrated circuitry. Fundamental material and technological problems, however, have prevented such developments for the past 35 years and persist to this very day.

Several decades of worldwide intensive research and development work in photonic devices have witnessed successes and failures of different material systems from glasses and gels and plastics to $LiNbO_3$ and $LiTaO_3$, from silicon and organic materials to III/V and II/VI semiconductors. These materials have satisfied, to a better or worse extent, certain applications requirements. However, there does not seem to exist a material system that would be as dominant in photonics as silicon is in electronics. The failed aspirations of many to develop an optical "equivalent" of the transistor, that is an all-optical switch of sorts, in a "photonic universal" material may simply reflect an excessive concern with the electronics analogy. It may very well turn out that photonics may never find one material system that will satisfy all possible needs.

Fortunately, the situation has been changing in recent years rather drastically, with the Internet providing most of the push for more bandwidth and new emerging application areas requiring higher speeds and more complex processing capabilities. Wavelength division multiplexing schemes requiring robust light processing functions triggered a new round of "integrated optics" research and development. Advances in material science and sophisticated technologies have been amazing, giving rise to new fields of nanotechnology and biophotonics that offer and indeed require capabilities for implementing optical and optoelectronic functionalities that were unthinkable several years ago.

Active optical waveguides are the light guiding structures that contain a material possessing optical gain and/or optical nonlinearity that leads to various active functionalities. Considering active photonic devices, as mentioned above, much remains to be accomplished; the whole area of photonic devices based on optical waveguides has almost exclusively been studied from the point of view of passive structure, except for semiconductor and fiber amplifiers and lasers. Even in those cases, though, the optical gain has traditionally been used either to generate coherent output or to provide linear signal gain. As a result, when a linear medium is used in a passive waveguide device, only simple functions such as wavelength multiplexing or optical filtering can be implemented. When a nonlinear medium is employed, the effects are either too weak (transparent crystals) or associated with large absorption losses (semiconductors near the band edge). However, optical waveguides remained the focus for many researchers for obvious reasons; waveguide configurations offer effective solutions to implementing many desired processing functions via efficient changes of refractive index due to diffractionless propagation of high power densities over long distances. The promise of combining effects of optical gain and optical nonlinearity in active optical waveguides is thus very attractive for future novel high performance and robust photonic devices.

4.2 OPTICAL GAIN AND NONLINEARITIES

4.2.1 Historical Perspective

Stimulated emission of light is a process whereby an atom or a molecule undergoes a transition from higher to lower energy states as a result of the presence of external stimulating photons and is key to the functionalities of both laser emission and optical gain devices. Although the idea of stimulated emission goes as far back as to Bohr and Einstein [1], the invention of the laser did not take place until the 1960s when Basov, Prokchorov, Schawlow, Townes, and Maiman [2] transformed it into reality. Similarly Gabor's idea of holography [3] did not materialize in a practical version until decades later, when it was made possible by the arrival of the laser. Kao and Hockham's [4] optical fiber as a new transmission medium was preceded by pioneering studies of Hondros and Debye [5].

Although much earlier, Shockley, Brattain and Bardeen did show, based on von Neumann's original idea conceived and calculations made in 1953, that semiconductors can be used for light amplification, it was the optical fiber that provided a tremendous stimulus for semiconductor laser research and development with the pioneering contributions of Kogelnik [6], Yariv [7], and others. Miller's [8] concept of integrated optics using for the first time optical waveguides of sorts conceived in the late 1960s and developed by some of the fathers of this field, Tamir [9], Tien [10], and others, evidently came too early from an applications point of view even though the fathers of modern nonlinear optics, Franken [11] and Bloembergen [12], excited many researchers over its potential in various applications. Seven of these early pioneers and more recently Kroemer and Alferov were awarded Nobel Prizes, thus significantly bringing photonics to the technology forefront and emphasizing its importance as a powerful enabling technology.

When Gibbs [13] showed bistable switching in a semiconductor and Chemla [14] successfully studied nonlinear optical effects in semiconductor quantum-well structures, many new exciting developments emerged, for example, bistability in waveguides proposed, studied and successfully developed by Stegeman [15]. Optical periodic configurations, first investigated by Rayleigh [16], then by both, the Bragg brothers [2] and Brillouin [17], and examined by Winful [18] in the nonlinear regime, were used for the first time for optical mixing in integrated optics by Normandin [19]. These works then led to various new photonic devices that exploited active optical waveguides, proposed and demonstrated by a number of other researchers including Cada [20, 21].

4.2.2 Optical Nonlinearities

Nonlinearity can lead to establishing active functionalities in waveguide photonic devices such as all-optical switching or optical bistability. In transparent regions of semiconductors or other crystals, the nonlinear behavior can be described via the polarization vector $\overline{P}$:

$$\overline{P} = \varepsilon_0 \left[\overline{\overline{\chi^{(1)}}}\,\overline{E} + \overline{\overline{\chi^{(2)}}}\,\overline{E}\,\overline{E} + \overline{\overline{\chi^{(3)}}}\,\overline{E}\,\overline{E}\,\overline{E} + \ldots \right] = (\varepsilon - \varepsilon_0)\,\overline{E} + \overline{P}_{NL} \qquad (4.1)$$

where ε_0 is the permittivity of vacuum, ε is the permittivity of the material, $\overline{\overline{\chi^{(i)}}}$'s are the material susceptibilities, $\overline{E}$ is the electric field vector, and $\overline{P}_{NL}$ is the nonlinear part of the polarization vector. Different orders of the nonlinearity then represent various known effects such as, for example, the electro-optic effect or the Kerr effect. These effects in turn lead to numerous applications such as electro-optic modulation, sum and difference frequency generation, of which the second-harmonic generation is an important special case, all-optical switching, four-wave mixing, soliton generation, and others [22].

In absorbing regions of semiconductors such as those appearing when the operating wavelength is close to the band edge, the Kerr-like effect is responsible for nonlinear behavior, which leads to intensity-dependent refraction and is a result of nonlinear resonant phenomena whereby the absorption and refraction are interdependent via Kramers–Kronig relations. An extensive review of resonant optical nonlinearities in semiconductors including a long list of references is found in Garmire [23]. Although the underlying physics leading to these effects is quite complex, one can often conveniently and approximately describe the refractive index changes as:

$$n(I) = n_0 + n_2 I \qquad (4.2)$$

where $n(I)$ is the intensity-dependent refractive index, I is the light intensity, n_0 is the linear refractive index, and n_2 is the nonlinear refractive index coefficient that depends on wavelength λ. Clearly, this can be viewed as a special case in Eq. (4.1) above when only the third-order nonlinearity is involved.

A classical example of an application of this effect is optical bistability [13]. Eq. (4.2) is valid only close to the band edge where the nonlinearity is drastically enhanced by several orders of magnitude due to the resonant generation of free carriers and when absorption is not saturated by a large intensity. As the operating wavelength moves away from the band edge, the nonlinearity enhancement weakens and the behavior follows more and more the pure Kerr effect. When the operating wavelength is further than double the band edge (half in energy), even two-photon absorption does not take place and the nonlinearity assumes a character of the third-order effect in Eq. (4.1) with a corresponding $\overline{\overline{\chi^{(3)}}}$ for a given material.

In addition to the known material optical nonlinear properties of compound semiconductors well described, for instance, in Haug [24], quantum wells, wires and dots as nanostructures have been of great interest. They exhibit remarkable properties nonexistent in their parent bulk compounds. Physical mechanisms can be engineered or tailored on a quantum mechanical level to accommodate device functionality requirements. Physics of such semiconductor nanostructures produces the consequences in optical and electro-optical properties, while several unique effects exist [25], e.g., quantum-confined Stark effect or phase-space absorption quenching, that can be and have been used in optical devices.

The quantum mechanical understanding of these effects is not trivial and is related to the strong excitonic resonances existing in these quantum structures. Their reduced dimensionality beyond the exciton Bohr radius forces an increase in exciton binding energy and that makes the structure very stable. This added stability causes exciton resonances to produce strong absorption that is, along with the associated refractive index, strongly dependent on the intensity of incident light [26]. However, when close to the excitonic resonance of a semiconductor quantum well, Eq. (4.2) above can be used to describe reasonably accurately the nonlinear refraction. Due to excitons, it is further enhanced by several orders of magnitude compared to the parent compound. For example in GaAs or InP quantum wells, a typical value is $n_2 \approx 10^{-4}\, cm^2/W$, which is about four orders of magnitude larger than in bulk.

Eqs. (4.1) and (4.2) apply generally to semiconductor nonlinear optical materials characterized by either the tensors $\overline{\overline{\chi^{(i)}}}$ that represent usually weak nonlinearities or by the nonlinear coefficient n_2 that can be large, especially in quantum well materials, but for the price of large absorption losses.

The use of optical waveguides for exploitation of nonlinearities is obviously advantageous. Diffractionless propagation over relatively long distances and high power densities achievable with less-than-milliwatt optical powers generated by laser diodes is easily obtainable in optical waveguides. Refractive index changes are thus enhanced owing to the optical field confinement over a guiding distance. Optical waveguides are also suitable for implementing the travelling wave interactions between guided waves, which offers a benefit of substantially improved large bandwidth modulation capabilities. Inherent optical amplification offered by semiconductors makes functional devices employing such capabilities distinctive and attractive. The possibility of integration in a common substrate is another very promising feature, especially if one considers integrating vertically various technological and application platforms, for example, the passive and active ones.

4.2.3 Optical Gain

Optical gain is a property that a material acquires by exciting its molecules, atoms, or carriers in large densities into higher energy states, resulting in amplification of incident optical power through stimulated emissions. Basic processes governing the absorption and spontaneous and stimulated emission of radiation in a material system are straightforward, although the theory and quantum physics behind them are quite complex. Einstein's coefficients and relations were the first expressions that related these three phenomena and led to the known ratio of the rate of stimulated emission over the rate of spontaneous emission, $1/[\exp(h\nu/kT) - 1]$, where $h\nu$ is the photon energy and kT is the thermal energy. This expression shows that in a system under thermodynamic equilibrium, spontaneous emission is a dominant mechanism for generating new photons.

Since the stimulated emission can amplify the incoming light in a unique coherent way as a result of interaction between the incoming (incident, stimulating) photons and atoms or carriers in an excited state, one needs to establish a non-equilibrium situation in a given material. This requires a population inversion whereby higher energy states (upper energy levels) are more occupied than lower ones. To achieve this condition, atoms or carriers must be excited from lower to upper levels by external means, for example, by an optical beam or electrical current. The process of excitation of a material is called pumping. Once a proper method of pumping is employed and a material is excited, light propagating in the material is amplified with a net optical gain following a well known exponential law, $p(z) = p_0\, e^{(g-\alpha)z}$, where p is the optical power along the length, z is the propagation direction along the waveguide, p_0 is the initial power, g is the material optical gain, and α is the material loss. An optical waveguide that contains a gain material is referred to as an active optical waveguide.

A typical example of an active waveguide is a silica optical fiber doped with erbium atoms. It uses process of optical pumping. A pump signal at, for example, $\lambda = 0.98\ \mu$m excites erbium ions and produces desired population inversion. The signal wave at $\lambda = 1.55\ \mu$m triggers stimulated emission and gets amplified as a result. The new generated photons have polarization, phase, and wavelength properties identical to those of the photons that triggered their generation. Because of the specific character of the energy level distribution of erbium ions in silica fiber, this process requires the energy of the pumping photon to be greater than that of the signal photon. An effect of stimulated Raman scattering is also employed in establishing optical amplification in silica fibers [33].

The presence of high carrier densities in a semiconductor material can lead to optical gains, such as those exploited in semiconductor optical amplifiers and lasers. They use process of electrical pumping. A forward-biased PN junction structure provides a means to inject electrons and holes into a common area (active or gain area) from each side of the junction. If large concentrations of the carriers build up in the active region, population inversion is established. Carriers in the active region also form an optical waveguide that is now active since it has optical gain. Signal photons passing through such an excited area — optical gain material — stimulate radiative recombinations of electrons and holes, resulting in coherent amplification of signal power. Some of the recombinations still occur spontaneously, leading to noise called amplified spontaneous emission. Similar noises also exist in optical fiber amplifiers.

The dependence of optical gain on carrier density is not linear, although linear approximations have been used extensively. More accurate approximations employ higher order functions, for example, quadratic functions, to better describe nonlinear behaviors. For quantum wells, the gains are reasonably approximated with logarithmic functions. Gain dependence on the wavelength (energy) defines the gain curve and is usually approximated by various functions, e.g., square and cubic, or is determined experimentally [26]. Generally, it increases and shifts to shorter wavelengths with increasing carrier density. It saturates with light intensity due to saturating carrier density, which effect can be advantageously employed in all-optical operations. More detailed discussion of this subject appears in a later section. Optical gain can be calculated theoretically; it is a complex quantum mechanical problem because gain depends on carrier density, gain medium, doping, carrier concentration, and many other parameters [27].

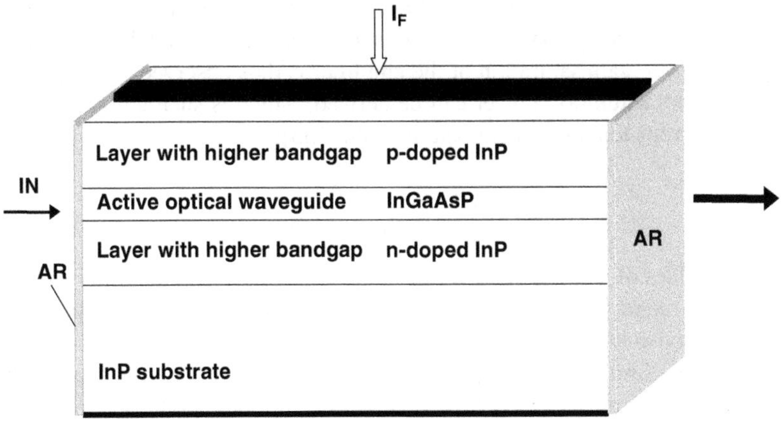

FIGURE 4.1 Semiconductor optical amplifier as active optical waveguide. AR = antireflection coating. I_F = driving electrical current.

After a semiconductor gain medium is placed in a waveguide formed, for example, by a semiconductor layer with a band gap energy lower (higher refractive index) than that of its surroundings, shown schematically in Figure 4.1 for the InP material system, one obtains an active optical waveguide that is basically a semiconductor optical amplifier. This was the original laser; a laser involves amplification rather than generation of light. What we now refer to as a laser is more of an optical oscillator than an optical amplifier. Light propagation in such an active waveguide can be described by travelling wave and rate equations, respectively:

$$\frac{dp}{dz} = (\Gamma g - \alpha)\, p \tag{4.3}$$

$$\frac{dN}{dt} = \frac{i_e}{qV} - \frac{N}{\tau} - \frac{(\Gamma g - \alpha)}{h\nu S} p \tag{4.4}$$

where Γ is the cross-section confinement factor of the propagating mode overlapping the active area, N is the carrier density dependent on time, t, i_e is the electrical injection current, q is the electron charge, V is the active region volume, S is the active area cross-section, and τ is the carrier lifetime that usually includes nonradiative recombinations due to defects, impurities and other traps, radiative recombinations due to spontaneous emission, and Auger processes. Eqs. (4.3) and (4.4) can be derived from using Maxwell's equations and employing the density matrix approach [28]. When N and p are both functions of t and z, the solutions provide the power and the carrier density distributions in the waveguide in both the steady-state and dynamic cases, including the relationship between the accumulated phase shift of the optical signal and the refractive index of the semiconductor optical amplifier.

Numerical simulations then yield the actual behavior of the semiconductor optical amplifier, especially the gain saturation phenomena [29]. For this case, if a saturating optical input is applied to the device, the carrier density decreases as a result of an increase in stimulated emissions, and the gain drops and shifts accordingly. By keeping the carrier density constant via a constant injection current while varying the input optical power, gain saturation of the amplifier is thus observed. An expression for the amplifier's saturation power can be derived, which is such an output power when the gain drops to half of its maximum value [33]:

$$p_{sat} = ln(2)\frac{h\nu S}{\Gamma\tau(dg/dN)} \tag{4.5}$$

Practical values of saturation power in semiconductor optical amplifiers are low, in the milliwatt range. This makes them basically unusable for in-line amplification in wavelength division-multiplexing optical fibers communication systems that carry tens to hundreds of milliwatts of optical power. Fiber amplifiers are more suitable; however, they suffer from relatively high noise due to amplified spontaneous emission. In comparison, Raman fiber amplifiers offer both high saturation power and relatively low noise while they exhibit a broad spectral bandwidth up to 50 *nm* that makes them very suitable for wavelength-division multiplexing-based system applications. The drawback is that they need high power lasers for pumping.

As mentioned, an optical gain possesses usable nonlinearities in addition to linear light amplifying properties. Various nonlinear mechanisms cause changes in the gain or refractive index. These nonlinear mechanisms include thermal effects and carrier effects. The thermal effects are not practically useful because of their slow response times on the order of microseconds or longer. The carrier effects are related to either interband or intraband transitions. The interband transitions are characterized by free-carrier lifetimes on the order of nanoseconds. The associated band-filling effect causes changes in the gain or the refractive index, leading to the saturation of optical gain with the optical input as discussed earlier. The intraband transitions cause nonequilibrium carrier distribution in the energy bands of the gain material, such as a spectral-hole burning or carrier heating effect. The relaxation time is on the order of hundreds of femtoseconds, which makes this nonlinear optical gain effect extremely attractive for construction of functional active waveguide devices [27].

Active optical nonlinearities as they relate to the gain material properties discussed above and as they demonstrate themselves in various desired or undesired functionalities, stem basically from both the dependence of refractive index on electron concentration (carrier density), dn/dN, and the dependence of material gain on carrier density, dg/dN. They are both responsible for a number of rather undesired effects in semiconductor communication lasers. dn/dN causes, for example, wavelength chirp, phase and frequency modulation, and spectral linewidth broadening. dg/dN affects threshold current, relaxation oscillations, and mode spectra [24]. These optical gain nonlinearities can, on the other hand, be turned into advantages. A thorough review is presented in Adams et al. [48]. Combined with optical signal gain, the nonlinearities can be useful in devices other than linear amplifiers and lasers. Wavelength conversion, optical flip-flopping, all-optical switching, all-optical oscillations, optical buffer memory, and other functionalities have been demonstrated [21]. Of several functional effects resulting from semiconductor optical amplifier nonlinearities, the three most common are cross-gain modulation, cross-phase modulation, and four-wave mixing [29].

Cross-gain modulation relies directly on the gain saturation effect. Two optical inputs are used; one is a modulated signal to be processed, the other is a continuous-wave optical signal. The modulated signal affects the carrier density and thus the gain because of its saturation characteristic. The continuous-wave signal experiences the same modulation at the output as a result. The output is inverted and may have some undesirable chirp imparted by large carrier density changes. The effect can be used for wavelength conversion.

Cross-phase modulation employs the change in refractive index associated with a change in carrier density controlled by the injection current of the semiconductor optical amplifier. The same arrangement is used with one modulated and one continuous signal, the former affecting the latter. Since gain is also a function of carrier density, the refractive index is related to gain. This relationship is often called the linewidth enhancement factor since its effect is observed in semiconductor lasers as spectral broadening of the lasing wavelength. Its value depends on the material and the device structure and is usually around 6 for InP-based structures at $\lambda = 1.55\ \mu m$. In principle, it is proportional to the ratio of the refractive index change with carrier density to the gain change with carrier density, $(dn/dN)/(dg/dN)$. The resulting refractive index change imparts a phase change on light travelling through the active waveguide, which is converted to intensity using an integrated optics interferometer, most often a Mach–Zehnder. Both these processes depend on carrier relaxation times that are on the order of hundreds of picoseconds at best. That limits the speed at which devices exploiting semiconductor optical amplifiers with these effects operate.

Four-wave mixing [49] is a coherent process in which two optical mode fields — the modulated signal and the continuous-wave signal — form a dynamic gain and refractive index grating in the semiconductor optical amplifier's waveguide. The grating period is determined by the beat frequency of the two optical fields. The modulated signal diffracts on the grating and generates a new signal that is its conjugate. Fast intraband effects, on the order of hundreds of femtoseconds as discussed earlier, such as carrier heating and spectral hole burning, can be made to dominate, thus allowing the four-wave mixing process to operate at very high speeds. However, it is a fairly weak and polarization-dependent effect that requires very involved experimentation. Differential cross-phase modulation, in which two mutually delayed pulses control two semiconductor optical amplifiers in an interferometer in that it can produce very short (on the order of picosecond) optical pulses is fast enough and more efficient than four-wave mixing while its implementation is less complex [50]. Even though much shorter pulses can be generated and high-speed data can be demultiplexed and processed in parallel with this technique, the repetition rate is still limited by the carrier lifetime.

4.3 ACTIVE FUNCTIONALITIES IN OPTICAL WAVEGUIDES

4.3.1 Introduction

Various active functionalities can be implemented exploiting the optical gain and nonlinearities effects discussed in the previous section. Waveguide configurations are best suited for such purposes due to their high control of light distribution.

If a high efficiency interaction between light and material is required, confinement of propagating modes and interacting carriers or electrons in a common region is the key. The double heterostructure semiconductor laser (concept illustrated in Figure 4.1) is a prime example of that principle [30] with its superb method of confining light in a waveguide that occupies the same volume as the injected interacting carriers trapped by a band gap energy difference. This device was the most important technical improvement on the original invention of the laser and it revolutionized the semiconductor laser industry by bringing to reality practical devices used in many applications today.

Efficient interactions can be obtained even without the optical amplification because of the light confinement in a waveguide and high light intensities achievable as a result. The prime example here is second harmonic generation with very high efficiencies and high output powers [31]. Bistable switching with a very high extinction ratio in nonlinear periodic multilayers [32] is another example of an efficient interaction when the distribution of light in a structure and in its constituting materials is precisely controlled. Other functionalities such as third harmonic generation, four-wave mixing, and self- and cross-phase modulation [33] have been demonstrated.

4.3.2 Harmonic Mixing

Harmonic mixing refers to nonlinear interaction of guided waves via the second-order nonlinearity, $\overline{\overline{\chi^{(2)}}}$. From Eq. (4.1), one can obtain the sum frequency (up-conversion) and the difference frequency (down-conversion) terms, respectively, as:

$$\begin{aligned} \overline{P}_{up} &\cong \chi^{(2)}\,\overline{EE} \\ \overline{P}_{down} &\cong \chi^{(2)}\,\overline{EE}^{*}, \end{aligned} \tag{4.6}$$

where $*$ means complex conjugation and $\cong$ is used to indicate equality except for a constant omitted for simplicity. This effect has been used extensively and successfully in, for example, periodically poled $LiNbO_3$ waveguides to generate visible light second-harmonic signals [34]. The phase matching problem has always been a crucial design issue if one intended to obtain high conversion efficiencies. Different techniques for achieving the phase match between the fundamental and sum frequency waves have been proposed and demonstrated, for example, periodic perturbation of the dielectric constant or boundary or a periodic modulation of the nonlinear optical properties of the waveguide material. Also in $LiNbO_3$, an off-axis cut was used to match mixed near-infrared (820 nm) and incoming (1550 nm) signals with a generated sum frequency (blue) light that, as a result, carried the information content of the incoming signal [35]. This functioned as an ultrafast optical mixer.

Nonlinear mixing of counter-propagating guided waves in a semiconductor waveguide can also lead to what one can consider nondegenerate four-wave mixing [36]. Two guided modes produce a nonlinear polarization that, in turn, acts as a source of sum frequency and difference frequency waves. If the orientation of the crystal's axis is in a correct relationship to the polarizations of the guided waves, the generated waves will travel out of the waveguide into both the substrate and the superstrate. Figure 4.2 illustrates a general situation. The beauty of this configuration is that no

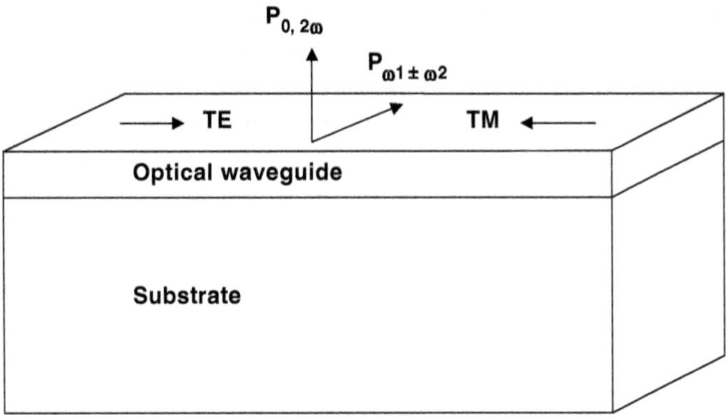

FIGURE 4.2 Harmonic mixing of guided modes in optical nonlinear waveguide.

phase matching condition needs to be satisfied, and thus the generated sum frequency (or second harmonic) is always produced once the counter-propagating waves are present in the waveguide.

Both GaAs and InP materials were shown to work well although the overall efficiencies are still low because of the value of $\overline{\overline{\chi^{(2)}}}$. A drastic improvement in the conversion efficiency can be achieved if one employs a multilayered $\lambda/4$ stack that matches the phase distribution of the generated wave. Green light from GaAs structures and red light from InP structures were observed and some interesting applications such as ultrafast autocorrelation, two-dimensional beam steering, and optoelectronic mixing [37] were proposed, demonstrated, or patented.

Eqs. (4.1) and (4.6) are more or less complex vector/tensor equations, depending on a given material's symmetry properties, the waveguide configuration, and guided mode polarizations. For example, referring to Figure 4.2, mixing TE and TM modes that are counter-propagating in a waveguide will produce a surface emitting sum frequency wave without a necessity of phase matching, as discussed above. On the other hand, mixing a TE mode with itself (co-propagation) in a III–V semiconductor waveguide produces a second-harmonic polarization component normal to the surface of the waveguide plane in the form [38]:

$$P_{2\omega} \cong 2d\, E_{TE}^{2}\, e^{i\,2\beta_{TE}\,z}\, e^{-i\,2\omega\, t} \tag{4.7}$$

where $2d = \overline{\overline{\chi^{(2)}}}$ is a scalar now representing one, preferably the strongest, element in the susceptibility tensor, $\overline{\overline{\chi^{(2)}}}$, with d being another symbol used extensively in the literature and related to the corresponding electro-optic constant via the permittivity tensor, usually tabulated for different materials [31]. β_{TE} is the TE mode's propagation constant, and ω is the angular optical frequency. This nonlinear polarization source generates the second harmonic traveling along the waveguide. Therefore, the spatial periodicity term, $e^{i\,2\beta_{TE}\,z}$, is extremely critical if one needs to extract the second harmonic at the end of the waveguide. Phase matching techniques and arrangements discussed above must be employed and they are not simple technical and/or technological tasks. In addition, absorption losses of the second-harmonic wave could be prohibitive in a semiconductor material such as GaAs or InP.

However, if one is interested in exploiting the mixing property of this nonlinear effect, which leads to an important and useful application of ultrafast optoelectronic mixing, phase matching is not an issue. Mixing two modes carrying different signals at ω_1 and ω_2 in this co-propagation regime yields the sum frequency polarization component in the form (see Figure 4.2):

$$P_{\omega_1+\omega_2} \cong 2dE_1^2 e^{i2\beta_1 z} e^{-i2\omega_1 t} + 2dE_2^2 e^{i2\beta_2 z} e^{-i2\omega_2 t} + 4dE_1 E_2 e^{i(\beta_1+\beta_2)z} e^{-i(\omega_1+\omega_2)t} \tag{4.8}$$

This nonlinear polarization generates three waves; the first two are the second harmonics and the third one is the sum frequency signal. A material with low absorption at ω_1 and ω_2 and high absorption at $\omega_1 + \omega_2$ configured as a waveguide as well as a PIN detector can yield a photocurrent directly proportional to the square of the generated nonlinear optical field; that is its power. The first two terms represent a DC while the third term is the desired optically mixed signal. The result is somewhat similar to the classical coherent detection scheme; however, the speed of this harmonic mixer is far superior to any photodetector. Because it exploits bound-electron nonlinear effects, the response times are on the order of much less than a picosecond, while the photodetector's mixing is always limited by the transport of free carriers, that is, realistically, by about a nanosecond.

Considering the down-conversion in Eq. (4.6), one can obtain similar expressions to Eq. (4.8) and find that in this case an electric field is readily produced in the direction perpendicular to the waveguide's plane. It can thus be collected with a pair of electrodes or a needle probe. The result is identical to the case of a photodetector used in a coherent detection system. The difference is that harmonic mixing is much faster, works in a broad wavelength range, and has no dark current because no free carriers are involved in the process. However, its conversion efficiency is very low compared to a photodetector. One can also examine a special case of optical rectification [39]. Eq. (4.6) yields, for a TE mode, the induced DC polarization in the direction perpendicular to the waveguide plane as (see Figure 4.2):

$$P_0 \cong 2d\, E_{TE}^2 \tag{4.9}$$

Knowing the mode power and the dimensions of the waveguide allows one to calculate the total dipole moment, the total charge, and the resulting DC voltage produced by the optical field in the waveguide. This voltage can be easily collected by electrodes deposited on the top of the waveguide or by a needle probe. With optical powers around a milliwatt in an InP-based ridge waveguide, voltages around a microvolt were measured [38]. This effect was advantageously used for monitoring the efficiency of coupling the light from a fiber into a waveguide.

Some interesting potential applications of harmonic mixers can be identified. In addition to harmonic detection and optical rectification coupling monitoring mentioned above, the optoelectronic mixing appears attractive. Fast mixing of electrical signals is not a trivial task and is always limited by the speed of electrical circuits that cannot match the speed of optics. Employing the principle discussed above, one could possibly implement a mixer with superb speeds well above what any electronics can handle. The low efficiency of the harmonic mixing concept can, in fact, be an advantage in all-optical fully transparent fiber networks. Extracting the high-speed clock of terabit data flexibly with respect to the bit rates and without affecting the data stream is a realistic possibility with the optical harmonic mixing scheme discussed. Optical gain inherent to III–V semiconductors can be used to compensate for low conversion efficiencies resulting in weak output signals, if required.

4.3.3 Field-Assisted Functionalities

All the electric field terms in Eq. (4.1) do not necessarily have to be contributors from modes in a waveguide, as is the case in, for example, sum and difference frequency generation or optical bistability discussed in the previous sections. Electro-optic modulation as one of the first practical uses of optical nonlinear effects [40] is an example; an external electric field controls the material's refractive index, which in turn modulates a guided wave. Sala offered a first comprehensive study of more general cases for third-order nonlinearity in isotropic media with applied external DC fields [41]. Other examples are the resonant optical nonlinearities in semiconductor quantum well materials that display large electroabsorption and associated electrorefraction known as the quantum-confined Stark effect [42]. When an external electric field is applied, the excitonic resonance weakens due to diminished electron–hole correlation of the exciton [26], and absorption, as well as refraction change, leading, in principle, to the same result: refractive index control with an external field via an optical nonlinearity.

This effect was used in the first waveguide vertical directional coupler [43] whereby a multiple-quantum well layer coupled nonlinearly and actively two waveguides. Figure 4.3 shows the configuration. The transfer of light from one guide to the other is controlled by the light intensities in the modes, thus effectively establishing

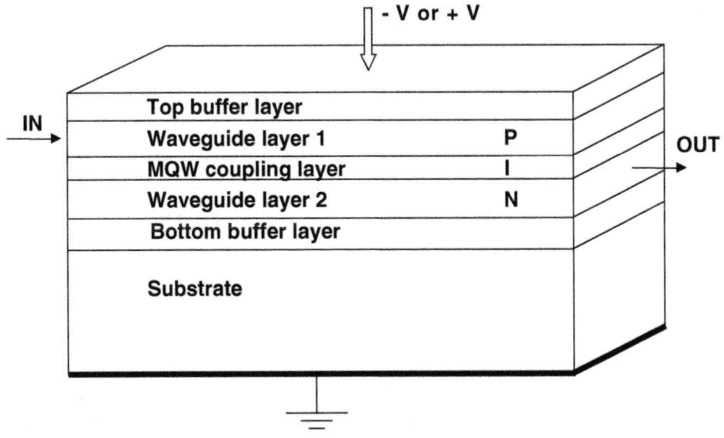

FIGURE 4.3 Multiple quantum-well vertical directional coupler as an active optical waveguide structure. MQW = multiple quantum-well material. V = control electrical voltage.

an operation of all-optical switching, or by applying reverse voltage via a PIN-doped structure, thus utilizing the quantum-confined Stark effect in the multiple-quantum well layer.

This is considered to be the first semiconductor ΔK switch, where ΔK is a change in the coupling coefficient between the waveguides due to the presence of the multiple-quantum well material. It is also possible to apply voltage in the reverse direction as a feedback from the detected optical output and obtain a self-electro-optic effect. A robust optical switching operation with high contrast and constant output power was demonstrated. This is recognized as the first waveguide self-electro-optic device [20]. Forward voltage, on the other hand, can turn the multiple-quantum well layer into an active optical waveguide, which can then provide gain as well as nonlinearity for switching. This subject still needs further investigation.

Maxwell's equations in SI units in a source-free region for non-magnetic non-time-dispersive media can be combined into a second-order partial differential equation with $\overline{P}_{NL}$ representing the given nonlinearity:

$$\nabla \times \nabla \times \overline{E} = -\mu_0 \varepsilon \frac{\partial^2 \overline{E}}{\partial t^2} - \mu_0 \frac{\partial^2 \overline{P}_{NL}}{\partial t^2} \tag{4.10}$$

For example, the fifth-order nonlinearity, $\overline{P}_{NL} = \varepsilon_0 \overline{\overline{\chi^{(5)}}}\,\overline{EEEEE}$, was recently exploited in designing an all-optical intracavity switching mirror for a pulsed laser [44]. The CdTe material that exhibits this high-order nonlinearity and has an attractive property of technological compatibility with several important substrates such as $LiNbO_3$, GaAs, and most importantly Si, was used. Using the developed model, a possible explanation was suggested for the experimentally known instabilities and chaotic dynamic behaviors in some optical bistable structures.

The third-order nonlinearity has been much more exploited in waveguide structures and devices, however; optical fibers and semiconductors including quantum wells are examples already discussed. This nonlinearity in silica glass of optical fibers or other isotropic materials with cubic symmetry is conventionally described in Cartesian coordinates by expanding the nonlinear polarization into its $x,\ y,\ z$ components as:

$$P_{NLi} \cong \sum_{i=x,y,z} \chi_{ijkl} E_j E_k E_l \tag{4.11}$$

Due to symmetry properties, most of the susceptibility components, χ_{ijkl}, are usually zero or equal to each other. This drastically simplifies the vector equation in Eq. (4.10) and leads to a scalar nonlinear wave equation whereby, for silica fiber, for example, and using Eq. (4.2), $n_2 = \frac{3\eta_0}{n_0^2 \varepsilon_0} \chi_{iiii}$, $i = x, y, z$, where η_0 is the impedance of free space, $\chi_{iiii} = \chi$, and a typical value is $n_2 \approx 10^{-16}\, cm^2/W$. The nonlinear polarization is then usually written as $\overline{P}_{NL} \cong \chi\, \overline{E} \left|\overline{E}\right|^2$, indicating an intensity-dependent field, which then enters Eq. (4.10) and yields the scalar nonlinear wave equation. This nonlinear wave (or Helmholtz) equation is solved analytically employing, for example, the slowly varying envelope approximation while assuming, practically correctly, that $n_0 \gg n_2 I$. The equation turns out to be identical to the well known Schrödinger equation in physics if a nonlinearity is introduced into it rather artificially; Schrödinger himself did not derive a nonlinear equation. In 1925, he formulated a linear motion equation for a free particle, which was then extended to the equation for the motion of a particle in a potential field. Nevertheless, for this similarity, the nonlinear wave equation that should really be properly called the nonlinear Helmholtz equation, has widely been called the nonlinear Schrödinger equation in the literature. One of its best known solutions describes the famous space soliton. The subject of optical solitons is treated thoroughly in the previous chapter of this book.

The third-order nonlinearity term in Eq. (4.1), $\overline{\overline{\chi^{(3)}}\overline{E}\overline{E}\overline{E}}$, does not necessarily have to be considered only as a refractive index change with the intensity of light. An externally applied electric field can play a role of one or more constituting field contributors, $\overline{E}$, in this term. This can lead to different application possibilities [45]. A number of situations can occur; for example, a linearly polarized mode field and an external field being either parallel or perpendicular to each other. Solving Eq. (4.10) for a field component and monochromatic light yields, in the first approximation, a field in the form:

$$E\,(t, z) \cong e^{-i(\omega t - k_0 n_{eff} z)} \tag{4.12}$$

where k_0 is the free space wave-vector, $n_{eff} = n_0 + \gamma\, n_2\, E_0^2$, E_0 is the external field, and γ is a constant dependent on the polarization relationship between the guided mode and the external field; it is largest when both polarizations are aligned. The complete solution in Eq. (4.12) contains a number of other terms corresponding to the DC, the self-phase modulation, and the second and third harmonic components. Especially interesting is that the second harmonic visible-light signal could be generated in a fiber with assistance from an external electric field. Considering that a long fiber can be used, this makes it an attractive alternative to coherent visible light sources.

If there were more than one signal in a waveguide at different optical frequencies, additional terms corresponding to cross-phase modulation and four-wave mixing would appear in the solution in Eq. (4.12). An assumption that $E_0 > E$ allows us to neglect those terms except for the largest one shown in Eq. (4.12). It is a reasonably practical condition since $E < V/\mu m$ for a milliwatt of optical power in a waveguide of a 10 μm cross-section while the external field could be as high as $E_0 >> V/\mu m$.

One can observe from Eq. (4.12) that the external field changes or modulates the optical mode phase. The change in phase is equivalent over a given distance, l, to a time delay, T_d, that is in this case variable and dependent on the externally applied field, E_0, as:

$$T_d = \frac{n_{eff}\, l}{c} = \frac{n_0\, l}{c} + \frac{\gamma\, n_2\, l\, E_0^2}{c} = \textit{fixed delay} + \textit{variable delay} \tag{4.13}$$

where c is the speed of light in vacuum. Therefore, as expected, the strength of the nonlinearity, the length of the delay line, and the applied external field determine the amount of time delay to be acquired. The most important aspect is indeed the external field controllability of the delay. This suggests an application as a programmable or tunable time delay element. Such an element is desirable for various optical systems, for example, those that employ photonic signal processing techniques for microwave applications, optical time-division multiplexing communication systems, systems of precise control of laser pulse time position in interferometry, scattering studies, and metrology [46].

If a linearly polarized mode has two orthogonal components and only one is aligned with the external field, similar analysis shows that the two orthogonal components will be affected differently. The difference is in the constant γ that assumes three times larger a value for the parallel component of the mode than for the orthogonal one. That yields a differential time delay, T_{dd}, between the two mode polarization components simply as:

$$T_{dd} = \frac{2}{3}\frac{\gamma\, n_2\, l\, E_0^2}{c} \tag{4.14}$$

This can be exploited in a variable differential delay device for polarization mode dispersion control. The differential delay in Eq. (4.14) is controlled by the external field, which can be generated by tracking the polarization mode dispersion and then fed back into the active delay device. In that manner, real time compensation can be achieved. In high performance 40 Gb/s and faster optical systems, this polarization mode dispersion is a critical problem that limits the ultimate speed of transmission. Its solution is challenging, and several approaches have been more or less successfully pursued, for example, electronic or optical control of polarization mode dispersion via a feedback, use of polarization maintaining fiber in a multiple stage polarization controller, or use of a variable differential group delay in an optical polarization mode dispersion compensator [47].

4.4 DYNAMICS IN ACTIVE OPTICAL WAVEGUIDES

4.4.1 INTRODUCTION

Dynamic behavior of active optical waveguides, that is semiconductor optical amplifiers and lasers or optical fibers with built-in gains, is a complex issue as can be deduced from the previous discussion of optical gain and its nonlinearities. The Von Neumann's original idea for exploiting stimulated emission in semiconductors was to amplify incoming light. Because the achievable gain per unit length was very small, a cavity was employed to increase manifold the interaction length, which led to the development of a semiconductor laser.

Fiber amplifiers, erbium-doped or Raman, were developed later with the same objective: to amplify incoming optical signals. The main driving force behind the development of optical amplifiers was the need for long-distance high-performance fiber communication systems and networks. Optical amplifiers can be used as in-line amplifiers to compensate for fiber losses and thus increase transmission distance. They can also be employed as booster amplifiers to increase the transmitted power right after the laser source or as front end preamplifiers just before the receivers. Their use in local area networks to offset coupling insertion and power splitting losses is also important.

As discussed, the nonlinearities associated with the optical gain effects limit the range of optical powers to be amplified, and possibly the dynamic range, and introduce undesirable distortions of the information-bearing signal. Optical transmission systems designers must deal with these severe limitations. In semiconductor optical amplifiers, the gain adjusts itself to the changing signal intensity as long as carriers with lifetimes on the order of hundreds of picoseconds can respond to its time variations. Common data rates on the order of gigahertz are affected by this. An amplifier that operates close to saturation conditions will limit the maximum output power and lead to signal distortion since different levels of signal intensity will experience different optical gains.

Polarization sensitivity is another critical problem. Since optical amplifiers are active waveguide devices, the mode propagation properties, especially mode polarizations, are pronounced and need to be dealt with if polarization-independent gain is required, as is the case for many applications. Gain nonlinearities also cause interchannel cross-talk in multichannel applications. However, as is often the case in practice, negative effects can be turned into useful functionalities. Many different applications in which optical gain nonlinearities were exploited somewhat positively have been proposed, demonstrated, and tested [29].

Perhaps the most studied application of gain nonlinearities in active optical waveguides is in wavelength converters. The principles of cross-gain and cross-phase modulation as well as four-wave mixing have been exploited. A use of a waveguide integrated Mach–Zehnder interferometer with semiconductor amplifiers in each arm, schematically sketched in Figure 4.4, demonstrated wavelength conversion at up to 40 Gb/s. Only one input (Input 1 in Figure 4.4) was used in this case.

In principle, the same interferometer can be used for high speed demultiplexing in optical time-division multiplexed systems. The continuous-wave input that serves in the wavelength converter as a light carrier for the signal to be converted on is replaced with a pulsed input that acts as a demultiplexing control signal, and similarly only one input is used. Every bit to be extracted requires a control pulse, and the whole integrated interferometer functions as a high speed optical AND gate. Using differential cross-phase modulation, a demultiplexer at 40 Gb/S was demonstrated; a full add–drop multiplexer at the same speed and using the same technique was reported.

This interferometer is also usable in implementing all-optical logic gates. Both inputs in Figure 4.4 are used in this case. All basic logic operations have been demonstrated including NOT, AND, OR, and particularly XOR which is important for parity checking and data encryption, at speeds of up to 20 Gb/s. A detailed review can be found in Stubkjaer [51]. Optical regenerators have been in the focus of researcher's

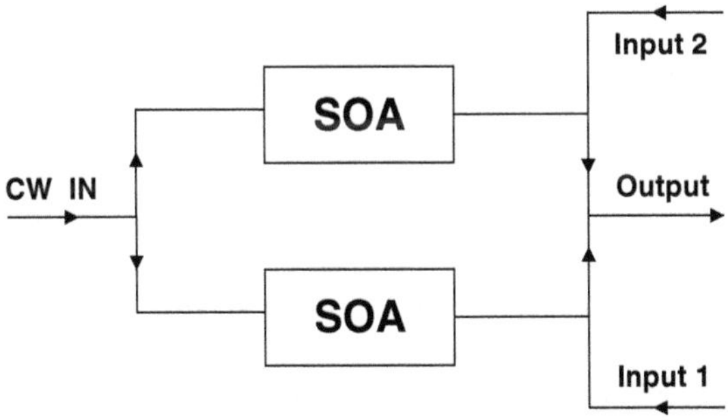

FIGURE 4.4 Active optical integrated waveguide Mach–Zehnder interferometer. SOA = semiconductor optical amplifier.

attention for some time; however, this functionality is still a domain of electronics. All-optical regeneration that performs all three operations, i.e., re-amplification (boosting signal strength), reshaping (improving signal-to-noise ratio), and retiming (removing pulse jitter), is the ultimate goal. Neither the erbium-doped fiber amplifier nor the Raman fiber amplifier can perform reshaping. The semiconductor optical amplifiers, on the other hand, possess that capability. Cross-gain modulation has poor conversion efficiency leading to signal extinction ratio degradation; however, cross-phase modulation performs quite well up to 40 Gb/s. The same interferometer has been exploited in setting up optical memory. Both single-bit flip-flop type and buffer type memories can be constructed employing cross-phase modulation. More detailed discussion is provided in the next section.

Bistable laser diodes as active optical waveguides with nonlinearities can also be used as memory elements since they exhibit the hysteresis required for the flip-flop operation. However, they usually suffer from functional and/or control problems that limit their practical use. In many cases investigated, the device is set optically but must be reset electrically. Even if optical reset is implemented, the optical flip-flops are wavelength-, power-, and polarization-sensitive — undesirable characteristics for applications. Some interesting results and more detail on this subject are discussed below.

Optical fibers as waveguides were also used in proposing and demonstrating optical memory functionality. Nonlinear Sagnac interferometers [52] have received considerable attention as possible functional devices. Originally, they employed fiber nonlinearity. However, because the nonlinearity of silica glass is very weak and required very long lengths, the semiconductor optical amplifier was later used, with the additional benefit of optical gain that turns the whole configuration into another interesting application of active optical waveguides. These designs are referred to as a semiconductor laser amplifier-in-a-loop mirror and a terahertz optical asymmetric demultiplexer. They are practically identical in construction and functionality, and operation at a bit rate of up to 1 Gb/s and an output extinction ratio of 15 dB was demonstrated. Although they can work as single-bit flip-flop type and buffer type memory, including full read and write capability, and as all-optical counters, half-adders and shift registers, the set-ups are complex and rather bulky and require a number of extra necessary components such as couplers, filters, and circulators. The systems are also bit rate-dependent and can handle only return-to-zero signals.

Another active optical waveguide arrangement for all-optical memory and logical operations is an ultrafast nonlinear interferometer [53]. It uses for its operation a semiconductor optical amplifier and exploits the birefringence of the polarization-maintaining fiber. A feedback is required in order to be able to use the output data as control pulses. Erbium-doped fiber amplifiers are necessary to provide feedback gain. Other hardware is needed, which makes the whole system very complex. Operation at a bit rate of 10 Gb/s with an extinction ratio of 20 dB was demonstrated. A number of other all-optical memory set-ups have been proposed and demonstrated, for example a coupled Mach–Zehnder interferometer with semiconductor optical amplifiers. Generally, however, until now, an all-optical device that would function as a true optical memory — a photon storage element of sorts — seems an elusive proposition despite the efforts spent on its possible invention. Electronics-based memories indeed dominate the field along with the ever-increasing storage capacity of the chip. However, it is still a long way to achieving the capacity of a human brain, which is estimated to be about 10^8Mbytes [54].

4.4.2 Oscillations and Microwave Generation

The integrated Mach–Zehnder interferometer with semiconductor optical amplifiers incorporated in the active waveguide configuration (Figure 4.4) is a flexible design that allows its numerous variations to be exploited in different applications. The all-optical flip-flop, wavelength conversion, regeneration, and memory functionalities were reviewed and discussed in the previous section. Generally speaking, the non-linear effect of optical gain saturation in a semiconductor optical amplifier in the interferometer can be used to perform all-optical signal processing, depending on the actual geometry and operating conditions. The first patented design [55] of an all-optical latching device suffered from severe limitations that made it unusable due to uncontrollable oscillations caused by a mismatch between the length of the reset pulse and the time delay of the feedback loop. It, however, pioneered a promising and novel idea of introducing optical feedback into an interferometer and exploiting it for its functionality. Details of such an arrangement are discussed below.

In its basic configuration, an interferometer contains a semiconductor optical amplifier in each of its arms. It requires a phase shift of π for complete switching, which in turn determines the required level of the injected electrical current or the input optical power, depending on whether electrical or optical control is desirable. As discussed previously, this is directly related to the phase delay imparted by a semiconductor optical amplifier on propagating modes in the active optical waveguides: the gain depends on carrier density, which affects the refractive index. Using basically Eqs. (4.3) and (4.4), one can model the whole interferometer and obtain different transfer functions [29], the characteristics of which can be tailored to the specific functionality required, for example wavelength conversion or all-optical logic. Generally, inverting and non-inverting functions can exist while an extremely high output extinction ratio, theoretically infinite, is possible. A real device is limited by spontaneous and amplified spontaneous emissions generated in the semiconductor optical amplifiers, and by variations in the gain and resulting phase shift due to electrical current bias drift.

The dynamic properties are determined by two mechanisms, namely the carrier lifetime and the travelling wave effect. Considering the carrier recombination rate, R, that includes all the effects up to the Auger recombination and the stimulated emission, one can express it as [28]:

$$R = AN + BN^2 + CN^3 + R_{st}N_{ph} \tag{4.15}$$

where A, B, C are recombination constants dependent on the process and the material N_{ph} is the photon density in the active region of the waveguide (i.e., the semiconductor optical amplifier), and R_{st} is the stimulated emission rate given by:

$$R_{st} = \frac{cg}{n - \lambda(dn/d\lambda)} \tag{4.16}$$

while realizing that both the refractive index, n, and the optical gain, g, are functions of the carrier density, N. It is clear from Eq. (4.15) that fast response times are achieved through both high carrier as well as photon densities. This is, in turn, accomplished by using large injection electrical currents and intense optical inputs.

The travelling wave effect contributes positively to the overall response of the interferometer by extending the bandwidth of its semiconductor optical amplifiers. When amplifiers are properly electrically biased and before the arrival of a pulse, a high carrier density is present. As the leading edge of an input pulse reduces the carrier density through stimulated emission, it is greatly amplified while the remainder of the pulse experiences a reduced or compressed optical gain. This occurs also in bistable laser diodes, where it is usually called gain quenching. More details regarding this effect and its use in semiconductor lasers are presented in the next section. The result of this phenomenon is that as the leading edge travels through the active optical waveguide section, it quickly depletes the carriers, thus causing a fast change in refractive index, and a leading-edge peak appears on the output pulse. At the trailing edge of the pulse, the reverse occurs; the carrier density recovers rather slowly, the pulse power drops to a very low level before recovering, and a negative peak appears on the output pulse. The net effect is that the high frequency content of the input optical pulse is amplified while it travels through the semiconductor optical amplifier, that is, the active section of the waveguides of the interferometer. This traveling wave effect increases with higher carrier densities and the length of the active optical waveguide sections. Although it extends the bandwidth, it distorts the shape of the pulses.

When an optical feedback as shown in Figure 4.5 is incorporated into an interferometer operating in a non-inverting mode, functionalities typical for nonlinear systems with positive feedback are obtained, for example, in this case, optical hysteresis, all-optical bistable switching or flip-flopping, or optically controlled latching. The dynamic behavior of such an interferometer demonstrates itself when the interferometer is used as an all-optical flip-flop. Three contributing factors shape the interferometer-with-feedback's transients. These are the transient behavior of the semiconductor amplifiers, the feedback delay time, and critical slowing down. The effects of semiconductor optical amplifier dynamics were discussed in the previous

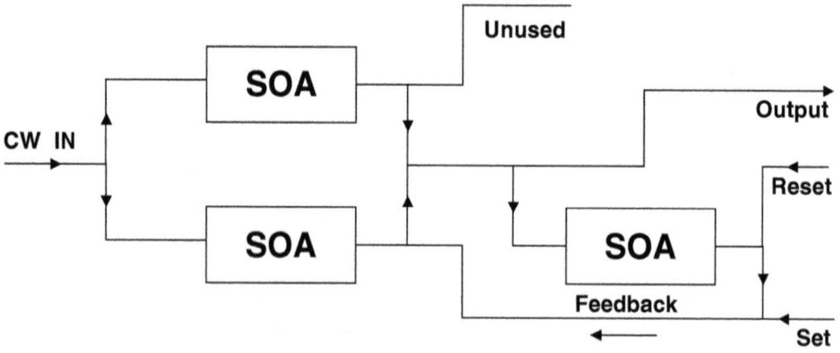

FIGURE 4.5 Active optical integrated waveguide Mach–Zehnder interferometer with optical feedback. SOA = semiconductor optical amplifier.

sections. The feedback delay can be controlled by the design and also, if another semiconductor amplifier is used within it to provide feedback gain and a control of the hysteresis width, by electrical current or additional optical control. Critical slowing down of the transition between bistable states occurs when the power of the switching optical signal remains close to a critical point. Critical points are points on the hysteresis curve corresponding to the switch-up and switch-down input optical powers.

Critical slowing down is a concern also in bistable laser diodes, and more detailed discussion on the subject is provided in the next section. Generally, its effects may be severe; the switching times of optical bistable devices may be longer by at least one order of magnitude than when critical slowing down is avoided by employing switching optical powers high enough to move the operating points quickly away from the critical points.

A combination of all three effects can produce a transient curve, as a response to an input optical square pulse, that carries signatures of these effects on different time scales that are thus observable in experiment. For example, critical slowing down creates the envelope of the transient since it is the slowest process. The feedback loop delay causes envelope ripples of the length of the feedback propagation time. Each ripple, in turn, has a shape determined by the dynamic characteristics of the semiconductor amplifiers. Ideally, for the best dynamic performance, one would want to avoid critical slowing down by using high switching optical powers, cutting the feedback length to a feasible minimum, and driving the semiconductor optical amplifiers hard [29].

A semiconductor optical amplifier incorporated in the feedback loop offers some benefits. It provides an extra parameter, optical gain that can be controlled externally, that helps to avoid undesirable oscillations when the interferometer is operated as a flip-flop. It also allows one to shape the transfer characteristic of the interferometer. It appears that if one would also incorporate a nonlinear active waveguide coupler in the feedback loop [56] to combine input optical powers, the transfer function could be shaped into a much more favorable and steeper curve having flat tops that can eventually improve the extinction ratio and switching times.

The flip-flop functions as a single-bit optical memory whereby the stored bit can be read at any time. It also operates as a buffer memory that remembers several bits at a time that, however, can only be read at regular intervals. The bit stream to be stored must be shorter than or equal to the optical delay of the feedback loop. The data bits then circulate around the feedback loop, being reamplified and reshaped by the interferometer during every pass. The reamplification and reshaping, which are the regenerative functionalities that improve the quality of pulses, represent the main benefits of using an active optical waveguide structure, that is, a semiconductor optical amplifier-based Mach–Zehnder interferometer with an active optical feedback.

The same interferometer with feedback (Figure 4.5) can be turned into an all-optical ring oscillator if the feedback is applied and the interferometer is operating in its inverting mode that can be established, for example, by adequate electrical bias of its semiconductor optical amplifiers or by proper design of the lengths of its arms. The device is self-starting and produces an output optical square wave when the transient effects are short compared to the fundamental frequency of the oscillator. That frequency is controlled by the feedback delay. The optical output square wave improves its quality after every cycle because of the regenerative properties of this interferometer with feedback, as described above. The only limitations are the speed of response and the optical noise in the system. The output optical power is determined by the transfer function of the interferometer. The oscillator can enter into a chaotic mode if the feedback optical gain is too high. This is related to the stability condition of the oscillator, which is a complex function of the interferometer parameters including the characteristics of its components [29].

Active optical waveguides can be employed in generating microwave and millimeter wave signals. The wideband wireless networks require a stable, narrow linewidth and high frequency RF carrier. Its frequency could be as high as tens of gigahertz and its linewidth needs to be as narrow as a few Hertz or even narrower. Due to high absorption in the air, the transmission distances are very short, resulting in a small coverage area and the need for an increased number of base stations. Since each base station then requires its own microwave hardware (RF oscillator, filters, other electronics), the cost is becoming prohibitive. In addition to the very expensive generation of the RF carrier, the electronic distribution is also expensive due to the cost of high-frequency coaxial lines. Therefore, a proposition was put forward to research and exploit an alternative solution, that is, to use the broadband capacities of well developed optical fibers for the distribution of the RF carriers generated in the central station over to the base stations [57].

There are two ways to generate and/or distribute an RF carrier via an optical carrier traveling in a fiber. The most straightforward means is laser modulation, either directly or via an external modulator. Direct current-driven modulation of a semiconductor laser is limited by its natural frequency that may range from several to a few tens of gigahertz. The effect of wavelength chirp known from high-performance optical communications systems may negatively affect the quality of the generated RF carrier. Commercially available external modulators operate at frequencies of up to several tens of gigahertz. Various techniques and variations thereof have been proposed and demonstrated with some success [58].

Active injection locking of a self-pulsating multisection distributed feedback semiconductor laser is another example of the direct modulation approach. Self-pulsation, when mixed on a fast photodetector, produces an RF signal with a broad linewidth because the phase noise of the optical modes is not correlated. Injecting a low-phase-noise electrical signal into one section of the laser can reduce the linewidth significantly by modulating laser gain. Optical injection locking is also possible [59]. The main advantage of these rather direct modulation methods is that they can offer RF linewidths in the range of less than a Hertz. While these approaches are simple in concept and relatively easy to implement, they suffer from some major limitations. The main problems are the cost because high-speed electronic modulation circuits are required and chromatic dispersion that can drastically reduce the resulting RF signal amplitude after it has been transmitted on the light carrier down the fiber [60].

Optoelectronic mixing of light outputs of two independent continuous wave lasers on a high-speed photodetector is a known method that generates an electrical signal whose frequency is proportional to the wavelength spacing of the two lasers. It is called optical heterodyning and it was successfully employed in the development of optical coherent receivers before the fiber amplifier existed and when every decibel increase in the optical fiber signal strength and its signal to noise ratio was critical.

The same idea can be used in optical distribution of RF carriers by sending two wavelengths down an optical fiber and generating the RF signal locally at the base station. The linewidth of such a remotely generated RF carrier is, however, one extremely critical parameter. Since the RF linewidth is basically given by the sum of linewidths of the two mixed lasers, two requirements must be precisely satisfied. The first is that extremely narrow linewidth lasers must be employed, and the second calls for mutual phase locking to have them coherently track each other's frequency drift. A typical linewidth of a semiconductor laser can be on the order of a megahertz at best — which is clearly not sufficient for an RF carrier expected to have a linewidth in the subhertz range.

External cavity laser arrangements have been implemented whereby the length of the cavity is much longer, which narrows the linewidth significantly. Gas and solid state lasers (e.g., Nd:YAG lasers) with much more distinct and well defined energy levels compared to semiconductors have inherently narrower linewidths. Employing such laser sources in an external cavity configuration, RF signals were obtained with a kilohertz linewidth [61]. As indicated above, further improvement of the RF linewidth can be achieved by employing an optical phase-lock loop that attempts to correlate dynamically in real time the phase noise of both lasers. The task is technically challenging; a frequency tunable laser is required. Typically, for semiconductor lasers that can be frequency tuned via their temperatures (e.g., 2 GHz/K), a subnanosecond loop delay is needed for proper operation of the optical phase-lock loop and that is not a trivial task. If functioning well, however, this technique can narrow the RF linewidth down to the Hertz range, limited then only by the RF reference. An additional benefit is the tunability over a range of tens of gigahertz [62].

When a semiconductor optical amplifier as an active optical waveguide is used in an optical cavity, lasing takes place as a result of the positive feedback and optical gain. Stimulated emission dominates the process, thus producing high quality coherent output. The main objective of a laser design is to obtain a single-mode output with as narrow a linewidth as possible. Since the gain curve is in most cases, except in vertical cavity surface-emitting lasers with very short cavities, much wider than the separation between the oscillating cavity modes, more than one mode can exist. This led to an idea of employing this inherent property of semiconductor lasers in optoelectronic mixing of the laser modes to generate microwave or millimeter-wave signals. Ideally, one would need two modes properly spaced and mixed on an optical nonlinearity to yield the beat signal.

The advantage is that except for a photodetector, no expensive high speed RF electronics equipment is required to implement this concept. The idea has a number of technical problems, however. The first is the linewidth quality of the modes; the second involves the stability and mutual coherence of both modes oscillating in a cavity they share; the third is the speed of response of the optical nonlinearity to be used; and the fourth is the conversion efficiency of optoelectronic mixing. All these issues are complex and require rigorous scientific treatment; some of them, for example, the stability and mutual coherence of laser modes in the same semiconductor cavity, are not fully understood yet [2].

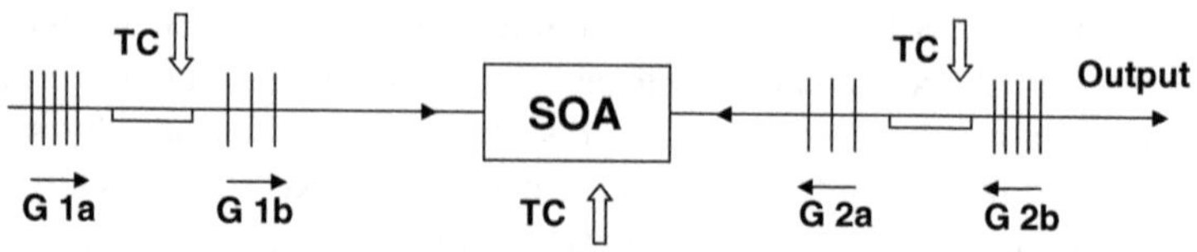

FIGURE 4.6 Tunable dual-fiber Bragg grating external cavity laser for microwave generation. SOA = semiconductor optical amplifier. G1a, G1b, G2a, G2b = fiber Bragg gratings. TC = temperature control.

A number of different set-ups implementing the idea of beating laser modes to generate an RF carrier have been proposed and demonstrated in recent years with various levels of success [63]. For example, a dual-mode semiconductor ring laser that employed an active optical waveguide in a form of a semiconductor optical amplifier as the gain medium was tested. The ring was built with an optical fiber containing a splitter to provide the ring feedback and output. A circulator was used to connect a dual-fiber-Bragg-grating that served as a two-wavelength-selective reflector in this external-cavity laser. An in-line version (Figure 4.6) was also implemented and tested; two dual-fiber-Bragg-gratings, one on each side of the active optical waveguide (gain medium), formed the lasing cavity. The reflection characteristics of the fiber Bragg grating determine which modes of the external cavity lase. They thus dictate the wavelength (frequency) difference that is used to generate the RF signal when mixed on a photodetector. A 40 GHz RF signal using gratings with a wavelength separation of 0.32 nm was produced. The stability was good, but the linewidth was too broad because of multiple RF frequencies present in the spectrum as a result of multimode laser operation due to too broad reflectivity spectra of the gratings.

Novel configurations based on the same idea have been studied recently [64]. One is a semiconductor optical amplifier dual-fiber-Bragg-grating external-cavity laser that offers external control of stability conditions and RF tunability. Similarly to Figure 4.6, a fiber Bragg grating is placed on each side of the gain chip. A space between the two gratings written in the fiber is used as a phase control via temperature tuning. This makes it possible to control and manipulate the relative standing wave positions of the cavity modes inside the chip. The gain competition between the modes is thus reduced, and reasonably stable dual-mode operation was obtained.

A relatively strong microwave beat signal was produced at around 30 GHz with a 3 dB linewidth of about 500 kHZ using fiber Bragg gratings with a 3 dB bandwidth of 0.1 nm. Tunability was about 100 MHZ/°C. Linewidth narrowing, overall stability, and tunability characteristics are still critical and problematic issues exist with these experimental set-ups.

An integrated optics version of any of these configurations would certainly improve the operating parameters, especially the stability. A full understanding of the interaction and the synergy between the two modes sharing the same semiconductor gain cavity must be acquired. The notion that the phase noise due to carrier density fluctuations is the same for modes oscillating in the same cavity [65] offers a promise that the noise coherence of the modes can be achieved under certain operating conditions, which would then lead to significantly reduced linewidths. Additional electronic linewidth narrowing circuitry can also improve the quality of the microwave signal to the levels required by practical wireless systems.

An interesting and innovative approach recently verified experimentally [66] uses an integrated optics technology on lithium niobate ($LiNbO_3$). An optical-waveguide-based Mach–Zehnder interferometer with a double-stub electrode structure acts as a high-frequency resonant-type modulator that has a narrowband modulation characteristic, about a few gigahertz. It is placed between two fiber-Bragg-grating filters, one of which (input) is phase shifted. The phase shift in the grating, which can be called a defect in a periodic structure if one uses the photonic crystal terminology, creates a transmission peak in the otherwise highly reflective characteristic. The input optical signal at the filter's transmission peak wavelength travels inside the cavity formed by both filters. It is modulated at the resonant modulator's frequency, which creates optical sidebands. The sidebands are reflected by the filters, and as the signal continues to travel forth and back in the cavity, it is modulated again on each pass.

New higher sidebands are generated as a result. The output filter is designed to transmit out the desired sideband that is in frequency an integer multiple of the fundamental modulation frequency. The delay between successive passes in this reciprocating process is designed to match the resonant modulation frequency, thus leading to high efficiency and high output power of the desired harmonic. A 50 GHz signal with a linewidth around 100 Hz was successfully generated. Incorporating optical gain into this integrated waveguide scheme seems to be a very promising improvement of the idea.

4.4.3 Bistability and All-Optical Switching

Optical bistability requires nonlinearity; the presence of optical gain enhances its performance parameters such as the extinction ratio and speed. Optical gain as a nonlinear effect can be exploited to establish bistable switching. It is interesting to note that optical bistability has been studied mostly in bulk materials and structures while its existence in active optical waveguides is considered as a somewhat undesirable phenomenon to be avoided or even eliminated.

The semiconductor laser (e.g., Figure 4.1 without antireflection coating) is in principle a generic structure of an active optical waveguide that has all the operating characteristics necessary for bistability and all-optical switching. However, since the main application area of semiconductor lasers has traditionally been in optical fiber communications and later in consumer electronic products such as CD players, the linear performance of the laser has been the focus of its research, design, development, and implementation. That is not to say that the behavior of semiconductor lasers has not been studied intensively and in detail; the overall perspective and objectives have been different. A vast research literature exists, indeed, related to gain and nonlinear behavior of laser systems; a good description and treatment can be found in Mandel [67].

Irrespective of a particular application, the most important operating parameters of a semiconductor laser are its linewidth and threshold. Both parameters affect switching performance and are thus of paramount importance for establishing efficient and fast active optical waveguide-based nonlinear devices. The pioneering work on distributed

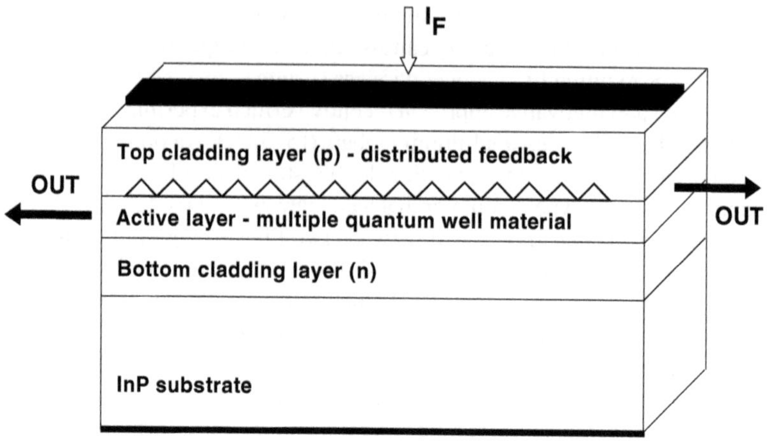

FIGURE 4.7 Conceptual sketch of a distributed feedback quantum-well semiconductor laser. I_F = driving electrical current.

feedback lasers [68] set the scene for tremendous improvements in these parameters. After quantum-well materials were advantageously introduced into their structures, distributed feedback lasers became the most important semiconductor laser structures with the best operating characteristics technically and technologically achievable [69]. Such a laser is shown in Figure 4.7. The optimal design is a very difficult task because large numbers of material and operating parameters are involved in the design and optimization processes; there could be tens of parameters, some of which may not be known exactly. Therefore, setting up systematic and reliable design and optimization procedures is an extremely challenging problem that has been tackled by a number of excellent researchers over the years. Some commercial software design programs achieve that task. However, a designer's experience combined with intelligent choices and decisions, often iterative while estimating certain parameters, taken as a part of the design procedure remains an unavoidable requirement.

One example of a somewhat systematic optimization methodology for modeling and design of a distributed feedback laser was described by Isenor et al. [70]. The novelty in this work lies in the fact that the model and the global optimization approach employed [71] are capable of addressing the issues of a laser functioning in the above-threshold operational regime. This enables us to analyze, model, design, and optimize crucial nonlinear characteristics such as multimode versus single-mode operation, wavelength stability, and linewidth that strongly affect laser performance.

Classical modeling techniques typically involve design evaluations at the laser's threshold; this does not represent an extremely challenging numerical problem. Extrapolation is then performed to study and design the operation parameters at the injection current of interest above threshold. The new methodology, on the other hand, makes it possible to maximize the laser's active waveguide (cavity) optical field flatness over a range of above-threshold injection currents directly. This leads to finding an optimally flat field solution in terms of structural design parameters while taking into account above-threshold operating conditions including optical nonlinearities.

Generally, two phenomenological yet fundamental effects dominate performance. One is the actual strength of the interaction between the forward and backward propagating waves in the active optical waveguide that determines the efficiency and thus the resulting threshold. The second effect is the uniformity of the optical power density and the carrier density distributions along the waveguide cavity. Less uniformity leads to the effect of spatial hole burning that in turn causes deterioration of the single-mode condition and the linewidth. Numerous analyses and designs of laser structures have been proposed and tested to deal with this difficult problem.

A novel comprehensive theoretical treatment that enables optimal design of active waveguide structures is given in Wang and Cada [72]. It is based on a new coupled power approach as opposed to the well known couple mode theory. A set of nonlinear coupled power equations that include optical gain and can treat the situations of index, gain, and complex coupling in the distributed-feedback structure is derived. It is also applicable to quantum-well lasers. Although the coupled power equations are in principle equivalent to the coupled mode equations, they provide a straightforward and insightful path toward analytical understanding of the optimal design of an active distributed-feedback waveguide structure employed in a semiconductor laser.

For amplitudes of forward and backward propagating modes, $E_F(z)$ and $E_B(z)$, respectively, the theory introduces four power density variables: the total photon power density along the cavity, $I_t(z) = E_F(z)\,E_F^*(z) + E_B(z)\,E_B^*(z)$, the net photon power density flux along the cavity, $I_d(z) = E_F(z)\,E_F^*(z) - E_B(z)\,E_B^*(z)$, the power density of the mutual interaction, and the phase relation between forward and backward propagating modes, $I_c(z) = 2\,E_F(z)\,E_B^*(z) = I_{cr}(z) + i\,I_{ci}(z)$. Four coupled differential equations that relate rates of change of these quantities to each other govern the distributions of these power densities along the cavity and the power exchange between the forward and backward propagating modes due to index, gain or complex coupling. Their close examination shows that the optical nonlinearity included in this description affects the mutual interactions of the modes and thus changes the net power density flux and the total power density of each mode. The phase shift plays an important role in determining the amount of power transferred between the forward and backward propagating modes and naturally confirms that controlling the phase shifts has a fundamental effect on overall performance.

From a physical point of view, the strongest mutual interaction leads to the minimum threshold of the laser — an extremely important operating parameter, as mentioned above. Such a condition is obtained when the forward and backward propagating modes are synchronized or their phase difference is at π along the cavity. From the definition of $I_c(z)$ thus follows that the lowest threshold is achieved when:

$$I_{ci}(z) = 0 \tag{4.17}$$

This beautifully simple condition for the lowest threshold in a distributed-feedback laser does, however, yield a complex requirement when applied. It includes the correct combination of laser parameters such as operating wavelength, coupling coefficients, phase difference between the gain coupling and the index coupling, and phase shifts.

Spatial hole burning involves a non-uniform carrier density distribution along the cavity resulting from a non-uniform optical power distribution along the same cavity.

Similarly, as in the case of the threshold, the power coupled theory yields a simple condition to be satisfied if a uniform distribution of the total power density along the cavity is desired:

$$\frac{dI_t(z)}{dz} = 0 \tag{4.18}$$

This is the second crucial condition that shows, after one examines in detail the set of differential equations governing all four power densities, that spatial hole burning can be completely eliminated by adjusting the phase shift and by properly choosing the variations of the coupling coefficients. Eqs. (4.17) and (4.18) thus represent a general principle that can be applied to optimally design an active optical waveguide distributed-feedback structure for a semiconductor laser with the lowest achievable threshold and minimum, if not completely eliminated, spatial hole burning. The conditions are general and apply also to multiple-phase shift structures and high power lasers. Quantum-well distributed-feedback structures can be studied with this approach as well. The equations are easily modifiable to take into account the differences between the bulk and quantum-well materials, such as the recombination mechanism, gain characteristic, coupling coefficient, and band structure confinement factor.

Because a semiconductor laser has optical gain and positive feedback, it is naturally suited for bistability and all-optical switching functionalities. The first proposal for a bistable injection laser diode was put forward by Lasher [73]. Many different bistable laser diodes were later proposed and demonstrated including using various physical nonlinear effects. A thorough review of this field is presented in Kawaguchi [74]. Conventional bistability in laser diodes is due to absorptive and dispersive nonlinearities. A use of saturable absorbers for example, whereby one of the mirrors acts as a nonlinear reflector, has been successfully demonstrated in system applications because of the stable operation achievable. However, because free carriers are involved in the process, the speed is limited to a nanosecond range. On the other hand, two-mode lasers display bistable switching as a result of the gain saturation phenomenon that is a much faster process.

In general terms, if two modes with intensities I_1 and I_2 and linear (unsaturated) gains g_1 and g_2 exist in a cavity, they are coupled via the nonlinearity of the gains. The nonlinearity or gain saturation is usually described using a widely accepted dependence in the form:

$$g(I) = \frac{g_0}{1 + I/I_{sat}} \tag{4.19}$$

where g_0 is the linear gain and I_{sat} is the saturation intensity. One can simplify the gain saturation description in Eq. (4.19) by linearizing it and assuming that the gain behaves as $g(I) \cong g_0(1 - \kappa I)$, with $\kappa = 1/(2I_{sat})$ being the saturation coefficient. The two-mode mutual interaction can then be described through the rates of change of the mode intensities as:

$$\begin{aligned} \frac{dI_1}{dt} &= g_1 I_1(1 - \kappa_{11} I_1 - \kappa_{12} I_2) \\ \frac{dI_2}{dt} &= g_2 I_2(1 - \kappa_{21} I_1 - \kappa_{22} I_2) \end{aligned} \tag{4.20}$$

where κ_{11}, κ_{22} and κ_{12}, κ_{21} are self-saturation and cross-saturation coupling coefficients, respectively. The solutions to Eq. (4.20) and their examination offer insight into all possible situations that can occur including the condition necessary for bistability. Well-known S-type bistability and pitchfork bifurcation curves can be obtained and explored for interesting features, for example, one or the other mode oscillating, both modes co-existing and competing for the gain, regular toggling between them (self-pulsation), or electrically or optically triggered bistable switching. The particular conditions that allow any of these cases to occur depend on the physical, electrical, optical, structural, and operational parameters of the laser; the processes are extremely complicated. Usually, in order to capture the rich dynamics, nonlinear coupled rate equations and the traveling wave equation similar to Eqs. (4.3) and (4.4) must be formulated for a particular situation. The set of equations is then analyzed and solved numerically. Involved and extensive simulations are required if one desires to acquire deep insight into the behavior of a bistable laser [27].

A number of bistable semiconductor laser structures and configurations have been studied theoretically and experimentally. External-cavity arrangements, two-segment stripe structures, Fabry–Perot and injection-locked lasers, side-injection light-controlled laser diodes, distributed-feedback and distributed-Bragg reflector lasers, vertical-cavity surface-emitting lasers, and many others with a wide range of operational parameters, stability, and promises for practical use [74] have been demonstrated.

Very fast polarization bistabilities, for example, are displayed by some of these structures. In this case, switching between TE and TM modes can be initiated with an external optical trigger. The modes bifurcate at a critical point at a certain driving current, while one of them takes over and starts to oscillate. When an optical trigger of the opposite polarization is injected into the cavity, the other mode starts to oscillate and remains in that polarization state. Since the effect apparently does not involve spontaneous recombination of free carriers, as opposed to some other bistable switching schemes, it is very fast indeed, on the order of several picoseconds.

Nonlinear systems are best described and understood by analyzing their stability conditions. Instabilities and nonlinearities in semiconductor lasers have traditionally been studied from a physical device point of view, including quantum mechanical principles and formulating and solving corresponding rate equations, as discussed above. Although this approach considers the carrier and photon densities distributed throughout the active and passive waveguide regions of the laser and thus leads to finding its operating conditions (e.g., conditions for bistability or self-pulsation), it is limited to numerical analyses due to inherent complexity of the problem. A nonlinear system approach is, however, possible and turns out to be very insightful and useful [75]. A stability analysis of bistable laser diodes as active optical waveguides can be performed starting from describing the laser system by the evolution equation:

$$\frac{d\overline{X}}{dt} = f(\overline{X}, \xi) \tag{4.21}$$

where $\overline{X} = [N_g, N_a, N_{ph}]$ is a state variables vector, N_g and N_a are the carrier densities in the gain and absorption regions, respectively, and N_{ph}, as before, is the photon density in the waveguide cavity. ξ is a control parameter, for example, the pump rate related to the driving electrical current or the optical pumping power. Eq. (4.21) describes the typical bistability curve for the photon density versus the pump rate. Setting $f(\overline{X_s}, \xi) = 0$, one can find $\overline{X_s}$, which represents either the on or the off state. As explained in discussions of semiconductor optical amplifiers, these points are called critical points and are related to the effect of critical slowing down.

Variation of $\overline{X}$ in the vicinity of the steady state solution $\overline{X_s}$ enables one to study its stability by linearizing Eq. (4.21) and then seeking the corresponding linear solutions. This leads to a third-order equation that yields three eigenvalues of the system. They have a general form as follows:

$$\lambda_1 = \nu_1(\overline{X_s}, \xi), \quad \lambda_2 = \nu_2(\overline{X_s}, \xi), \quad \lambda_3 = \nu_3(\overline{X_s}, \xi) \tag{4.22}$$

with $\nu_i (i = 1, 2, 3)$ being the eigenvalue problem functions determined by the actual material and structural parameters. The eigenvalues are generally complex while their real parts characterize the relaxation rate of the system toward its on and off states and their imaginary parts determine the frequency of the damped oscillations of the system. The on and off states are stable if the corresponding real parts of $\lambda_i (i = 1, 2, 3)$ are negative; one or more positive real parts imply instability. The presence of the critical points is signaled by vanishing of the real part of at least one eigenvalue, which means the divergence of the relaxation time; in other words, the relaxation time approaches infinity. The effect associated with this situation, already mentioned above, is called critical slowing down and it can occur in both the on and off states.

The stability analysis based on this approach indicates that the maximum switch-on and switch-off times are inversely proportional to the minimum value of the real part of the corresponding eigenvalue. It thus qualitatively explains the numerical and experimental results reported in a number of papers [75], whereby the switch-on transients are characterized by the relaxation rate with oscillation and the switch-off transients are determined only by the relaxation rate. The self-pulsation emerges from the on state at the switch-off pump rate corresponding to the Hopf bifurcation point on the bistability curve, which is when the complex conjugate eigenvalue pair for the on state, $\lambda_1 = \lambda_2^*$, changes the sign of its real part, that is when $Re\{\lambda_1\} = Re\{\lambda_2\} = 0$.

The analysis is quite general and can thus be applied with more or less minor modifications to any bistable laser diode including quantum-well distributed-feedback structures. Again, the rate equations as discussed above have to be formulated for a given structure and then the stability analysis is performed. The optical gain function $g(N)$ appearing in the rate equations significantly affects the properties of bistable laser diodes. The widely used linear approximation is not accurate enough for cases of bistability and all-optical switching when the carrier densities are well above the threshold carrier density (transparency density). A quadratic approximation, for instance, of the form $g(N) \cong aN^2 + bN + c$, where a, b, and c are constants determined by the material and operating conditions (doping, carrier concentration, temperature, wavelength), is more appropriate. The adoption of this approximation and its inclusion in the rate equations expose the significance of the a constant, for example. It basically controls the pump threshold and the frequency of the self-pulsations.

The rate equations are complicated coupled nonlinear differential equations that are solved numerically. The stability analysis, however, enables us to examine them analytically, at least in the first approximation, and thus gain important insight into the laser's nonlinear behavior. This leads to establishing the linear stability matrix of the on and off states, which in turn yields the three eigenvalues, $\lambda_1, \lambda_2, \lambda_3$. Examination of the characters of the eigenvalues provides an understanding of bistable, all-optical switching, and self-pulsation conditions by introducing two key parameters, namely the ratio of the carrier lifetime to the photon lifetime, δ, and the normalized spontaneous emission coefficient, σ.

Despite the complexities, one can still, analytically in an asymptotic approximation, obtain the expressions for the eigenvalues and thus draw some conclusions regarding the bistable points. In the case of a two-section laser (active and passive optical waveguides), for example, for the off state, all three eigenvalues are real and negative, meaning that this state is always stable up to the switch-up pump rate. The rate itself depends mainly on the gain profile and the ratio of the length of the gain region (active waveguide section) to the whole cavity (waveguide) length; it is independent of the carrier lifetime in the absorption region (passive waveguide section).

The existence of the off state then depends on the spontaneous emission from the gain region, that is, on one of the key parameters, σ, conveniently introduced. The eigenvalues for the on states come out as a complex conjugate pair for the first two, as mentioned earlier, while the third one is real and negative. This means that the on state may become unstable and a periodic solution can emerge (self-pulsation). Physically this means that when the carrier lifetime is about the same over the whole length of the laser cavity, the on state is stable. It becomes unstable when the carrier lifetime of the passive waveguide section (absorption region) is shorter than that of the active waveguide section (gain region); bistability will be lost at that point and self-pulsations may occur. Clearly, the other key parameter, δ, introduced above plays a determining role in this case. It also determines the width of the hysteresis, if any exists.

One can conclude that the two key parameters are useful tools in analyzing bistable switching and self-pulsation, while turn-on and turn-off dynamics are characterized by two different sets of eigenvalues. The switch-off performance is drastically affected by the carrier lifetime in the absorption region because the turn-off process is governed by recovery of absorption, which saturates due to electrons overfilling the conduction band. Therefore, the turn-off time can be reduced by decreasing the carrier lifetime in the passive waveguide section. The switch-on performance is affected by the carrier lifetime in the gain region, that is, the active waveguide section. This time decreases with increasing the pump rate by, for example, a higher driving electrical current of the laser diode.

The self-pulsation frequency itself is a complex function of a number of parameters. An approximate analytical expression was derived for the first time in the form [75]:

$$f_{sp} = \frac{1}{2\pi}\sqrt{\frac{\beta}{\tau_{cg}\tau_{ph}}} \tag{4.23}$$

where τ_{cg} is the carrier lifetime in the gain region and τ_{ph} is the photon lifetime. The function β depends on the carrier densities in the gain and absorption regions, the photon density in the cavity, the gain profile constants, and the ratio of the gain region length to the total cavity length. Eq. (4.23) clearly indicates that in order to increase the self-pulsation frequency one needs to reduce the photon lifetime, reduce the carrier life time in the gain region, or increase the photon density via a higher pump rate or higher optical gain. This favors quantum-well lasers that have high optical gains and short carrier and photon lifetimes compared to other structures. Typically, with$\tau_{cg} \approx\ 1\ ns$ and $\tau_{ph} \approx\ 1\ ps$, Eq.(4.23) suggests $f_{sp} \approx 10\,GHz$. Self-pulsations in the gigahertz range have been observed, although even frequencies around 60 GHz were reported, most probably due to two-mode beating-type oscillations [76].

True all-optical switching in which "light controls light" as stated in the early days of optical bistability, has been an elusive goal for decades. Although simple, straightforward, and attractive conceptually, it has become a physical, technical, and technological nightmare. Physically, effects that can lead to all-optical switching exist, but an optical nonlinear phenomenon that is usually weak or too lossy must be employed. Optical nonlinearities including nonlinear optical structures and devices have been studied extensively for a long time. Some remarkable results have been obtained in bulk [13] as well as in semiconductor multiple quantum wells [14] and in waveguides [15]. Technically and technologically, hundreds of new and novel structures and devices have been proposed, patented, and tested, but no practical devices emerged and "saw the light of day," except perhaps for frequency-doubling devices (e.g., laser pointers).

The main reason seems to have been the underestimation of the fact that the optical nonlinearities of materials used in device implementations are either too weak (transparent crystals) or associated with large absorption losses (semiconductors near the band edge). The weak nonlinearities must be enhanced in an accumulative way by making the interaction lengths long, while the nonlinear effects coupled with optical losses require some way of amplification or resonance to compensate for such losses. Even in cases when a resonance effect can be utilized, optical bistability using semiconductor nonlinearities has not materialized in the form of practical devices. The required bistable switching energies and achievable speeds, limited in most cases by the carrier recombination time, cannot match the capabilities of electronics. Analogy with electronics, whereby an "optical transistor" was thought to be "the" all-optical device of the future, does not work in photonics, which is not suitable for cascading devices to implement complex and robust functions, as is the case with electronics. Robustness and speed in optics lie in a natural parallelism that has not yet been fully exploited.

True all-optical switching means that the control light signal directly affects the controlled light signal. It is sometimes confused with optical transparent switching in which an electrical control signal is derived from the optical information-bearing signal without converting it to the electrical domain, and then it is used to affect the switching via, for example, the electro-optic effect. Because the information-bearing signal remains in the optical domain, it is sometimes incorrectly termed all-optical switching, although the actual switching is performed electrically. The prime example is high-speed $LiNbO_3$-based or InP-based optical switching components.

True all-optical switching without optoelectronic conversions is possible in both passive and active waveguides. In passive waveguides, third-order nonlinearities, either resonant such as in semiconductors or non-resonant such as in optical fibers, have been employed, as explained earlier. Switching power is the main drawback due to the weakness of these nonlinear effects. Even when quantum-well materials with nonlinearities several orders of magnitude higher than those in their bulk counterparts are used, the overall figure of merit, switching energy required per bit of information, can by no means match the performance of electronics. In active waveguides, the situation is more encouraging since optical gain can lead to enhanced nonlinear effects in terms of efficiency, with the additional benefit of the speed of stimulated emission.

It is thus reasonable to believe that the semiconductor laser [77] as a generic active optical waveguide structure offers the most promising potential for the development of true all-optical devices. A number of theoretical and experimental studies of various semiconductor laser diodes have been conducted [78, 79]. The majority of the work, however, dealt with optical switch-on and electrical switch-off systems that are not truly all-optical operations.

A recent demonstration of a real all-optical flip-flop [80] employs a nonlinear directionally coupled waveguide as a means of resetting a bistable laser diode. The device is set in the same way as other configurations using a saturable absorber. To reset it, a reset pulse injected into the coupled waveguide that runs in parallel to the active waveguide of the laser acts to increase the coupling and thus extract the light from the laser cavity. This brings the absorber back to its high (unsaturated) absorption state, which resets the flip-flop. Operation at 2 GHz was achieved.

True all-optical switching dynamics including experimental results and a theoretical model were first reported by Zhou et al. [81]. A two-section strained quantum-well distributed-feedback laser (Figure 4.8) performed set–reset (flip-flop) operations with an optical trigger from another distributed-feedback laser. Light pulses directly controlled the light output. The interaction employed to affect this functionality is a complex physical phenomenon that involves photons injected into an active optical waveguide (laser cavity), carrier densities, and stimulated emission. Nonlinear effects of absorption saturation and gain quenching are responsible for the switching.

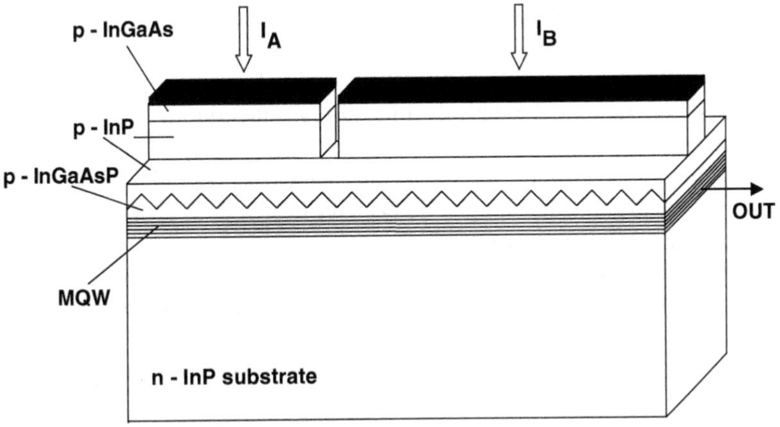

FIGURE 4.8 Two-section strained quantum-well distributed-feedback laser for all-optical switching. MQW = multiple-quantum-well material. I_A, I_B = driving electrical currents.

The laser consisted of a passive and an active optical waveguide. When kept close to the lasing threshold by electrical bias until an optical trigger pulse arrived, it switched into the high (lasing) output as a result of the trigger photons saturating the absorption in the passive waveguide section of the laser. That section thus acted as a saturable absorber mirror for the active waveguide section. The laser stayed in the high lasing state even when the triggering pulse was no longer present because there was enough optical gain in the whole structure to keep the absorption saturated. When the next optical trigger identical to the first one arrived, it depleted the excited carriers injected from the DC supply in the active waveguide section and thus quenched the optical gain. Because the gain was insufficient to overcome absorption losses in the passive waveguide section, the laser switched back to its low state and stayed there until the next triggering pulse arrived. Clearly, the optical injection used for the control in this case did not involve the same physical processes as the electrical injection.

While an electrical trigger injects free carriers directly into the PN junction, the optical trigger creates the free carriers through absorption of photons in the PN junction. Most importantly, the electrical injection is usually uniform along the length of the cavity because the electrical strip contact covers it, thus creating a uniform distribution of photon density unless spatial hole burning occurs. The optical trigger, on the other hand, does not automatically generate or deplete carriers uniformly along the cavity. A more involved model is required to correctly describe the dynamics of this all-optical switching system that includes non-uniform distribution of the photon density along the laser cavity [81].

The basic rate and traveling-wave equations (4.3) and (4.4) can be used as starting points. However, they must be extended to describe the carrier densities in both the passive (absorption) and active (gain) waveguide sections, and the photon density in both sections. The resulting nonlinear first-order differential equations are time and space varying, and are mutually coupled. Their solution is difficult and a numerical algorithm must be used. The results are very interesting, showing that the non-uniformity of the photon density distribution enables us to explain the effect of the reset operation triggered by the same optical pulse used to effect the set operation. One can conclude that the physical mechanism behind these two inherently different switching dynamics (set and reset) is the existence of a positive feedback in absorption saturation during the set operation and a negative feedback in gain quenching during the reset operation. Alternatively, one observes that during the set operation the absorption saturation in the passive waveguide section is supported by an increase in the optical output (positive feedback) and during the reset operation the drop in the optical output due to gain quenching holds back the decrease in carrier density in the active waveguide section (negative feedback).

The switching speed (actual turn-on and turn-off times) is very high. The optical generation (set) and depletion (reset) of carriers are inherently fast processes since carrier generation from the valence to the conduction band and carrier recombination via stimulated emission are on very short time scales — shorter than a picosecond. The fundamental limitation, however, is the carrier recovery time that in effect limits the maximum repetition frequency of all-optical switching. The reason is that the absorption recovery in the passive waveguide section (absorption region) requires a minimum pulse width for the reset operation, and the gain recovery in the active waveguide section (gain region) causes an additional delay for the set operation. With nanosecond carrier recovery times, only hundreds of megahertz in repetition frequency are possible. A realistic and feasible reduction in the carrier recovery time suggests repetition frequencies in the gigahertz range.

A two-mode two-section laser is not only usable for very high speed, self-pulsation, power bistability, or all-optical switching. As discussed earlier, it can also be employed in the generation of microwave signals for wireless communication systems by beating its two modes on an optical detector. In addition, it displays a very interesting and promising characteristic of bistable wavelength switching. If both sections of a laser are biased to oscillate, one deals with two active optical waveguides coupled nonlinearly. Each section has its own resonant wavelength and, depending on the actual operating conditions, namely the electrical bias on each section, one or both modes can oscillate. The non-uniform carrier density distribution and most importantly the non-uniform photon distribution along both sections determine the actual optical output.

Pronounced non-uniformity of density distributions occurs when the lasing sections are biased well above the threshold, usually several times the threshold value. A phase diagram shows various situations, depending on the bias on the two sections. A bistability region exists in the phase diagram; the output can be switched between the two wavelengths of the two active waveguide sections. Since the switching occurs while the laser is lasing, it is very fast because of stimulated emissions, although gain competition between the modes can slow it down. At the same time, the output power is practically constant for any of the two wavelength bistable states. This suggests interesting applications in optical wavelength rather than power logic [82].

Wavelength bistability and wavelength all-optical switching in this type of laser are complex processes that involve interplay of several participating nonlinear optical effects. The non-uniformity of the carrier and photon densities that lead to the non-uniform distribution of gain and refractive index along the active waveguides is one important contributing phenomenon. Other effects include forward and backward wave coupling, nonlinear gain in the two sections and its saturation, and nonlinear coupling between the modes.

Including all these factors into a model is a complicated process. Two approaches are possible. One involves the classical use of rate equations expanded now to incorporate the two modes, each with a different gain. One of the approximations of the gain functions should be used, namely the quadratic or the saturated one for a bulk-material-based laser, or the logarithmic one for a quantum-well laser. Four differential nonlinear coupled equations result, describing carrier and photon densities in each section. The traveling-wave equations must then be built in by allowing the densities, gain, and refractive index to vary along the cavity length. Although this method is relatively simple to implement and may yield reasonable results, it is more or less intuitive in that it lets the user choose the longitudinal functions for the non-uniformity of the variables and thus is not always reliable and accurate [83].

A more rigorous approach is employing coupled-power equations [72] to find proper longitudinal distributions [84] instead of guessing them. Spontaneous emission, shown to be important in the dynamics of bistable lasers [75], must be included in the coupled-power equations re-derived for this case. Each mode is represented by its own set of equations. The two rate equations must also be included in the construction of a complete model, one for each carrier density in each section. The quantities in the rate equations and the photon densities described by the coupled-power equations are now functions of the longitudinal variable.

This method eventually requires eight coupled power equations for each active waveguide section of the laser. With the two rate equations, a total of eighteen nonlinear differential, mutually coupled equations is required for a complete model. It is a complex system to solve, which is analytically impossible. Advanced numerical approaches are thus needed; however, results correctly describe the dynamics including the wavelength bistability.

In conclusion, active optical waveguides offer most in terms of possible and realistic functionalities that will be required if all-optical switching, logic, computations, processing, and even memory are to become realities. Recent incredible advances in photonic crystal materials [85] discussed in more detail in another chapter of this book are very promising. Combined with strong optical nonlinearities of quantum-well and quantum-dot materials and arranged into active waveguiding configurations with strong light confinement, rich periodic structure-based dispersive properties, and inherent optical gain, they will certainly drive progress toward true high performance all-optical devices. Semiconductor nanotechnologies will play a crucial role in their implementations.

4.5 CONCLUSIONS AND FUTURE

The extraordinary success of integrated active electronics resulted from a general approach whereby an as-clean-as-possible common crystal material (silicon) is grown first. It is then processed by introducing "dirty" impurities in a controlled manner to modify its electronic properties. Integrating such artificial alterations onto a basic material then establishes all the various functionalities the integrated chip offers. Once the human mind put software into the chip as an added intelligence, the electronic chip became a powerful technological means for enabling progress in almost all aspects of our life. Since photonics does not have and may never have a silicon-type material, a different approach may be necessary. Rather than attempting to find an ideal universal material for implementing photonic devices and integrated configurations, we may need to focus on various materials with unifying structural properties, i.e., two- and three-dimensional periodicities that are natural properties of many materials.

Because periodic structures offer almost infinite numbers of possible light propagation properties, they may serve as a universal "material" for photonic devices of the future.

Two- and three-dimensional periodic structures, made in various material systems chosen depending on particular areas of applications, could then serve as a "starting material" of sorts — as silicon serves the electronics field. Such common photonic materials or rather their structuralities, modified by introducing optical nonlinear and optical amplifying regions, will then represent the specific device functionalities built into their very structures. An analogy with biology rather than electronics comes to mind. Biological materials such as DNA molecules structure themselves right from the start to the desired functionalities and have the built-in genetic intelligence to do so. The self-assembly, self-organization, and adaptation abilities of biological materials are truly miraculous and expose quite clearly our own, still primitive level of making and building useful structures and devices.

Photonic band gap principles date back to the early work in periodic structures done by Lord Rayleigh in 1887 [16] and others who followed him. Recent advances in photonic band gap materials [86] may facilitate development along these original ideas because they offer tremendous opportunities for design and implementation of future photonic devices. Microstructured and nanostructured systems with photonic band gaps allow, in principle, modification of interactions between light and matter. If such modification can be performed dynamically, for example, through active nonlinear optical properties of the material, new functionalities may become realistic.

Photonic crystals possess properties that allow intrinsic technological integratability of basic device configurations that may lead to monolithic integration of optical generation, detection, and control. This can lead to scalability that, in turn, leads to efficiency and robustness in terms of functionality, an evolutionary synergy so valuable in monolithic electronic integration. Obtainable functions include all-optical operations such as switching, wavelength conversion, clock extraction, retiming, reamplification, and reshaping functions, wavelength logic, ultra-fast optical and optoelectronic mixing and tunable heterodyning, computational functions, and others that can drastically enhance future communication, computing, processing, and sensing systems in their functionality and flexibility. Fully three-dimensional photonic crystals with complete photonic band gaps that incorporate tunable defects (active and nonlinear) may bring about invention of a photon-based rather than electron-based storage element. True optical storage is one of the biggest challenges in photonics; successful implementation of a device of this sort would exert a revolutionary impact on optical communications and quantum computing and produce benefits in biomedicine and many other fields.

Full exploitation of wavelength has not yet been achieved; wavelength is used only for transmission channel carriers in optical communications systems. Its utilization as, for instance, intelligence added to transmitted bits of information could significantly influence the future development of algorithms and protocols in optical telecommunications. The color bit concept [87] whereby every bit consists of several wavelengths that represent an address of the bit's destination is indeed an interesting one. Such exploitations of wavelength as a functional parameter in processing and intelligent operations (self-addressing, logic) can tap into the potential of photonics well beyond the present state of the art. The new photonic devices employing novel material structures and active nonlinear optical waveguides can then facilitate not only information technology, communications, data processing and computing, but also sensing, biomedical and life sciences, manufacturing, health care, environmental sustainability, and many other applications areas, demonstrating the inherent pervasiveness of photonics.

The goal of researching and developing such active integrated photonic configurations will, of course, require novel approaches to modeling, designing, and evaluation of these devices. The philosophy for design of future photonic devices may be based on the fact that optical gain can turn optical waveguide and periodic nanostructured elements into multifunctional devices with low insertion losses and high operating speeds, especially where nonlinearities inherent to the material or the gain process are utilized. A basic generic structure with optical gain, for example the one employed in semiconductor optical amplifiers and lasers, can form an elementary designer building block for the research and development in integrated active photonics that uses active optical waveguides.

As a conclusion, a common material or structural platform that can support in a compatible and technologically viable way the integration of various active photonic devices with complex functionalities including optical memory into complicated configurations is the primary long-term challenge of research in this area. If such research contributes successfully to solving this extremely challenging problem, a demonstration of such a powerful technology would be an achievement as significant as the invention of the electronic integrated circuit for which Kilby received the Nobel Prize in 2000.

ACKNOWLEDGMENTS

The author wishes to thank all his graduate students and postdoctoral fellows without whose creative minds and hard work many of the results presented here would not have been obtained. Collaborations with esteemed colleagues from various institutions are also acknowledged. The institutions include Nortel Networks, the National Research Council and the Communications Research Center of Canada, Ecole Polytechnique Federale Lausanne of Switzerland, Siemens AG of Germany, the Lebedev Institute of Physics of the Russia Federation, the Polytechnic University of New York, United States, the Polytechnic University of Madrid, Spain, and the Optical Technologies Division (now Agilent) of the Center for Research in Telecommunications of Telecom Italia, Italy. Financial support from these organizations and from the Natural Sciences and Engineering Research Council of Canada is greatly appreciated.

REFERENCES

1. A. Einstein, Die quanten Theorie der Strahlung, *Phys. Leit*, 18, 121 (1917).
2. B.E.A. Saleh and M.C. Teich, *Fundamentals of Photonics*, John Wiley & Sons, New York (1991).
3. D. Gabor, *Proc. Roy. Soc. A*, 197, 454 (1949).
4. K.C. Kao and G.A. Hockham, *Proc. IEEE*, 113, 1151 (1966).
5. D. Hondros and P. Debye, *Annal Physik*, 32, 465 (1910).
6. H. Kogelnik and C.V. Shank, *Appl. Phys. Lett.*, 18, 152 (1971).
7. A. Yariv, *Proc. Esfahan Symp.*, August 29, 1971.
8. S.E. Miller, *Bell Syst. Tech. J.*, 48, 2059 (1969).
9. T. Tamir, *Integrated Optics*, Springer-Verlag, New York (1975).
10. P.K. Tien, *Appl. Opt.*, 10, 2395 (1971).
11. P.A. Franken et al., *Phys. Rev. Lett.*, 7, 118 (1961).

12. N. Bloembergen et al., *Phys. Rev.*, 127, 1918 (1962).
13. H.M. Gibbs et al., *Appl. Phys. Lett.*, 25, 451 (1979).
14. D.S. Chemla and D.A.B. Miller, *J. Opt. Soc. Am.*, B2, 1155 (1985).
15. G.I. Stegeman, *IEEE J. Quant. Electr.*, 18, 1610 (1982).
16. Lord Rayleigh, *Phil. Mag.*, 24, 145 (1887).
17. L. Brillouin, Dover Publications, New York (1953).
18. H.G. Winful et al., *Appl. Phys. Lett.*, 35, 379 (1979).
19. R. Normandin et al., *IEEE J. Quant. Electr.*, 27, 1520 (1991).
20. M. Cada et al., Optical nonlinear devices, *Int. J. Nonlinear Opt. Phys.*, 3, 169 (1994).
21. M. Cada, Nonlinear optical devices, *Opt. Pura Aplic.*, 38, 1, (2005).
22. Y.R. Shen, *Principles of Nonlinear Optics*, John Wiley & Sons, New York (1984).
23. E. Garmire, Resonant optical nonlinearities in semiconductors, *IEEE J. Sel. Topics Quant. Electr.*, 6, 1094 (2000).
24. H. Haug, *Optical Nonlinearities and Instabilities in Semiconductors*, Academic Press, New York (1988).
25. D.A.B. Miller, Physics and applications of quantum wells in optics, *Tech. Dig.*, 4, 196 (1989).
26. C. Weisbuch and B. Vinter, *Quantum Semiconductor Structures*, Academic Press, New York (1991).
27. H. Kawaguchi, *Bistabilities and Nonlinearities in Laser Diodes*, Artech House, Boston (1994).
28. G.P. Agrawal and N.K. Dutta, *Long Wavelength Semiconductor Lasers*, Van Nostrand Reinhold, New York (1986).
29. R. Van Dommelen, Active Mach–Zehnder interferometers with feedback, Ph.D. thesis, Dalhousie University, Halifax, Canada (2005).
30. H.C. Casey and M.B. Banish, *Heterostructure Lasers*, Academic Press, New York (1978).
31. A. Yariv, *Quantum Electronics*, John Wiley & Sons, New York (1989).
32. J.J. He and M. Cada, Optical bistability in semiconductor periodic structures, *IEEE J. Quant. Electr.*, 27, 5 (1991).
33. R.P. Khare, *Fiber Optics and Optoelectronics*, Oxford University Press, New York (2004).
34. M.M. Fejer, Nonlinear optical frequency conversion, *Phys. Today*, 47, 25 (1994).
35. M. Cada and J. Ctyroky, in *Guided Wave Optoelectronics*, T. Tamir, Ed., Plenum Press, New York (1995).
36. R. Normandin and G.I. Stegeman, Nondegenerate four-wave mixing in integrated optics, *Opt. Lett.*, 4, 58 (1979).
37. M. Cada, Optical nonlinear devices, International Symposium on Microwave and Optical Technologies, Ostrava, Czech Republic (2003).
38. M. Cada, Optical harmonic mixers, *IEEE J. Quant. Electr.*, 31, 269 (1995).
39. M. Bass et al., Optical rectification, *Phys. Rev. Lett.*, 9, 445 (1962).
40. P. Yeh, *Introduction to Photorefractive Nonlinear Optics*, John Wiley & Sons, New York (1993).
41. K.L. Sala, Nonlinear refractive index phenomena in isotropic media subjected to a DC electric field: exact solutions, *Phys. Rev. A*, 29, 4 (1984).
42. D.A.B. Miller et al., Band-edge electroabsorption in quantum well structure: the quantum-confined Stark effect, *Phys. Rev. Lett.*, 53, 2173 (1984).
43. M. Cada et al., Nonlinear guided waves coupled nonlinearly in a planar GaAs/GaAlAs multiple-quantum-well structure, *Appl. Phys. Lett.*, 49, 13 (1986).

44. M. Cada, Switching mirror in the CdTe-based photonic crystal, *Appl. Phys. Lett.*, 87, 1 (2005).
45. M. Cada and M. Qasymeh, Light propagation in external-field-assisted nonlinear waveguides, *IEEE J. Quant. Electr.*, July, 2006.
46. R. Ramaswami and K.N. Sivarajan, *Optical Networks*, Morgan Kaufman, New York (2002).
47. R. Noe et al., Polarization mode dispersion compensation at 10, 20, and 40 Gb/s with various optical equalizers, *J. Lightwave Technol.*, 17, 9 (1999).
48. M.J. Adams et al., Nonlinearities in semiconductor laser amplifiers, *Opt. Quant. Electr.*, 27, 1013 (1995).
49. G. Agrawal, Population pulsations and nondegenerate four-wave mixing in semiconductor lasers and amplifiers, *J. Opt. Soc. Amer. B*, 5, 147 (1988).
50. K. Tajima, All-optical switch with switch-off time unrestricted by carrier lifetime, *Jpn. J. Appl. Phys.*, 32, L1746 (1993).
51. K.E. Stubkjaer, Semiconductor optical amplifier-based all-optical gates for high-speed optical processing, *IEEE J. Sel. Topics Quant. Electr.*, 6, 1428 (2000).
52. M. Eiselt et al., Semiconductor laser amplifier in a loop mirror, *J. Lightwave Technol.*, 13, 2099 (1995).
53. R.J. Manning et al., 10 Gbit/all-optical regenerative memory using single SOA-based logic gate, *IEEE Electr. Lett.*, 35, 158 (1999).
54. H. Moravec, When will computer hardware match the human brain? *J. Evol. Technol.*, 1 (1998).
55. K.B. Roberts, Optical detection and logic devices with latching, U.S. Patent 5,999,284 (1999).
56. J. De Merlier et al., Experimental demonstration of all-optical regeneration using an MMI-SOA, *IEEE Photonics Technol. Lett.*, 14, 660 (2002).
57. J. Cooper, Fiber-radio for the provision of cordless/mobile telephony services in the access network, *Electr. Lett.*, 26, 2054 (1990).
58. J.J. O'Reilly et al., Optical generation of very narrow line width millimeter wave signals, *Electr. Lett.*, 28, (1992).
59. K.Y. Lau, Narrow-band modulation of semiconductor lasers at millimeter wave frequencies by mode locking, *IEEE J. Quant. Electr.*, 26, (1990).
60. U. Gliese et al., Chromatic dispersion in fiber optic microwave and millimeter-wave links, *IEEE Trans. Microwave Theory Tech.*, 44, (1966).
61. G.J. Simonis and K.G. Purchase, Optical generation, distribution and control of microwaves using laser heterodyne, *IEEE Trans. Microwave Theory Tech.*, 38, (1960).
62. R.T. Ramos and A.J. Seeds, Delay, linewidth and bandwidth limitations in optical phase lock-loop design, *Electr. Lett.*, 26, (1990).
63. S. Taylor, Optical generation of millimeter waves, M.A.Sc. thesis, Dalhousie University, Halifax, Canada (2004).
64. M. Cada et al., Photonic transmitters for optical distribution of microwaves, *Proceedings of 10th International Symposium on Microwave and Optical Technology*, Fukuoka, Japan, p. 175 (2005).
65. C.H. Henry, Theory of the linewidth of semiconductor lasers, *IEEE J. Quantum Electr.*, 14, (1982).
66. T. Kawanishi, High-speed optical modulators and photonic sideband management, *Proceedings of 10th International Symposium on Microwave and Optical Technology*, Fukuoka, Japan, p. 716 (2005).
67. P. Mandel, *Theoretical Problems in Cavity Nonlinear Optics*, Cambridge University Press, Cambridge, U.K. (1997).

68. H. Kogelnik and C.V. Shank, Coupled-wave theory of distributed feedback lasers, *J. Appl. Phys.*, 43, 2327 (1972).
69. H. Guafouri-Shiraz and B.S.K. Lo, *Distributed Feedback Laser Diodes*, John Wiley & Sons, New York (1996).
70. G. Isenor et al., *A Global Optimization Approach to Laser Design: Optimization and Engineering*, Vol. 4, Kluwer Academic Publishers, New York, p. 177 (2003).
71. J.D. Pinter, *Computational Global Optimization in Nonlinear Systems*, Lionheart Publishing, Atlanta (2001).
72. J.Y. Wang and M. Cada, Analysis and optimum design of distributed-feedback lasers using couple-power theory, *IEEE J. Quant. Electr.*, 36, 52 (2000).
73. G.J. Lasher, Analysis of a proposed bistable injection laser, *Solid State Electr.*, 7, 707 (1964).
74. H. Kawaguchi, Bistable laser diodes and their applications: state of the art, *IEEE J. Sel. Topics Quant. Electr.*, 3, 1254 (1997).
75. J.Y. Wang et al., Dynamic characteristics of bistable laser diodes, *IEEE J. Sel. Topics Quant. Electr.*, 3, 1271 (1997).
76. B. Sartorius et al., 12–64 GHz continuous frequency tuning in self-pulsating 1.55-μm multiquantum-well DFB lasers, *IEEE J. Sel. Topics Quant. Electr.*, 1, (1995).
77. G.P. Agrawal, *Semiconductor Lasers: Past, Present, and Future*, AIP Press, New York (1997).
78. M. Jinno and T. Matsumoto, Nonlinear operations of 1.55-μm wavelength multi-electrode distributed-feedback laser diodes and their applications for optical signal processing, *J. Lightwave Technol.,* 10, 448 (1992).
79. G.H. Duan et al., Modeling and measurement of bistable semiconductor laser, *IEEE J. Quant. Electr.*, 30, 2507 (1994).
80. M. Takenaka and Y. Nakano, Realization of an all-optical flip-flop using directionally coupled bistable laser diode, *IEEE Photonic Technol. Lett.*, 16, 45 (2004).
81. J. Zhou et al., All-optical bistable switching dynamics in 1.55-μm two-segment strained multiquantum-well distributed-feedback lasers, *J. Lightwave Technol.*, 15, 342 (1997).
82. R. Van Dommelen and M. Cada, Wavelength logic: research report, Canadian Institute for Photonic Applications (2001).
83. J. Kinoshita, Modeling of high-speed DFB lasers considering the spatial hole burning effect using three rate equations, *IEEE J. Quantum Electr.*, 30, 929 (1994).
84. R. Van Dommelen, Bistable distributed feedback laser diodes, M.A.Sc. thesis, Dalhousie University, Halifax, Canada (1999).
85. S.G. Johnson and J.D. Joannopoulos, *Photonic Crystals: The Road from Theory to Practice*, Kluwer, Boston (2002).
86. E. Istrate et al., Behavior of light at photonic crystal interfaces, *Phys. Rev. B*, 71, (2005).
87. M.Cada, Color bit concept, private communication (1989).

5 Wavelength Dispersive Planar Waveguide Devices: Echelle and Arrayed Waveguide Gratings

Pavel Cheben

CONTENTS

5.1 DIFFRACTION GRATING: A RETROSPECTIVE

Few devices invented by man have contributed more to our understanding of the physical world than the diffraction grating. The first experiments with diffraction gratings in the 1820s by Joseph von Fraunhofer proved they were extraordinary tools for studying the world at both microscopic and cosmological scales. Using gratings, he discovered dark lines in light emitted by several substances and observed similar lines in light from the sun and other stars,[1] now known as Fraunhofer's lines.

A practical use of the grating came soon afterward when Fraunhofer used the absorption lines as markers for precise measurements of the refractive indices of glasses used in achromatic lenses made in an optical workshop in Munich. To make his gratings, Fraunhofer also built the first ruling engine, arguably the most precise machine in existence then. He also derived and verified the grating equation and found that groove shape influences the intensity distribution in different diffraction orders. Now known as the blazing effect, the effect was mathematically described many decades later by Rayleigh[2] and experimentally studied by Wood[3] in 1910. Fraunhofer's pioneering work[1,4] laid the foundations of diffraction grating technology, but the making of the first diffraction grating is credited to American astronomer David Rittenhouse[5] who in 1786 made a simple grating by wrapping a wire around the threads of two fine pitch screws. Interestingly, it took more than a century after the discovery of diffraction by Francesco Grimaldi in 1665 to learn what happened when Grimaldi's single aperture was repeated many times to form a grating.

By 1850, F.A. Norbert began to provide scientists including Ångström and Rayleigh with gratings superior to those of Fraunhofer. He was the world's main source of diffraction gratings until his death in 1881. In the 1870s, Rayleigh showed theoretically that gratings can outperform prisms in their ability to resolve spectral lines, and several gratings with spectral resolutions superior to those of the most powerful prisms were made by L.M. Rutherfurd in New York. In the 1870s, Cornu realized and later demonstrated that a planar grating would focus diffracted light if ruled with variable groove spacing. This concept of "chirping the grating" is fundamental in the fields of diffractive optics,[6] group velocity dispersion compensation in optical communications,[7] ultra-fast light pulse generation and shaping,[8] and broadband multilayer dielectric mirrors.[9]

A new era in diffraction grating technology started in the 1880s when Henry A. Rowland at John Hopkins University made gratings about ten times larger and more accurate than any previously made. Rowland's ruling engines incorporated advanced concepts such as temperature control and vibration isolation and were the most important grating sources for the world's scientific community for nearly 50 years.[10] Rowland's discovery that gratings can be formed on a concave (rather than plane) substrate and hence simultaneously disperse and focus light[11] was a milestone that led to the development of ultraviolet (uv) and x-ray spectroscopy.

Another important discovery was a grating mount, now named after Rowland, that eliminates defocus and primary coma while keeping spherical aberration small. A concave grating in a Rowland circle mount is used in both modern spectroscopic instruments, namely in Paschen–Runge, Eagle, Abney, and Johnson–Onaka configurations,[12] and in advanced waveguide echelle grating[13,14] and arrayed waveguide grating (AWG) demultiplexers[15] used in wavelength division multiplexed (WDM) communications. Finally, the availability of new and precise spectral data in the 1880s, most of which were obtained with gratings made by Rowland, was the key for fundamental discoveries on the nature of spectral lines by Rydberg,[16] Balmer,[17] Kayser,[18] Runge and Pashen,[19] and the discovery of Zeeman's[20] effect. It is well known that these spectral studies paved the way toward Bohr's theory of the hydrogen atom, opening the era of atomic physics and quantum mechanics.

In the early years of the 20th century, Albert Michelson of the University of Chicago made key contributions to grating manufacture. He suggested controlling the position of the grating grooves by the interferometer that he invented. His idea was put into practice four decades later when frequency stable light sources and electronic servo systems were made available by Harrison[21] at the Massachusetts Institute of Technology (MIT). Michelson's first engine was upgraded several times and is still operational nearly 100 years after it was built (at Thermo RGL in Rochester, New York). This machine covers the widest range of groove spacings (20 to 10,800 grooves/mm) of any ruling engine. Michelson's second engine was developed further at MIT and it is known as the MIT A engine. These extraordinary engines reliably performed fabrication tasks with a precision in the nanometer range more than a century before the word *nanofabrication* came into vogue.

Since Gabor's discovery of holography[22] in 1948, interferometric recording has also been used to make diffraction gratings. In this technique, a spatial interference pattern formed by two intersecting coherent light beams is holographically recorded in a high resolution photosensitive material. The first interference gratings of spectroscopic quality were produced in 1967 by Rudolph and Schmahl[23] in Germany and Labeyrie and Flamand[24] in France. However, the first interference gratings were made in the 1890s. These were Lippmann's color photographs,[25] some of which still exist and can be found in the London Science Museum. Lippmann's photographs are Bragg-type reflection gratings recorded by two counter-propagated beams.

Interference gratings were also produced by Wiener[26] as a side product of his classical experiment proving that the electric field vector is the photographically active one, and also produced later by Cotton.[27] Michelson considered making his gratings interferometrically,[28] but did not pursue the idea because no intense coherent light source and photosensitive material of a sufficiently fine granularity existed then.

Recently, a technique called scanning beam interference lithography (SBIL) was developed at MIT.[29] It combines the interferometric technique with the mechanical scanning typical for conventional ruling engines. In SBIL, gratings are written by a high precision scan of a photoresist-coated substrate with a small area interference pattern formed by two narrow uv beams with wavefronts of high quality. By scanning the fringe pattern, nonlinear fringe distortion due to the residual wavefront errors can be averaged out in the scan direction and further reduced by overlapping adjacent scans. Grating groove precision approaching 1 nm has been reported across a 300 mm substrate and grating fabrication time was reduced from days to minutes compared to ruling engines.

An important interferometrically recorded grating known as fiber Bragg grating was discovered in 1978 by Hill et al.[30] Light propagating in the Ge-doped core of an optical fiber was allowed to interfere with the weak reflected beam from the end facet of the fiber. This can be regarded as the waveguide version of the reflection-type interferometric grating discovered by Lippmann[25] and later developed by Denisyuk[31] in the context of reflection holograms. Meltz et al.[32] showed that fiber gratings can be recorded by two ultraviolet beams transversally illuminating the photosensitive waveguide core, thus allowing the grating period to be adjusted by changing the angle between the two beams. Lemaire et al.[33] found that refractive index grating modulation can be increased by two orders of magnitude (up to $\Delta n \sim 0.01$) by soaking a Ge-doped fiber prior to grating recording in hydrogen or deuterium at a high pressure. These findings[30,32,33] are important milestones on the way to remarkably successful fiber grating technology.[34] Today fiber gratings are used in many applications,[35] for example, as wavelength filters, dispersion compensators, gain equalizers in erbium-doped silica fiber amplifiers, WDM (de)multiplexers and add/drop filters, mirrors in fiber lasers and amplifiers, mode and polarization converters, and fiber optic sensors.

The interferometric technique is also commonly used for making waveguide gratings that provide wavelength-selective feedback required for single longitudinal mode operation in distributed Bragg reflector (DBR) and distributed feedback (DFB) lasers. These lasers have distributed grating corrugations placed either directly in the laser waveguide cavities (DFB) or at the ends of the cavities that act as Bragg mirrors (DBR), hence providing wavelength-selective feedback. They were first conceived, theoretically studied,[36] and experimentally demonstrated[37] by Kogelnik and Shank at Bell Telephone Laboratories. The first DBR laser[37] was made in a poly(methylmethacrylate) (PMMA) host doped with rhodamine 6G dye. The first DFB laser[38] was made in a dichromated gelatin (DCG) host doped with the same dye.* The Bragg gratings were recorded holographically in the respective hosts. One advantage of PMMA and DCG is that they are photosensitive and virtually grainless, and for this reason they are often used as high resolution media for recording holograms[39] and also in photography.*

Modern semiconductor DFB lasers were developed during the 1980s[40,41] and today they are widely used in WDM communication systems. In DFB lasers, the gratings are often recorded holographically in thin photoresist layers and then transferred by etching to the waveguide layers underneath. The grating period required for first order Bragg diffraction in InP-based lasers at a telecommunication wavelength of 1.5 μm is $\sim$ 240 nm. Gratings of this pitch can be produced interferometrically

* DCG was used in a pioneering photography experiment by Fox Talbot in 1852, and later by Lippmann.[25]

using an inexpensive He-Cd laser with a wavelength of 442 or 325 nm in a single step over the whole wafer, hence avoiding the need for making the gratings by costly and time-consuming electron-beam writing or the still uncommon deep uv stepper projection lithography.

Essential to these practical developments were advances in diffraction grating theories, particularly in electromagnetic theories of gratings[42] that are also suitable for rigorous simulations of waveguide echelle gratings discussed in the next section. Approximate methods such as coupled-mode[43,44] and transfer matrix[45] analyses have often been used for waveguide gratings.

A brief overview of grating theories can be found in an excellent comprehensive textbook on diffraction gratings by Loewen and Popov.[46] Recently there has been a shift in favor of numerical methods, particularly with advances in the finite difference time domain (FDTD) technique[47] first proposed by Yee[48] in 1966. The FDTD is applicable for complex grating geometries including photonic bandgap structures and can calculate polarization effects.

Most of the gratings mentioned are weakly modulated and periodic in one dimension, for example, parallel shallow grooves in a metal coating of a substrate of a spectroscopic grating, a shallow waveguide corrugation in a DFB laser, or a small modulation of refractive index in the fiber Bragg gratings. Indeed, such periodic structures can be extended into two or three dimensions and include large grating modulations.

In an analogy to the fiber Bragg grating mirror, a multilayer plane mirror can be formed by periodically altering thin layers of materials with different refractive indices ($n_1, n_2 = n_1 + \Delta n$) and optical thickness $\lambda/4$, yielding a period $\lambda/2$, so that the light partially reflected by different layers collectively adds in phase resulting in the mirror effect. As the bandwidth over which the mirror effect is obtained increases with Δn, an omnidirectional mirror effect can be expected over a large wavelength bandwidth or *bandgap* by forming regular three-dimensional arrays of different materials with large Δn. Such *photonic crystal* or *photonic bandgap* grating-like structures have been extensively studied[49,50] after they were suggested in the late 1980s.[51,52] The beginning of this new field came surprisingly late considering the amazing variety of similar structures that catch our attention almost daily in inorganic and organic nature,[53] the existence of readily applicable models in solid state physics (theorems of Floquet[54] and Bloch[55] dating back to 1883 and 1928, respectively), and even in optics. The latter includes a study by Rayleigh of light reflection in some multilayered minerals,[56] deducing a solution for waves in a one-dimensional periodic media,[57] and showing that such media exhibit an effect we now call bandgap.

Despite important advances in our understanding of photonic crystals, difficulties are still significant in fabricating practical photonic crystal devices, even when we use modern tools like electron beam lithography and advanced dry-etch processes developed in the microelectronics industry. Remarkable advances have been achieved in making photonic crystals in optical fibers, namely 0.3 dB/km loss in solid core fibers,[58] 1.7 dB/km in hollow core fibers,[59] and continuous fiber lengths of 100 km.[58]

For making planar waveguide photonic crystal devices, silicon-on-insulator (SOI) is a natural platform. Several research groups are working on reduction of waveguide propagation losses in photonic crystal SOI slab waveguides,[60] and losses as small as 8 dB/cm have recently been reported by IBM.[61] Among many potential applications of photonic crystals,[50,62] those that appear particularly attractive explore their peculiar dispersion properties and slow down the light,[63,64] enhance the optical fields in microcavities,[65] or have the ability to enhance or suppress the spontaneous emission of light[66] by Purcell's effect.[67] Interesting new devices may arise by arranging photonic crystals into a conventional diffraction grating configuration[68] or by combining the photonic crystals with AWGs. The latter has been suggested by Martínez et al.[69] who demonstrated that dispersion of an AWG can be markedly enhanced by placing in the waveguide array a triangular prism-like region of photonic crystal waveguides with a large group index.

In the post-cold war period, the main force driving technical innovation in photonics has been the telecommunications industry. Since the mid 1990s, the information capacity of optical fiber transmission links has been vastly increased by using wavelength division multiplexing (WDM).[70,71] Each fiber carries many different wavelengths and each wavelength carries a different data stream.*

Diffraction gratings are essential to many devices used in optical communication networks, including wavelength stabilized lasers, optical amplifiers, wavelength converters, multiplexers and demultiplexers, add–drop filters, dispersion compensators, and gain equalizers.[72]

In the following sections we will discuss the waveguide echelle grating (EG) and arrayed waveguide grating (AWG) devices. Similar to ordinary diffraction gratings, these devices can spatially separate (demultiplex) or combine (multiplex) light of different wavelengths — the two key functions of WDM optical communication systems.

5.2 WAVEGUIDE ECHELLE GRATINGS

5.2.1 Fundamentals and Background

Before proceeding to the discussion of modern waveguide EG devices, we will briefly recall a few simple but useful grating formulas. Consider light incident on a grating surface as shown in Figure 5.1a. We assume the incident light is an optical mode guided in the slab waveguide (see Figure 5.2a) of effective index n_{eff}, and the grating is formed through the waveguide layers in a direction perpendicular to the waveguide plane. Most of the arguments to follow apply also to free-space gratings. Each grating facet represents a small source reflecting and transmitting incident light according the laws of reflection and refraction, and simultaneously acts as a narrow diffraction slit, thus determining the envelope of the far-field diffraction pattern, also called the *slit function SF* (Figure 5.1b).

---

* For example, WaveStar™ 400G system deployed in 1999, supports 40 WDM channels at 10 Gb/s per channel, hence with a bit rate of 400 Gb/s per fiber that is equivalent to 5,160,000 voice channels.

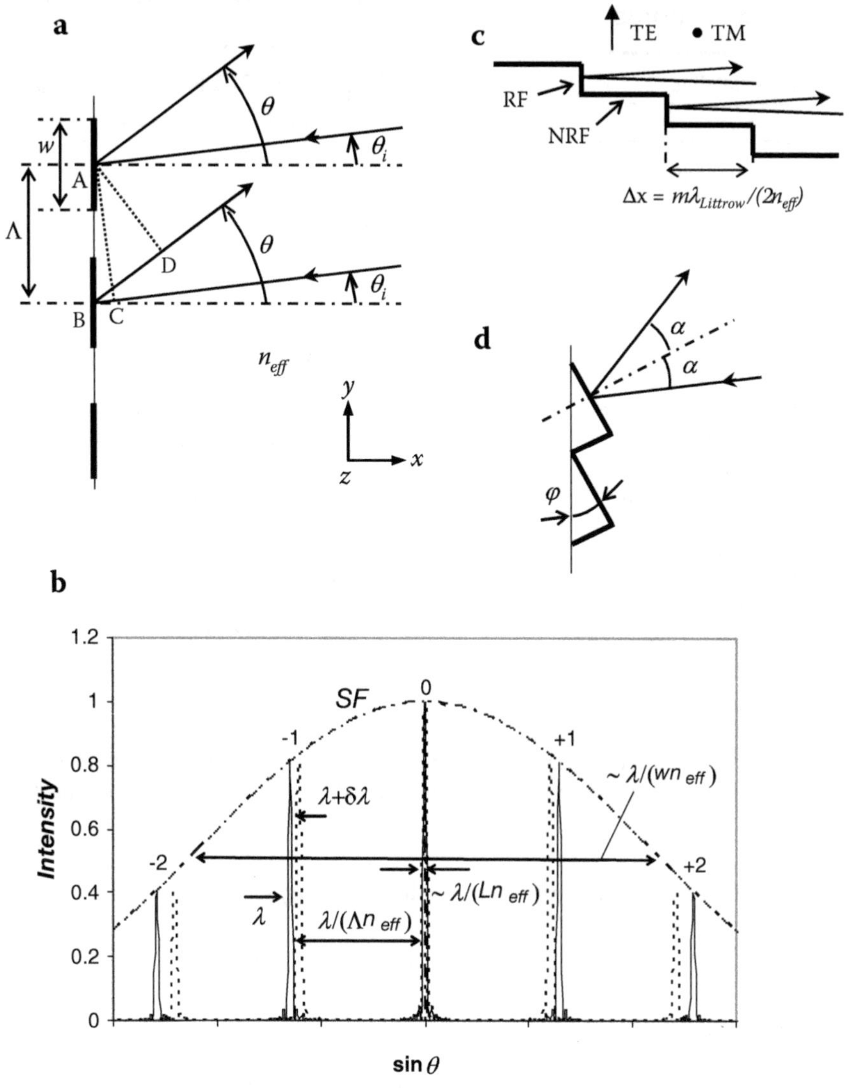

FIGURE 5.1 a) Diffraction grating schematics in reflection geometry. b) Far field intensity distribution of the light diffracted by a grating. c) Echelle grating used in near-Littrow incidence; the electric field directions of the TE and TM polarizations are indicated; RF and NRF are reflecting and non-reflecting facets, respectively. d) Blazed diffraction grating.

Gratings are often used in reflection geometry with the grating surface typically covered by a reflective metal layer so that each facet acts as a small mirror. The light reflected by different facets will add in phase, hence producing interference maxima in those directions θ for which the phase difference between the light diffracted by

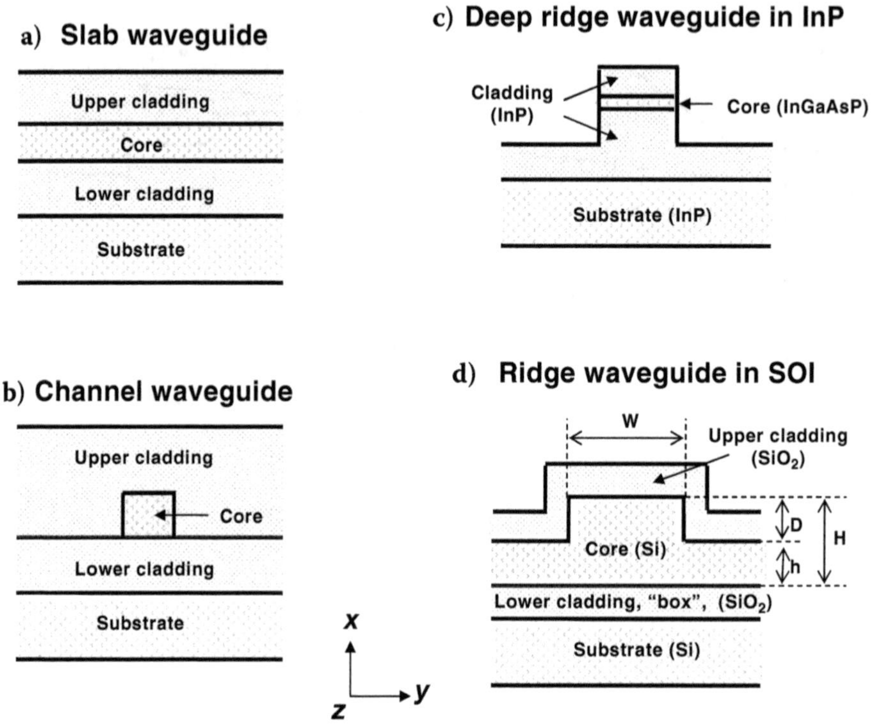

FIGURE 5.2 Examples of different types of optical waveguides. a) Slab waveguides; b) buried channel waveguide; c) deep ridge waveguide in InP; and d) ridge waveguide in silicon-on-insulator.

two adjacent facets is $2\pi m$, as expressed by the *grating equation*:*

$$\sin\theta + \sin\theta_i = \frac{m\lambda}{n_{eff}\Lambda}, \quad m = 0, \pm 1, \pm 2, \ldots \tag{5.1}$$

where θ_i is the angle of incidence, Λ is the grating period (pitch), λ is the wavelength measured in vacuum, and m is the diffraction order.

For a grating of a finite width $L = N\Lambda$, the normalized far field intensity distribution $I(\theta)$ can be expressed from the Fresnel–Kirchhoff diffraction formula, as a product of the interference function IF and the slit function SF:

$$I(\theta) = SF \cdot IF = \left(\frac{\sin\beta}{\beta}\right)^2 \left(\frac{\sin N\gamma}{N\sin\gamma}\right)^2 \tag{5.2}$$

where $\beta = (\pi n_{eff} w \sin\theta)/\lambda$, $\gamma = (\pi n_{eff}\Lambda \sin\theta)/\lambda$, and N is the number of facets each of width w. Normal incidence ($\theta_i = 0$) is assumed. The slit function SF appears

* The sign convention adopted in Eq. 5.1 assumes positive angles in the sense indicated in Figure 5.1a.

as the envelope of the interference function *IF* (see Figure 5.1b).The maxima of *IF* occur within the *SF* envelope at $\gamma = m\pi$, as predicted by Eq. 5.2.

For a given incident angle, wavelength, and grating period, the grating equation can be satisfied for different orders *m* providing $\sin\theta < 1$. The orders for which $\sin\theta > 1$ are called evanescent orders. Their wavevector component along the grating normal is imaginary, hence their amplitude vanishes at distances larger than a few wavelengths from the grating surface. The evanescent orders must be taken into account in waveguide gratings and in rigorous calculations of diffraction efficiency or polarization effects in electromagnetic theories of gratings.

For small wavelength-to-period ratios (λ/Λ), many orders may exist. Gratings working in high orders are called echelle gratings* or EGs (Figure 5.1c) and they are attractive because of their large dispersion and resolution. They were first proposed by Wood[73] and extensively studied by Harrison[74] who noticed that the resolution of a grating does not depend on the number of grooves but on the width of the grating. Harrison proposed to solve the problem of diamond wear† by ruling coarse gratings (large Λ) and using them in high orders.

The energy of light diffracted by the grating in a specific direction can be increased by inclining the grating facets by a so-called blazed angle (φ; see Figure 5.1d) such that the diffraction angle coincides with the reflection angle relative to the facet normal. This is called *blazing* the grating. The blazing shifts the far-field envelope so that it peaks in the desired direction θ compared to $\theta = -\theta_i$ when the grating is not blazed. Grating diffraction efficiency close to 100% is expected in the scalar approximation for an order *m* diffracted in or close to the blazed direction θ. By differentiation of Eq. 5.1, the grating *angular dispersion* is obtained:

$$d\theta/d\lambda = m/(n_{eff}\Lambda\cos\theta)$$

The grating *resolution*, $R = \lambda/\Delta\lambda$, also called *resolving power*, is the ability of a grating to separate adjacent wavelengths. Because the angular half-width of the far-field diffraction pattern by a grating of aperture $L = N\Lambda$ is $\Delta\theta \sim \lambda/(Ln_{eff}\cos\theta)$,‡ two peaks of wavelengths λ and $\lambda + \Delta\lambda$ are considered resolved if they are separated by at least this amount (Rayleigh criterion), the resolving power is obtained from Eq. 5.1:

$$R = mN = L(\sin\theta_i + \sin\theta)n_{eff}/\lambda = OPD_{max}/\lambda \qquad (5.3)$$

where $OPD_{max} = N\Lambda n_{eff}(\sin\theta_i + \sin\theta)$ is the optical path difference between the extreme rays on the grating. In an AWG (see Section 5.3), OPD_{max} in Eq. (5.3) is the optical path difference between the longest and shortest waveguides in the phase array.

It is worth noting that the resolution of a grating-based instrument equals the number of wavelengths contained in the optical path difference between the rays

---

* From French *échelle* (ladder), because of their staircase-like teeth profile.

† When the best diamond crystal is used to rule grating, the diamond wear typically limits the maximum accumulated travel to less then 20 km. A 12-inch grating of a period 400 nm would require a travel of more than 200 km.

‡ $\Delta\theta$ is defined as separation between the peak and first adjacent minimum of *IF*.

originating from the two extremes of the grating aperture. From the above formulas it follows that working in high diffraction orders is beneficial for maximum resolution and dispersion, and both EGs and AWG devices benefit from this fact. Working in high orders, however, comes at a cost of a reduced *free spectral range* (*FSR*), which is the difference between two wavelengths diffracted into the same direction in successive orders, i.e., $\theta(m, \lambda) = \theta(m \pm 1, \lambda \mp FSR)$. From Eq. 5.1, FSR $= \lambda/m$. EGs with resolutions exceeding 10^6 are used in stellar or atomic spectroscopy. Wavelength-to-period ratios as small as 0.001 and orders up to $m \sim 1500$ have been used in astronomy,[75] while in EGs and AWGs for WDM applications, $m \sim 20$ is typically sufficient for (de)multiplexing of wavelength channels separated at 100 GHz. To avoid the spatial overlap of different wavelengths in successive orders when the wavelength range exceeds the *FSR*, a cross-dispersion prism is often inserted in high resolution free-space echelle spectrometers. Otherwise the *FSR* sets the limit of the maximum available spectral range as is the case with waveguide EGs and AWGs.

Modern waveguide EG devices often use a configuration first suggested by Eagle in 1910.[76] The Eagle mounting (see Figure 5.3a) uses the Rowland configuration with the angles of incidence and diffraction nearly equal, similar to the case of Littrow mounting for plane gratings (Figure 5.1c). For a Rowland circle configuration* in the coordinates of Figure 5.3a, the following relation for grating linear dispersion is obtained by differentiating the grating equation, assuming $\sin\theta_i \sim \sin\theta$, and including the effect of waveguide effective index dispersion $dn/d\lambda$:[77]

$$D = \frac{dy}{d\lambda} = \frac{4r \sin\theta}{\lambda}\left(1 - \frac{\lambda}{n}\frac{dn}{d\lambda}\right) \tag{5.4}$$

where r is the radius of the Rowland circle. Residual aberrations in the Rowland circle configuration can be further reduced by shifting the position of the grating teeth from the exact Rowland geometry, a procedure called stigmatization.[78] Designs with two stigmatic, i.e., aberration-free, wavelengths λ_1 and λ_2 are often used in waveguide EG demultiplexers. The stigmatic correction can be understood as follows: The light of wavelengths λ_1 and λ_2 that originates in the input waveguide I, upon diffraction by a concave grating G, is brought into focus at points O_1 and O_2 where the output waveguides for receiving the wavelengths λ_1 and λ_2 are located (see Figure 5.3a). The imaging of the input located at I onto the outputs at O_1 and O_2 will be aberration-free if the phase of the light that originates from arbitrary adjacent grating teeth located at $G_i \equiv (x_i, y_i)$ and $G_{i+1} \equiv (x_{i+1}, y_{i+1})$ will arrive at the focal points O_1 (for λ_1) and O_2 (for λ_2) with a phase difference of 2πm or, expressed in terms of the path differences:

$$IG_i + G_iO_1 - (IG_{i+1} + G_{i+1}O_1) = \frac{m\lambda_1}{n_1}, \tag{5.5}$$

$$IG_i + G_iO_2 - (IG_{i+1} + G_{i+1}O_2) = \frac{m\lambda_2}{n_2}, \tag{5.6}$$

* In the Rowland configuration, the grating is located on an arc of radius 2r tangent to the Rowland circle of radius r. The spectrum is formed along an arc on the Rowland circle.

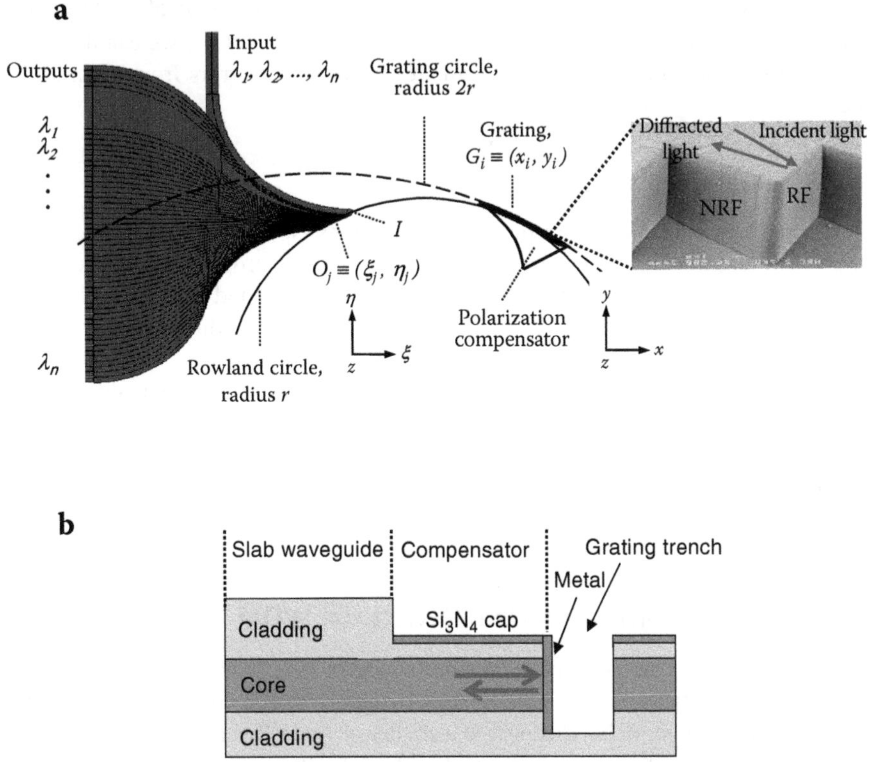

FIGURE 5.3 a) A schematics of echelle grating demultiplexer in Eagle geometry. b) Cross-section of the polarization compensator. Waveguide birefringence is modified in the section with thinned waveguide cladding and high index (Si_3N_4) overlayer.

where n_1 and n_2 are the effective indices along the path IGO at wavelengths λ_1 and λ_2, respectively. For a grating with N teeth, the above equations can be reduced to a system of $(2N)$ linear equations from which, using simple algebra, the unknown coordinates of the grating teeth (N of x_i and N of y_i) can be found.

Once the exact position $G_i \equiv (x_i, y_i)$ of the grating teeth required for the stigmatic correction is found, aberrations can be further reduced by allowing for small variations $\Delta\xi_j$ and $\Delta\eta_j$ in the output waveguide positions from their initially assumed position (ξ_j, η_j) at the Rowland circle to minimize the phase errors, where the index j denotes the j-th output waveguide. From Figure 5.3a it is observed that parameter $\Delta\xi_j$ is primarily responsible for focusing and $\Delta\eta_j$ for frequency (wavelength) responses of individual channels. The light at wavelength λ_j will arrive at an arbitrary output channel j located at $O_j \equiv (\xi_j, \eta_j)$ from two adjacent grating teeth G_i and G_{i+1} with a phase error of:

$$\Delta\Phi_j^i = \frac{2\pi}{\lambda_j} n_j (IG_i + G_iO_j - IG_{i+1} - G_{i+1}O_j) - 2\pi m \tag{5.7}$$

For the two stigmatic wavelength channels λ_1 and λ_2, $\Delta\Phi_1^i = \Delta\Phi_2^i = 0$ for any grating facet G_i, as explained above. For every output channel λ_j, we can define a functional representing the sum of the squares of the phase errors from all grating facets:

$$F_j = \sum_{i=1}^{N} \alpha_i \left|\Delta\Phi_j{}^i\right|^2 \tag{5.8}$$

F_j can be minimized with respect to the variations $\Delta\xi_j$ and $\Delta\eta_j$ in the position O_j of each output waveguide, and the weight factor α_i is proportional to the light intensity at different facets. Using the calculated values of $\Delta\xi_j$ and $\Delta\eta_j$, the optimized position $(\xi_j + \Delta\xi_j, \eta_j + \Delta\eta_j)$ of the output waveguides with reduced aberrations is found.

The outlined aberration reduction steps would obviously be extremely difficult to implement with conventional (bulk optics) diffraction gratings, in contrast to planar waveguide EGs in which the optimized positions of the grating teeth and the output channels can easily be produced on the lithographic mask and then transferred on the chip.

5.2.2 Early Monolithic Grating Devices for WDM

Attempts were made to assemble miniature spectrometers from separate elements (i.e., collimated optics and bulk gratings in a miniature mount connected with input and output optical fibers), but this proved a challenge for WDM applications. Different components must be precisely aligned with each other, often at submicrometer tolerances, and this alignment must be preserved at packaging and then for many years during device operation. Monolithic solutions are preferred because of compactness, ease of alignment with optical fiber arrays, stability, and cost advantages. One of the first monolithic devices[79] was demonstrated in the late 1970s. It used a gradient index (GRIN) lens as a collimator attached at one end to the input and the output fibers. A miniature diffraction grating was attached to the other end of the GRIN lens, with a small prism inserted to provide a tilt of the grating with respect to the beam propagation direction, resulting in a miniature Littrow mounting. The first waveguide demultiplexer using a Rowland configuration was demonstrated by Tangonan et al.[80] who bonded a flexible grating sheet on a curve-polished edge of an ion-exchanged planar waveguide. Similar devices were reported in the early 1980s.[81–83] To avoid time-consuming curve edge polishing, a demultiplexer with a reflection-type chirped (focusing) grating attached to a flat chip edge,[84] and a device using a collimating lens and ordinary grating attached to a chip edge[85] were reported. However, performance of these early devices was insufficient for practical WDM application. For applications with less demanding specifications, such as spectroscopy and sensing, compact miniature spectrometers based on bulk optics elements have been successfully used.*

The first demultiplexers based on gratings etched in a slab waveguide were demonstrated in the early 1990s in various waveguide platforms: SiO_2 glass waveguides

* See, for example, http://www.oceanoptics.com/.

on Si substrate (Siemens AG),[86,87] InGaAs/AlGaAs/GaAs (NRC Canada),[88,89] and InGaAsP/InP (Siemens AG,[90,91] Bellcore,[92,93] and NRC Canada.[94] Bellcore also reported a multiwavelength laser integrating an EG demultiplexer within the laser cavity.[95] AT&T later demonstrated a similar device using an AWG as the wavelength-selective element.[96] Most of these early EG devices used reflective gratings in a Rowland configuration similar to that used in state-of-the-art waveguide EG devices (Figure 5.3a). Some used a so-called flat field design[97] with a spectrum formed along a straight line (rather than a curve) directly at the chip end face[86,87] or a transmission waveguide grating configuration.[98] The main problems with these early devices were their large losses and polarization sensitivities. Large polarization-dependent peak wavelength shifts of the demultiplexed channels were caused by slab waveguide birefringence, and the polarization-dependent loss (PDL) was due mainly to intrinsic polarization dependency of grating diffraction efficiency (see Section 5.2.3.2). These problems were resolved by arrayed waveguide grating (AWG) devices[99] that soon served as the dominant optical mux/demux technology.

Several problems observed with waveguide EGs made many skeptical about their usefulness as WDM mux/demux devices. For example, a book on diffraction gratings by Loewen and Popov[100] states: "Otherwise, planar waveguide gratings serve for this application* mainly as an intellectual exercise or when efficiency is not important . . . These reasons are enough to deprive truly planar gratings of practical usefulness for dense wavelength demultiplexing." Obviously this statement no longer applies.

5.2.3 Advanced Planar Waveguide Echelle Grating Devices

Recently, planar waveguide echelle grating devices were developed in silica-on-silicon[13,101] by Optenia Inc. and InGaAsP/InP[14] by MetroPhotonics Inc., with performances comparable to arrayed waveguide gratings and significantly smaller sizes. The schematic of a 48-channel EG device developed by Optenia Inc. is similar to Figure 5.3a. This was the first device that minimized to the level acceptable for WDM applications the grating loss, the PDL, and the polarization-dependent spectral shift. The device was fabricated in SiO_2 glass waveguides on a silicon substrate, i.e., a silica-on-silicon waveguide platform. The SiO_2 waveguide core layer was doped with a small amount of phosphorus to provide a refractive index increase of $\Delta n = 0.012$ in the core compared to the cladding. All layers were grown by plasma-enhanced chemical vapor deposition (PECVD) on silicon substrates. The input and output waveguides were buried channel waveguides with $\sim 5 \times 5\,\mu m^2$ cross-section; the slab waveguide had a 5-μm thick core surrounded by the upper and lower cladding SiO_2 layers. The input and output waveguides joined the slab waveguide along the Rowland circle.

The light from the input waveguide was coupled into the slab waveguide where it diverged in the waveguide plane and illuminated the grating. The light was diffracted backward from the concave grating and different wavelengths were focused at

* Referring to WDM.

different positions along the Rowland circle where the output waveguides were arrayed with a spacing of approximately 14 μm to give a 100-GHz spacing between adjacent demultiplexed channels according to Eq. (5.4). The grating was formed by etching a $\sim$ 10-μm deep vertical trench (90° within $\pm 1°$) through the waveguide layers (see Figure 5.3b) using an anisotropic etching process based on C_4F_8 and argon gases. As will be discussed later, the grating verticality is essential for achieving acceptable device loss and crosstalk. The grating teeth have a 6-μm wide reflective facet (*RF*, Figure 5.3a) and a 10.7 μm long nonreflective facet (*NRF*, Figure 5.3a), resulting in a grating order of $m = 20$, as it follows from the constructive interference condition: for order m in near-Littrow incidence, $2L_{NRF} \sim m\lambda_c/n_{eff}$, where $\lambda_c =$ 1550 nm is the demultiplexer central wavelength, n_{eff} is the effective index of the slab waveguide fundamental mode, and L_{NRF} is the length of the nonreflecting facet.

The device has a Gaussian passband spectrum, and the size of the unpackaged device is 18×17 mm^2. The measured adjacent channel crosstalk is better then 35 dB, the PDL is less than 0.2 dB across the entire C-band, and the polarization-dependent wavelength shift is less than 10 pm. These parameters are comparable to AWG devices, but device size is significantly smaller. Typical fiber-to-fiber insertion loss was $\sim$ 4 dB – about 2 dB larger than those for the best AWGs.[102] This loss penalty was partly due to waveguide propagation loss and it is expected to be further reduced by using optimized low-loss silica-on-silicon platforms.

It was also demonstrated that the 48-channel 100-GHz EG demultiplexer can be scaled up to 256-channels with a 25-GHz spacing simply by increasing the focal length of the grating from 7.5 mm (48-channel device) to 20 mm (256-channel device).[13] The 256-channel device has adjacent channel crosstalk of 30 dB, insertion loss of 10 dB, and a polarization-dependent wavelength shift less than 10 pm. Using an AWG on a waveguide platform with the same index step, a comparable 256-channel device would occupy about ten times the area of the EG device.

We will now discuss the critical issues in waveguide EG devices for WDM applications.

5.2.3.1 Slab Waveguide Birefringence

In dielectric slab waveguides, the two distinct sets of linearly polarized solutions of Maxwell equations are known as the transverse electric (TE) and transverse magnetic (TM) modes. TE modes have the electric field polarized in the plane of the waveguide (along the y axis in Figure 5.2a). Electric field distributions (E_y) and the propagation constants β of different TE modes can be found as a solution of the wave equation subject to boundary conditions at the interfaces between the waveguide core and the upper and lower claddings. The other field components are $H_y = 0$, $H_x = -\beta E_y/(\omega\mu_o)$, $H_z = -(i/\omega\mu_0)(dE_y/dx)$, where $k = \omega(\varepsilon_0\mu_0)^{1/2} = \omega/c$ is the modulus of the wave vector, ε_o and μ_o are the vacuum dielectric permittivity and magnetic permeability, $\omega = kc$ is the light angular frequency, and c is the speed of light in vacuum.

TM modes have the magnetic field polarized in the plane of the waveguide (y axis). H_y is found by solving the wave equation, and other field components are $E_y = 0$, $E_x = \beta H_y/(\omega\varepsilon_o n^2)$, and $E_z = (i/\omega\varepsilon_0 n^2)(dH_y/dx)$. Each of the TE and TM modes has a distinct propagation constant β, hence a distinct mode effective index $n = n_{eff} = k/\beta$.

Slab waveguides in the EG devices are often designed to be single mode for TM and TE polarizations. Because optical fibers deployed in telecommunication networks do not preserve light polarization, it must be assumed that the light coupled from the fiber to the demultiplexer chip does not have a defined polarization state and both TM and TE polarized fundamental modes can be excited in the slab waveguide.

Because TE and TM modes have different propagation constants (β_{TE} and β_{TM}) and effective indices ($n_{TE} = k/\beta_{TE}$ and $n_{TM} = k/\beta_{TM}$), they will propagate in slightly different directions $\theta_{TE} \neq \theta_{TM}$ (see Eq. 5.1) upon diffraction by the grating, and will arrive at different focal positions near the Rowland circle. This results in a wavelength shift* of $\Delta\lambda \sim \lambda\Delta n/n_g$ between demultiplexer spectra for TE and TM polarized light, where $\Delta n = n_{TE} - n_{TM}$ is the waveguide birefringence, and $n_g = n - \mathrm{d}n/\mathrm{d}\lambda$ is the group index. Since this birefringence originates from the different slab waveguide geometries along the x and y axes and hence different boundary conditions for light polarized along these axes, it can be referred to as waveguide geometrical birefringence.

The wavelength shift $\Delta\lambda$ can cause interchannel crosstalk, and even if $\Delta\lambda$ is small compared to channel separation, it can produce significant PDL over the channel passband. This difference in the mode propagation constants will also result in a small difference in the propagation times for the TE and TM modes, or polarization mode dispersion (PMD), but because of the small chip size this effect is negligible at the modulation frequencies presently used in WDM systems.

We have assumed isotropic refractive indices for the core and cladding layers. This assumption is, however, not realistic for a glass waveguide deposited on a silicon substrate that often has large stress-induced refractive index anisotropy. This is because of the high temperature anneals required to form silica films for optical waveguides. Both plasma-enhanced chemical vapor deposition (PECVD) and flame hydrolysis deposition (FHD) require high temperature anneals. Temperatures of 900 to 1100°C are used in PECVD to consolidate and homogenize the glass and to quench strong absorption lines often present near 1480 nm and 1510 nm associated with overtones of Si–H and Si–N bonds.† Even higher temperatures are required for FHD because glass is formed from fine ($\sim$0.1 μm in size) silica particles that need to be sintered into a homogeneous film at 1100 to 1300°C, similar to the vertical axial deposition process used for making optical fiber preforms in which FHD has its origin. As the wafer cools to room temperature, the different thermal expansion coefficients of the Si wafer ($3.6{\cdot}10^{-6}\mathrm{K}^{-1}$) and of the SiO_2 glass ($5.4{\cdot}10^{-7}\mathrm{K}^{-1}$) layers make the silicon wafer contract more than the glass. This creates a large compressive in-plane stress in the glass that in turn modifies the glass refractive index via the elasto-optic effect. The refractive index change induced by stress σ for light polarized along the i axis is:

$$\Delta n_i = -C_{ij}\sigma_j \tag{5.9}$$

* In the literature also known as PD-λ, or PDW, i.e., polarization dependent wavelength.

† Hydrogen and nitrogen originate from SiH_4 and N_2O gases used in the PECVD process.

where C_{ij} is the stress-optic tensor.[103] In a material with an initially isotropic refractive index n, only elements C_{11} and C_{12} are nonzero, and are related to the strain optic tensor p_{ij} by $C_{11} = (n^3/2E)(p_{11} - 2\mu p_{12})$ and $C_{12} = (n^3/2E)(p_{12} - \mu p_{12} - \mu p_{11})$, where μ is the Poisson's ratio and E is the Young's modulus of the material.

In glass slab waveguide layers grown on a silicon substrate, it can be assumed that stress in the direction normal to the layers is negligible ($\sigma_3 = \sigma_\perp = 0$) because the layers are constrained by the silicon substrate only in the in-plane direction for which $\sigma_1 = \sigma_2 = \sigma_{||}$. The stress-induced refractive index changes in the waveguide plane $\delta n_{||}$ and perpendicular to the waveguide plane $\delta n_\perp$ are therefore $\delta n_{||} = -(C_{11}+C_{12})\sigma_{||}$ and $\delta n_\perp = -2C_{12}\sigma_{||}$, and the resulting stress-induced birefringence is $\Delta n_s = \delta n_{||} - \delta n_\perp = (C_{12} - C_{11})\sigma_{||}$. The effective indices of the modes with electric field parallel (TE polarization) and perpendicular (TM polarization) to the waveguide plane will hence be shifted by $\delta n_{||}$ and $\delta n_\perp$, respectively, and the waveguide effective index birefringence $\Delta n = \Delta n_{TE} - \Delta n_{TM}$ will be correspondingly shifted by Δn_s.

A stress-induced birefringence of $\Delta n_s = -1.3 \cdot 10^{-3}$ is predicted from Eq. 5.9 for a compressive stress of $\sigma_{||} \sim -300$ MPa typically found in PECVD silica glass on silicon and the following parameters for fused silica: $p_{11} = 0.121$, $p_{12} = 0.270$, $E = 70$ GPa, and $\mu = 0.17$.[104] Measurements in our laboratory on silica-on-silicon slab waveguides from various vendors confirmed birefringence values ranging from $-4 \cdot 10^{-4}$ to $-1 \cdot 10^{-3}$, about two orders of magnitude larger than the waveguide geometrical birefringence in the absence of stress.

The slab waveguide birefringence was measured with an accuracy of $\sim 10^{-6}$ by a technique proposed by Janz by which birefringence is obtained from the measured period $\Lambda = 2\pi/(\beta_{TM} - \beta_{TE})$ of the fringe pattern arising from the coherent superposition of Rayleigh scattered light originating from the TE and TM modes observed at 45° from the waveguide plane normal.[105,106] A stress birefringence of $-1 \cdot 10^{-3}$ causes a polarization dependent wavelength shift of $\Delta\lambda = 1$ nm, which is in a demultiplexer with 100 GHz (~ 0.8 nm) channel spacing obviously intolerably large.

In silica-on-silicon AWG devices, the wavelength shifts are usually controlled to a level of $\Delta\lambda < 0.01$ nm. Unfortunately, techniques used for reducing polarization sensitivity in AWGs often cannot be applied for EG devices. For example, inserting a half-wave plate in the middle of the phased array is often used in AWGs (see Section 5.3.2.3)[107] but there is no such symmetry plane for inserting a half-wave plate along the light path in an EG device. In AWGs based on glass waveguides, in-plane stress can be balanced by applying stress in the direction parallel to the buried channel vertical sidewalls (Figure 5.2b) using a cladding material with a thermal expansion coefficient (TEC) matched to the silicon substrate.[108,109] This can be achieved by including B and P dopants in silica glass (BPSG) cladding. This stress balancing obviously cannot be applied to slab waveguides (because they do not have vertical sidewalls). Using a waveguide core material such as BPSG with TEC matched to the Si substrate[110,111] can eliminate TEC mismatch-induced stress not only in buried waveguides but also in slab glass waveguides. However, BPSG with TEC matched to Si has higher optical loss and may degrade with time.

The polarization compensator shown in Figure 5.3b is an effective solution to the problem of polarization-dependent wavelength shift in glass waveguide EG devices. The compensator consists of a prism-shaped region (Figure 5.3a) in which the waveguide birefringence is modified.[13,112] TE and TM polarized light is refracted at the compensator prism boundary at slightly different angles to compensate for $\Delta\theta = \theta_{TE} - \theta_{TM}$ in the unmodified slab waveguide, so that both polarizations eventually arrive at the same position at the Rowland circle.

Waveguide effective index and birefringence can be modified by partially etching the waveguide layers, as demonstrated by He et al.[113] We later applied this technique to silicon-on-insulator (SOI) AWGs.[114] However, to fully compensate for birefringence in glass waveguides, all the cladding and even part of the core may need to be etched away. This introduces a loss penalty due to mode mismatch between the compensator and the unmodified slab waveguide regions. This problem was resolved by etching only a part of the cladding, leaving approximately 1 μm of the cladding over the core, and capping the thinned cladding by a compensating overlayer with a high refractive index[115] that can be a few hundred nanometer thick film of silicon nitride with $n \sim 2$,[13,112] as shown in Figure 5.3b. The TE and TM mode evanescent tails interact differently with the high index layer, causing the TE effective index to increase relative to the TM index, hence increasing the birefringence in the compensator region. Because the mode shapes in the compensator and the unmodified waveguide are similar, the insertion loss and PDL penalty are negligible. Using this technique, the polarization-dependent wavelength shifts were routinely reduced from about 0.5 nm to 0.01 nm.

This birefringence modification by a thin capping layer of high refractive index can be regarded as an extension of the concept of form birefringence.[116] It has been known for more than a century[117] that a specific birefringence can be created by combining materials of different refractive indices at scales smaller than the wavelength of light. For example, interleaving thin layers of two materials of thickness a and b and refractive indices n_a and n_b results in different effective indexes for light polarized parallel and perpendicular to the layers: $n_{\parallel} \sim [(an_a^2 + bn_b^2)/(a+b)]^{1/2}$ and $n_{\perp} \sim n_a n_b[(a+b)/(an_b^2 + bn_a^2)]^{1/2}$, first derived by Ritov[118] and consistent with recent results by Gu and Yeh[119] and Lalanne and Hugonin.[120] Form birefringence has been used for phase matching wavelength conversion processes in waveguides,[121] polarization compensation in silicon microphotonic devices,[122] and glass buried ridge waveguides.[123] Form birefringence is also found in inorganic (e.g., mica) and organic (e.g., tobacco mosaic virus crystals)[124] materials.

5.2.3.2 Polarization Dependence of Grating Diffraction Efficiency

Calculation of diffraction efficiency of echelles is considered one of the most intractable problems in the electromagnetic theory of gratings[125] because of the numerical difficulties posed by the large number of diffraction orders, deep triangular grooves with high blaze angles, and small λ/Λ ratios. It was often assumed in the past that because of the low λ/Λ ratios of EGs, polarization effects played a minor role and scalar theories were sufficient for predicting the behavior of echelles with sufficient accuracy. Significant divergences between theory and experimental data were often incorrectly attributed to experimental uncertainties rather than limits of the scalar approach.[125] The insufficiency of the scalar approach was confirmed in the first theoretical (electromagnetic) and experimental investigations of EGs performed by Loewen et al.[126] Now it seems commonly accepted that boundary integral equation methods (BIMs)[125–127] are well suited for the simulation of echelles.

The impact of an EG profile on demultiplexer performance was studied with scalar[128] and vector diffraction theories.[129,130] In contrast to the vector theories, scalar theories failed to show the influence of profile imperfections on polarization sensitivity. It was theoretically shown[130] that polarization-dependent loss can be reduced to less than 0.3 dB for a 32-nm bandwidth by rounding the grating corners with a specific radii. However, this is achieved at the cost of an increased diffraction loss and increased fabrication challenges when making the grating with corners of specific radii.

The problem of polarization dependency of diffraction efficiency in metallized waveguide echelle gratings was resolved by eliminating the metal on the nonreflecting facets (Figure 5.3a, inset).[13] This can be achieved, for example, by removing metal from the nonreflecting facets of a fully metallized grating (e.g., by a directional reactive ion etch) or by metal deposition only on the reflecting facets. The physics behind the polarization dependency of diffraction efficiency in metallized gratings concerns the different boundary conditions for TE and TM polarizations at different facets (see Figure 5.1c). At the reflecting facet, both TE and TM polarizations have electric fields parallel to the facet, so that the same boundary conditions (continuous tangential component of the electric field when crossing the dielectric–metal boundary) apply for TE and TM waveguide modes and total electric field in the waveguide near the metallized boundary is zero (incoming $\boldsymbol{E}$, reflected -$\boldsymbol{E}$), as required for a perfectly conducting metal. The situation is different at the nonreflecting facet. The boundary condition (continuous, hence zero tangential electric field when crossing the boundary) is automatically satisfied for the TE mode but not for the TM mode. For the latter, the boundary condition is satisfied by forming a disturbance that can be interpreted as a surface current that excites various diffraction orders.

This was studied for a waveguide EG grating demultiplexer by Delâge and Dossou[131] who solved Maxwell's equation in the region close to the grating using a finite element method (FEM), and matched this numerical solution near the grating surface to a Rayleigh expansion[132] into plane waves that describes the grating diffraction in the far field but its rigor fails in close proximity to the grating surface.[133] To obtain a good matching of the two solutions, more than thirty orders are needed, including eight evanescent orders (i.e., with imaginary propagation constants). This energy transfer into other diffraction orders for TM polarization is the main cause of PDL in metallized EG devices and it can be alleviated by avoiding metallization on nonreflecting facets.[13,134] This conclusion was confirmed by Shi et al.[135] using the rigorous coupled wave theory.[136,137] The PDL problem in gratings with both facets metallized can also be alleviated by inserting a dielectric spacer layer between the waveguide core and the metal, as proposed by Xu et al.[138]

Another possibility is avoiding metal completely by using total internal reflection (TIR) in retroreflecting grating teeth.[139–141] Because this requires two reflections per tooth and the number of corners is doubled, the losses due to grating imperfections such as verticality error, sidewall roughness, and corner rounding are increased. Furthermore, in glass waveguides, the critical angle for TIR at the glass–air interface is ~42° ($n = 1.5$), so that TIR is effective only in a limited range of diffraction angles near Littrow incidence. This reduces the wavelength range compared to waveguides with high refractive indices (e.g., InP) in which TIR can accommodate a wider range of angles.

Still another possibility to minimize the polarization dependence is to overcoat the grating teeth with a multilayer dielectric mirror with alternating high and low refractive index layers, each of optical thickness $\lambda/4$. Dielectric multilayer coatings are often used in deep-uv or x-ray gratings because no good reflectors below 100-nm wavelength are available. In the case of the so-called Bragg–Fresnel grating,[142] the grating is etched directly in a multilayer substrate. At optical wavelengths, however, a problem arises because of resonance anomalies in diffraction efficiency caused by excitation of the modes guided in the multilayer dielectric stack.[143] This is further exacerbated by grating roughness that can phase match the incident wave and the waveguide modes in the multilayer. These resonance anomalies can be alleviated by reducing the grating period, but the latter obviously limits the operating order for gratings.

5.2.3.3 Grating Loss

The ability to etch grating sidewalls with a minimum deviation from the vertical is essential in reducing insertion losses. A simple estimate of loss incurred upon reflection on a nearly vertical sidewall etched in the waveguide at an angle of $90° + \alpha$ can be obtained from the overlap integral of the fundamental waveguide mode with the same mode tilted by 2α upon reflection:[77] $\varsigma(\alpha) = \left|\int E^2(z)\exp\left[i\beta z\sin(2\alpha)\right]dz\right|^2$, where z is in the direction perpendicular to the waveguide plane (see Figure 3a), β is the mode propagation constant, and E is the normalized electric field amplitude of the mode. In glass waveguides with an index step of $\Delta n = 0.012$, a grating wall offset of 1° from the vertical produces a coupling loss of about 0.5 dB from the incident mode to the reflected mode. Evaluating this loss for the same grating made in a silicon-on-insulator waveguide with an index step of $\Delta n \sim 2$ shows that keeping the excess loss below 0.5 dB still demands nearly vertical sidewalls ($\alpha < 1.5°$). The ability to etch nearly vertical sidewalls is thus essential, irrespective of the chosen waveguide platform. Another issue that must be considered is the influence of the verticality error on the coupling to the higher order waveguide modes producing ghost peaks deteriorating the crosstalk.

Reactive ion etching (RIE) is commonly used for vertical etches. In RIE, etch profiles can be controlled by balancing the chemical (isotropic) and physical (anisotropic) etch components. For a given type of RIE reactor, the etch profile depends on the process chemistry, pressure, temperature, ion density, and ion energy.[144] The etching uses halogens (e.g., Cl_2, Br_2), halogenated (e.g., HBr), or fluorocarbon (e.g., C_4F_8, CF_4, CHF_3) etchant gases, depending on the material to be etched. For the relatively deep etches needed in glass waveguide EG devices ($\sim$10 μm), high etch rates are preferred.

To improve anisotropy of the etch and hence sidewall verticality, a second gas can be added to passivate the sidewall by forming a nonvolatile etch by-product. The latter is preferentially removed from the horizontal areas by directional ion bombardment, but not from the vertical sidewall, hence protecting the latter from chemical etching and resulting in a vertical sidewall.

High etch rates can be achieved by increasing the ion energy that determines the physical etching component, but this is not desirable because it can damage the mask (e.g., mask edge faceting and reduced etching selectivity) and the sample. Using high energy ions also makes it difficult to control the sample temperature that also determines the sidewall angle.[145]

Higher ion densities also result in higher etch rates. Both the anisotropic nature and the verticality of the etch are improved if the ions have low energy and are incident normal to the surface. The directionality of ion trajectories can be improved by using low pressure processing since this increases the mean free path of the plasma species and decreases the number of collisions. Low pressure also increases the volatility of the etch by-products that can, if re-deposited on the sample, contaminate the etched surface. ICP (inductively coupled plasma) RIE systems, in which plasma excitation is provided by an independent radio frequency (RF) source, can provide low and controllable ion energy and high ion density that is adjustable independently from ion energy, and operate at low pressures. These are regarded as desirable conditions for deep vertical etching. ICP RIE techniques for etching deep vertical grating facets in both glass and InP waveguide EG devices have been developed and are discussed in Chapter 7 by Lamontagne.

Reflectivity reduction due to mirror imperfections such as metal absorption and sidewall roughness also contributes to insertion loss. A reflectivity higher than $R = 0.9$ near normal incidence, hence less than a 0.5-dB loss penalty compared to a perfect mirror, has been demonstrated in trenches deeply etched in glass slab waveguides and metallized with Al.[13]

Yet another source of grating loss is the rounding of teeth corners. Light is dispersed by the rounded corners into a wide range of angles and may propagate in the slab waveguide as stray light or unwanted orders that, if intercepted by the output waveguides, increase the noise floor and contribute to the crosstalk. Because corner rounding affects the TE and TM polarizations differently,[130] it is also a source of PDL. The rounding effect can be alleviated by reducing the ratio between the rounded and straight areas of the grating teeth by increasing the length of the facets, thus working in higher orders. Longer facets, however, result in a narrower far field envelope (slit function, Eq. 5.2 and Figure 5.1b). The increased insertion loss nonuniformity or roll-off should be kept within a fraction of a decibel across the operating wavelength range of a demultiplexer. The corner rounding is due to lithographic resolution limits and the limited fidelity with which the mask pattern can be transferred to the waveguide by etching. The lithography limit can be improved by a slight modification of the corner shape on the etch mask. Such etch bias correction techniques used specifically shaped small compensatory patches (called serifs) near the grating corners.

5.2.3.4 Influence of Grating Teeth Positions on Crosstalk and Passband

If an error is introduced in the facet position from its ideal location, the phase relationship between the light reflected from different facets is altered. The distortion of the optical phase and intensity distribution along the focal line will cause crosstalk and widen the passband. The main source of errors in the grating facet positions is mask pixelation. For a mask writing machine with a step size of 100 nm, the pixelation error in the position of each facet is ±50 nm, and the resulting crosstalk is about 20 dB for a demultiplexer in InP.[146] A pixelation-limited crosstalk of less than 45 dB has been calculated for the same waveguide platform for a step size of 5 nm;[146] the latter can be achieved with modern e-beam machines or steppers.

The EG crosstalk is largely governed by the phase errors $\delta\phi$ related to the facet position errors δx by $\delta\phi = 2k\delta x = 4\pi n_{eff}\delta x/\lambda$. Therefore using a waveguide platform with a lower index can tolerate a larger pixelation error for the same crosstalk. This results in an increased tolerance (approximately two times) to errors in facet positions for EG devices in glass ($n \sim 1.5$) compared to InP ($n \sim 3.18$). Compared to AWGs, it is an advantage of EGs that the main source of their phase errors results from geometrical errors in the positions of the facets. In an AWG, the relative phase of the light propagating in different waveguides depends on the geometrical length of the arrayed waveguides,[147] waveguide cross-section dimensions, refractive index and thickness of waveguide layers, and stress (see Section 5.3.2.1). The influence of these imperfections increases with the size of the waveguide array and with decreasing channel spacing, hence EGs appear to be better choices over AWGs for WDM systems with large numbers of densely spaced channels, for example, hundreds of channels spaced at 25 or 10 GHz.

The grating is preferably written in one field. Fields sizes on the order of 5 cm^2 are readily achieved with stepper projection lithographs. For earlier EG devices with patterns exposed by stitching several (N_s) smaller fields, stitching errors must be considered. An error in stitching accuracy results in an EG spectrum with satellite peaks separated from each other by approximately FSR/N_s with a magnitude of the peaks decreasing with distance from the main peak.[148]

Small and controllable changes of the grating teeth positions can be used for reducing device aberrations (stigmatization, see Section 5.2.1) and also for widening and flattening the channel passband. By intentionally introducing phased shifts at the grating via displacing (dithering) the grating teeth by a small and controlled amount, a specific phase and intensity distribution can be created along the focal line (Rowland circle). The same effect is achieved in an AWG by dithering the lengths of the waveguides in the phased array.

For example, a top-hat or a double peak intensity distribution can be created that, after convolution with the output waveguide mode profile, results in a flatter and a wider passband compared to the original Gaussian profile. It is desirable to have the (de)multiplexer transmittance constant inside the specified passband. The passband must be wide enough to provide tolerance to laser wavelength drifts or variations in passband peak positions due to manufacturing errors, temperature changes, or polarization-dependent wavelength shifts. The edges of the passband must be steep to avoid crosstalk with adjacent channels. This also helps to widen the cumulative passband when cascading multiple filters. This is particularly important in flexible advanced optical networks through which optical signals pass, compared to conventional point-to-point WDM networks, more frequently through several stages of (de)multiplexers without being regenerated.

A typical requirement based on network engineering parameters is to ensure a 1-dB bandwidth no narrower than 50% of the channel spacing.[149] Various passband flattening and widening techniques applicable for EG and AWG devices are known, including sinc function amplitude modulation in the dispersive element producing a top-hat distribution in the focal region,[150] input waveguides with parabolic tapers[151] or multimode interference couplers,[152] interleaved gratings,[153,154] cascaded gratings and Mach–Zehnder interferometers,[155] gratings with amplitude modulation and phase dithering,[156,157] and gratings with phase dithering only.[158,159] The advantage of the latter is that it can be implemented simply by dithering the grating facet positions. The passband is flattened, broadened, and its transitions steepened, and the chromatic dispersion can also be minimized. The chromatic dispersion is determined by the group delay $\tau(\omega) = -d\varphi/d\omega = (\lambda^2/(2\pi c))(d\varphi/d\lambda)$, where $\varphi(\omega)$ is the device frequency phase response for light of angular frequency ω. The quoted chromatic dispersion $D = d\tau/d\lambda$, typically measured in picoseconds per nanometer, is the maximum group delay difference within the passband. D is often larger in EGs than in AWG devices due to symmetry considerations.[160]

A typical AWG has symmetric intensity distribution across the phased array and as such it pertains to a subclass of finite impulse response (i.e., with no feedback) digital filters known as linear phase filters. Such filters produce linear phase responses irrespective of the amplitude response (the latter can be, for example, a top-hat function), pertaining to a category of non-minimum phase (non-MP) filters. In contrast to non-MP filters, minimum phase (MP) filters have the amplitude response that uniquely determine the phase response and vice versa. The two are related by the Hilbert transform (Kramers–Krönig relation). In an AWG with symmetrical field distribution (non-MP filter), group delay is nearly flat within the channel passband; hence $D \sim 0$. The condition of symmetric intensity distribution is not satisfied in EGs because the grating facets are located on an angled curve and the facet widths vary asymmetrically with respect to the central facet (hence there is no symmetry plane), so that EGs are MP filters with varying group delays within the channel passband, hence $D \neq 0$. Furthermore, passband flattening techniques tend to induce coupling between the amplitude and phase responses (see Section 5.3.2.5).[160]

The facet dithering technique[159] was demonstrated in a 40-channel EG device in InP with a measured passband of 41 GHz at -1 dB, more than double the Gaussian passband (18 GHz at –1 dB) of a similar device.

5.2.4 Current Trends in Waveguide Echelle Grating Devices

In addition to the previously discussed 40- and 256-channel silica-on-silicon waveguide demultiplexers, various other EG-based devices have been proposed by Optenia Inc.,[13,101] including an integrated double pass equalizer,[161] an optical performance monitor,[162] and an optical add–drop multiplexer.[163] MetroPhotonics Inc. developed similar InP-based EG devices for real-time optical power monitoring[164] and dynamic channel equalization[165] operating on a 40-channel 100-GHz spacing WDM frequency grid.

The optical power monitor[164] integrates an EG demultiplexer built into a passive slab waveguide, schematically similar to that shown in Figure 5.3a, with an evanescently coupled *pin* photodetector array in an active receiver waveguide section. The passive section has a 0.6-μm thick transparent InGaAsP core (with a bandgap wavelength of $\lambda_g = 1\mu m$) sandwiched between 1-μm thick upper and lower InP claddings. The active section comprises a *pin* photodiode formed by vertically stacking a bottom *n*-doped InP layer, a thin intrinsic GaInAs absorption layer, and an upper *p*-doped InP layer. These layers are grown atop a transparent waveguide being common with the passive section that assures an efficient mode coupling between the two sections.[166] The evanescent tail of the mode is absorbed in the absorption layer as it propagates along the active waveguide section, and the generated photocurrent is collected by the reverse-biased pin diode.

The small overlap of the mode with the absorption layer (hence a small absorption per unit length) is compensated by elongating the active waveguide portion at a cost of increased *RC* time constant and reduced operation speed. This is not a concern for WDM optical power monitoring or control where high speed is not required. All the layers are grown by metal organic chemical vapor deposition (MOCVD). Both the passive and active sections are single mode waveguides with matched TE and TM modes with efficient wavelength-independent coupling between the two sections.

Voltage-controlled light attenuation in the absorption layer of EG devices was also demonstrated based on the Franz–Keldysh (bulk materials) and the quantum-confined Stark effect (quantum wells). The first devices were polarization-dependent as a result of different confinement factors Γ of the TE and TM modes in the absorption layer, with $\Gamma_{TE} \sim 1.6\Gamma_{TM}$. Densmore et al. suggested[167] eliminating the polarization dependence by replacing the bulk absorption layer with a tensile-strained quantum well layer with a higher absorption for TM compared to the TE mode. In a voltage-controlled attenuation device, the bandgap of the absorption layer is chosen to be slightly above the photon energy. Applying the external electric field perpendicular to the absorption layer increases the rate of interband transitions (light absorption) by orders of magnitude in the absorption layer. In the EG devices, electrically controlled absorption can be used for optical power control, equalization, and power stabilization. Accuracy of optical attenuation of better than 0.1 dB and a fast response time of ~ 40 ns have been reported.[167]

An InP-based 32-channel waveguide EG WDM data receiver[168] with a photodetector array optimized for high speed operation, a -3 dB bandwidth exceeding 1.25 GHz and a triplexer chip[169] with a waveguide EG integrated with high speed photodetectors have been demonstrated. The triplexer receives up to 1.7 GHz video and 2.5 Gb/s data in the 1550- to 1560-nm and the 1480- to 1500-nm wavelength ranges, respectively, and transmits and monitors data at up to 2.5 Gb/s in the 1280- to 1340-nm range.

A waveguide EG comb filter working in a high order of $m \sim 2000$ with a free spectral range of 100 GHz has been reported by LNL Optenia.[170,171] Comb filters are used in WDM networks to reduce cumulative crosstalk and improve adjacent channel isolation and they also can be employed as stable frequency references. The large orders result in very long (hundreds of microns) grating facets and if the facets are straight, the angle of incidence necessarily varies along the facet, resulting in a reduced diffraction efficiency and asymmetric spectral response. Blazing can be optimized for any position at the facet by curving the facets along a family of ellipses with foci located at the input and the central output waveguides, as shown by Packirisamy and Delâge.[172] Although the comb filter reported by Bidnyk et al.[170] was fabricated in a planar silica waveguide with low index contrast ($\Delta n = 0.75\%$), the total die areas including the input and output waveguides were only 0.57 cm^2 and 2.4 cm^2 for a Gaussian and flat-top passband filters, respectively.

A waveguide EG demultiplexer has also been demonstrated on a silicon-on-insulator (SOI) platform,[173] but the crosstalk (16 dB), on-chip loss (–10 dB) and polarization-dependent wavelength shift ($\Delta\lambda \sim 0.13$ nm) must be substantially improved before envisioning a similar device for practical applications. Because of a very large refractive index difference between the silicon waveguide core (n ~ 3.476 at 1550 nm) and the SiO_2 cladding (n ~ 1.5), the slab SOI waveguide is multimode unless the core thickness is reduced to $H \sim \lambda/[2(n_{core}^2 - n_{clad}^2)^{1/2}] \sim 0.25\mu$m.

Because the higher order slab modes have different propagation constants, they arrive at different positions at the demultiplexer focal line, degrading the crosstalk. Reduction of multimode effects is crucial for SOI-based EGs.[174] Excitation of higher order modes at the junction between the input and the slab waveguides and at the grating trench must be reduced to –30 dB or less, unless efficient solutions to mode filtering in multimode slab waveguides are developed. Using a thin single mode SOI slab waveguide is a possibility but problems with high coupling loss between the fiber and the chip and high propagation losses in SOI waveguides of submicron cross-sections must first be resolved.

A waveguide grating spectrometer has also been demonstrated in a high index contrast silicon oxynitride (SiO_xN_y) platform.[175] The device used a transmission grating, also called a grating prism or *grism*, with a diffraction efficiency of > 60%, a spectral range of $\lambda = 350$ to 650 nm, and a resolution of 9 nm. In comparison with reflection gratings, transmission gratings have larger grating facets,[176] but since the phase delay factor is smaller than in reflection, dispersion is decreased.

The significant size advantage of EG compared to AWG devices allows for higher integration densities, more functions per chip, and a larger number of spectral channels with denser frequency spacing. This makes waveguide EG devices promising not only for WDM applications, but also in other fields, particularly spectroscopy, sensing, and future wavelength multiplexed optical interconnects.

5.3 ARRAYED WAVEGUIDE GRATINGS

The problems that plagued early waveguide EG devices were circumvented by M.K. Smit's invention in 1988 of the arrayed waveguide grating (AWG),[99] also known as the phasar (phased array) or waveguide grating router (WGR). The AWG soon became a key device in WDM optical communication systems for performing functions such as wavelength multiplexing and demultiplexing, wavelength filtering, signal routing, and optical crossconnects, among others. Numerous review articles and textbooks describe AWG theory, fabrication, and applications.[15,177–181] In the limited space of this section, after a brief explanation of AWG operation principles, we will discuss recent developments in AWG technology, with an emphasis in Section 5.4 on compact AWG devices that use waveguides with large refractive index contrasts.

5.3.1 AWG Principle

The AWG operates similar to the microwave-phased array antennas often used in space communications and satellite tracking. By controlling the phase relationship between the beams emitted by the individual elements of the array, antenna radiation direction can be changed without mechanical movements.

Figure 5.4a is a schematic of an AWG demultiplexer. Light comprising different wavelengths is coupled from an optical fiber to the input channel waveguide (Figure 5.2b) where it propagates toward the input combiner — a slab waveguide (Figure 5.2a) that confines the light in a direction normal to the waveguide plane. In the in-plane direction, light propagates as it would in free space, diverging to illuminate the waveguide array. The field profile at the junction between the input waveguide and the slab waveguide is typically Gaussian,* so that the divergence half-angle α in the input slab waveguide can be estimated from the Gaussian diffraction formula as $\alpha = 2\lambda/(\pi w n_{eff})$ where w is the Gaussian beam waist (mode diameter) in the input channel waveguide. α is the angle between the light propagation direction and the Gaussian far field asymptote of the $1/e^2$ irradiance surface.

The light couples from the input combiner into the array of waveguides† that start along an arc centered at the input waveguide–combiner junction, with a radius equal to the focal length of the combiner. This ensures that the light injected from the central input waveguide arrives at the beginning of each of the arrayed waveguides with the same phase. For a waveguide array with a constant length difference ΔL between the adjacent waveguides and effective index n_{eff}, the light arrives to the end of the array with a phase difference between the adjacent waveguides of $\Delta\varphi_{AWG} = 2\pi\,\Delta L n_{eff}/\lambda$. This phase shift is $2\pi m$ for demultiplexer central wavelength

$$\lambda_c = n_{eff}\Delta L/m \tag{5.10}$$

resulting in constructive interference of order m between the wavelets emerging from the waveguide array output aperture. Because the latter is curved along an arc centered at the focal point where the central output channel waveguide joins the output combiner, the phase front emerging from the waveguide array bears the same curvature, hence converging toward the central output waveguide. For a wavelength λ, the maximum interference condition is satisfied in direction θ with respect to the coupler axis when:

$$n_{eff,\,s}\Lambda \sin\theta + n_{eff,\,a}\Delta L = m\lambda \tag{5.11}$$

* The non-Gaussian input field is often used for passband flattening; see Sections 5.2.3.4 and 5.2.3.5.

† Coupling from the slab waveguide to the waveguide array is the most significant source of loss in an AWG, because of the mismatch between the field distributions of the slab waveguide and the arrayed waveguides. At the slab-array interface, the continuous field in the slab waveguide couples to a segmented field of an array of waveguides. In a simplified picture, a part of the light that falls into the gaps between the array waveguides is lost, but it may also propagate in the array as stray light deteriorating crosstalk. This source of loss is absent in waveguide echelle grating devices.

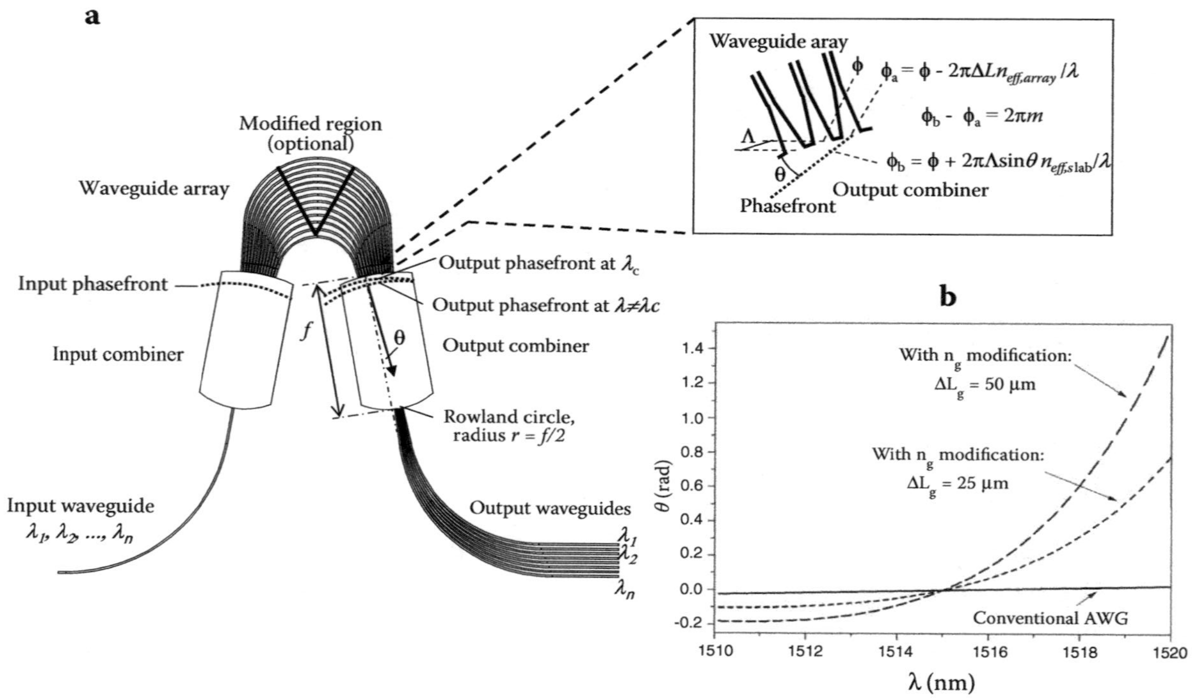

FIGURE 5.4 a) A schematics of an AWG demultiplexer. Phase conditions at the interface between the waveguide array and the output combiner are shown in the inset. b) Dispersion enhancement in an AWG with a triangular region with modified group index. In the example shown here the dispersion enhancement was calculated for a triangular region with photonic crystal waveguides.

as is evident from phase conditions shown in the inset of Figure 5.4a. Equation 5.11 is the equivalent of the grating equation (Eq. 5.1) where Λ is the waveguide pitch at the interface of the waveguide array and the output coupler and $n_{eff,\,s}$ and $n_{eff,\,a}$ are the effective indexes of the slab and array waveguides, respectively. In Eq. (5.11), we assumed that light enters the input coupler through the central input waveguide, so that the wavefront in the input coupler is not tilted with respect to the coupler axis. When other than a central input waveguide is used, the incident angle $\theta_i \neq 0$ and the $\sin\theta_i$ term needs to be included in Eq. (5.11) as in Eq. (5.1).

The effect of the change in mode index with wavelength is included in a modified interference order $M = m(n_{g,a}/n_{eff,\,a})$, where $n_{g,a}$ is the group index of the arrayed waveguides. The angle θ of the diffracted beam in the output coupler is then given by:

$$\sin\theta = \frac{(\lambda - \lambda_c)M}{n_{eff,\,s}\Lambda} \tag{5.12}$$

Although it may not seem so at first sight, the AWG operating principle is identical to that of the waveguide EG. In an EG, the light interference that produces a wavelength-varying propagation direction of diffracted light is due to the phase difference $\Delta\varphi_{EG} = 4\pi\,\Delta x n_{eff}/\lambda$ between the light reflected by the adjacent facets arranged in a staircase-like fashion with a step Δx (Figure 5.1c). In an AWG, the required phase difference $\Delta\varphi$ is obtained by propagating the light in N waveguides of varying lengths with a constant length difference ΔL between adjacent waveguides, so that $\Delta\varphi_{AWG} = 2\pi\,\Delta L n_{eff}/\lambda$, where n_{eff} is the effective index of a mode in an arrayed waveguide. The analogy with the EG is even more obvious for a reflection type AWG, first proposed by Dragone,[182] for which $\Delta\varphi_{AWG,\,R} = 4\pi\,\Delta L n_{eff}/\lambda$.

By differentiating Eq. 5.12, the AWG linear dispersion in the focal region is $\mathrm{d}x/\mathrm{d}\lambda = f\,\mathrm{d}\theta/\mathrm{d}\lambda = f n_{g,\,a}\Delta L/(\lambda_c n_{eff,\,s}\Lambda) = fM/(n_{eff,\,s}\Lambda)$ for a coupler of length f. This dispersion yields constant channel spacing* if the receiver waveguides are equidistantly spaced along the focal curve. To minimize the aberrations as explained in the EG section, the focal curve lies on the Rowland circle of radius $f/2$, tangent to the arc contouring the array output aperture of radius f.

Following the steps outlined in Section 5.2.1, the $FSR = \lambda_c/M$ as in an ordinary grating. Similar parallels with grating construction also apply for other AWG parameters. For example, the far field pattern is given by Eq. 5.2 (see Figure 5.1b) with the slit function SF arising from diffraction from a single arrayed waveguide with mode width w at the slab–array interface. Hence, changing the waveguide width by tapering the waveguides near the slab–array interface has the same effect as varying the grating facet width, namely the loss penalty (roll-off) when approaching the peripheral output channels increases with w. Details on AWG design and theory can be found for example in Smit and van Dam[15] and Muñoz et al.[183]

AWGs can be regarded as generalized Mach–Zehnder interferometers with multiple unbalanced arms formed by waveguides of different lengths. The first Michelson-type AWG[184,185] used two interleaved waveguide arrays (Figure 5.7a).

---

* AWGs with unequally spaced waveguides are not common but are useful for suppression of four-wave mixing crosstalk in dispersion-shifted fibers. See Forghieri, F. et al., *IEEE Photon. Tech. Lett., 6, 754, 1994.*

This device benefits from a markedly increased light throughput, also known as the *étendue* advantage of a Michelson interferometer.*

5.3.2 State-of-the-Art AWGs

Among various targets of current AWG research,[186] the following topics appear particularly relevant: miniaturization, increasing channel numbers, decreasing channel spacing, reduction of insertion loss, crosstalk and chromatic dispersion, flattening and widening the passband, elimination of polarization and temperature sensitivity, and improving spectral tuning capabilities including advanced add–drop, cross-connect, and filtering functions, and integration with photodetectors, lasers, modulators, variable optical attenuators, optical amplifiers, switches, and other photonic elements. We will briefly discuss some of the achievements and challenges related to these targets.

Most of the AWGs deployed in WDM optical communication systems use low index contrast glass waveguides that are usually $6 \times 6\,\mu m^2$ in cross-section, with core-cladding refractive index differences of $\Delta n \sim 0.75\%$.† The advantages of these conventional waveguides include low losses (< 0.02 dB/cm[187]) and good mode matching with standard optical fibers, yielding an insertion loss of ~ 2 dB or less for 100-GHz spaced 40-channel Gaussian passband AWGs. However, the minimum bend radius (typically defined as the radius corresponding to a bending loss of 0.05 dB per $90°$ bend) is large, for example $r \sim 5$ mm for 0.75% Δn. This yields prohibitively large devices for channel counts over 100, e.g., chip size would be more than 100 cm^2 for a 25-GHz 200-channel AWG that is difficult to fabricate. This is mainly because it is difficult to assure that the fluctuations in waveguide parameters and the resulting phase errors (see Section 5.3.2.1), are small enough over such large areas.

A cascaded fiber-connected two-stage 1080-channel 25-GHz demultiplexer covering the full range of S, C, and L bands was reported in 0.75%-Δn waveguides.[188] It uses a primary 10-channel 2.5-THz AWG as a first stage and each of the resulting 10 wavelength bands is further demultiplexed at the second stage by one of the 10 secondary 200-channel 25-GHz AWGs, each made on a separate 4-inch wafer. Waveguide platforms with larger Δn and still low losses of 0.05 dB/cm for 1.5% Δn (bending radius $r = 2$ mm) and 0.09 dB/cm for 2.5% Δn ($r = 1$ mm) have been developed for large scale AWGs.[189] A demultiplexer with 400 channels spaced at 25 GHz in the C and L bands with more than 1000 waveguides in the phased array, chip loss in the range of 3.8 to 6.4 dB, adjacent crosstalk of -20 dB, and a chip size of 12×6 cm^2 was fabricated on a 1.5% Δn platform.[190] Since it was impossible to lay out such a large circuit on a 4-inch Si wafer, a 6-inch Si wafer was used.

Several methods have been developed[189] to reduce the connection loss due to mode mismatch between optical fibers and high Δn silica-on-silicon waveguides.

* Large *étendue* (light throughput) is an intrinsic property of a Michelson interferometer that was first noted by P. Jacquinot (*J. Opt. Soc. Am.*, 44, 761, 1954). The Jacqinot étendue advantage and the Fellgett multiplex advantage (Fellgett, P.B., Ph.D. thesis, University of Cambridge, 1951) constitute the main reason Fourier transform Michelson interferometers dominate the field of infrared spectroscopy.

† Glass (silica-on-silicon) AWGs have been commercially produced since 1994.

Coupling loss is about 2 dB per point when standard (SMF-28) fiber is bonded to a 1.5% Δn waveguide with a $4.5 \times 4.5\ \mu m^2$ core. A 1.5% Δn waveguide was connected to a high numerical aperture (NA), small core fiber with a mode size matching that of the waveguide. The other end of the high-NA fiber was spliced with a standard SMF fiber. By thermally treating the joint, the core doping of the high-NA fiber diffuses into the cladding, and the core diameter and the mode size are increased to match those of the SMF fiber. This yielded a low coupling loss of 0.2 dB/point. An on-chip integrated spot size converter using both vertical and lateral tapering was also demonstrated with a coupling loss of 0.2 dB/point.[189]

A solution to reduce waveguide bend radius without increasing the Δn of the silica-on-silicon platform is etching air trenches at the waveguide bends to locally increase the index contrast.[191,192] Alternatively, turning mirrors may be used instead of waveguide bends.[193] In the future, we can expect a gradual abandoning of the low index contrast silica-on-silicon platform in favor of waveguide platforms with high index contrast, such as InP, silicon-on-insulator, or silicon oxynitride, as discussed in Section 5.4. This trend is also expected for emerging AWG applications such as pulse train generation, sensing, spectroscopy, metrology, and optical interconnects.

5.3.2.1 Phase Errors

The influence of phase and amplitude errors on AWG response has been analyzed in various models,[194,195] including the effects of the errors on the coupling between the array waveguides.[196] The coupling can be described by the Green's function[197] representing the response of the waveguide array when light is coupled into a single waveguide in the uncoupled region of the array. Deterioration of the crosstalk due to phase and amplitude errors has been associated with breaking the symmetry of the Green's function.[196]

One of the difficulties in large-scale AWGs is crosstalk deterioration caused by phase errors arising from variations in the arrayed waveguide width, thickness, material composition, and stress. Because the influence of such errors increases with the size of the waveguide array, the effect can be severe for densely spaced AWGs (25 GHz and less) with large numbers of channels. As the phase errors are related to the effective index fluctuations by $\delta\varphi = 2\pi\,\delta n_{eff} L/\lambda$, in a glass AWG with an average arrayed waveguide path length of $L \sim 40$ mm, effective index changes on the order of $\delta n_{eff} \sim 1 \cdot 10^{-6}$ produce phase errors of $\delta\varphi \sim 1.6$ rad.

Because the electric field distribution in the focal region is the Fourier transform (Fraunhofer diffraction pattern) of the field profile at the slab–array interface, phase errors over larger areas (low spatial frequencies) of the waveguide array produce peaks or side lobes close to the main peak that mostly affect crosstalk between the adjacent channels. The crosstalk can be reduced by adjusting the phase delays in the individual arrayed waveguides, but this is a rather tedious task. The phase φ_i can be measured for each waveguide i of the array by low coherence interferometry.[198,199] A light source with a low coherence length, typically a LED, is split into two branches of a Mach–Zehnder interferometer, one inserted with an AWG and the other with a variable delay line. The phases and the amplitudes of light passing through different arrayed waveguides are found by scanning the optical delay line. The coherence length of the broadband source is selected to be shorter than the arrayed waveguide length increment ΔL. The interference due to a particular waveguide is observed only when the optical path length through the variable delay line approaches the light path length through the AWG via that particular waveguide. This allows separate measurements of the phase errors for each waveguide.

Alternatively, the phase and amplitude errors can be found by placing the AWG in one arm of a Mach–Zehnder interferometer (that can be conveniently formed on the same chip as the AWG) and scanning the input wavelength.[200] The phase and amplitude errors of different waveguides can be extracted from different Fourier components of the signal measured at the output of the Mach–Zehnder interferometer.

Phase errors have been corrected by depositing a stress-applying amorphous silicon (a-Si) film on the upper cladding to modify the arrayed waveguide effective index via the elasto-optic effect, and then trimming the a-Si patch length over different waveguides via uv laser ablation.[201] Alternatively,[202] the phase correction can be achieveds by uv-induced refractive index changes in the glass by the same mechanism used for making fiber Bragg gratings.[34] All the waveguides are exposed at the same time by using a metal mask with different opening lengths over each waveguide i that are proportional to the phase errors $\delta\varphi_i$ to be compensated. Because the phase errors and the effective index sensitivity to uv light are different for TE and TM modes, two masks must be made for each AWG device. This compensation technique was used to improve the crosstalk in large-scale AWGs.[181]

5.3.2.2 Higher Order Waveguide Modes

Because of their different propagation constants, the higher order waveguide modes in the arrayed waveguides cause crosstalk-degrading ghost peaks and changes in the passband shape and center wavelength. This is of particular concern for waveguides in high index contrast platforms as they may support many modes. Even when the single mode condition is achieved in these waveguides (see Section 5.4.2), higher order leaky modes can propagate to the end of the device.

The propagation loss for these leaky modes should be sufficiently large, typically > 50 dB/cm.[203] Higher order modes can be excited by misalignment of the input fiber or by abrupt changes in waveguide parameters, for example, at the junctions between straight, curved, slab, and channel waveguides.

If an input other than the central input waveguide is used to inject light into the input combiner of an AWG, the wavefront is tilted in the combiner, increasing higher order mode excitation at the combiner–array interface. A simple solution to filter out the higher order modes was demonstrated in an SOI AWG demultiplexer by choosing bend radii small enough for the higher order modes to radiate out at the bends, but large enough to keep the bend loss negligible for the fundamental mode.[204] Another solution is to insert mode filters in the waveguide array, for example, a 1×1 multimode interference (MMI) coupler[205] in each of the arrayed waveguides. A properly designed 1×1 MMI can provide more than 20 dB higher order mode suppression over a relatively large bandwidth ($\sim$ 100 nm), as demonstrated in an InP-based AWG channel selector.[206]

5.3.2.3 Polarization Sensitivity

From Eq. 5.10 we can observe that an effective index birefringence Δn_{eff} in the arrayed waveguides produces a polarization-dependent shift of $\Delta\lambda = \Delta n_{eff}\Delta L/m$ in the demultiplexer central wavelength. This polarization-dependent wavelength shift was discussed in Section 5.2.3.1, including the birefringence compensating techniques of in-plane stress balancing by applying stress in the direction parallel to the buried channel vertical sidewalls[108,109] and by using waveguide layers with thermal expansion coefficients (TECs) matched to the Si substrate.[110,111] Other polarization compensation techniques include:

1. Choosing the $FSR = \Delta\lambda$ to make the mth order TE spectrum overlap with the TM spectrum of the $m - 1$ order[207,208]
2. Modifying waveguide birefringence over a triangular region of the array (see Figure 5.4a) which acts as a prism, producing polarization-dependent wavefront tilt similar to the prism slab compensator shown in Figure 5.3a[209,210]
3. Eliminating TEC mismatch by using silica instead of a silicon substrate[211]
4. Making stress releasing grooves beside the arrayed waveguides[212]
5. Changing core width[213]
6. Using raised-strip waveguides[213,214]
7. Inserting a half-wave plate in the middle of the phased array[107]

In the latter technique pioneered by NTT, polarization independence is achieved by converting TE polarization to TM and vice versa in the middle of the array so that the overall phase accumulation through the array is made independent of input polarization. When a polyimide half-wave plate with a thickness of 15 μm is inserted in an 18-μm wide groove and fixed with a uv-curable adhesive, a low excess loss of 0.4 dB is obtained for a conventional silica (Δn = 0.75%) AWG. However, with increasing Δn, it becomes more difficult to bridge the groove due to increasing divergence of the free-propagating beam in the groove. This demands techniques other than the half-wave plate. For example, stress balancing by dopant-rich glass[111] was used to reduce the polarization wavelength shift to <0.01 nm in 1.5% Δn AWGs.[216]

Compensating the effect of array waveguide effective index birefringence may not be sufficient for complete elimination of polarization-dependent wavelength shifts in AWGs. From Eq. 5.12 it follows that polarization insensitivity is achieved when the factor $M/n_{eff,\,s} = m[n_{g,\,a}/(n_{eff,\,a}n_{eff,\,s})]$ is polarization-independent. Advanced polarization compensation techniques may also need to include effects of birefringence in the arrayed waveguide group index and in the slab waveguide effective index. The latter has recently been demonstrated by NTT.[217] A small polarization-dependent wavelength shift exists even if the waveguide array is perfectly birefringence compensated by a $\lambda/2$ plate. While $\Delta\lambda$ vanishes near the central output channel, it linearly increases toward the peripheral channels. The origin of this wavelength shift is in the slab waveguide (combiner) birefringence. Although the wavefront of the central output channel passes normal to the array–slab interface, the tilt of the wavefront increases as one moves from the central channel toward the peripheral channels. In a birefringent slab waveguide, a polarization-dependent refraction at the array–slab boundary, in turn, causes a TE–TM wavelength shift to linearly increase from the central channel toward the peripheral channels. This channel-dependent $\Delta\lambda$ was eliminated in a silica-based AWG by placing a half-wave plate across the output combiner slab waveguide.[217]

5.3.2.4 Temperature Sensitivity

From Eq. 5.10 it follows that the AWG central wavelength changes with temperature as $d\lambda_c/dT = \lambda_c[(1/n)dn/dT + \alpha]$, where α is the linear thermal expansion coefficient. In glass AWGs, $d\lambda_c/dT$ is of the order of 0.01 nm/°C, determined mainly by the temperature dependence of silica glass refractive index $dn/dT \sim 1.1 \cdot 10^{-5}$/°C (we ignored temperature-induced stress causing refractive index changes via elasto-optic effect; in silica glass, $\alpha \sim 10^{-7}$/°C). Active temperature stabilization by a heater or Peltier cooler is often used, but it requires continuous power consumption of several watts and temperature control electronics. This can be avoided with an athermal design, i.e., with substantially reduced temperature sensitivity.

Various athermal designs have been demonstrated, including a moving fiber coupler,[218] attaching the die to a bimetal plate[219] or metal stress bar,[220] moving the output coupler by a compensating metal plate,[221] and using a reflective AWG with a compensatory mirror assembly that tilts with temperature.[222] These designs use mechanically moving parts attached to the chip to compensate for temperature-induced changes in the position of the focal spot.

Static athermalization was demonstrated by using waveguides with $(1/n)dn/dT = -\alpha$ (see equation above for $d\lambda_c/dT$), but this requires unconventional waveguides, e.g., TiO_2 waveguide core and SiO_2 cladding on a Si substrate.[223] One practical athermalization technique involves creating a prism-like region in the waveguide array[224,225] or in the slab combiner,[226] often segmented into multiple triangular grooves to reduce bridging distance and loss, and filled with a material with different dn/dT compared to the material outside the groove.

Silicone with a large negative dn/dT ($-37\cdot10^{-5}$/°C) is often used. Being conceptually similar to the polarization compensating prism that provides polarization-dependent wavefront tilt as explained in Section 5.2.3.1, the temperature compensating prism provides a temperature-dependent wavefront tilt as a consequence of the varying refraction of the prism with temperature. For given dn/dT coefficients of the materials inside and outside the prism, the shape of the prism can be found such that the light of a given wavelength arrives, regardless of temperature, to the same focal position. Using the prism technique, a temperature-dependent wavelength shift has been reduced by 20 times compared to an AWG without a compensator, for the temperature range $T = 0$ to 65°C, with an excess loss of 0.7 dB.

5.3.2.5 Chromatic Dispersion

A typical AWG has a symmetric intensity distribution across the waveguide array, and as such its chromatic dispersion D is negligible (see Section 5.2.3.4). However, in a practical AWG this symmetry is disturbed by phase and amplitude errors that are randomly distributed in the arrayed waveguides. This increases chromatic dispersion.

Because the errors increase with decreasing channel separation (increasing the size of the waveguide array), the chromatic dispersion increases similarly: $D \propto 1/\Delta f^2$, where Δf is the channel frequency spacing. D can easily exceed user specifications (typically $D < 10$ ps/nm) for 50 GHz and smaller channel spacing. This problem is further exacerbated in devices with flat spectral responses because they tend to exhibit coupling of the phase response with the amplitude response, resulting in a larger D that may not be tolerable for high data-rate systems. In addition to the phase errors in the waveguide array, D is also proportional to the wavefront curvature at the junction of the input waveguide and the slab combiner.[227]

In a popular passband flattening technique using a parabolic waveguide horn at the end of the input waveguide,[151] the phase wavefront is deformed because of a phase difference between the 0th and 2nd modes that in turn increases the chromatic dispersion. It was recently shown[228] that by a judicious design of the horn, the phase front can be flattened and chromatic dispersion reduced to 3.2 ps/nm compared to 20.9 ps/nm for an AWG with a conventional parabolic taper. It has also been demonstrated[229] that chromatic dispersion can be reduced by using an AWG with a hyper-Gaussian passband of order n and transmission $T(\lambda) = 0.5^{\xi(\lambda)}$, where $\xi(\lambda) = |(\lambda - \lambda_c)/(\Delta\lambda_{3dB}/2)|^n$ and $\Delta\lambda_{3dB}$ is the filter 3-dB bandwidth. $D < 0.7$ ps/nm was achieved in a 50-GHz AWG with a flat and wide hyper-Gaussian passband. Other passband flattening techniques are discussed in Section 5.2.3.4.

5.4 AWGS AND SPECTROMETERS IN WAVEGUIDE PLATFORMS WITH HIGH INDEX CONTRAST

Strong mode confinement in waveguides with high refractive index differences between the core and the cladding makes it possible to achieve mode sizes of submicrometer dimensions and bend radii as small as a few micrometers, therefore yielding ultracompact devices. We will briefly discuss AWG devices implemented on InGaAsP/InP and SOI platforms, although encouraging results have also been reported with the silicon oxynitride (SiO_xN_y) waveguide platfom.[230]

5.4.1 InGaAsP and InP

AWGs made in InP have been intensely investigated[231–233] not only because of the advantage of monolithic integration with electronics and active elements but also for their miniaturization advantages. An InGaAsP waveguide core gives a large accessible bandgap wavelength range ($\lambda_g \sim 0.92$ to 1.65 μm) of quaternaries lattice-matched to the InP cladding and substrate. Passive and active elements can then be integrated on the same substrate, the former requiring a material with $\lambda_g < \lambda$ while the latter $\lambda_g \geq \lambda$ where λ is the operation wavelength. A compact 64-channel 50-GHz InP-based AWG occupying an area of 3.6×7 mm^2, which is $\sim 1/100$ that of the silica-based AWG, was demonstrated in the late 1990s,[234] and an extremely small 4×4 channel 400-GHz AWG of size 230×330 μm^2 has recently been reported.[235]

Many photonic circuits integrating AWGs with photodiodes, lasers, modulators, switches, or semiconductor optical amplifiers (SOA) have been reported, including WDM receivers, channel monitors, optical cross-connects, add–drop multiplexers, channel selectors and equalizers, and multiwavelength lasers. A review of these complex devices is beyond the scope of this chapter and can be found, for example, in Yoshukini[231] and Leitjens.[232]

For illustration purposes, we choose two recent examples. A multichannel modulator has been demonstrated with a channel spacing of 25 GHz and 80-Gb/s throughput (10 Gb/s $\times$ 8 channels) by monolithically integrating an InGaAsP/InP AWG with InGaAlAs-InAlAs tensile strain multiquantum-well electro-absorption modulators and InGaAsP semiconductor optical amplifiers.[236] A 10-channel WDM transmitter was reported by Infinera. It comprises an array of 10 DFB lasers, electro-absorption modulators, variable optical attenuators, and optical power monitors, all integrated with a 10-channel AWG multiplexer.[237]

The InP AWG devices often use deep-ridge waveguides with thin (0.3 to 0.5 μm) waveguide cores (see Figure 5.2c) etched by RIE (using Br_2–N_2 chemistry[238]) in waveguiding layers grown by MOCVD or molecular beam epitaxy (MBE). The thin core facilitates integration with active devices, as the core thickness and mode size are similar to those used in semiconductor laser technology. By deep etching, a strong lateral confinement is achieved by vertical semiconductor–air waveguide boundaries allowing for small bending radii ($\sim$ 100 μm or smaller). However, strict control of waveguide width is necessary to make the waveguides free of birefringence.

The birefringence also strongly depends on the bandgap wavelength λ_g of the core. The birefringence sensitivity to the waveguide width errors rapidly increases with the difference between the bandgap wavelengths of the InGaAsP core and the InP upper and bottom cladding layers. For an InGaAsP core with $\lambda_g = 1.05\ \mu$m, the fabrication tolerance of the waveguide width is about $\pm 0.1\ \mu$m,[231] which can be reproducibly achieved with modern dry etching technology.

Mitigation of polarization sensitivity using the prism-like sections with a modified birefringence either in the waveguide array[210] or in the slab combiner[113] was reported. In deep-etched ridge InP waveguides, waveguide loss on the order of $\sim$ 0.5 dB/cm can be obtained, which is larger than in silica-on-silicon waveguides, but this is compensated by small device size. Loss can be completely eliminated, even net gain achieved, by integration with semiconductor optical amplifiers.

Because of strong confinement in the array waveguides, the excess loss due to field mismatch at the slab–array interfaces is large (a 2.8-dB excess loss for a gap of 0.3 μm)[234] but it can be reduced by optimizing the slab–array interface, e.g., by decreasing the gap between the arrayed waveguides. An excess loss as low as 0.4 dB has been achieved[239,240] by using a double etch process. This involves deep etch waveguides in the array adiabatically tapered to shallow etched regions near the area where the array waveguides join the input and output combiners. An original technique has been proposed to reduce the field mismatch loss by creating interference fringes in the couplers by launching two mutually inclined beams in the input combiner using two input waveguides. The field mismatch loss is reduced when the multiple peaks of the interference fringes exactly match the positions of the arrayed waveguides at the slab–array interface.[241] However, based on AWG imaging properties, it follows that splitting the light into two input waveguides produces a split (double) image in the output focal plane with a 3-dB split loss, demanding a solution for merging the two images.

A fiber-to-chip coupling loss of 2 dB was reported for an InP AWG with an integrated spot-size converter, with a 1-dB coupling fiber position tolerance of 2 μm. For an overview of spot-size converters for InP-based devices, see Moerman et al.[242]

Because of the large refractive index of InGaAsP/InP, phase errors are increased compared to glass waveguides, but devices with crosstalk lower than 30 dB have been demonstrated[231] and encouraging results were also recently reported on crosstalk reduction by 1×1 MMI mode filters.[205] A temperature-insensitive InGaAsP/InP design has also been demonstrated.[231] Athermalization was achieved by using InGaAsP cores with high and low dn/dT quaternary compositions of bandgap wavelengths of 1.3 and 1.1 μm. A triangular prism with high dn/dT waveguides was inserted into the waveguide arrays of low dn/dT waveguides.

5.4.2 Silicon-on-Insulator (SOI)

Silicon has been the dominant platform for the microelectronics industry for several decades and it appears that it will remain so in the foreseeable future. Benefit from the integration of mainstream microelectronic technology with photonics has been the main driving force in the emerging field of silicon photonics.[243–246] In intrinsic single crystal bulk silicon, light with wavelength longer than the Si bandgap wavelength ($\lambda_g = 1.107\,\mu$m) can propagate with a low loss. The transparency range thus includes the 1.3- and 1.55-μm telecommunication windows. Particularly attractive is the high refractive index of Si, with $n \sim 3.5$ at 1.55μm. When silicon is used as the waveguide core and is surrounded, for example, by a SiO_2 cladding of $n \sim 1.5$, an index step of $\Delta n \sim 2$ is obtained. In such waveguides, light is highly confined in the core which can have cross-sections as small as $\sim$ 250 nm and bending radii can be reduced to a few micrometers. Ultra-compact planar waveguide devices can thus be made in Si.

The first silicon photonic devices were realized on SiGe platforms.[247–249] However, the disadvantages of SiGe waveguides include large stresses built in during the heteroepitaxial growth of waveguide layers and relatively low index steps.[250] Recent silicon photonic devices are almost exclusively realized on an SOI platform.[251] SOI is a composite substrate with a top single crystal Si layer separated from the Si wafer by a thin SiO_2 oxide (buried oxide or *box*) layer.* The top Si layer acts as the waveguide core and the buried oxide as the bottom cladding, providing strong vertical confinement. The latter is an important advantage of SOI as opposed to InP, which has a large index contrast in the lateral direction but not in the vertical direction.

---

* SOI was developed in the 1970s as a niche substrate technology for military and space applications demanding materials with improved radiation hardness. The main reasons for using SOI in microelectronic applications include reduced parasitic substrate capacitance and low leakage current due to the insulating SiO_2 bottom oxide layer, leading to increased speed and lower power consumption of MOSFET devices. With the advent of Smart-Cut technology (Bruel, M., *Nucl. Instrum. Meth. B*, 108, 313, 1996) in the 1990s that significantly improved wafer quality, SOI entered the mainstream of ultralarge scale integration (ULSI) integrated circuits (ICs).

Ridge waveguides (Figure 5.2d) are often used in SOI instead of channel waveguides (Figure 5.2b). This is because the single mode condition demands, due to large index contrast, a cross-section of $\sim 0.32 \times 0.32\,\mu m^2$ for a channel waveguide (Figure 5.2b) with the Si core buried in SiO_2. With current fabrication technology, the loss due to light scattered by the sidewall roughness is still rather large for such waveguides (often called photonic wires).* In addition, because of a large mode mismatch, fiber-to-waveguide coupling becomes a serious problem. Various solutions to reduce sidewall roughness-induced loss have recently been demonstrated. Thermal oxidation after reactive ion etching (RIE) preferentially consumes asperities on the waveguide surface, resulting in a residual *rms* roughness of ~ 2 nm and a loss reduction from 32 to 0.8 dB/cm for a 0.5-μm wide SOI waveguide[252] and deep uv lithography and RIE yield losses as low as 2.4 dB/cm for $0.5 \times 0.33\ \mu m^2$ waveguides.[253] Alternatively, a wet etch can be used after conventional RIE to smoothen the roughness, or etching can be avoided altogether by forming the SiO_2 waveguide cladding via oxidation of the Si, for example, using the LOCOS (local oxidation of silicon) process, as we suggested.

In contrast to SOI channel waveguides of large cross-sections that support many modes, a single mode condition can be achieved in SOI ridge waveguides of large cross-sections. The single mode condition has often been approximated by the following formula:[248] $W/H \leq 0.3 + r/\sqrt{1 - r^2}$ and $r = h/H \geq 0.5$, where h is the slab waveguide height in the lateral (etched) regions and W and H are the ridge width and height, respectively† (Figure 5.2d). This condition can be intuitively understood as the cut-off of the higher order ridge modes as their effective indices approach the effective index of the fundamental mode in the lateral slab waveguide regions. This makes the higher order modes leak from the ridge to the lateral slab regions so that after a sufficiently long propagation distance only the fundamental mode remains in the ridge. More details on SOI waveguides are included in Chapter 6 of this book by Knights and Jessop.

The first AWG in SOI was demonstrated in 1997.[254] Because of the large waveguide cross-section and the small stress in the Si waveguide core, a polarization-dependent wavelength shift as small as 0.04 nm was obtained without any compensation technique. However, SOI with a thick (5 μm) Si layer was used and the bend radius was 2 cm, resulting in a large device size of $2.7 \times 2.7\ cm^2$ for this four-channel AWG. Similarly, SOI AWG devices commercialized by Bookham

---

* According to a model by F.P. Payne and J.P.R. Lacey (*Opt. Quantum Electron.*, 26, 977, 1994), the scattering loss coefficient is $\alpha \sim \sigma^2\kappa/(nk_0d^4)$ where σ is *rms* sidewall roughness, k_0 is the free space wavevector, n and d are the refractive index and width of the waveguide core, respectively, and κ depends on the details of the waveguide geometry and the statistical distribution of the roughness. Although this is only an approximate model, the important prediction is the inverse fourth power dependence of the scattering loss on waveguide dimensions. This is a good approximation for waveguide widths down to ~ 300 nm. When the waveguide dimensions are reduced further, the loss due to sidewall roughness begins to decrease (Grillot, F. et al., *IEEE Photon. Technol. Lett.*, 16, 1661, 2004) as a consequence of mode delocalization from the core, i.e., the effective mode index is reduced and the mode begins to expand into the cladding. For a rigorous scattering analysis of microphotonics waveguides, see the recent work of T. Barwicz and H.A. Haus based on the volume current method (*J. Lightwave Technol.*, 23, 2719, 2005).

† This Soref's formula is a useful guideline for designing single mode waveguides but should not be regarded as a rigorous condition (see Lousteau, J. et al., *J. Lightwave Technol.*, 22, 1923, 2004 and Chan, P.P. et al., *J. Lightwave Technol.*, 23, 2103, 2005).

Technology[255] have thick Si waveguide core layers, resulting in comparatively large devices.

The first miniature SOI AWG was reported by Pearson et al.[204] This eight -channel 200-GHz multiplexer/demultiplexer has an array of 100 waveguides designed at an interference order of $m = 49$ and an overall chip size of 5×5 mm^2. We used similar designs in our subsequent AWG devices, typically made on SOI with $\sim 2\ \mu$m thick Si with a typical on-chip loss of $\sim$ 6 dB, crosstalk below 20 dB, and a polarization-dependent wavelength shift as small as 0.02 nm with birefringence compensation. An overview of these devices can be found in Cheben et al.[256]

Currently the smallest reported AWG[257] uses single mode Si photonic wire waveguides of 0.32 μm thickness and 0.45 μm width and has a size of $70 \times 60\ \mu$m^2. The device is designed with a free spectral range of 85 nm and a channel spacing of 11 nm; the measured crosstalk is 13 dB and the estimated on-chip loss $\sim$ 1 dB. The performance of similar ultra-compact AWG devices is expected to rapidly improve with advances in nanofabrication technologies. Compact mux/demux devices based on large dispersion in photonic crystals, using the super-prism effect have also been reported.[258–260]

5.4.2.1 Waveguide Birefringence

In the course of silicon photonic research in our laboratory,[251,256,261,262] one important direction has been the mitigation of polarization sensitivity in SOI AWGs. The geometrical birefringence of a ridge waveguide can be eliminated by adjusting the waveguide width-to-depth ratio.[263,264] However, when the waveguide size is reduced to about 2 μm or smaller, the birefringence due to waveguide geometry can become large and highly sensitive to changes in waveguide dimensions. The birefringence is often further increased by the stress present in the core as a result of the upper cladding.

The cladding is usually an SiO_2 layer with an internal stress of several hundreds of MPa. The stressed cladding exerts a force on the Si ridge that modifies the refractive index of the waveguide via the elasto-optic effect. The compressive stress of the oxide film results in a compressive stress in the in-plane direction and a tensile stress in the out-of-plane direction within the Si core (Figure 5.5a), hence the birefringence. Xu et al. demonstrated that the stress-induced birefringence in Si ridge[265,266] and photonic wire waveguides[267] can be used to eliminate the waveguide birefringence.

Birefringence-free operation was achieved for waveguides with different cross-section geometries by adjusting the SiO_2 cladding thickness[265] or by modifying stress in the oxide film by controlling deposition conditions or with thermal anneals,[268] as shown in Figure 5.5b. The technique was experimentally demonstrated on a compact nine-channel 200-GHz AWG demultiplexer fabricated in SOI with a 2.2-μm Si layer,[265] and this technique was shown to be effective for designing polarization-insensitive SOI ring resonators.[269]

The polarization-dependent wavelength shift ($\Delta\lambda = \lambda_{TM} - \lambda_{TE}$) in the AWG spectra was reduced from –0.54 nm in the absence of a cladding, to 0.02 nm after depositing a 0.8-μm PECVD cladding oxide with in-plane compressive stress of $\sigma \sim -320$ MPa. The corresponding decrease in modal birefringence ($\Delta n = n\Delta\lambda/\lambda$)

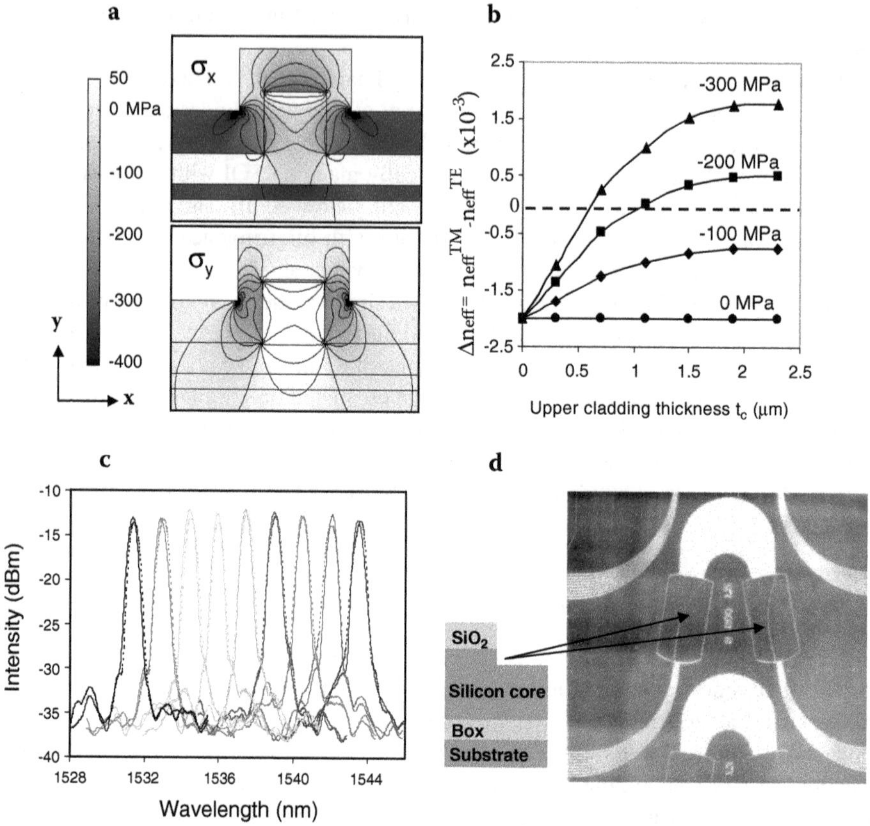

FIGURE 5.5 a) Anisotropic stress distribution in an SOI ridge waveguide clad with SiO_2 film in compressive stress. b) Variation of SOI ridge waveguide birefringence as a function of SiO_2 cladding thickness for different values of stress in the SiO_2 film. c) The measured spectrum of the AWG with polarization dependent wavelength shift compensated by 0.8-mm thick PECVD cladding oxide with in-plane compressive stress of $\sigma \sim -320$ MPa. d) A polarization compensated SOI AWG using a compensator with partially etched Si waveguide core in the input and the output combiners.

is from $1.2 \cdot 10^{-3}$ to $4.5 \cdot 10^{-5}$.[265] The measured spectrum of the compensated AWG is shown in Figure 5.5c. Polarization-compensated SOI AWGs were also demonstrated by partially etching the Si waveguide core to form prism-like compensators in the input and output slab waveguide combiners (Figure 5.5d), with $\Delta\lambda$ reduced from 2.2 to 0.04 nm.[114] Compensators with inverted symmetries were used to avoid total internal reflection at the compensator boundaries. One drawback of etching the core is the insertion loss penalty on the order of 2 dB (depending on the etch depth required for a designed compensator strength) arising from the mode mismatch at the boundary between the etched compensator and non-etched slab waveguide regions.

Instead of partially etching the silicon core, it was shown that the mode mismatch loss can be reduced to ~ 0.2 dB by depositing over the core in the compensator region a thin (~ 200 Å) SiO_2 film capped by an a-Si layer (~ 1 μm, n_{a-Si} ~ 3.45).[122] Because the TE mode penetrates deeper than the TM mode into the high index a-Si capping layer through the thin oxide layer, the TE effective index is increased compared to the TM. This double layer cladding structure is an example of the form birefringence discussed in Section 5.2.3.1.

5.4.2.2 Fiber-to-Waveguide Coupling

Because of a large mode size disparity, the optical coupling between an optical fiber and an SOI waveguide with a small cross-section is very inefficient. The mode diameter of a standard SMF-28 fiber is 10.4 μm measured at 1/e^2 intensity. In order to match this large fiber mode to an SOI waveguide mode with an area typically two orders of magnitude smaller, mode transforming structures in both the in-plane and out-of-plane directions must be used.

Input waveguides adiabatically tapered in both directions are conceptually simple, but the out-of-plane tapering requires gray-scale lithography, which is not yet a standard technique in the industry. Grating[270] and photonic crystal[271] couplers have been demonstrated, but their fabrication is demanding. An interesting approach is use of an inverse tapered waveguide[272,273] that adiabatically narrows down to a width of about 100 nm or smaller as the waveguide approaches the coupling facet. The waveguide effective index is reduced by narrowing the waveguide, which causes the mode to expand and eventually match that of the fiber.

Delâge et al.[274] recently demonstrated a waveguide coupler formed by depositing on top of a 0.5-μm thick SOI waveguide a 4-μm thick gradient index (GRIN) amorphous silicon (a-Si) layer with a refractive index in the proximity of the waveguide that is close to that of the Si waveguide and then gradually (quadratically) decreases toward the upper surface of the a-Si layer (Figure 5.6a and b). This structure acts as an asymmetric GRIN lens.* A wide input field launched into the coupler periodically converges into focus in the high refractive index region, and if the a-Si structure is terminated at one of the focal positions, the light is effectively coupled into the thin Si waveguide. A coupling loss of ~ 0.3 dB has been predicted for an ideal quadratic profile. A coupler with the continuous gradient index profile replaced by a few discrete layers with calculated coupling losses of 2.9, 1.5, and 0.6 dB for one, two, or three layers, respectively, has been proposed.[274] An approximately four-fold increase in coupling efficiency has been measured for a coupler with a single a-Si layer on an SOI with a 0.8-μm thick Si waveguide.[275]

A subwavelength grating fiber-to-waveguide coupler has also been proposed.[276] The mode is expanded by changing the duty ratio $r = (\Lambda - d)/\Lambda$ of a subwavelength grating of a period $\Lambda \sim 300$ nm etched in the Si core, where d is the width of trenches

---

* GRIN waveguide coupler can be regarded as a multimode interference (MMI) device. The modes of an ideal GRIN lens with quadratic index profile $n^2(x) = n_0^2\,(1 - x^2/x_0^2)$ are hermite Gaussian polynomials with propagation constants approximately spaced at $\Delta\beta \sim 1/x_0$, hence with a period of input field reconstruction of $\Delta z = 2\pi x_0$ along propagation direction z. As in a quarter-pitch GRIN lens, the input field profile will be brought into focus at $\Delta z/4 = \pi x_0/2$.

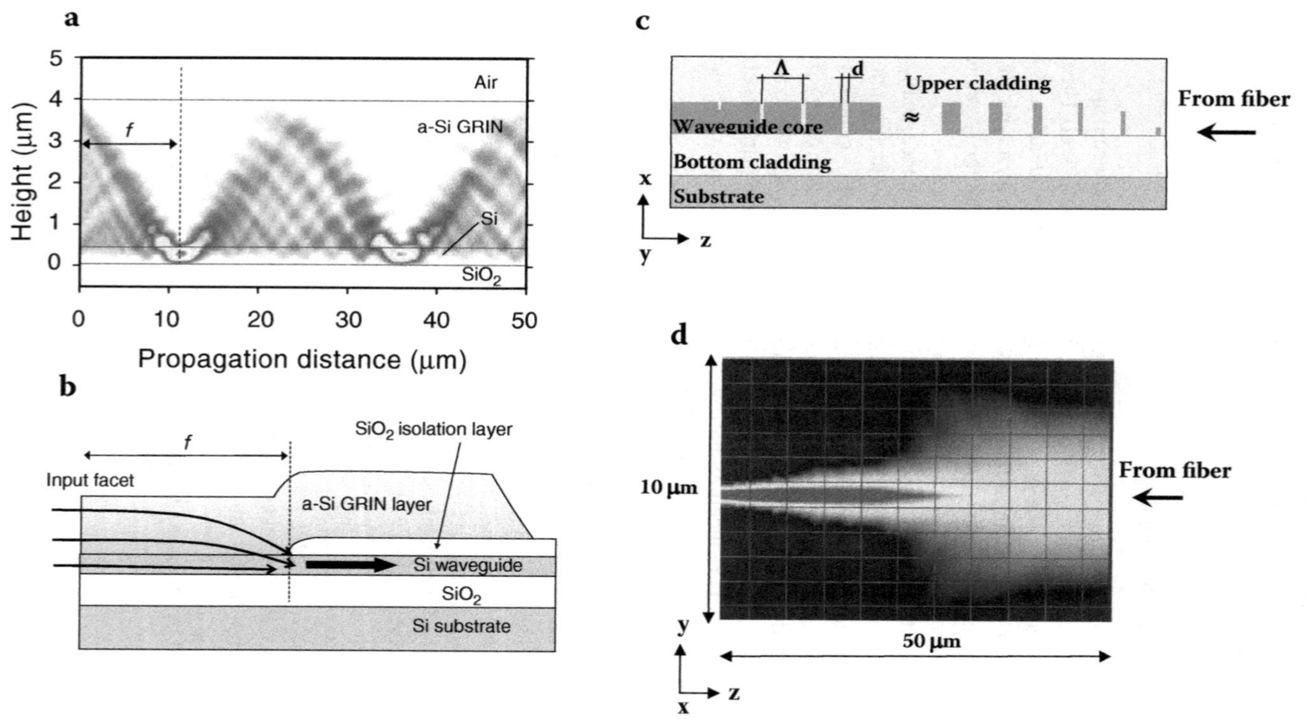

FIGURE 5.6 a) Light coupling to a thin SOI waveguide using GRIN effect. b) Schematics of an input coupler using a-Si GRIN layer. c) Schematics of a sub-wavelength grating fiber-to-waveguide coupler. d) FDTD simulation of light coupling from an SMF-28 fiber to a thin SOI waveguide using the sub-wavelength grating coupler.

etched in the Si core that are eventually filled with a SiO_2 cladding (Figure 5.6c). The grating duty ratio is gradually changed from $r \sim 0.1$ at the coupler end facing the fiber to $r \sim 10$ at the other end of the coupler. Within the effective medium approximation, the effective index of the mode in the taper increases with duty ratio. At the same time, diffraction by the grating is negligible because the grating period Λ is less than the first order Bragg grating period $\Lambda_{Bragg} = \lambda/(2n_{eff})$. As an increasing portion of the Si core is etched away and replaced by SiO_2 when approaching the coupler end facing the fiber, an effective index close to that of an SiO_2 waveguide is obtained at the fiber–chip interface. Hence reflection loss at the waveguide-fiber junction can be very small if index matching fluid is used. The FDTD calculations shown in Figure 5.6d predict a coupling efficiency of more that 75% between a SMF-28 fiber and the fundamental mode of a 0.3-μm thick SOI waveguide,[276] and further optimization appears possible.

5.4.2.3 Waveguide Spectrometers

Examples of emerging AWG applications include optical interconnects, spectroscopy, metrology, chemical and biological sensing, medical instrumentation, and space-based sensing. High resolution spectroscopies (e.g., Raman or Fourier transform infrared) are needed for the development of new detection platforms in genomics and health-related applications. High resolution techniques currently require the use of large and expensive spectroscopic instruments because the resolution of miniature spectrometers is still rather limited.

Based on AWGs in a high index contrast platform such as SOI, compact waveguide spectrometers can be made with many channels and large spectral resolutions.[277] An important advantage of SOI in this application is that it allows for very narrow slit-like rectangular waveguides at the Rowland circle that are required to maximize spectral resolution, similar to free-space spectrometers that require narrow input slits for high resolution. In a recently reported 100-channel SOI microspectrometer[277] with a designed resolution of 0.08 nm at a 1.5-μm central wavelength and a chip size of 8×8 mm^2, the shallow etch input and output ridge waveguides were tapered down to 0.7-μm wide rectangular (slit-like) waveguides deep etched down to the buried oxide near the input and output Rowland circles. Adiabatic tapering with low loss and negligible mode conversion was used.[277]

The reduction of the input waveguide width below $w \sim 0.7$ μm may be impractical because the increasing angular width of the diffraction pattern requires combiners with low f numbers for which aberrations can become difficult to control.

The fundamental limit to the minimum waveguide width is due to the mode delocalization effect. When the waveguide width is reduced to a critical value on the order of the wavelength of the light in the waveguide, the mode effective index is reduced and the mode starts expanding from the core into the cladding. This demands increasing separation between the neighboring waveguides at the Rowland circle to avoid crosstalk between the adjacent spectral channels. According to calculations by Delâge et al.,[77] the minimum waveguide separation d along the focal curve in order to achieve a crosstalk level below 40 dB is $d \sim 0.73\lambda\Delta n^{-0.55}$, which for SOI waveguides ($\Delta n \sim 2$) gives $d \sim 0.75$ μm for $\lambda \sim 1.5$ μm.

5.4.2.4 Michelson-Type AWG: Étendue Advantage

The need for narrow input waveguide apertures to achieve high resolution can be circumvented by arranging an AWG in a Michelson instead of a Mach–Zehnder configuration.[184] This can be achieved by interleaving two AWGs of different orders (AWG_1 of orders M_1 and AWG_2 of order M_2) in a reflective configuration, as shown in Figure 5.7a. From Eq. (5.12), it can be observed that the angle between the two wavefronts emerging from the corresponding waveguide arrays changes with the wavelength as $d\theta/d\lambda = d\theta_2/d\lambda - d\theta_1/d\lambda \sim (M_2 - M_1)/(n_{eff,s}\Lambda)$, resulting in Fizeau interference fringes with a wavelength-dependent period $p(\lambda)$ in the slab waveguide combiner where the two wavefronts overlap.

Monochromatic light of wavelength λ produces sinusoidal fringes of period $p(\lambda) = \lambda/(2n_{eff,\,s}\sin(\theta/2))$, whereas for arbitrary input spectral density $B(\lambda)$, the light intensity $I(x)$ as a function of position x along the interference pattern is:

$$I(x) = \int_0^{\infty} B(\lambda)\left(1 + \cos\frac{2\pi x}{p}\right) d\lambda \tag{5.13}$$

Once $I(x)$ is measured, the spectrum $B(\lambda)$ can simply be retrieved by Fourier transformation; $I(x)$ can be, for example, measured by sampling the field with an array of receiver waveguides. For a large number of receiver waveguides, it may be advantageous to redirect the light from the waveguides out of the waveguide plane by 45° waveguide mirrors and intercept the two-dimensional array of free-propagating beams by a photodetector array, as shown in Figure 5.7b and discussed in Section 5.4.2.6.

The main advantage of this spectrometer configuration is that, as a Michelson-type interferometer, it permits a markedly larger input aperture size and hence a larger light-gathering capability than conventional AWG or EG devices, without degrading resolution. A 60-fold increase in the input aperture size has been demonstrated in a simulation of a Michelson AWG spectrometer fabricated in an SOI platform.[184]

5.4.2.5 AWG Dispersion Enhancement by Waveguide Group Index Modification

An interesting possibility for increasing AWG dispersion and resolution without increasing the size of the waveguide array is to modify the waveguide group index within a triangular prism region of the waveguide array, similar to that used for polarization or temperature compensation. When the waveguide group index is modified within the triangular region of the waveguide array (Figure 5.4a), the AWG dispersion equation becomes:[69]

$$\sin\theta = \frac{(\lambda - \lambda_c)(M + M')}{n_{eff,\,s}\Lambda} = \frac{(\lambda - \lambda_c)(n_g\Delta L + \Delta n_g \Delta L_g)}{\lambda_c n_{eff,\,s}\Lambda} \tag{5.14}$$

where $M = mn_g/n_{eff}$ is the order of a conventional AWG with length increment ΔL between the adjacent waveguides, $M' = m\Delta n_g/\Delta n_{eff}$ is the order enhancement factor due to group index modification, ΔL_g is the length difference between the adjacent waveguides with modified group index, and Δn_g and Δn_{eff} are respective

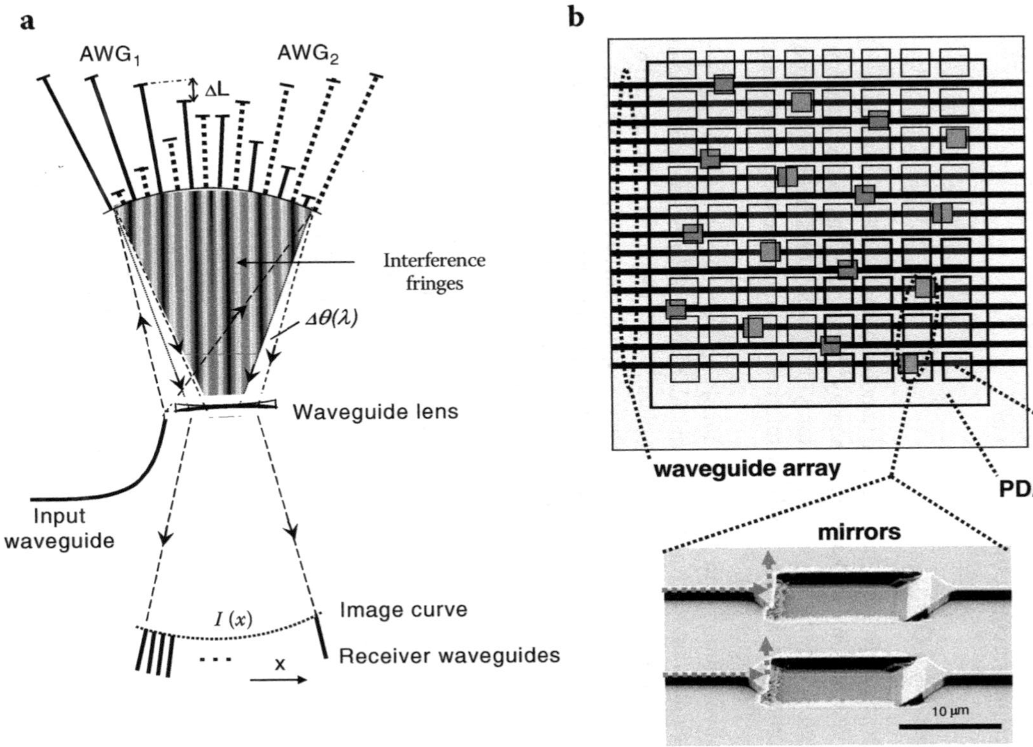

FIGURE 5.7 a) AWG in a Michelson configuration with two interleaved waveguide arrays. b) Optical read-out of a waveguide array. The light guided in a one dimensional array of output waveguides is converted by 45° waveguide mirrors into a two-dimensional (2D) array of free-propagating beams intercepted by a receiving device, for example a photodetector array (PDA).

changes in group and effective indexes caused by waveguide modification within the triangular region of the array. Group index can easily be modified by changing the waveguide cross-section, for example, the waveguide width.[278] Very large Δn_g and hence a large dispersion enhancement can be obtained over a limited wavelength range near the edges of the stop bands of various structures, including gratings, resonators, or photonic crystals. A 300-fold increase in group index has recently been reported in SOI photonic crystal waveguides.[63] Figure 5.4b shows the AWG dispersion enhancement effect calculated by Martínez al.[69] for an SOI AWG using the group index measurement by Notomi et al.[279]

5.4.2.6 Optical Read-Out

Since AWG spectrometers may have several hundred output channels, reading out the channels with conventional packaging technology[280] that relies on aligned attachment to the chip edge of a linear photodetector array or optical fibers held in a v-groove assembly becomes prohibitively costly. A solution is to integrate the photodetectors on the microspectrometer chip.[164]

An alternative solution is to divert the signals guided in a one-dimensional array of output waveguides into a two-dimensional (2D) array of free-propagating beams out-coupled in a direction normal to the chip plane, where they can be intercepted by a 2D photodetector array.[281,282] This out-of-plane coupling can be achieved by inserting a 45° waveguide mirror in each waveguide and staggering the mirrors into a 2D array matching the pixel spacing of the photodetector array (Figure 5.7b). At the angled Si–cladding interface, the waveguide mode undergoes total internal reflection (TIR). Being based on TIR, the mirror performance is broadband, polarization-independent, and nearly lossless for mirrors with high surface quality. Finite difference time domain (FDTD) simulations of 45° mirrors in a 1.5-μm thick SOI predict a coupling loss of 0.14 dB with a negligible PDL, for a mirror incorporating a 193-nm thick Si_3N_4 antireflection coating on the upper surface.

We have fabricated such mirrors by chemically assisted ion beam etching (CAIBE) with oblique ion incidence through an opening in a Ni hard mask deposited on the sample, yielding a mirror facet angle of $45 \pm 1°$ and a negligible excess loss. Because a mirror occupies an area of approximately $5 \times 5\ \mu m^2$, in principle, many thousands of mirrors can be arrayed in a 1-mm^2 chip area. This makes this simple technique attractive not only for optical read-out of waveguide spectrometers and biosensor arrays but also for dense off-chip interconnects in VLSI devices where, as it has been argued,[283] there is no known physical solution other than optics to the VLSI interconnect bottleneck problem.

Other methods have also been used for off-chip coupling. These include gratings in surface-emitting horizontal cavity semiconductor lasers,[284] gratings for coupling between planar waveguides and optical fibers,[285,286] beam shaping outcoupling elements and waveguide holograms,[287] and photonic crystal couplers.[288,289] Polarization dependency and a limited spectral range are often characteristic of these structures.

A broadband operation has recently been reported in a grating coupler with a subwavelength grating (SWG) mirror in SOI.[290] A grating with a period $\Lambda_g = \lambda_0/n_{eff}$ is formed by partially etching the silicon waveguide core. The first order diffraction at a wavelength λ in the proximity of λ_0 yields two out-of-plane diffraction orders propagating in approximately opposite directions out of the waveguide plane. Their respective angles in the superstrate (air) α_a and in the Si substrate α_s with respect to the waveguide plane normal are: $\alpha_{a,s} = \arcsin[(n_{eff}k - K)/(n_{a,s}k)]$, where $K = 2\pi/\Lambda_g$ is the modulus of the grating vector, $k = 2\pi/\lambda$ is the free-space propagation constant, and n_a and n_s are the refractive indices of the air and the substrate, respectively. These two diffraction orders are recombined by reflecting the upward propagating order by an SWG mirror. The mirror[291] comprises a periodic array of amorphous silicon ($n \sim 3.46$, with a period Λ_{swg} and width $d_{swg} = r\Lambda_{swg}$, r is the duty ratio) deposited on a low refractive index SiO_2 spacer separating the SWG mirror from the Si waveguide core. For light diffracted by the waveguide grating and incident normally on the SWG mirror, all the diffraction orders are evanescent (suppressed) for $\Lambda_{swg} < \lambda/n_{SiO_2}$.

The SWG mirror function can be intuitively understood based on the fact that the portion of light transmitted through each a-Si pillar is phase shifted π with respect to the light transmitted in the gaps between the pillars, the two parts canceling each other. This results in suppressed transmission through the SWG layer, the latter acting as a mirror. The optical thickness of the SiO_2 spacer is chosen so that the resulting out-coupled co-propagating beams add in-phase modulo 2π. The FDTD calculations predict the coupling efficiency to be larger than 90% over a broad wavelength range of 1520 to 1580 nm for a coupler length of 80 μm. The SWG mirror obviates the need for blazing the waveguide grating and deposition of multilayer dielectric or metal mirrors. The metallic mirror can induce a substantial loss in thin SOI waveguides with a thin dielectric cladding spacer, where the latter is preferred to maximize the wavelength bandwidth when interferometrically combining the two diffracted beams.

ACKNOWLEDGMENTS

I would like to thank Siegfried Janz, Dan-Xia Xu, André Delâge, Boris Lamontagne, Adam Densmore, Marie-Josée Picard, Edith Post, Winnie Ye, María Luisa Calvo, Oscar Martínez Matos, Jose Rodrigo, Stoyan Tanev and Milan Sejka for many insightful discussions and valuable comments on this manuscript, and the Institute for Microstructural Sciences, National Research Council of Canada, for continuous support. Discussions with my former colleagues at Optenia Inc. are also gratefully acknowledged.

REFERENCES

1. Fraunhofer, J., Kurtzer Berich von den Resultaten neuerer Versuche über die Gesetze des Lichtes, und die Theorie derselben, *Gilberts Ann. Phys.*, 74, 337–378, 1823.
2. Lord Rayleigh, On the manufacture and theory of diffraction gratings, *Phil. Mag.*, Series 4, 47, 193–205, 1874.
3. Wood. R., The echelette grating for the near infra-red, *Phil. Mag.* XX, Series 6, 770–778, 1910.

4. Fraunhofer, J., Über die Brechbarkeit des electrishen Lichts, *K. Acad. d. Wiss. zu München*, 61–62, 1824.
5. Rittenhouse, D., An optical problem proposed by F. Hopkinson and solved, *J. Am. Phil. Soc.*, 201, 202–206, 1786.
6. Cindrich, I., Ed., *Diffractive and Holographic Optics Technology*, Proceedings of the International Society for Optical Engineering, vol. 2152, SPIE, Bellingham, 1994.
7. Agraval, G.P., *Lightwave technology*, Wiley, New York, 2004, chap. 2.
8. Steinmeyer, G., Sutter, D.H., Gallmann, L., Matuschek, N., and Keller, U., Frontiers in ultrashort pulse generation: Pushing the limits in linear and nonlinear optics, *Science*, 286, 1507–1512, 1999.
9. Kuhlmann, T. et al., New multilayer coatings for EUV optics, in *Fraunhofer IOF annual report*, 2001, p. 36.
10. Harrison, G.R. and Loewen, E.G., Ruled gratings and wavelength tables, *Applied Optics*, 15, 1744–1747, 1976.
11. Rowland, H., Preliminary notice of results accomplished on the manufacture and theory of gratings for optical purposes, *Phil. Mag. Suppl.*, 13, 469–474, 1882.
12. Hutley, M.C., *Diffraction gratings*, Academic Press, London, 1982, chap. 7.
13. Janz, S. et al., Planar waveguide echelle gratings in silica-on-silicon, *Photon. Technol. Lett.*, 16, 503–505, 2004.
14. Tolstikhin, V.I. et al., Monolithically integrated optical channel monitor for DWDM transmission systems, *J. Lightwave Technol.*, 22, 146–153, 2004.
15. Smit, M.K. and van Dam, C., Phasar-based WDM devices: Principles, design and applications, *IEEE J. Select. Topics Quantum Electron.*, 2, 236–250, 1996.
16. Rydberg, J., Recherche sur la constitution des spectres d'émission des éléments chimiques, *Comp. Rend.*, 110, 394–400, 1890.
17. Balmer, J., Notiz über die Spectrallinien des Wasserstoffs, *Wied. Ann.*, 25, 80–87, 1885.
18. Kayser, H. and Runge, C., Über die Spectren der Elemente: I-VII Abschnitt, *Physik. Abh. Königlichen Acad. Der Wiss. zu Berlin*, S1, 1–16, 1890.
19. Runge, C. and Pashen, F., Ueber das Spectrum des Heliums, *Astrophys. J.*, 3, 4–28, 1895.
20. Zeeman, P., On the influence of magnetism on the nature of light emitted by a substance, *Phil. Mag.*, 43, 226–239, 1897.
21. Harrison, G.R., Production of diffraction gratings: I. Development of the ruling art, *J. Opt. Soc. Am.*, 39, 413–426, 1949.
22. Gabor, D., A new microscopic principle, *Nature*, 161, 777, 1948.
23. Rudolph, D. and Schmahl,. G., Verfahren zur Herstellung von Röntgenlinsen und Beugungsgittern, Umschau, 67, 225, 1967.
24. Labeyrie, A. and Flamand, J., Spectrographic performance of holographically made diffraction gratings, *Opt. Commun.*,1, 5, 1969.
25. Lippmann, M.G., La photographie des couleurs, *Comptes Rendus Hebdomadaires des Séances de l'Académie des Sciences*, 112, 274–275, 1894.
26. Wiener, O., Stationary light waves and the vibration direction of polarized light, *Annalen der Physik*, 40, 203–243, 1890.
27. Cotton, A., Gratings obtained by photographing stationary waves, *Bulletin des Sciences de la Société Francaise de Physique*, 71–73, 1901.
28. Michelson, A.A., *Studies in Optics*, Univ. Press, Chicago, 1927.
29. Konkola, P. et al., Nanometer-level repeatable metrology using the Nanoruler, *J. Vac. Sci. Technol.*, B21, 3097–3101, 2003.

30. Hill, K.O., Fujii, Y., Johnson, D.C., and Kawasaki, B.S., Photosensitivity in optical fiber waveguides: Application to reflection filter fabrication, *Appl. Phys. Lett.*, 32, 647–649, 1978.
31. Denisyuk, Yu. N., Photographic reconstruction of the optical properties of an object in its own scattered radiation field, *Sov. Phys. Dokl.*, 7, 543–545, 1962.
32. Meltz, G., Morey, M.M., and Glenn, W.H., Formation of Bragg gratings in optical fibres by a transverse holographic method, *Opt. Lett.*, 14, 823–825, 1989.
33. Lemaire, P.J., Atkins, R.M., Mizrahi, V., and Reed, W.A., High pressure H_2 loading as a technique for achieving ultrahigh UV photosensitivity and thermal sensitivity in GeO_2 doped optical fibers, *Electron. Lett.*, 29, 1191–1193, 1993.
34. Hill, O.K. and Metlz, G., Fiber Bragg grating technology fundamentals and overview, *J. Lightwave Technol.*, 15, 1263–1276, 1997.
35. Giles, C.R., Lightwave applications of fiber Bragg gratings, *J. Lightwave Technol.*, 15, 1391–1404, 1997.
36. Kogelnik, H. and Shank, C.V., Coupled-wave theory of distributed feedback lasers, *J. Appl. Phys.*, 3, 2327–2335, 1972.
37. Kogelnik, H. and Shank, C.V., Stimulated emission in a periodic structure, *Appl. Phys. Lett.*, 18, 152–154, 1971.
38. Kaminow, I.P., Weber, H.P., and Chandross, E.A., Poly(methyl methacrylate) dye laser with internal diffraction grating resonator, *Appl. Phys. Lett.*, 18, 497–499, 1971.
39. Cheben, P., Holographic recording materials for optical data storage, in *Advanced Optics*, Calvo, M. L., Ed., Ariel Ciencia, Barcelona, 2002, chap. 10.
40. Akiba, S., Usami, M., and Utaka, K., 1.5 μm λ/4-shifted InGaAsP/InP DFB lasers, *J. Lightwave Technol.*, 5, 1564–1573, 1987.
41. Buus, J., *Single frequency semiconductor lasers*, SPIE Press, Bellingham, 1991.
42. Petit, R., Ed., *Electromagnetic theories of gratings*, Springer-Verlag, Berlin, 1980.
43. Kogelnik, H., Theory of optical waveguides, in *Guided-wave optoelectronics*, Tamir, T., Ed., Springer-Verlag, New York, 1990.
44. Yariv, A., Coupled-mode theory for guided-wave optics, *IEEE J. Quantum Electron.*, QE-9, 919–933, 1973.
45. Yamada, M. and Sakuda, K., Analysis of almost-periodic distributed feedback slab waveguides via a fundamental matrix approach, *Appl. Opt.*, 26, 3474–3478, 1987.
46. Loewen, E.G., and Popov, E., *Diffraction gratings and applications*, Marcel Dekker, New York, 1997.
47. Taflove, A. and Hagness, S., *Computational Electrodynamics: The Finite-Difference Time Domain Method*, Artech House, Boston, 2000.
48. Yee, K.S., Numerical solution of initial boundary value problems involving Maxwell's equation in isotropic media, *IEEE Transactions on Antennas and Propagation,* AP-14, 302–307, 1966.
49. Joannopoulos, J.D., Villeneuve, P.R., and Fan, S., Photonic crystals: putting a new twist on light, *Nature*, 386, 143–149, 1997.
50. *Proceedings of the NATO Advanced Study Institute on Photonic Crystals and Light Localization*, Kluwer Academic Publishers, Dordrecht, 2001.
51. Yablonovitch, E., Inhibited spontaneous emission in solid state physics and electronics, *Phys. Rev. Lett.*, 58, 2059, 1987.
52. John, S., Strong localization of photons in certain disordered dielectric superlattices, *Phys. Rev. Lett.*, 58, 2486, 1987.
53. Vukusic, P. and Sambles, J.R., Photonic structures in biology, *Nature*, 424, 852–855, 2003.

54. Floquet, G., Sur les équations différentielles linéaires à coefficients périodiques, *Ann. École Norm. Sup.*, 12, 47–88, 1883.
55. Bloch, F., Über die quantenmechanik der electronen in kristallgittern, *Z. Physik*, 52, 555–600, 1928.
56. Lord Rayleigh, On the remarkable phenomenon of crystalline reflexion described by Prof. Stokes, *Philosophical Magazine*, 26, 256–265, 1888.
57. Lord Rayleigh, On the maintenance of vibrations by forces of double frequency, and on the propagation of waves through a medium endowed with a periodic structure, *Philosophical Magazine*, 24, 145–159, 1897.
58. Kurokawa, K. et al., Penalty-free dispersion managed soliton transmission over 100 km low loss PCF, in *Proc. OFC*, PDP21, 2005.
59. Mangan, B.J. et al., Low loss (1.7 dB/km) hollow core photonic bandgap fiber, *Proc. OFC*, PDP24, 2004.
60. Prather, D.W., Photonic crystals, an engineering perspective, *Optics and Photonics News*, 16–19, June 2002.
61. Dulkeith, E., McNab, S.J., and Vlasov, Y.A., Mapping the optical properties of stab-type two-dimensional photonic crystal waveguides, *Phys. Rev. B.*, 72, 115102, 2005.
62. Scherer, A., Photonic crystals for confining, guiding, and emitting light, *IEEE Transactions on Nanotechnology*, 1, 4–11, 2002.
63. Notomi, M. et al., Extremely large group-velocity dispersion of line-defect waveguides in photonic crystal slabs, *Phys. Rev. Lett.*, 87, 253902–1, 2001.
64. Vlasov, Y.A. et al., Active control of slow light on a chip with photonic crystal waveguides, *Nature*, 438, 65–69, 2005.
65. Scherer, A. et al., Photonic crystals for confining, guiding, and emitting light, *IEEE Trans. Nanotechnol.*, 1, 4–11, 2002.
66. Lodahl, P. et al., Controlling the dynamics of spontaneous emission from quantum dots by photonic crystals, *Nature*, 430, 654–657, 2004.
67. Purcell, E.M., Spontaneous emission probabilities at radio frequencies, *Phys. Rev.*, 69, 681, 1946.
68. Maystre, D., Photonic crystal diffraction gratings, *Opt. Express*, 8, 209–216, 2001.
69. Martínez, O.M. et al., Arrayed waveguide grating based on group index modification, to be published in *Journal of Lightwave Technol.*, 24, 1551–1552, 2006.
70. Kogelnik, H., High-capacity optical communications: personal recollections, *IEEE J. Select. Topics in Quant. Electron.*, 6, 1279–1286, 2000.
71. Keiser, G.E., A review of WDM technology and applications, *Optical Fiber Technology*, 5, 3–39, 1999.
72. Giles, C.R., Lightwave applications of fiber Bragg gratings, *J. Lightwave Technol.*, 15, 1391–1404, 1997.
73. Wood, R.W., The use of echelette grating in high orders, *J. Opt. Soc Am.*, 37, 733–737, 1947.
74. Harrison, G.R., The production of diffraction gratings II: The design of echelle gratings and spectrographs, *J. Opt. Soc. Am.* 38, 522–528, 1949.
75. Bach, K. G., Bach, B. W., and Bach, B. W., Jr., Large ruled monolithic echelle gratings, *SPIE Proc.*, 4014, 118–124, 2000.
76. Eagle, A., On a new mounting for a concave grating, *Astrophys. J.*, 31, 120, 1910.
77. Delâge, A. et al., Recent developments in integrated spectrometers, in *Proc. of IEEE ICTON'04*, 78–83, 2004.
78. März, R., *Integrated optics design and modeling*, Artech House, London, 1994.
79. Tomlinson, W., Wavelength multiplexing in multimode optical fibers, *Appl. Opt.*, 16, 2180–2185, 1977.

80. Tangonan, G.L. et al., Planar multimode devices for fiber optics, presented at the Int. Conf. Integrated Opt. and Optical Fiber Commun., 21–5, Amsterdam, The Netherlands, Sept. 17–19, 1979.
81. Watanabe, R. and Nosu, K., Slab waveguide demultiplexer for multimode optical transmission in the 1.0–1.4 μm wavelength region, *Appl. Opt.*, 19, 3588–3590, 1980.
82. Fujii, Y., and Minowa, J., Optical demultiplexer using a silicon concave diffraction grating, *Appl. Opt.*, 22, 974–978, 1983.
83. Yen, H.W. et al., Planar Rowland spectrometer for fiber-optic wavelength demutiplexing, *Opt. Lett.*, 6, 639–641, 1981.
84. Suhara, T., et al., Integrated-optic wavelength multi- and demultiplexers using a chirped grating and a ion-exchanged waveguide, *Appl. Opt.*, 21, 2195–2198, 1982.
85. Voges, E., Multimode planar devices for wavelength division multiplexing and demultiplexing, presented at the Int. Conf. Integrated Opt. and Optical Fiber Commun., 29A1-5, Tokyo, Japan, June 27–30, 1983.
86. Clemens, P.C., März, R., Reichelt, A., and Schneider, H. W., Flat-field spectrograph in SiO_2/Si, *IEEE Photon. Technol. Lett.*, 4, 886–887, 1992.
87. Clemens, P.C., Heise, G., März, R., Reichelt, A., and Schneider, H. W., 8-Channel optical demultiplexer realized as SiO_2/Si flat-field spectrograph, *IEEE Photon. Technol. Lett.*, 6, 1109–1111, 1994.
88. Fallahi, M. et al., Demonstration of grating demultiplexer in GaAs/AlGaA suitable for integration, *Electron. Lett.*, 28, 2217–2218, 1992.
89. Fallahi, M. et al., Grating demultiplexer integrated with MSM detector array in InGaAs/AlGaAs/GaAs for WDM, *IEEE Photon. Technol. Lett.*, 5, 794–797, 1993.
90. Cremer, C. et al., Grating spectrometer in InGaAsP/InP for dense wavelength division multiplexing, *Appl. Phys. Lett.*, 59, 627–629, 1991.
91. Cremer, C. et al., Grating spectrograph integrated with photodiode array in InGaAsP/InGaAs/InP, *IEEE Photon. Technol. Lett.*, 4, 108–110, 1992.
92. Soole, J.B.D. et al., Monolithic InP/InGaAsP/InP grating spectrometer for the 1.48-1.56 μm wavelength range, *Appl. Phys. Lett.*, 58, 1949–1951, 1991.
93. Soole, J.B.D. et al., Integrated grating demultiplexer and pin array for high-density wavelength division multiplexed detection at 1.5 μm, *Electron. Lett.*, 29, 558–560, 1993.
94. He, J.-J., Monolithic integrated wavelength demultiplexer based on a waveguide Rowland circle grating in InGaAsP/InP, *J. Lightwave Technol.*, 16, 631–638, 1998.
95. Soole, J.B.D., Multistripe array grating integrated cavity (MAGIC) laser: A new semiconductor laser for WDM applications, *Electron. Lett.*, 28, 1805–1807, 1992.
96. Zirngibl, M. et al., 12-frequency WDM laser based on a transmissive waveguide grating router, *Electron. Lett.*, 30, 701–702, 1994.
97. McGreer, K.A., A flat-field broadband spectrograph design, *IEEE Photon. Technol. Lett.*, 7, 397–399, 1995.
98. Ojha, S.M. et al., Demonstration of low loss integrated InGaAsP/InP demultiplexer device with low polarization sensitivity, *Electron. Lett.*, 29, 805–807, 1993.
99. Smit, M.K., New focusing and dispersive component based on optical phased array, *Electron. Lett.*, 24, 385–386, 1988.
100. Loewen, E.G. and Popov, E., *Diffraction gratings and applications*, Marcel Dekker, New York, 1997, p. 337.
101. Janz, S. et al., The scalable planar waveguide component technology: 40 and 256-channel echelle grating demultiplexers, in *Tech. Dig. Integrated Photonic Research*, Vancouver, Canada, July 2002, p. IFE1-1.

102. Dixon, M. et al., Performance improvements in arrayed waveguide-grating modules, in *Proc. SPIE*, 4640, 79–92, 2002.
103. Huang, M., Stress effects on the performance of optical waveguides, *Int. J. Solid Structures* 40, 1615, 2003.
104. *CRC handbook on laser science and technology*, CRC, Boca Raton, 1971.
105. Janz, S., Cheben, P., Dayan, H., and Deakos, R., Measurement of birefringence in thin-film waveguides by Rayleigh scattering, *Opt. Lett.*, 28, 1778–1780, 2003.
106. Janz, S. and Cheben, P., Birefringence measurement, International PCT application WO 03/034019 A1, World Intellectual Property Organization, publication date 24 April 2003.
107. Takahashi, H., Hibino, Y., and I. Nishi, Polarization-insensitive arrayed waveguide grating wavelength demultiplexer on silicon, *Opt. Lett.*, 17, 499, 1992.
108. Ojha, S.M. et al., Simple method of fabricating polarization insensitive and very low crosstalk AWG grating devices, *Electron. Lett.*, 34, 78, 1998.
109. Kilian, A. et al., Birefringence free planar optical waveguide made by flame hydrolysis deposition (FHD) through tailoring of the overcladding, *J. Lightwave Technol.*, 18, 193, 2000.
110. Chun, Y.Y. et al., Birefringence reduction in a high boron-doped core silica-on-silicon planar optical waveguides, *J. Korean. Phys. Soc.*, 29, 140, 1996.
111. Suzuki, S., Polarization insensitive arrayed waveguide gratings using dopant-rich silica-based glass with thermal expansion adjusted to Si substrate, *Electron. Lett.*, 33, 1173, 1997.
112. Janz, S. et al., Method for polarization birefringence compensation in a waveguide demultiplexer using a compensator with a high refractive index layer, International PCT application WO 03/023465, World Intellectual Property Organization, publication date 20 March 2003.
113. He, J.-J. et al., Integrated polarization compensator for WDM waveguide demultiplexers, *IEEE Photon. Technol. Lett.*, 11, 224–226, 1999.
114. Cheben, P. et al., Birefringence compensation in silicon-on-insulator arrayed waveguide grating devices, in *Silicon-based and hybrid Optoelectronic II*, Robbins, D. J., Trezza, J. A., and Jabbour, G. E., Eds., Proceedings of SPIE, 3953, 11–18, 2000.
115. Cheben, P. et al., Method of polarization compensation in grating- and phasar-based devices by using over-layer deposited on the compensating region to modify local slab waveguide birefringence, International PCT application WO 02/097490, World Intellectual Property Organization, publication date 5 December 2002.
116. Born, M. and Wolf, E., *Principles of Optics*, 7th ed., Cambridge University Press, Cambridge, 1999, pp. 837–840.
117. Lord Rayleigh, On the influence of obstacles arranged in rectangular order upon the properties of a medium, *Phil. Mag.*, 34, 481, 1892.
118. Rytov, S.M., Electromagnetic properties of a finely stratified medium, *Soviet Physics JETP*, 2, 466-475, 1956.
119. Gu, C. and Yeh, P., Form birefringence dispersion in periodic layered media, *Opt. Lett.*, 21, 504–506, 1996.
120. Lalanne, P. and Hugonin, J.-P., High-order effective-medium theory of subwavelength gratings in classical mounting: application to volume holograms, *J. Opt. Soc. Am. A.*, 15, 1834–1851, 1998.
121. Fiore, A. et al., Huge birefringence in selectively oxidized GaAs/AlAs optical waveguides, *Appl. Phys. Lett.*, 68, 1320–1320, 1996.

122. Cheben, P. et al., Birefringence compensation in silicon-on-insulator planar waveguide demultiplexers using a buried oxide layer, in *Photonics Packaging and Integration III*, Heyler, R.A., Robbins, D.J., and Jabbour, G.E., Eds., Proceeding of SPIE, 4997, 181–197, 2003.
123. Wörhoff, K. et al., Birefringence compensation applying double core waveguide structures, *IEEE Photon. Technol. Lett.*, 11, 206, 1999.
124. Wilkins, M.H.F., Stokes, A. R., Seeds, W. E., and Oster, G., Tobacco mosaic virus crystals and three-dimensional microscopic vision, *Nature*, 166, 127–129, 1950.
125. Goray, L.I., The modified integral method and real electromagnetic properties of echelles, in *Diffractive and Holographic Technologies for Integrated Photonic Systems*, Sutherland, R. L., Prather, D. W., Cindrich, I., Eds., Proceedings of SPIE, 4291, 13–24, 2001.
126. Loewen, E.G. et al., Echelle: Scalar, electromagnetic, and real-groove profiles, *Appl. Opt.*, 34, 1707, 1995.
127. Kleemann, B.H., Mitreiter, A., and Wyrowski, F., Integral equation method with parametrization of grating profile, theory and experiments, *J. Mod. Optics*, 43, 1323, 1996.
128. Deri, R.J., Kallman, J.S., and Dijaili, S.P., Quantitative analysis of integrated optic waveguide spectrometers, *IEEE Photon. Tech. Lett.*, 6, 242–244, 1994.
129. Graf, U.U. et al., Fabrication and evaluation of an etched infrared diffraction grating, *Appl. Opt.*, 33, 96–102, 1994.
130. Chowdhury, D., Design of low-loss and polarization-insensitive reflection grating-based planar demultiplexers, *IEEE Journ. Selec. Topics Quant. Electron.*, 6, 233–239, 2000.
131. Delâge, A. and Dossou, K., Polarization dependent loss calculation in echelle gratings using finite element method and Rayleigh expansion, *Optical and Quantum Electronics*, 36, 223–238, 2004.
132. Lord Rayleigh, On the dynamical theory of gratings, *Proc. Royal Soc.* (London) A79, 399–416, 1907.
133. Loewen, E.G., and Popov, E., *Diffraction gratings and applications*, Marcel Dekker, New York, 1997, pp. 373–374.
134. Xu, D.-X. et al., Echelle gratings with low polarization dependent loss (PDL) using metal coating on the reflective facets only, International PCT patent application WO 03/046619 A1, World Intellectual Property Organization, publication date 5 June 2003.
135. Shi, Z., He, J-J., and He, S., Waveguide echelle grating with low polarization-dependent loss using single-side metal-coated grooves, *IEEE Photon. Tech. Lett.*, 16, 1885–1887, 2004.
136. Moharam, M.G. and Gaylord, T.K., Rigorous coupled-wave analysis of metallic surface-relief gratings, *J. Opt. Soc. Am. A*, 3, 1780–1787, 1986.
137. Lalanne, P., Highly improved convergence of the coupled-wave method for TM polarization, *J. Opt. Soc. Am. A*, 13, 779–784, 1996.
138. Xu, D.-X, et al., Metallised echelle grating with reduced polarization dependence using dielectric spacer layers, International PCT application WO 03/046624 A1, World Intellectual Property Organization, publication date 5 June 2003.
139. McGreer, K.A. et al., Advanced grating based WDM demultiplexer, Proc. of SPIE, 2918, 92, 1996.
140. Erickson, L. et al., Using a retro-reflecting echelle grating to improve WDM demux efficiency, in *Digest of the IEEE/LEOS Summer Topical Meeting*, 82–83, 1997.

141. Janz, S. et al., Right angle corner retroreflectors, International PCT application WO 02/097482 A2, World Intellectual Property Organization, publication date 5 December 2002.
142. Aristov, V.V., Erko, A.I., and Martinov, V.V., Principles of Bragg-Fresnel multilayer optics, *Rev. Phys. Appl.*, 23, 1623–1630, 1988.
143. Mashev, L. and Popov, E., Diffraction efficiency anomalies of multicoated dielectric gratings, *Opt. Commun.*, 51, 131–136, 1984.
144. Humpreys, B. and Koteles, E., Fabrication challenges for enabling metropolitan WDM network technologies, *Compound Semiconductor*, 87–94, 2001.
145. Lamontagne, B., Method for deep and vertical dry etching of dielectrics, International PCT application WO 02/097874 A1, World Intellectual Property Organization, publication date 5 December 2002.
146. He, J.J., Monolithic integrated waveguide demultiplexer based on a waveguide Rowland circle grating in InGaAsP/InP, *J. Lightwave Technol.*, 16, 631–638, 1998.
147. Chen, W. et al., The role of photomask resolution on the performance of arrayed-waveguide grating device, in *Proc. SPIE*, 4087, 283–292, 2000.
148. Koteles, E., Integrated planar waveguide demultiplexers for high density WDM applications, *Fiber and Integrated Optics*, 18, 211–244, 1999.
149. Dixton, M. et al., Performance improvements in arrayed waveguide-grating modules, *Proc. SPIE*, 4640, 79–92, 2002.
150. Okamoto, K. and Yamada, H., Arrayed waveguide grating multiplexer with flat spectral response, *Opt. Lett.*, 20, 43–45, 1995.
151. Okamoto, K. and Sugita, A., Flat spectral response arrayed waveguide grating multiplexer with parabolic waveguide horns, *Electron. Lett.*, 32, 1661–1662, 1996.
152. Soole, J.B.D. et al., Use of multimode interference couplers to broaden the passband of wavelength-dispersive integrated WDM filters, *IEEE Photon. Technol. Lett.*, 8, 1340–1342, 1996.
153. Rigny, A., Bruno, A., and Sik, H., Multigrating method for flattened spectral response wavelength multi/demultiplexer, *Electron. Lett.*, 33, 1701–1702, 1997.
154. Ho, Y.P., Li, H., and Chen, Y.J., Flat channel-passband wavelength multiplexing and demultiplexing devices by multiple Rowland circle design, *IEEE Photon. Technol. Lett.*, 9, 342–344, 1999.
155. Thompson, G.H.B., An original low-loss and passband flattened SiO_2 on Si planar wavelength demultiplexer, in *Tech. Dig. Optical Fiber Conference*, 77, 1998.
156. Dragone, C., Efficient techniques for widening the passband of a wavelength router, *J. Lightwave Technol.*, 16, 1895–1906, 1998.
157. Kamalakis, T. and Sphicopoulos, T., An efficient technique for the design of an arrayed waveguide grating with flat spectral response, *J. Lightwave Technol.*, 19, 1716–1725, 2001.
158. Delâge, A. et al., Method of creating a controlled flat pass band in an echelle or waveguide grating, International PCT patent application WO 02/097484 A1, World Intellectual Property Organization, 5 December 2002.
159. He, J.J., Phase-dithered waveguide grating with flat passband and sharp transitions, *IEEE Journ. Select. Top. Quant. Electron.*, 8, 1186–1193, 2002.
160. Lenz, G., Optimal dispersion of optical filters for WDM systems, *IEEE Photon. Technol. Lett.*, 10, 567–569, 1998.
161. Janz, S. et al., Integrated double pass equalizer for telecommunications network, International PCT application WO 02/098026 A1, World Intellectual Property Organization, publication date 5 December 2002.

162. Pearson, M. et al., Optical performance monitor, International PCT application WO 03/024011 A2, World Intellectual Property Organization, publication date 20 March 2003.
163. Janz, S. et al., Integrated optical add-drop muliplexer using optical waveguide mirrors and multiplexer/demultiplexer, International PCT application WO 02/098038 A1, World Intellectual Property Organization, publication date 5 December 2002.
164. Tolstikhin, V. et al., Monolithically integrated optical channel monitor for DWDM transmission systems, *J. Lightwave Technol.*, 22, 146–153, 2004.
165. Tolstikhin, V. et al., Monolithically integrated InP-based dynamic channel equalizer using waveguide electroabsorptive attenuators-photodetectors, European Conference on Optical Communications (ECOC 2002), paper P2.21, Copenhagen, September 2002.
166. Tolstikhin, V. et al., Single-mode vertical integration of p-i-n photodetectors with optical waveguides for monitoring in WDM transmission systems, *IEEE Photon. Tech. Lett.*, 15, 843–845, 2003.
167. Densmore, A. et al., Integrated electroabsorption attenuator-photodetector for optical power control in WDM transmission systems, *IEEE J. Select. Top. Quant. Electron.*, 8, 1435–1444, 2002.
168. Densmore, A. et al., DWDM data receiver based on monolithic integration of an echelle grating demultiplexer and waveguide photodiodes, *Electronic Letters*, 41, 766–767, 2005.
169. Tolstikhin, V., InP-based photonic integrated circuit triplexer for FTTP applications, presented at 31st European Conference on Optical Communications, Glasgow, UK, 25–29 September 2005.
170. Bidnyk, S. et al., Planar echelle grating DWDM comb filters, Optical Fiber Communication Conference (OFC 2004), paper TuL2, 2004.
171. Bidnyk, S. et al., Planar comb filters based on aberration-free elliptical grating facets, *J. Lightwave Technol.*, 23, 1239, 2005.
172. Packirisamy, M. and Delâge, A., Planar waveguide echelle grating device with astigmatic grating facets, U.S. Patent and Trademark Office, Provisional patent No. 09/986,828.
173. Wang, W., Etched-diffraction-grating-based planar waveguide demultiplexer on silicon-on-insulator, *Opt. and Quantum Electron.*, 36, 559–566, 2004.
174. Dai, D. and He, S., Reduction of multimode effects in a SOI-based etched diffraction grating demultiplexer, *Opt. Commun.*, 247, 281–290, 2005.
175. Sander, D. and Müller, J., Self-focusing phase transmission grating for an integrated optical microspectrometer, *Sensors and Actuators A*, 88, 1–19, 2001.
176. Sander, D., Blume, O., and Müller, J, Transmission gratings with SiON-slab-waveguides, *Apl. Opt.*, 35, 4096–4101, 1996.
177. Takahashi, H., Suzuki, S., and Nishi, I., Wavelength multiplexer based on SiO_2-Ta_2O_5 arrayed-waveguide grating, *J. Lightwave Technol.*, 12, 989–995, 1994.
178. Parker, M.C. and Walker, S.D., Design of arrayed-waveguide gratings using hybrid Fourier-Fresnel transform techniques, *IEEE J. Select. Topics Quantum Electron.*, 5, 1379–1384, 1999.
179. Special issue on arrayed grating routers/WDM mux/demuxs and related applications/uses, *IEEE J. Select. Top. Quantum Electron.*, 8, 1087–1214, 2002.
180. Okamoto, K., *Fundamentals of Optical Waveguides*, Academic Press, London, 2000, chap. 9.
181. Hibino, Y., Recent advances in high density and large scale AWG multi-demultiplexers with higher index contrast silica based PLCs, *IEEE J. Select. Topics Quantum Electron.*, 8, 1090–1101, 2002.

182. Dragone, C., Efficient reflective multiplexer arrangement, U.S. Patent 5 450 511, September 12, 1995.
183. Muñoz, P., Pastor, D., and Capmany, J., Modeling and design of arrayed waveguide gratings, *J. Lightwave Technol.*, 20, 661–674, 2002.
184. Cheben, P., Powell, I., Janz, S., and Xu, D.-X., Wavelength-dispersive device based on a Fourier-transform Michelson-type arrayed waveguide grating, *Opt. Lett.*, 30, 1824–1826, 2005.
185. Cheben, P., Powell, I., Janz, S., and Xu, D.-X., Wavelength dispersive Fourier-transform spectrometer, U.S. patent application 11/221,925, September 9, 2005.
186. Smit, M.K., Progress in AWG design and technology, in Proc. of WFOPC'05 (IEEE/LEOS), 26–31, 2005.
187. Hida, Y. et al., A 10-m long silica-based waveguide with a loss of 1.7 dB/m, in *Proc. Integrated Photonics Research,* paper IthC6, 1995.
188. Takada, K. et al., A 25-GHz-spaced 1080-channel tandem multi/demultiplexer covering the S-, C-, and L-bands using arrayed waveguide grating with Gaussian passband as primary filter, *IEEE Photon. Technol. Lett.*, 14, 648–650, 2002.
189. Hibino, Y., Recent advances in high-density and large-scale AWG multi/demultiplexers with higher index-contrast silica-based PLCs, *IEEE J. Select. Top. Quantum Electron.*, 8, 1090–1101, 2002.
190. Hida, Y. et al., 400-channel 25-GHz spacing arrayed-waveguide grating covering a full range of C- and L-bands, in *Proc. OFC2001*, paper WB2, 2001.
191. Yamauchi, J., Ikegaya, M., and Nakano, H., Bend loss of step-index slab waveguides with a trench section, *Microw. Optical. Technol. Lett.* 5, 251–254, 1992.
192. Popović, M. et al., Air trenches for sharp silica waveguide bends, *J. Lightwave Technol.*, 20, 1762–1772, 2002.
193. Suzuki, T. and Tsuda, H., Ultrasmall arrowhead arrayed-waveguide grating with V-shaped bend waveguides, *IEEE Photon. Technol. Lett.*, 17, 810–812, 2005.
194. Muñoz, P. et al., Analytical and numerical analysis of phase and amplitude errors in the performance of arrayed waveguide gratings, *IEEE J. Select. Top. Quantum Electron.*, 8, 1130–1141, 2002.
195. Maru, K. et al., Statistical analysis of correlated phase error in transmission characteristic of arrayed-waveguide gratings, *IEEE J. Select. Top. Quantum Electron.*, 8, 1142–1148, 2002.
196. Klekamp, A. and Münzner, R., Imaging errors in arrayed waveguide gratings, *Opt. and Quant. Elecron.*, 35, 333–345, 2003.
197. Pizzato, F., Perone, G., and Montrosset, I., Arrayed waveguide grating demultiplexers: a new efficient numerical analysis approach, *Proc. SPIE*, 2620, 198, 1999.
198. Takada, K. et al., Measurement of phase error distributions in silica-based arrayed-waveguide grating multiplexers by using Fourier transform spectroscopy, *Electron. Lett.*, 30, 1671–1672, 1994.
199. Chen, W. et al., Improved techniques for the measurement of phase error in waveguide based optical devices, *J. Lightwave Technol.*, 21, 198–205, 2003.
200. Saida, T., Arrayed-waveguide grating with built-in interferometer for measuring amplitude and phase distribution, *IEEE Photon. Technol. Lett.*, 17, 1659–1661, 2005.
201. Yamada, H. et al., Statically-phase-compensated 10 GHz-spaced arrayed-waveguide grating, *Electron. Lett.*, 32, 1580–1582, 1996.
202. Takada, K. et al., Beam-adjustement-free crosstalk reduction in 10 GHz-spaced arrayed-waveguide grating via photosensitivity under UV laser irradiation through metal mask, *Electron. Lett.*, 36, 60–61, 2000.

203. Koktohu, M. et al., Control of higher order leaky modes in deep-ridge waveguides and application to low-crosstalk arrayed waveguide gratings, *J. Lightwave Technol.*, 22, 499–508, 2004.
204. Pearson, M.R.T. et al., Arrayed waveguide grating demultiplexer in silicon-on-insulator, in *Proc. SPIE*, 3953, 11–18, 2000.
205. Vazquez, C. et al., Multimode interference filter to solve degradation on couplers common-mode rejection, *LEOS Annu. Mtg. Conf. Proc.*, 326–327, 1998.
206. Kikuchi, N., Monolithically integrated 100-channel WDM channel selector employing low-crosstalk AWG, *IEEE Photon. Technol. Lett.*, 16, 2481–2483, 2004.
207. Vellekoop, A.R. and Smit, M.K., Four-channel integrated-optic wavelength demultiplexer with weak polarization dependence, *J. Lightwave Technol.*, 9, 310–314, 1991.
208. Zirngibl, M. et al., Polarization-independent 8 × 8 waveguide grating demultiplexer on InP, *Electron. Lett.*, 29, 201–202, 1993.
209. Takahashi, H. et al., Polarization-insensitive arrayed-waveguide wavelength demultiplexer with birefringence compensating film, *IEEE Photon. Tech. Lett.*, 5, 707–709, 1993.
210. Zirngibl, M., Joyner, C.H., and Chou, P.C., Polarization compensated waveguide grating router on InP, *Electron. Lett.*, 31, 581–582, 1995.
211. Suzuki, S., Innoue, Y., and Ohmori, Y., Polarization-insensitive arrayed-waveguide grating multiplexer with SiO_2-on-SiO_2 structure, *Electron. Lett.*, 30, 642–643, 1994.
212. Nadler, C.K. et al., Polarization insensitive, low-loss, low-crosstalk wavelength multiplexer modules, *IEEE J. Sel. Top. Quant. Electron.*, 5, 1407-1412, 1999.
213. Inoue, Y. et al., Novel birefringence compensating AWG design, Tech. Dig. OFC 2001, Paper WB4-1, Anaheim, 2001.
214. Bissessur, H. et al., Polarization-independent phased-array demultiplexer on InP with high fabrication tolerance, *Electron. Lett.*, 31, 1372–1373, 1995.
215. Kasahara, R., Birefringence-compensated silica-based waveguide with undercladding ridge, *Electron. Lett.*, 38, 1178–1179, 2002.
216. Hibino, Y., Silica-based planar lightwave circuits and their applications, *MRS Bull.*, 365–371, May 2003.
217. Hida, Y., Inoue, Y., and Kominato, T., Elimination of PD-λ at all output ports of silica-based AWG, *Electron. Lett.*, 40, 1118–1119, 2004.
218. Heise, G., Shneider, P.C., and Clemens, P.C., Optical phased array filter module with passively compensated temperature dependence, in *Proc. 24th Eur. Conf. Opt. Commun.*, Madrid, 319–320, 1998.
219. Ooba, N. et al., Athermal silica-based arrayed-waveguide grating multiplexer using bimetal plate temperature compensator, *Electron. Lett.*, 36, 1800–1801, 2000.
220. Soole, J.B.D. et al., Athermalised monolithic VMUX employing silica arrayed waveguide grating multiplexer, *Electron. Lett.*, 39, 1318–1319, 2003.
221. Saito, T. et al., 100-GHz 32-ch athermal AWG with extremely low temperature dependency of central wavelength, *Tech. Dig. OFC*, 1, 57–58, 2003.
222. de Peralta, L.G. et al., Control of central wavelength in reflective-arrayed waveguide-grating multiplexers, *IEEE J. Quant. Electron.*, 40, 1725–1731, 2004.
223. Hirota, H. et al., Athermal arrayed-waveguide grating multi/demultiplexers composed of TiO_2-SiO_2 waveguides on Si, *IEEE Photon. Technol. Lett.*, 17, 375–377, 2005.
224. Inoue, Y. et al., Athermal silica-based arrayed-waveguide grating multiplexer, *Electron. Lett.*, 33, 1945–1946, 1997.
225. Kamei, S. et al., A 1.5%-Δn athermal arrayed-waveguide grating multi/demultiplexer with very low loss groove design, *IEEE Photon. Technol. Lett.*, 17, 588–590, 2005.

226. Maru, K. et al., Super-high-Δ athermal arrayed waveguide grating with resin-filled trenches in slab region, *Electron. Lett.*, 40, 374–375, 2004.
227. Marhic, M.E. and Yi, X., Calculation of dispersion in arrayed waveguide grating demultiplexers by a shifting-image method, *IEEE J. Sel. Top. Quant. Electron.*, 8, 1149–1157, 2002.
228. Kitoh, T. et al., Low chromatic-dispersion flat-top arrayed waveguide grating filter, *Electron. Lett.*, 39, 1116–1118, 2003.
229. Fondeur, B. et al., Ultrawide AWG with hyper-Gaussian profile, *IEEE Photon. Technol. Lett.*, 16, 2628–2630, 2004.
230. Horts, F. et al., Compact 1-to-8, double 1-to-4 or four-fold 1-to-2 space switch based on beam steering by a waveguide array, in *Proc. IEEE Electron. Comp. Technol. Conf.*, 922, 1999.
231. Yoshikuni, Y., Semiconductor arrayed waveguide gratings for photonic integrated devices, *IEEE J. Sel. Top. Quant. Electron.*, 8, 1102–1114, 2002.
232. Leijtens, X., Developments in photonic integrated circuits for WDM applications, *Proc. SPIE*, 5247, 19–25, 2003.
233. Kohtoku, M., Semiconductor arrayed waveguide gratings for integrated photonics devices, *Electron. Commun. Jpn.*, 83, 18–26, 2000.
234. Kohtoku, M. et al., InP-based 64-channel arrayed waveguide grating with 50 GHz channel spacing and up to –20 dB crosstalk, *Electron. Lett.*, 33, 1786–1787, 1997.
235. Barbarin, Y. et al., Extremely small AWG demultiplexer fabricated on InP by using a double-etch process, *IEEE Photon. Technol. Lett.*, 16, 2478–2480, 2004.
236. Suzaki, Y. et al., Monolithic integrated eight-channel WDM modulator with narrow channel spacing and high throughput, *IEEE J. Sel. Top. Quant. Electron.*, 11, 43–49, 2005.
237. Nagarajan, R. et al., Large-scale photonic integrated circuits, *IEEE J. Sel. Top. Quant. Electron.*, 11, 50–65, 2005.
238. Oku, S., Shibata, Y., and Ochiai, K., Controlled beam dry etching of InP by using Br_2-N_2 gas, *J. Electron. Mater.*, 25, 585–591, 1996.
239. Herben, C.G.P. et al., Low-loss and compact phased array demultiplexer using a double etch process, *Proc. ECIO '99*, 211–214, 1999.
240. Zhu, Y.C. et al., A compact phasar with low central channel loss, in *Proc. ECIO '99*, 219–222, 1999.
241. Suzuki, K. et al., Reduction in the diffraction loss of an arrayed-waveguide grating by use of an interference fringe between slab and arrayed waveguides, *Opt. Lett.*, 30, 2400–2402, 2005.
242. Moerman, I., van Daele, P.P., and Demeester, P.M., A review on fabrication technologies for the monolithic integration of tapers with III–V semiconductor devices, *IEEE J. Sel. Top. Quant. Electron.*, 3, 1308–1320, 1997.
243. Soref, R.A., Silicon-based optoelectronics, *Proc. IEEE*, 81, 1687–1706, 1993.
244. Pavesi, L., Will silicon be the photonic material of the third millennium? *J. Phys. Condens. Matt.*, 15, 1169–1196, 2003.
245. Reed, G. and Knights, A.P., *Silicon Photonics: An Introduction*, John Wiley & Sons, Chichester, 2004.
246. Pavesi, L. and Lockwood, D.J., Eds., *Silicon Photonics*, Springer-Verlag, Berlin, 2004.
247. Soref. A.R., Navamar, F., and Lorenzo, J.P., Optical waveguiding in a single crystal layer of germanium silicon grown on silicon, *Opt. Lett.*, 15, 270–272, 1990.
248. Soref, R.A., Schmidtchen, J., and Petermann, K, Large single-mode rib waveguides in GeSi-Si and Si-on-SiO_2, *IEEE J. Quant. Electron.*, 27, 1971, 1991.

249. Schleppert, B. et al., Integrated optics in silicon and Si-Ge-heterostructures, *J. Lightwave Technol.*, 14, 2311–2322, 1996.
250. Janz, S. et al., Optical properties of pseudomorphic $Si_{1-x}Ge_x$ for Si-based waveguides at the $\lambda = 1300$-nm and 1550-nm telecommunications wavelength bands, *IEEE J. Sel. Top. Quant. Electron.*, 4, 990–996, 1998.
251. Xu, D.X. et al., Silicon-on-insulator (SOI) as a photonic platform, *12th Int. Symp. SOI Technol. Dev., ECS Proc.*, 207–218, 2005.
252. Lee, K.K. et al., Fabrication of ultra-low loss Si/SiO_2 waveguides by roughness reduction, *Opt. Lett.*, 26, 1888–1890, 2001.
253. Dumon, P. et al., Low loss photonic wires and ring resonators fabricated with deep UV lithography, *IEEE Photon. Technol. Lett.*, 16, 1328–1330, 2004.
254. Trinh, P.D. et al., Silicon-on-insulator (SOI) phased-array wavelength multi/demultiplexer with extremely low-polarization sensitivity, *IEEE Photon. Technol. Lett.*, 9, 940–941, 1997.
255. Bozeat, R.J. et al., Silicon based waveguides, in *Silicon Photonics*, Pavesi, L. and Lockwood, D.J., Eds., Springer-Verlag, Berlin, 2004.
256. Cheben, P. et al., Scaling down photonic waveguide devices on the SOI platform, in *SPIE Proc.*, 5117, 147–156, 2003.
257. Sasaki, K., Ohno, F., Motegi, A., and Baba, T., Arrayed waveguide grating of 70×60 μm^2 size based on Si photonic wire waveguides, *Electron. Lett.*, 41, 801–802, 2005.
258. Kosaka, H., Superprism phenomena in photonic crystals, *Phys. Rev. B*, 58, R10096, 1998.
259. Wu, L. et al., Superprism phenomena in planar photonic crystals, *IEEE J. Quant. Electron.*, 38, 915–918, 2002.
260. Baba, T., Photonic crystal light deflection devices using the superprism effect, *IEEE J. Quant. Electron.*, 38, 909–914, 2002.
261. Janz, S. et al., Microphotonic elements for integration on the silicon-on-insulator waveguide platform, *J. Sel. Top. Quant. Electron.*, in press, 2006.
262. Janz, S. et al., Microphotonics, current challenges and applications, in *Frontiers in Planar Lightwave Circuit Technology, Design, Simulation, and Fabrication*, Janz, S. et al., Eds., Nato Science Series II, Vol. 216, 2006, chap. 1.
263. Vivien, L. et al., Polarization-independent single-mode rib waveguides on silicon-on-insulator for telecommunication wavelengths, *Opt. Commun.*, 210, 43, 2002.
264. Whiteman, R.R. et al., Recent progress in the design, simulation and fabrication of small cross-section silicon-on-insulator VOAs, *Proc. SPIE*, 4997, 146, 2003.
265. Xu, D.X. et al., Eliminating the birefringence in silicon-on-insulator ridge waveguides by use of cladding stress, *Opt. Lett.*, 29, 2384–2386, 2004.
266. Ye, W.N. et al., Birefringence control using stress engineering in silicon-on-insulator (SOI) waveguides, *J. Lightwave Technol.*, 23, 1308–1331, 2005.
267. Xu, D.X. et al., Prospects and challenges for microphotonic waveguide components based on Si and SiGe, in *Electrochem. Soc. Proc.*, 7, 619–633, 2004.
268. Xu, D.X. et al., Control and compensation of birefringence in SOI waveguides, in *Proc. LEOS IEEE Annu. Mtg.*, Vol. 2, 590–591, 2003.
269. Xu, D.X., Janz. S., and Cheben, P., Design of polarization-insensitive ring resonators in SOI using cladding stress engineering and MMI couplers, *IEEE Photon. Technol. Lett.*, 18, 343–345, 2006.
270. Masanovic, G.A., Passaro, V.M.N., and Reed, G.T., Dual grating-assisted directional coupling between fibers and thin semiconductor waveguides, *IEEE Photon. Technol. Lett.*, 15, 1395–1397, 2003.

271. Taillert, D. et al., A compact two-dimensional grating coupler used as a polarization splitter, *IEEE Photon. Technol. Lett.*, 15, 1395–1397, 2003.
272. Almeida, V.R., Panepucci, R.R., and Lipson, M., Nanotaper for compact mode conversion, *Opt. Lett.*, 28, 1302–1304, 2003.
273. Lee, K.K. et al., Mode transformer for miniaturized optical circuits, *Opt. Lett.*, 30, 498–500, 2005.
274. Delâge, A. et al., Graded-index coupler for microphotonic SOI waveguides, in *Proc. SPIE*, 5577, 204–212, 2004.
275. Delâge, A. et al., Monolithically integrated symmetric graded and step-index couplers for microphotonic waveguides, *Opt. Express*, 14, 148–161, 2006.
276. Cheben, P. et al., Subwavelength waveguide grating for mode conversion and light coupling in integrated optics, *Opt. Express*, 14, 4695–4702, 2006.
277. Cheben, P. et al., A 100-channel near-infrared SOI waveguide microspectrometer: design and fabrication challenges, *SPIE Proc.*, 5644, 103–110, 2005.
278. Martínez, O.M. et al., A wavelength demultiplexer based on waveguide broadening in silicon-on-insulator platform, *SPIE Proc.*, 5622, 885–891, 2004.
279. M. Notomi et al., Extremely large group-velocity dispersion of line-defect waveguides in photonic crystal slab, *Phys. Rev. Lett.*, 87, 253902–1, 2001.
280. Kato, K., Inshii, M., and Inoue, Y., Packaging of large-scale planar waveguide circuits, *IEEE Trans. Comp. Packag. Manufac. Technol.*, B21, 121, 1998.
281. Cheben, P. et al., Optical off-chip interconnects in multichannel planar waveguide devices, U.S. Patent Application 60/481,828, December 24, 2004.
282. Lamontagne, B. et al., Fabrication of out-of-plane micro-mirrors in silicon-on-insulator planar waveguides, presented at 12th Can. Semiconductor Technol. Conf., August 2005, Paper TH1.2, *J. Vac. Sci. Technol.,* submitted, 2006.
283. Miller, D.A.B., Optical interconnects to silicon, *IEEE J. Sel. Top. Quant. Electron.*, 6, 1312, 2000.
284. Evans, G.A. et al., Grating surface emitting lasers, in *Surface Emitting Semiconductor Lasers and Arrays*, Evans, G.A. and Hammer, J.M., Eds., Academic Press, New York, 1993, chap. 4.
285. Taillaert et al., An out-of-plane grating coupler for efficient butt coupling between compact planar waveguides and single-mode fibers, *IEEE J. Quant. Electron.*, 38, 949-955, 2002.
286. Masanovic, G.Z., Passaro, V.M.N., Reed, B.T., Coupling optical fibers to thin semiconductor waveguides, *Proc. SPIE*, 4997, 171–180, 2003.
287. Larsson, A. et al., Grating coupled surface emitters: integrated lasers, amplifiers, and beam shaping outcouplers, *Proc. SPIE*, 3626, 190–200, 1999.
288. Ziolkowski, R.W. and Liang, T., Design and characterization of a grating-assisted coupler enhanced by a photonic-band-gap structure for effective wavelength-division demultiplexing, *Opt. Lett.*, 22, 1033–1035, 1997.
289. Mekis, A. et al., Two-dimensional photonic crystal couplers for unidirectional light output, *Opt. Lett.*, 25, 942–944, 2000.
290. Cheben, P. et al., A broad-band waveguide grating coupler with a subwavelength grating mirror, *IEEE Photon. Technol. Lett.*, 18, 13–15, 2006.
291. Mateus, C.F.R. et al., Ultrabroadband mirror using low-index cladded subwavelength grating, *IEEE Photon. Tech. Lett.*, 16, 518–520, 2004.

6 Silicon Waveguides for Integrated Optics

Andrew P. Knights and Paul E. Jessop

CONTENTS

The development of optical functionality on a silicon platform holds great promise for the widespread deployment of integrated circuits that combine photonics and microelectronics. Such devices would exert direct impacts in many hi-tech arenas such as microelectronics, telecommunications, and the biological, chemical, and mechanical sensor industries. In this chapter, we attempt to outline the current status of and future prospects for silicon-based photonics. We begin with an explanation of why silicon will be a dominant optical platform in the coming decade. Next we review the physics that underpins optical waveguiding in silicon. Following this background material, we describe some applications and specific devices (both passive and active) that have been demonstrated successfully.

6.1 INTRODUCTION: WHY SILICON?

The choice of silicon as a substrate for the fabrication of optical waveguides follows naturally from its domination of microelectronics. To begin then, we briefly review the histories of silicon electronic devices and integrated circuits (ICs).

The historical development of the solid state microelectronics industry is punctuated by fundamental scientific discoveries, the invention of complex device structures, and the development of ingenious fabrication equipment and technology that allows large-scale production of transistors with dimensions on a submicron scale with unit costs of a tiny fraction of a cent [1]. The initial impetus for semiconductor device research in the first half of the 20th century was the need for alternatives to vacuum tubes as energy-efficient switches. Toward the end of 1947, Bell Telephone Laboratories employees John Bardeen, Walter Brattain, and William Schockley demonstrated a point contact transistor that led directly to the invention of the bipolar transistor and their receipt of the Nobel Prize for physics. This truly ground-breaking research yielded the first solid state switch and the search for methods of miniaturizing such devices began.

The Bell solid state transistor was fabricated in polycrystalline germanium, not in silicon. However, during the 1950s, silicon was shown to possess several properties that made it preferable to germanium for device fabrication. Not least is the ease with which a high quality oxide (SiO_2) may be formed on its surface. This oxide provides an almost perfect passivation layer. It is highly electrically resistive, is resistant to most chemical etching (but may be selectively removed using hydrofluoric acid), and provides an efficient diffusion barrier to common silicon dopants (and thus may be used as a doping mask during a diffusion doping process). Further, in comparison to germanium, silicon is better suited to power handling and operation at elevated temperatures.

In 1958, while working at Texas Instruments (TI), Jack Kilby demonstrated that it was possible to fabricate a resistor, capacitor, and transistor using single-crystal silicon [2]. This technological landmark led directly to the first truly integrated circuit (fabricated using the TI germanium platform) and its importance was recognized by the award of a Nobel Prize to Kilby in 2000. In the almost five decades since Kilby's early work, the microelectronics industry has witnessed reductions in individual device size and concomitant increases in chip functionality. This trend led to the doubling of device density approximately every 24 months, roughly in line with the prediction of Gordon Moore in 1962. Moore's law has more or less remained relevant, forming the motivation for the International Technology Roadmap for Semiconductors (ITRS) in 1993 (then known as the National Technology Roadmap for Semiconductors). The roadmap is a needs-driven document that assumes that the industry will be dominated by complementary metal oxide semiconductor (CMOS) silicon technology [3]. In fact, the MOSFET transistor forms the basic element of many standard products such as high speed MPU, DRAM and SRAM.

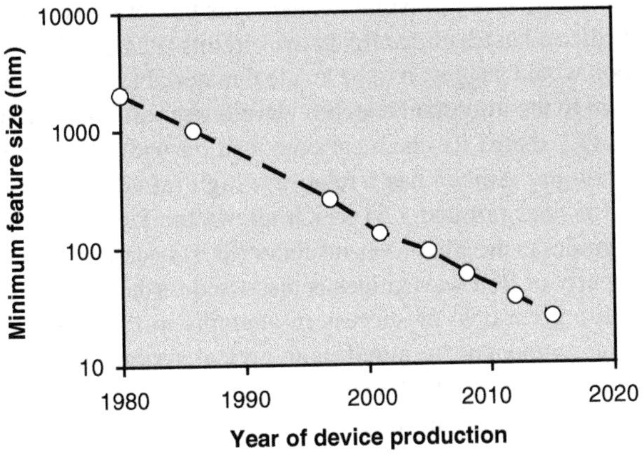

FIGURE 6.1 Indication of the reduction in minimum devices dimension of silicon integrated circuits.

The continued shrinkage of device dimension is indicated by the minimum feature size found on the surface of a CMOS device. This trend is summarized in Figure 6.1. In the early 1960s, the minimum feature size was on the order of tens of microns. This value was driven below one micron in the 1980s, while today we may anticipate the size of the smallest feature to be less than 100 nm.

It is clearly impossible for device dimensions to continue to shrink forever. Ultimately such a trend would see the requirement for features with dimensions smaller than those of individual atoms. Even before this fundamental limit is approached though, devices will become so small that quantum effects will inhibit their classical operation. As minimum feature size approaches the point of transition between classical and quantum behavior, the roadmap outlines a number of serious obstacles to the continuation of Moore's law. For example, concomitant with the shrinking of lateral dimensions has been the reduction in the thickness of the MOS transistor's insulating gate oxide. We are now approaching the point where the gate oxide is required to be so thin that charge carriers can tunnel through the insulating layer.

The integration of photonic and electronic functionality is seen as a solution to another roadblock — that associated with the increasing power demands of metal–dielectric interconnects. More generally, photonics promises an increase in computing speed and efficiency for microprocessors by reducing the response time of global interconnects to a point where the switching speed of the transistor is dominated by the gate delay. The silicon roadmap considers optical interconnects as primary options for replacing metal–dielectric systems [3]. It concedes that although this option has many advantages, there exist several clear areas requiring significant research. Associated with any optical interconnect system are a number of individual elements such as emitters, waveguides, and detectors. The roadmap highlights the importance of the delays associated with *all* elements even though the signal travels at the speed of light.

It is important to note that no single material possesses the optimum properties for each individual device found in an IC, but silicon contains a base material from which *all* the required devices can be fabricated. This flexibility is also important in the fabrication of silicon-based, planar lightwave circuits (PLCs). Many of the optical properties of silicon would suggest it to be an ideal material for PLC fabrication. It is virtually transparent to the important telecom wavelengths around 1550 nm. Further, silicon dioxide (SiO_2) shares its chemical composition with glass fiber, providing a degree of compatibility. Silicon has a relatively high refractive index around 3.5, compared to that for fiber (around 1.5) which allows the fabrication of submicron dimensional waveguides in the silicon-on-insulator (SOI) system. This so-called high-index contrast property of SOI waveguides is discussed further in Section 6.2.

There has been a great deal of success in attempts to monolithically fabricate the vast majority of components required in an optical system such as waveguides, modulators, (de)multiplexers, and detectors on a silicon substrate. Several examples of these successes are outlined in Sections 6.3 and 6.4. However, the outstanding limitation of silicon in the photonics arena is its indirect band gap that prevents the straightforward formation of efficient optical sources. Indeed, the search for a silicon-based optical source forms the greatest challenge to the widespread dominance of silicon in opto-electronics. This chapter includes a brief overview of two approaches to the fabrication of such a source in Section 6.4.

6.2 SILICON WAVEGUIDING: WHAT'S DIFFERENT ABOUT HIGH INDEX CONTRAST?

6.2.1 SILICON-ON-INSULATOR

Silicon has a band-gap energy of 1.12 eV, which places its absorption band edge at a wavelength of 1.1 μm. For wavelengths shorter than this, silicon is highly absorbing and is an important photonic material for photodetectors and for CCD and CMOS imaging. For wavelengths longer than 1.1 μm, including the most important optical communications bands, high purity silicon is transparent, suitable for use as an optical waveguide material. A number of different materials combinations can be used to provide the necessary refractive index contrast between a silicon waveguiding layer and its substrate. The reduction in refractive index due to free carriers is sufficient to allow moderately doped silicon to serve as a substrate for an undoped silicon waveguide layer [4]. When silicon-on-sapphire became readily available for electronics, it also found use as an optical waveguide material [5]. There has also been considerable interest in using SiGe alloy layers as optical waveguides, with pure silicon serving as the substrate material [6]. More recently, however, the vast majority of work on silicon waveguides has focused on the use of silicon-on-insulator (SOI), in which the substrate is a thin amorphous layer of SiO_2 that optically isolates a single-crystal silicon layer from an underlying bulk silicon wafer.

SOI technology has been developed over the past three decades, motivated primarily by the enhanced device speed and reduced power consumption that are possible when thin SiO_2 layers are used for electrical isolation in silicon MOSFET devices. A variety of methods have been developed using ion implantation and/or wafer bonding techniques to produce isolated thin silicon layers with the crystalline perfection demanded for microelectronic device use [7]. This technology has matured to the point where SOI is widely used in many current-generation integrated circuits. SOI is now a readily available, relatively low cost material that can be exploited for other applications, including micro-electro-mechanical systems (MEMS) and optical waveguiding. SiO_2 layer thicknesses are typically in the range of hundreds of nanometers to a single micron. For electronics applications, the silicon layer might be as thin as 100 nm; however, thicknesses up to several microns are available and are often desirable for waveguide applications.

6.2.2 Single Mode Condition

For most integrated optical devices, it is essential that the waveguides support only a single optical mode. One way to ensure single-mode operation is to control the refractive index contrast, Δn, between a waveguide layer and its substrate. This is a critical parameter that determines the maximum thickness for which a slab waveguide will be single-mode. In III–V semiconductor epilayer waveguides, for example, Δn is normally made quite small (<0.05) so that the waveguide will support only one optical mode, but will also have dimensions that are sufficiently large (>1 μm) to couple conveniently to conventional optical fibers. For SOI waveguides, Δn is fixed and it is large ($n_{Si} - n_{SiO_2} \approx 3.5 - 1.46$). This means that for an SOI slab waveguide to support only a single mode, the silicon layer must be kept to a thickness of less than about 250 nm. As will be discussed below, such small waveguide dimensions are preferred for certain applications. However, such "nanophotonic" waveguide devices demand relatively complex fabrication methods and make input and output coupling difficult. In general, it is desirable to have available a waveguide technology in which the waveguides can have dimensions on the order of a few microns while maintaining single mode operation.

Although SOI slab waveguides with silicon thickness on the order of microns will be highly multimode, it has been shown that a properly designed rib waveguide can be effectively single-mode even with rib widths and silicon thickness of several microns. This was first recognized by Soref et al. [8], with further refinements of the analysis subsequently provided by Pogossian et al. and others [9,10]. The explanation for this somewhat counterintuitive result is based on the effective index method for approximating the solution for rib waveguide modes. One takes the "vertical mode" shape within the rib to be the same as that of a slab waveguide with the same thickness but with infinite lateral extent. The two-dimensional mode profile is then taken to be the product of this vertical mode and a horizontal mode that is calculated by solving for the modes of a fictitious slab waveguide oriented perpendicular to the substrate. The fictitious waveguide is symmetric. Its core index is the effective index of the rib's vertical mode and its cladding index is the effective index vertical mode solution for the thinner slab waveguide regions on either side of the rib. If the slab waveguide calculations corresponding to the rib waveguide thickness and the thickness of the side regions generate multimode solutions (as will be the case for thick SOI), then the effective index approach will generate multiple solutions, one for each possible combination of a vertical mode times a horizontal mode. However, for relatively shallow etch depths, the lowest order vertical mode in the slab regions beside the rib will have a larger effective index than all of the rib's vertical modes except for the lowest order one. As a result, for all but the lowest order rib mode, power in the rib waveguide will leak into the slab regions on either side. The rib waveguide will only support the propagation of its lowest order mode. Regardless of the silicon thickness, it is possible to design a rib waveguide structure such that only one of the many modes resulting from an effective index calculation will not leak away. This requires that the rib etch depth be kept below a certain critical value that depends on the aspect ratio of the rib. This is neatly summarized in a simple expression for the single mode condition [8]:

$$\frac{\mathrm{W}}{\mathrm{H}} \leq 0.3 + \frac{r}{\sqrt{1 - r^2}} \tag{6.1}$$

where W and H are the width and silicon thickness, respectively, of the rib structure, and r is the ratio of the silicon thickness beside the rib to the thickness of the rib itself. Figure 6.2 shows the good agreement between this expression and experimental observations of the transition between single-mode and multimode operation.

6.2.3 Nano-Optics

Although SOI's large refractive index contrast imposes design constraints on single-mode rib waveguides, it can also be considered an advantage. It makes possible core sizes much smaller than one micron, which in turn makes possible very compact integrated optical devices with waveguide bending radii as small as a few microns. Silicon layer thicknesses below about 250 nm are required for slab waveguides to be unconditionally single-mode. When two-dimensional waveguides are patterned in such thin silicon, a guide with a width of a few hundred nanometers is typically defined by etching away the surrounding silicon all the way down to the SiO_2 substrate and then depositing a layer of SiO_2 or a transparent polymer to embed the silicon. The result is a "photonic wire" surrounded on all sides by a low index medium.

The small dimensions of photonic wires are conceptually very attractive for on-chip optical interconnects and integrated optical devices with small footprints. However, they are very demanding in terms of fabrication technology. Conventional optical lithography (with illumination wavelengths of 300 nm or more) cannot provide the required resolution. Most of the research reported to date in this area makes use of electron beam lithography. This provides the necessary accuracy and is ideal for proof of concept, but it is not well suited for mass production because it is a serial process. For this, nanophotonic waveguides using the available tools of advanced CMOS fabrication based on deep ultraviolet (UV) lithography must be fabricated [11].

Two major obstacles must be overcome before silicon photonic wires become widely used. One is their high level of propagation loss and the other is the difficulty in coupling to and from the much larger modes of a typical optical fiber. Propagation losses result mainly from scattering due to sidewall roughness. Although state-of-the-art fabrication technologies can result in surface roughness that is small on an absolute scale, the small dimensions of silicon photonic wires exaggerate the roughness on a relative scale. Furthermore, the small core sizes of photonic wires result in a significant fraction of the mode power propagating in the cladding and hence a mode amplitude at the waveguide–cladding interface that is quite large compared to micron-sized waveguides.

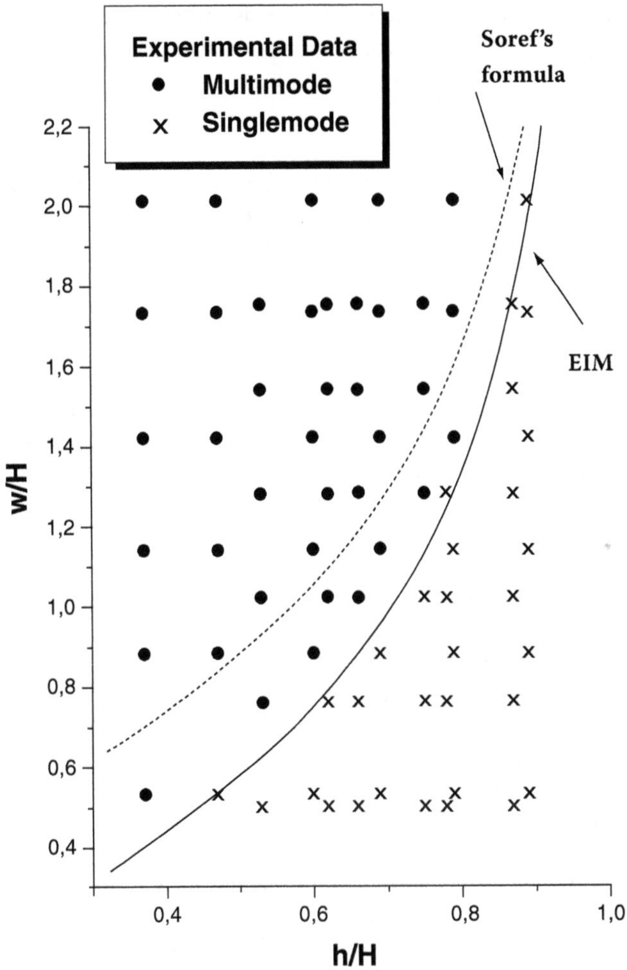

FIGURE 6.2 Plots of theoretical curves for the single-mode limit, and a matrix representing experimental observations of multimode and single-mode conditions. The circles and crosses correspond to multimode and single-mode waveguides, respectively. (From Reference [9], ©IEEE1998.)

Recent studies [11,12] have reported the fabrication of silicon wire structures with sidewall roughnesses of 5 nm or smaller. This resulted in propagation losses as low as 2.4 dB/cm for waveguide widths of 400 to 500 nm. It was observed that losses increased substantially for narrower waveguides. Although this loss level is larger than for many competing waveguide material systems, it is low enough to be used in practical devices with the small footprints associated with nanophotonics. Further reduction of losses can be expected using partial oxidization of the silicon wire to smooth out surface roughness. Lee et al. [13] reported a reduction in the propagation loss of a 500 nm wide strip waveguide from 32 to 0.8 dB/cm using this method.

They also reported similar loss reduction using anisotropic wet etching to reduce the sidewall roughness associated with conventional reactive ion etching.

6.2.4 Photonic Crystals

Photonic crystals [14] are structures in which there is a periodic modulation of the dielectric constant in one, two, or three dimensions. They can be fabricated by etching arrays of air holes in otherwise uniform transparent materials. For wavelengths commensurate with periodicity, the propagation of electromagnetic waves in such a structure is profoundly affected by the effects of interference from the periodically repeated dielectric interfaces. This is directly analogous to the way in which the periodic electrostatic potential in a crystal lattice determines the solutions for the wave functions of the electrons that propagate within the crystal. Silicon is the ideal material for fabrication of photonic crystals because of the advanced state of silicon processing technology and the large refractive index contrast that exists at an air–silicon interface.

In a photonic crystal, the familiar dispersion relation, $\omega = (c/n)k$, is replaced by a very complex, direction-dependent ω *vs.* k band structure like that of an electron in a crystal lattice. For certain combinations of wavelength and propagation direction, electromagnetic waves cannot propagate through the structure. For certain periodic structures, including a hexagonal array of air pores, it is possible to design photonic crystals with complete photonic band gaps, meaning that within a range of wavelengths lightwave propagation is forbidden for all directions and all polarizations. In other words, the photonic density of states is zero.

In order to exploit photonic crystals for functional devices, it is far easier to make use of two-dimensional rather than three-dimensional structures. Figure 6.3 shows a two-dimensional photonic crystal fabricated in SOI [15]. An array of holes is etched straight through the silicon layer to define a band gap with respect to lightwave propagation within the plane of the silicon layer. In this example, the underlying SiO_2 layer has been etched away to form a free-standing silicon membrane. A waveguide is formed by the introduction of a lattice defect that, for example, could be a single row of missing pores. Light is guided within the defect row, confined vertically by the large index step, as in a conventional SOI waveguide. Lateral confinement, however, is due to the fact that in-plane propagation outside of the defect row is forbidden for wavelengths that fall within the photonic band gap. Even at abrupt bends in the waveguide path, there is no loss due to in-plane scattering as would occur in a conventional waveguide.

The propagation losses in photonic crystal waveguides tend to be dominated by vertical out-of-plane scattering and are typically tens of decibels per centimeter. However, the best results are on the order of 0.1 dB/cm [16]. The problems associated with coupling light to and from optical fibers are much the same as with photonic wires and similar spot-size conversion techniques are employed to overcome them.

While a simple row of missing pores in a two-dimensional photonic crystal can form a low loss waveguide, a line of only a few missing pores can form an ultra-compact optical resonator that can serve as a basis for a variety of devices that are ideally suited for optical integration. Notomi et al. [16], for example, have

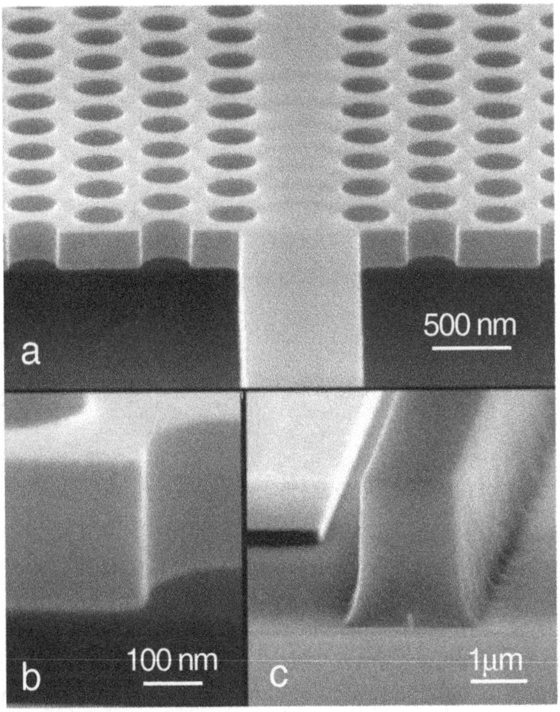

FIGURE 6.3 SEM micrographs of a) photonic crystal membrane waveguide with strip access waveguide. b) Sidewall profile showing, ~ 90 angle sidewalls. c) End of a silicon taper tip. (From Reference [15].)

demonstrated narrow-band transmission filters and 3- and 4-port channel-drop devices based on resonant tunneling in silicon photonic crystals.

6.3 PASSIVE DEVICES

6.3.1 Arrayed Waveguide Gratings

The arrayed waveguide grating (AWG) has become a very important device for use in advanced fiber optic telecommunications systems, and its fabrication technology is now very highly developed. Its function is to multiplex or demultiplex large numbers of closely spaced wavelength channels that are transmitted simultaneously in a single optical fiber. This volume contains a dedicated chapter explaining in detail the principles of operation and potential applications of the AWG; however, its importance to the development of silicon-based planar lightwave circuits requires some specific explanation in this chapter.

Figure 6.4 illustrates the design of an AWG. When operating as a demultiplexer, the multi-wavelength input from a fiber is coupled into a single input rib waveguide on the integrated optical chip. This waveguide feeds light into a planar waveguide

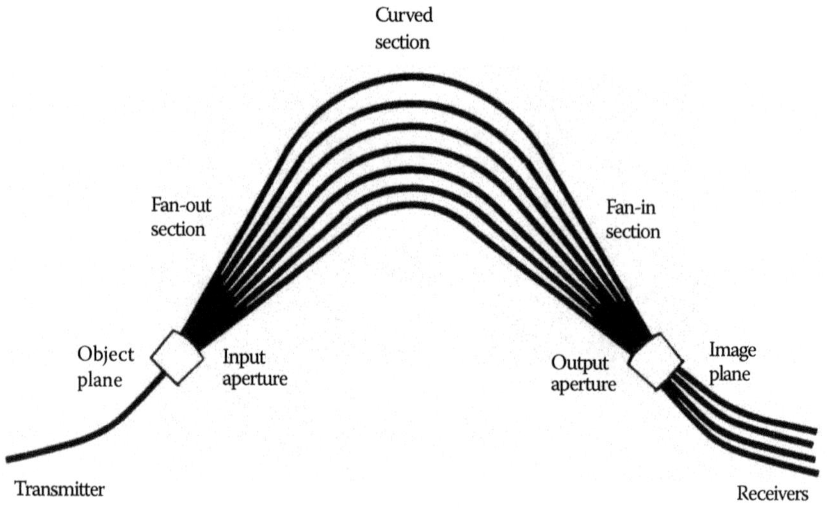

FIGURE 6.4 Layout of an Arrayed Waveguide Grating (AWG).

section (free propagation region) where it spreads out in the waveguide plane due to the absence of any lateral confinement. The free propagation region ends at the input to an array of single-mode rib waveguides that deliver their lightwave signals to a second free propagation region that is normally a mirror image of the first. The waveguide array follows a curved path such that there is a fixed path length difference between adjacent array elements. Thus, lightwaves that enter the waveguide array in phase with each other exit it with a constant non-zero phase shift between adjacent array elements. The multiple inputs to the second free propagation region act like the multiple elements of a phased array radio antenna. Their signals interfere so as to form a narrow beam at the far side of the free propagation region. For one particular wavelength, the precise location of this output beam is determined by the phase shift between radiating elements. It can be made sufficiently narrow so that it couples to only one of the many output waveguides. Since the channel-to-channel phase shift introduced by the curved waveguide array is wavelength-dependent, different wavelengths can be made to couple into different output waveguides.

It has now become routine to make these devices with such precision that the center wavelengths assigned to the output channels can be accurately matched to the established wavelength channels on the ITU grid. The majority of commercial AWG devices are currently fabricated using silica glass as the waveguide material. SOI poses some particular design problems for AWGs. However, recent studies have shown that these can be overcome, making SOI an alternative that offers the usual advantages in terms of device size and higher levels of photonic integration.

One of those problems is the fact that for practical silicon layer thicknesses (>1 μm), the free propagation region is highly multimode. The rib waveguides that deliver light to this region can be made single mode, but if they couple significantly to the higher order modes in the planar region, the resulting interchannel cross-talk will be unacceptable for any practical device. Pearson et al. [17] carried out a detailed

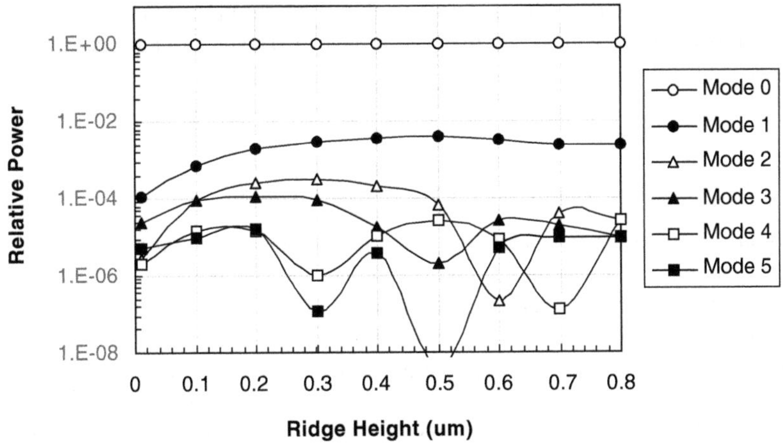

FIGURE 6.5 Power coupled to slab waveguide modes from a single-mode input waveguide in SOI. (From Reference [17].)

calculation of the mode shape in the rib waveguides and its overlap integral with the multiple modes of the free propagation region. This was used to calculate the distribution of the input power among the many modes. The results are shown in Figure 6.5 as a function of the etch depth used to define the input rib waveguide for the case of 1.5 μm thick silicon, for which the slab waveguide supports six modes at a wavelength of 1550 nm. The coupling to the higher order modes was found to be very small, with more than 99% of the power coupling to the fundamental mode. The same calculation applies to the coupling out of the free propagation region and into the output rib waveguides. The small amount of power in the higher order modes will couple to the single mode output rib waveguide with an efficiency that is well below 1%. The combined effects of the input and output coupling mean that the higher order modes are suppressed and do not contribute significantly to cross-talk.

Another potential problem for silicon-based AWGs is the birefringence inherent in the rib waveguides that will result in undesirable polarization dependence and cross-talk. The degree of birefringence depends on the rib's etch depth and also on its shape (vertical walls formed by reactive ion etching versus sloped walls from wet chemical etching). Pearson et al. showed that for vertically etched ribs, it is possible to find a combination of rib width and etch depth for which the birefringence vanishes [17]. In the interest of polarization independence, they chose to use this rib structure even though it was in violation of the single mode condition described above. Multimode rib waveguides could be tolerated because the curved waveguides in the array acted very effectively as mode filters. This is because the bending loss in curved waveguides tends to make a very abrupt transition from being negligibly low to being extremely large as the bending radius is decreased. The radius where this transition occurs is much smaller for the lowest order mode than it is for higher order modes. By designing the bending radii for the waveguide array to be small enough to extinguish the higher order modes yet large enough not to perturb the lowest order mode, the usual single mode criterion is relaxed. Additional polarization compensation techniques that have been developed for silica glass and InP AWGs have also been applied successfully to SOI devices [17].

6.3.2 Waveguide Integrated Grating Structures

Bragg grating structures are widely used in optical fibers and integrated optics for a variety of passive and active device functions. They consist of a waveguide structure in which there is a small periodic variation of the refractive index as in the case of fiber Bragg gratings or a corrugation of the waveguide surface as in devices such as distributed feedback lasers. A Bragg grating is highly reflecting in a narrow band of wavelengths centered at twice the grating period, Λ, but has negligible effect on the propagation of light at other wavelengths. In SOI rib waveguides with large cross-sections, there is an additional complication as observed and explained by Murphy et al. [18]. Transmission spectra show the usual resonant dip in intensity when the matching condition, $\Lambda = \lambda_0/(2N_{eff})$, is met. However, additional dips occur at shorter wavelengths due to the grating-assisted coupling of the forward propagating fundamental mode to backward propagating leaky modes.

An interesting example of a SOI Bragg grating device is the demonstration by Bookham Technology of a narrow-band laser in which a III–V semiconductor optical amplifier is aligned by hybrid integration techniques with a SOI rib waveguide that contains a Bragg grating [19]. A laser cavity is formed by a highly reflecting coating at one end of the amplifier chip and by the grating at the other end. This mimics the more commonly used technique of using a fiber Bragg grating as an external cavity reflector to lock a laser's operation to a predetermined wavelength. However, it has advantages in terms of manufacturability and permitting higher levels of integration. The Bookham device also incorporated a monitor photodiode and passive fiber alignment features into the integrated optical circuit. Jones et al. [20] of Intel Corp. demonstrated a tunable external cavity laser using a SOI Bragg grating. They made use of thermal tuning of the grating wavelength to effect the laser output.

Electrical tuning of the Bragg grating can be accomplished by controlling the free carrier concentration in the rib waveguide. This is most easily done by creating p^+ and n^+contact regions on either side of the rib to form a *p-i-n* diode. Current injection in the diode will change the effective index and thereby shift the peak reflection wavelength (dynamic modification of refractive index in silicon waveguides is explained in detail in Section 6.4). Optical modulation at frequencies in excess of 1GHz has been shown to be possible using this method [21]. Electrically tunable gratings have also been shown to be useful for wavelength switching in SOI optical add–drop multiplexers [22].

Waveguide gratings are also important as a means of coupling light in and out of SOI rib waveguides that often have dimensions that are too small for efficient direct butt coupling to optical fibers. A beam of light in air that is incident on a waveguide grating at an angle ϕ will be coupled into the guide if:

$$\Lambda = \lambda_0/(N_{eff} - \sin\varphi) \tag{6.2}$$

Equation (6.2) also defines the angle of the output beam that a grating will cause to emerge from a waveguide. Careful optimization of the grating shape and etch depth is required in order to achieve efficient coupling. Ang et al. achieved 84% output coupling from a 1 μm thick waveguide using dry etching techniques with a tilted substrate mount to create a blazed grating structure [23].

Grating coupling has also been used to enhance the efficiency of surface illuminated silicon photodetectors [24]. There is growing interest in including short wavelength silicon detectors in standard CMOS circuitry. This includes (but is not limited to) SOI as the substrate material. In conventional detectors, the quantum efficiency is limited by the fact that the absorbing silicon layer is typically thinner than the $1/e$ absorption length for the wavelengths of interest. A large fraction of the input light passes through the detector's active region without being absorbed. By using a grating coupler to redirect the input light into the thin silicon waveguiding layer, the lightwave path in the silicon is greatly increased, thereby increasing the overall absorption. Csutak et al. demonstrated a four-fold increase in detector efficiency using a waveguide grating coupler [24].

The critical wavelength dependence of waveguide grating devices makes them suitable for a variety of sensor applications. One class of chemical sensors and biosensors is based upon the measurement of changes in φ in (6.2) to measure small chemically induced changes in the refractive index of the cover layer that alter N_{eff}. Bragg grating structures have wavelength dependences that are highly sensitive to changes in the strains and/or temperatures seen by the gratings. A good example of a sensor based on this effect is the integrated optical Michelson interferometer described by Tsao et al. for temperature monitoring [25].

6.3.3 Polarization Control

Standard single-mode optical fiber normally delivers randomly polarized light to an integrated optical device. As a result, the device should ideally be designed for polarization-independent operation. Since this may not be possible, a need exists for components that can separate TE and TM polarizations and components that can convert one to the other.

As previously described, it is possible to achieve zero birefringence in SOI waveguides. Silicon is not an inherently birefringent material, and SOI is not subject to the stress-induced birefringence present in some material systems due to thermal expansion coefficient mismatch between the waveguide layer and the substrate. However, SOI waveguides possess the usual birefringence associated with the different N_{eff} values for the TE and TM solutions. In a slab waveguide, the birefringence $(n_{TE} - n_{TM})$ is always non-zero and positive. For a SOI rib waveguide, it is usually positive. However, as shown in a detailed analysis by Dai and He [26], relatively narrow waveguides have a range of etch depths for which the birefringence becomes negative. With very precise control over the rib width and depth, it is possible to fabricate rib waveguides with zero birefringence. The results of their calculations are shown in Figure 6.6, where t and r are the rib width and the thickness of the lateral slab regions relative to the rib height. The combinations of r and t that produce zero

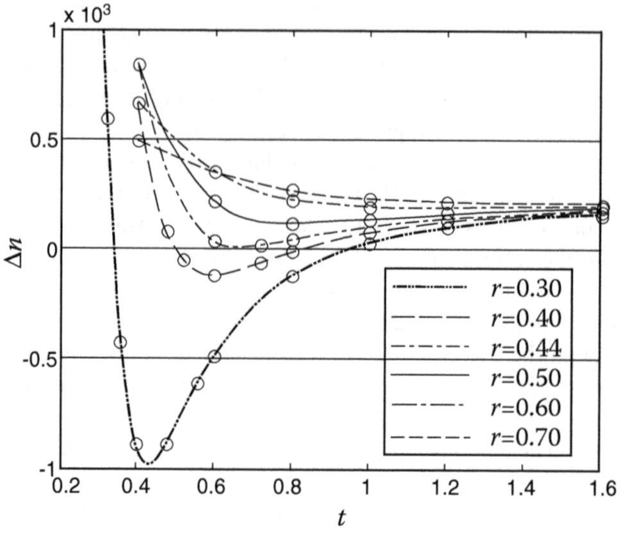

FIGURE 6.6 Birefringence Δ**n** of a SOI rib waveguide as **t** varies for different values of **r**. (From Reference [26].)

birefringence are not necessarily compatible with the single mode condition described above. However, there is a small operating range in which the two conditions can be met simultaneously.

Polarization splitters are devices that direct the TE and TM waves that copropagate in a single waveguide into two separate waveguides. They are used in many optical devices for communications where the device function is highly polarization-dependent. Kiyat et al. designed and fabricated compact polarization couplers based on a simple dual-channel coupler design [27]. An alternative design makes use of lateral Bragg reflectors to cross-couple one selected polarization between two adjacent waveguides [28].

A polarization converter or rotator is a device that can convert light in a TE polarized waveguide mode to a TM mode. This requires an integrated optical analogue of a half wave plate with its fast and slow axes oriented at ±45° relative to the substrate plane. This is generally difficult to achieve because the conventional rib waveguide geometry invariably results in fundamental modes that are polarized parallel or perpendicular to the substrate. In order to rotate the polarizations of the fundamental modes, the basic waveguide geometry must be rotated. This can be accomplished by fabricating an asymmetric rib in which one sidewall is vertically etched by reactive ion etching and the other sidewall is defined by an anisotropic wet etch that exposes the <111> crystal plane of silicon to form a sloped surface at an angle of 54.7° relative to the <100> oriented substrate. A detailed calculation of the mode structure of such an asymmetric waveguide shows that, for certain combinations of rib width and etch depth, the orientation of the waveguide axes can be set to ±45°, without violating the single mode conditions. Figure 6.7 shows the predicted optical axis rotation obtained from a full vectorial finite element mode solver calculation [29].

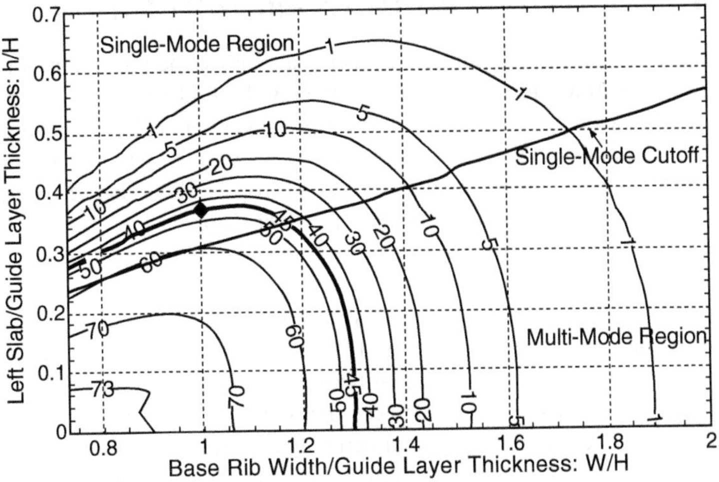

FIGURE 6.7 Optical-axis rotation angle contours for asymmetric SOI waveguides. The single mode cutoff curve indicates the geometries that guide only the two fundamental polarized modes. The contour for an optical-axis rotation of 45° is emphasized and the parameters chosen for fabrication are denoted by the symbol ◊. (From Reference [29].)

To achieve the desired TE–TM polarization conversion, it is not only necessary to fabricate a waveguide with its mode axes rotated by $\pm 45^{\circ}$. The length of the slanted waveguide section must be L_{π} such that the accumulated phase shift between the modes is 180°. Brooks et al. fabricated polarization converters based on these calculations [30]. Figure 6.8 shows an image of one of their asymmetric waveguides. For this structure, the optimum device length, L_{π}, was 2.6 mm.

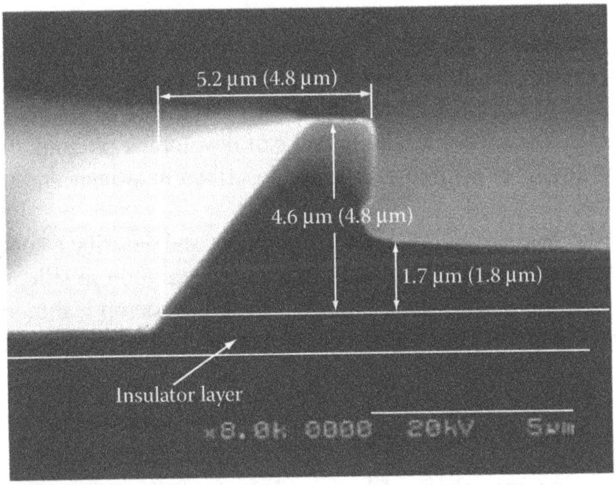

FIGURE 6.8 Scanning electron microscope image of waveguide cross-section. Optimal dimensions are included in brackets. (From Reference [30].)

6.4 PHOTONIC AND MICROELECTRONIC INTEGRATION

One of the most significant advantages of using silicon for the formation of optical waveguides is the ability to monolithically integrate both photonic and microelectronic functionalities in one seamless process flow. Potential applications appear almost endless. Examples include the integration of waveguide optical detectors and operational amplifiers; light sources and drive electronics; diode-based modulators and switches; and read-out circuitry for optical sensor chips.

The development of silicon optoelectronics has been the subject of significant academic-based research following the pioneering work of Richard Soref in the 1980s and 1990s. His 1993 paper remains relevant to the goals and general approach still employed at the present time [31]. Since Soref wrote his outstanding review, numerous journal articles and two books on silicon photonics [32,33] have been published. Further, waveguide devices that combine electronic and photonic functionalities have been and remain commercially available. Bookham Technology (although now concerned primarily with non-silicon products) pioneered the development of integrated silicon photonic components, demonstrating clearly that such an approach to integrated optics fabrication was viable [34]. Bookham manufactured a range of products including a solid state variable optical attenuator (VOA), and a (de)multiplexer with integrated VOAs. At present, Kotura, a U.S.-based company, uses silicon opto-electronic integrated circuit technology (SOEIC) to manufacture four- and eight-channel VOAs — the Ultra VOA arrays. The operation of this device is described in a comprehensive paper by Wenhua Lin and Tom Smith available at the Kotura website [35]. Attenuation is controlled dynamically using free carrier absorption via an integrated *p-i-n* diode. This technique is described further in Section 6.4.1. The Kotura paper also provides a clear explanation of the advantages of photonic and electronic integration in silicon, and suggests how devices might evolve, e.g., the combination of a VOA with an integrated power monitor would provide a waveguide device capable of power leveling via a monolithically integrated feedback mechanism.

Of some significance is the relatively recent emergence of Intel Corp. as a leader in the development of silicon photonics. The research group led by Mario Paniccia has made a number of impressive breakthroughs — some of which are described in detail in Sections 6.4.1 and 6.4.3. The benefits of developing photonic functionality and an outline of Intel's goals in this area are described in articles and white papers available at its website [36].

It is beyond the scope of this chapter to document and describe a complete list of all recent developments in photonic and electronic integration in silicon and hence we discuss only the areas of greatest interest. These include optical modulators and switches, sub-band-gap detection and the development of an integrated optical source.

6.4.1 MODULATORS AND SWITCHES

The modulation of an optical signal is one of the basic functions required in any photonic circuit, particularly those used for telecommunications. Modulation implies an induced change in an optical field, for example, amplitude or phase. This is achieved via a change in the complex refractive index (n') of the waveguide given by:

$$\Delta n' = \Delta n + \Delta ik \tag{6.3}$$

where Δn is the change in the real part of the refractive index and Δik relates to an increase (or decrease) in the optical extinction coefficient.

In general, changes in the refractive index of materials can be induced by the application of an electric field. This results in a change in the real component (electrorefraction) or the extinction coefficient (electro-absorption). The modification of n' may be such that it is a linear or second order function of the applied field so that:

$$\Delta n' = P(\lambda)E + K(\lambda)E^2 \tag{6.4}$$

where E represents the applied field and P and K are wavelength-dependent constants.

The linear and second order modifications of refractive indices are more generally known as the Pockels and Kerr effects respectively. The Pockels effect causes birefringence in crystals that lack inversion symmetry and forms the basis of operation of modulators formed in non-centro-symmetric materials such as lithium niobate and gallium arsenide. For elemental semiconductors such as silicon, demonstration of the Pockels effect is not possible. In contrast, the refractive index of silicon does exhibit a change in response to an applied electric field of quadratic form. However, the Kerr effect for silicon is extremely weak. It was quantified by Soref and Bennett [37] for a wavelength of 1300 nm to be approximately $\Delta n = 10^{-4}$ for an applied field of 10^6 V/cm (a value above that corresponding to the electrical breakdown of silicon).

For materials that are centro-symmetric and for which only a small Kerr response is observed, alternative methods are required to achieve significant changes to the refractive index and hence allow the formation of monolithically integrated modulators. In the following three subsections, we describe approaches used successfully to achieve dynamic modifications of the refractive indices of silicon waveguides.

6.4.1.1 Thermo-Optic Effect

The change in refractive index of a material as a function of its change in temperature is described as the thermo-optic effect. It has been known for some time that this effect is relatively large in silicon, for instance, according to Fischer et al. [38]:

$$\delta n/\delta T = 2 \times 10^{-4} K^{-1} \tag{6.5}$$

for silicon where $\delta n/\delta T$ is the thermo-optic coefficient (note that this refers to the real component of refractive index; for a positive change in temperature, an increase in refractive index is observed). The application of a thermal load to a waveguide causing an increase in temperature of only 5K results in a change in refractive index of 1×10^{-3}. This is significantly larger than, for instance, the maximum available change in index related to the Kerr effect.

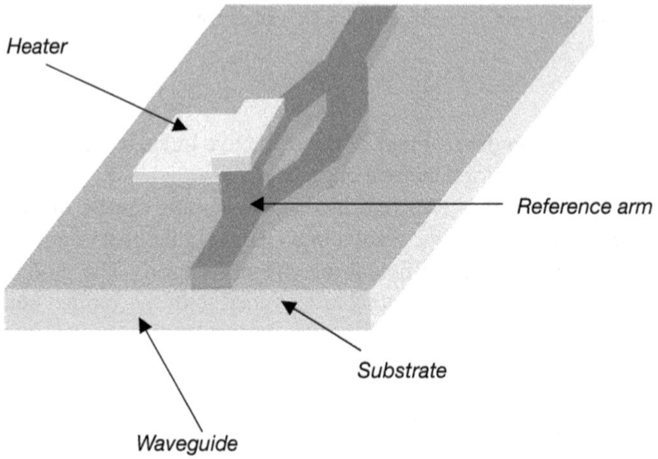

FIGURE 6.9 Mach-Zehnder interferometer based on the thermo-optic effect.

The thermo-optic effect has been successfully exploited in devices that work on the principle of phase modulation. As early as 1991, Treyz demonstrated a Mach–Zehnder (MZ) interferometer formed using SOI waveguides that permitted the dynamic attenuation of light with a wavelength of 1319 nm [39]. The principle of operation is shown schematically in Figure 6.9. The incoming signal is split equally between the two arms of a y junction. In one arm, the refractive index is controlled using an integrated heater. This could simply consist of a metal electrode deposited on the surface of the chip. By careful control of the thermal loading, the refractive index may be altered to induce a phase shift in the optical signal in the heater arm, as compared to that in the reference arm. At the point of signal recombination, at the second y junction, the phase difference between the otherwise identical signals can be controlled to vary between 0 and π. Subsequent interference of the signals is thus controlled to vary from totally constructive to totally destructive.

Fischer et al. reported a MZ-based modulator with low insertion loss that supported single mode propagation. For a 1300 nm wavelength, the on–off ratio was determined to be 13 dB for TE polarization [38]. Maximum attenuation was achieved for thermal power of 150 mW. Their key result is reproduced here as Figure 6.10.

Although simple in principle, any device relying on (de)coupling is susceptible to polarization dependence. Fischer et al. [38], for example, noted that for TM polarization the maximum attenuation was closer to 9 dB, which in practice is the true limit of performance. Another disadvantage associated with thermo-optic signal switching is the relatively low speed, a result of reliance on thermal diffusion mechanisms to turn the device from the on to the off state. In Fischer et al. [38], the device was noted to switch at 60 kHz, while Treyz [39] noted that it was close to only 20 kHz. The switching speed can be increased by limiting the thickness of the buried oxide layer in the SOI structure; oxide has a thermal conductivity approximately two orders of magnitude smaller than that for silicon. This increases the amount of thermal loading required to achieve a π phase change.

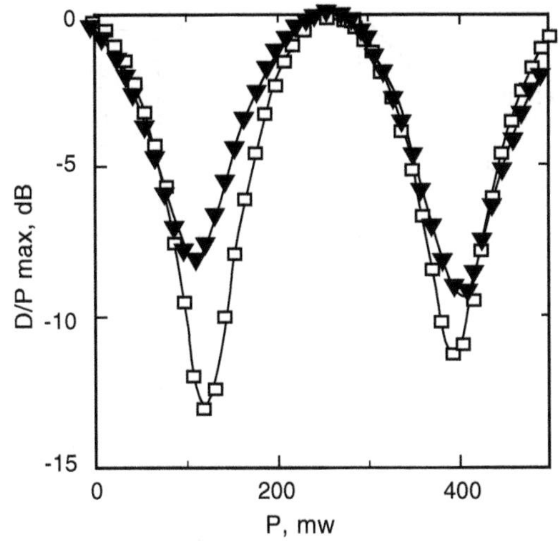

FIGURE 6.10 Normalized optical power versus electrical switching power for the Mach-Zehnder device reported in reference [38]. Open squares and closed triangles represent TE and TM polarization respectively. (©IEEE 1994.)

6.4.1.2 Carrier Injection

An alternative to the thermo-optic effect for dynamic control of the refractive index of silicon (and the only viable electro-optic method), is the use of injected charge carriers. The optical properties of silicon are strongly affected by the presence of free charges — namely electrons and holes. In the same work in which they considered the effect of applied electric field, Soref and Bennett also determined the change in the refractive index of silicon as a function of free carrier concentration [37]. In a rigorous treatment, they extracted a large range of experimental values of optical absorption from the research literature. Using the Kramers–Kronig relationship, they subsequently calculated values for Δn versus carrier concentration. These were compared to theoretical relationships obtained from the classical Drude model where:

$$\Delta n = -(e^2\lambda^2/8\pi c^2\varepsilon_0 n)[\Delta N_e/m_{ce}{}^* + \Delta N_h/m_{ch}{}^*) \tag{6.6}$$

$$\Delta\alpha = -(e^3\lambda^2/4\pi c^3\varepsilon_0 n)[\Delta N_e/m_{ce}{}^{*^2}\mu_e + \Delta N_h/m_{ch}{}^{*^2}\mu_h) \tag{6.7}$$

e is the electronic charge; α is the absorption coefficient; ε_0 is the permittivity constant; λ is the wavelength; n is the unperturbed, real-part of the refractive index; $m_{ce}{}^*$ is the effective electron mass; $m_{ch}{}^*$ is the effective hole mass; ΔN_e is the change in electron concentration; and ΔN_h is the change in hole concentration. The resulting plots of refractive index modification versus carrier concentration for a wavelength of 1300 nm are reproduced in Figure 6.11. The authors noted the reasonable agreement between experimental values and theory for both Δn and $\Delta\alpha$. Holes were found to be more efficient in perturbing the refractive index, especially for high carrier concentrations, e.g., for $\Delta N = 10^{17}$ cm^{-3}, where Δn (holes) $= 3.3\Delta n$ (electrons). The most

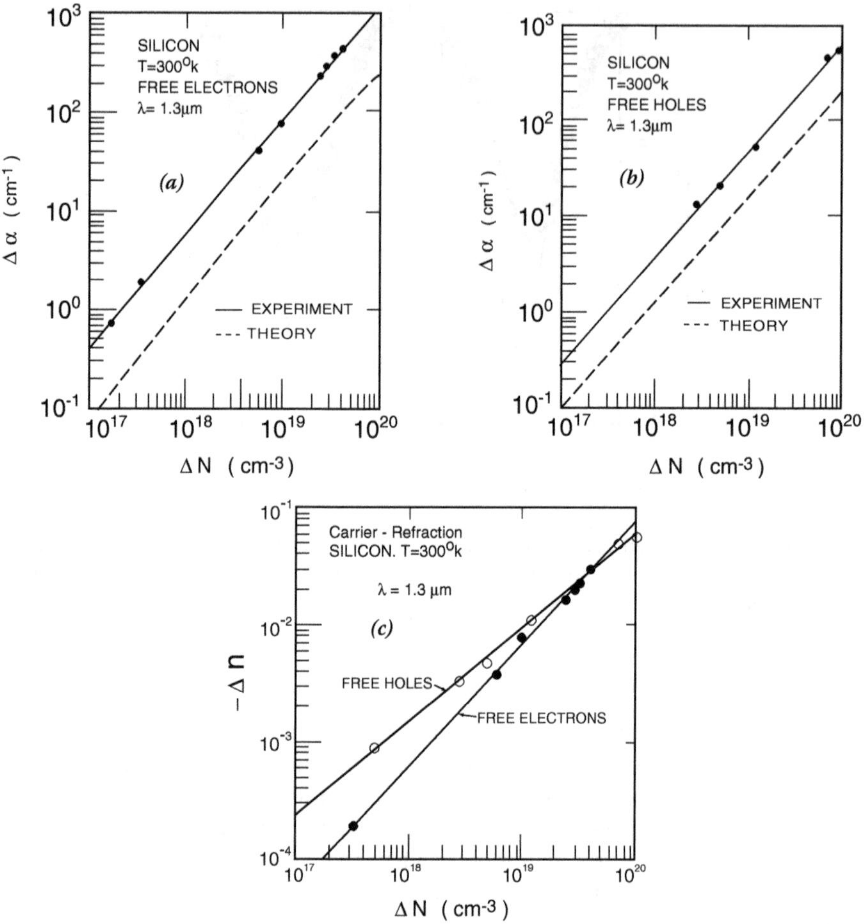

FIGURE 6.11 Change in optical properties of silicon as a function of free carrier concentration at a wavelength of 1300nm-(a) change is absorption versus free electron concentration; (b) change in absorption versus free hole concentration; (c) change in real component of the refractive index versus hole and electron concentration. (From Reference [37], ©IEEE 1987.)

significant results of this work were summarized in the following four empirical relationships that accurately predict the change in refractive index in silicon as a function of carrier concentration for the two important telecommunication wavelengths of 1300 nm and 1550 nm:

$$\Delta n(1300nm) = \Delta n_e + \Delta n_h = -[6.2 \times 10^{-22} \Delta N_e + 6.0 \times 10^{-18} (\Delta N_h)^{0.8}] \quad (6.8)$$

$$\Delta \alpha(1300nm) = \Delta \alpha_e + \Delta \alpha_h = 6.0 \times 10^{-18} \Delta N_e + 4.0 \times 10^{-18} \Delta N_h \quad (6.9)$$

$$\Delta n(1550nm) = \Delta n_e + \Delta n_h = -[8.8 \times 10^{-22} \Delta N_e + 8.5 \times 10^{-18} (\Delta N_h)^{0.8}] \quad (6.10)$$

$$\Delta \alpha(1550nm) = \Delta \alpha_e + \Delta \alpha_h = 8.5 \times 10^{-18} \Delta N_e + 6.0 \times 10^{-18} \Delta N_h \quad (6.11)$$

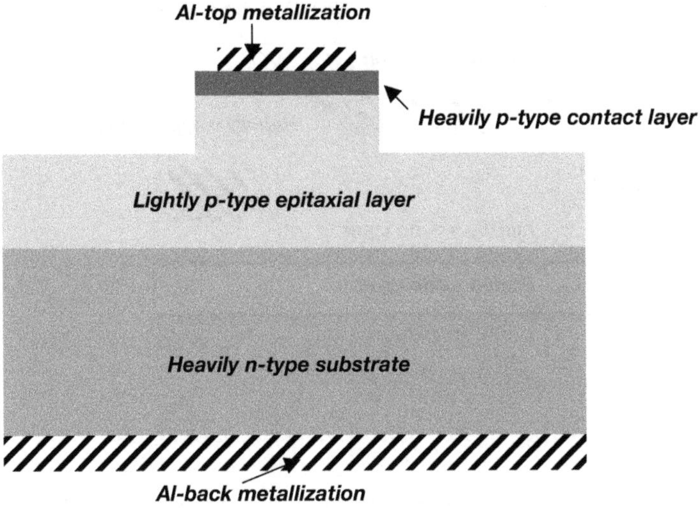

FIGURE 6.12 Monolithically integrated, vertical injection p-i-n diode and silicon waveguide. (From Reference [40].)

The accuracy of these expressions is such that Equations (6.7) through (6.10) are used universally during the design of electro-optic silicon modulators and switches.

Early demonstrations of the use of free carrier injection to modulate an optical signal in a silicon waveguide were made by Treyz, May, and Halbout. One device design was based upon a silicon-on-silicon waveguide [40], with vertical confinement provided by the contrast in doping of a heavily n-type substrate and an epitaxially grown thin film consisting of a very lightly doped p-type layer of 7.7 μm, and a heavily doped p-type contact layer of 0.5 μm thickness, shown schematically in Figure 6.12. After etching a rib structure into the p layer to provide lateral optical confinement, aluminum metallization was performed to the device backside and the top of the rib. In forward bias, the p-i-n structure floods the optical waveguide with free carriers, the concentration of which depends on the current density. The authors were thus able to control optical attenuation of a 1319 nm signal confined within the waveguide. This structure was one of the earliest reported variable optical attenuators fabricated in silicon. For a device length of 500 μm, and current density of 3×10^3 Acm^{-2}, corresponding to injected carrier concentrations approaching 10^{18}cm^3, an attenuation of >20 dB/cm was measured. The switching speed of the device was estimated to be in the megahertz range, consistent with free carrier recombination rates in silicon, and represented a considerable improvement on devices that rely on the thermo-optic effect.

Treyz, May, and Halbout also integrated a p-i-n diode with one arm of a MZ structure [41], similar to that shown in Figure 6.9. Phase shifting was controlled via the injection of carriers rather than by the thermo-optic effect. The p-i-n diode was identical in design to that in Figure 6.12, and 500 μm in length. They were able to induce a $\pi/4$ phase shift in one arm of the MZ by injecting a current density of 1.6 Acm^{-2} corresponding to a carrier concentration of around 6×10^{17} cm^{-3}.

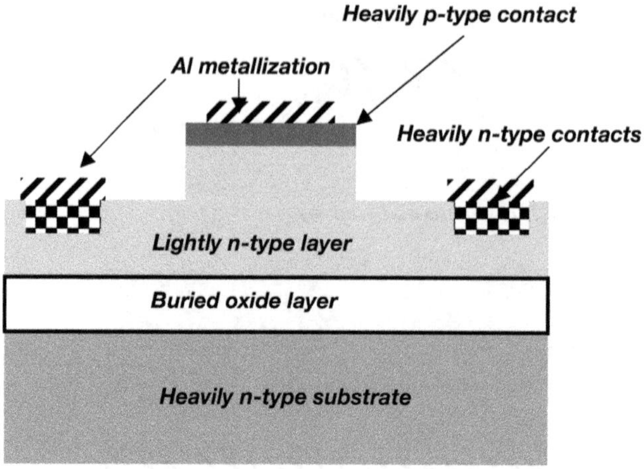

FIGURE 6.13 Lateral injection p-i-n diode, integrated with a silicon-on-insulator waveguide. (After Reference[42].)

A MZ carrier injection device was applied to an SOI waveguide by Zhao et al. [42]. The SOI structure prevents the fabrication of a through-wafer *p-i-n* diode; hence they designed a lateral injection diode with a *p*-type contact on top of the rib and *n*-type contacts on either side introduced via ion implantation of boron and arsenic, respectively. Their diode design is reproduced schematically in Figure 6.13. For a device length of 816 μm, the authors observed a 98% extinction ratio at an applied forward bias of 0.95 V for an optical signal of 1300 nm wavelength. Electron irradiation of the device at an energy of 14 MeV introduced an unreported concentration of carrier recombination centers. These would be expected to increase the modulation bandwidth of the device at the expense of decreasing the modulation efficiency. Unfortunately, no supporting report on either of these properties was provided.

A comprehensive treatment of the integration of a *p-i-n* diode and SOI waveguide was provided by Reed and Knights [33] following several reports by Reed's research group on various device geometries (for example, see Hewitt and Reed [43]). Of some importance, Reed describes the design of an asymmetrically doped lateral *p-i-n* structure that has commercial popularity (formerly with Bookham and presently with Kotura) as the basis for a VOA technology described in the introduction to this section. The asymmetric design does not require the doping or metallization of the rib, thus avoiding significant performance issues related to polarization-dependent passive loss in the off state. A schematic of Reed's design from Reference [43] is described by Figure 6.14. Using high level commercial device simulation software analysis [44], Reed described the impacts of several device parameters on efficiency and switching speed, i.e., doping concentration, dopant displacement from the rib, depth of dopant, and shape of the rib base. In general, it was determined that moving the dopant regions closer to the rib base reduced both switching time and increased device efficiency. Also, diffusing the dopant close to the buried oxide layer greatly increased device efficiency. With optimized parameters, a π phase shift could be

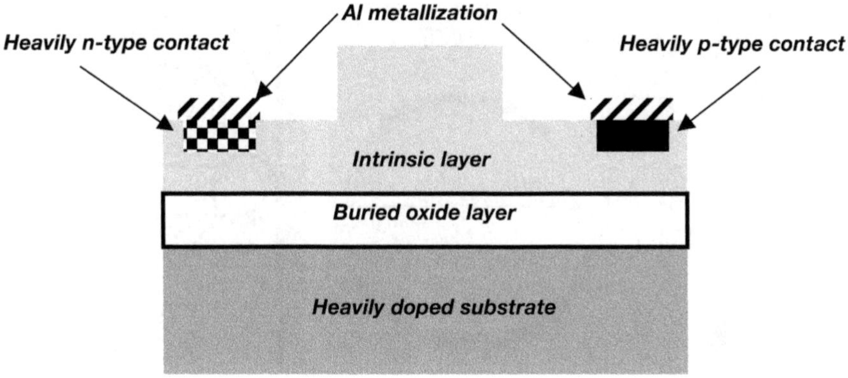

FIGURE 6.14 Lateral injection, asymmetric p-i-n diode, integrated with a silicon-on-insulator waveguide. (After Reference [43].)

achieved in a 500 μm long device for a current of <10 mA. Further, it was confirmed that carrier injection MZ devices fabricated in SOI with cross-sections of several square micrometers whose switching is dominated by carrier recombination, were limited in bandwidth to tens of mega hertz. Reed's work was of significance because it demonstrated the important role of device simulation while clearly describing the limits on the performance of large cross-section optoelectronic SOI devices.

The ongoing quest to increase the switching speeds of carrier injection devices has thus focused on the reduction of waveguide size beyond submicron dimensions. Of some note was a recent report by the group of Michal Lipson at Cornell University who integrated a carrier injection diode and a micro-ring resonator passive structure formed with a strip waveguide 250 nm in height and 450 nm wide. The ring had a diameter of 12 μm [45]. An electron micrograph of the device is reproduced in Figure 6.15. Inside the ring the silicon is doped heavily p-type while a halo of n-type doping is implanted around the outside of the ring. This design is a significant improvement on previous carrier injection devices in two important respects: (1) the waveguide dimensions are submicron and hence the silicon volume into which carriers fill or escape is relatively small; (2) the light-confining resonant structure enhances the effect of any refractive index change induced by free carriers.

Reproduced from their report is Figure 6.16. The main panel shows the transmission spectra of the ring resonator for diode forward applied biases of 0.54 V (subturn on bias), 0.87 V, and 0.94 V for light of wavelength around 1574 nm. For the subturn on bias, the spectra exhibit a 15 dB drop in transmission through the resonant structure. The electron–hole pair density in the cavity increases as the forward bias on the diode increases, resulting in a shift in the resonant wavelength of the ring structure. The inset in Figure 6.16 shows the transfer function of the device, i.e., the transmission of an optical signal of 1573.9 nm as a function of applied bias. In effect, for this wavelength, the device operates as a highly efficient modulator with a 15 dB extinction ratio. The modulation rate of the device was demonstrated to be in excess of 1 Gb/s.

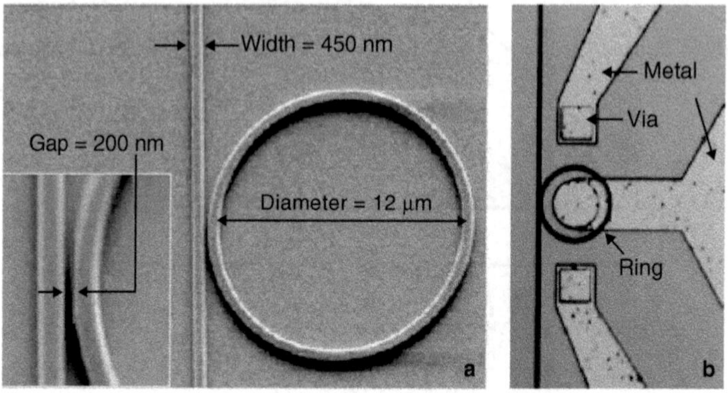

FIGURE 6.15 a) Top-view scanning electron microscope image of a micro-ring-resonator coupled a waveguide with a close-up view of the coupling region. b) Top-view microscope image of the same ring resonator after the integration of a central p-doped region, and a halo-doped n-type region. (From Reference [45].)

6.4.1.3 Carrier Depletion

In contrast to carrier injection devices, those that work on the principle of carrier depletion are not limited in bandwidth by the recombination of the injected carriers. Several electro-optic modulator designs relying on carrier depletion were outlined by

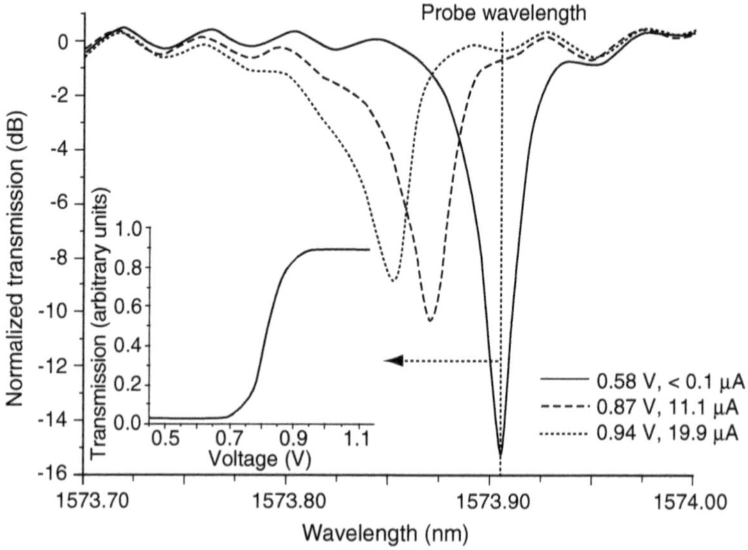

FIGURE 6.16 The main panel shows the transmission spectra of the ring resonator shown in Figure 6.6, for forward diode bias voltages of 0.58V, 0.87V, and 0.94V. The inset shows the transfer function of the modulator for light with a wavelength of 1573.9 nm. (From Reference [45].)

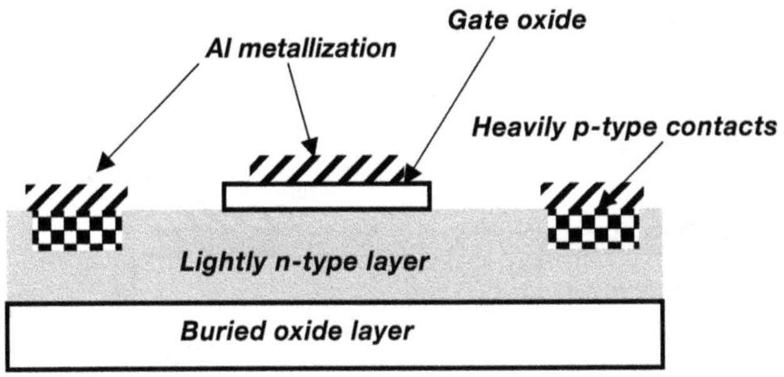

FIGURE 6.17 Schematic representation of a p-channel MOSFET integrated with a strip loaded waveguide. (After Reference [46].)

Giguere et al. [46]. Reproduced in Figure 6.17 is the authors' design for a modulator that resembles the structure of a *p*-channel MOSFET. As is the case for carrier injection modulators, optical confinement in the vertical direction is provided by a buried oxide such as that available in SOI substrates. An oxide strip on the wafer surface provides lateral confinement via strip loading, and acts as the gate oxide for the transistor structure. The applied gate voltage alters the carrier concentration below the gate region (the optical mode confinement region).

For a negative gate voltage, holes are attracted beneath the gate oxide. The effect is enhanced if a second gate electrode is added to the bottom buried oxide. The authors predicted a refractive index change of 10^{-4} for a change in gate voltage from zero to −4 V. They also summarized the operation principle of and the restrictions imposed by using such a design: (1) the gate voltage is able to modulate the drain source current (as expected in any MOSFET); (2) insulated gate structures designed to modulate electron and hole concentrations must be thin waveguides because the maximum depletion layer thickness decreases with increasing carrier concentration; and (3) unlike carrier injection modulators, the modulated carrier concentrations are not uniformly distributed in the silicon waveguide.

Despite proposals for designs for carrier depletion modulators in the late 1980s and early 1990s, the first significant attempt to fabricate a high bandwidth, high efficiency depletion modulator was not reported until 2004. This resulted mainly from the dominance of telecommunication applications in the intervening years and the lack of requirement for waveguides structures with dimensions smaller than 1 μm. Large waveguide dimensions are incompatible with the second design requirement outlined by Giguere.

In 2004, the Intel Corp. research team led by Mario Paniccia reported a depletion modulator capable of operation in the gigahertz regime [47]. A representation of their device is reproduced in Figure 6.18. The waveguide structure consists of a lightly doped *n*-type slab and a lightly doped *p*-type *poly-Si* region separated by a thin oxide that acts as the insulating gate during modulation. Simulation and measurement confirmed that the device propagated a single optical mode at a wavelength of 1550 nm. It was noted however, that the gate oxide induced a strong polarization effect on the waveguide and hence all results were reported for TE polarization only. Several

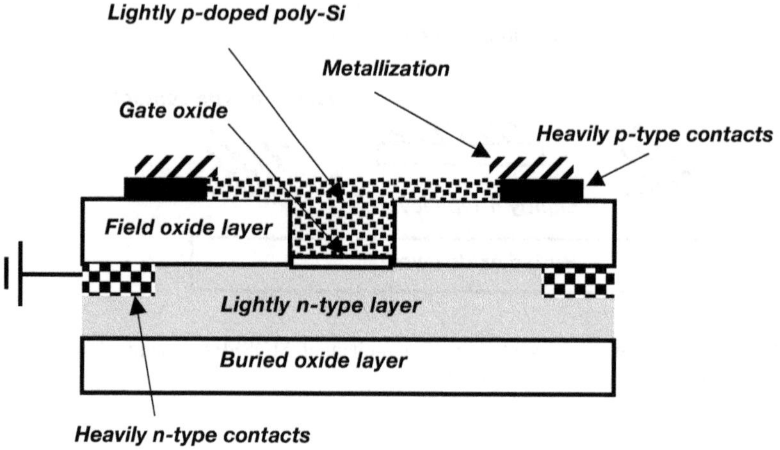

FIGURE 6.18 Carrier depletion, electro-optic modulator after Liu et al. (From Reference [47].)

modulators were measured, varying in length from 1 to 8 mm. When the *poly-Si* was biased positively, a small accumulation layer of free charge was induced on either side of the gate oxide. This charge induces a change in the waveguide refractive index. By placing the modulator in one arm of a MZ structure, the authors were able to quantify the phase change of the waveguide versus applied bias. Their results are reproduced as Figure 6.19.

The significant result of the work reported by the Intel group was the modulation bandwidth of the device. For the first time, an all-silicon optical waveguide fabricated using standard monolithic processing technology could be modulated at a rate greater than 1 GHz. In fact, as shown in Figure 6.20 reproduced from the same paper, the 3 dB bandwidth of the devices was greater than 3 GHz.

Using an approach similar to that of Hewitt and Reed for carrier injection devices [43], members from the same Intel group reported recently on methods to increase the bandwidth of their depletion device beyond 10 GHz [48]. The key step to improving the bandwidth relies on optimized doping profile engineering in the n-type slab and p-type *poly-Si* regions and the reduction in the gate oxide thickness to 6 nm. The simulation results are reproduced in Figure 6.21.

6.4.2 Sub-Bandgap Detection

Optical detectors fabricated using silicon device technology based on p-n and p-i-n diodes have been available since the concept of silicon integrated circuits was conceived. Coincident with the silicon bandgap value of around 1.1 eV, silicon detectors are usually marketed with peak sensitivity around 700 nm with suitability for the short haul telecommunications wavelength of 850 nm — incompatible with long haul telecommunication wavelengths in the infrared. This presents a contradiction for the fabrication of highly integrated silicon photonic circuits. Clearly, it is

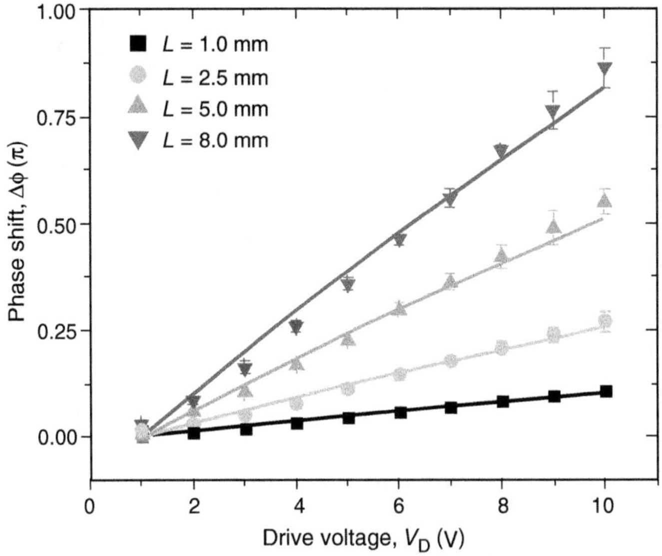

FIGURE 6.19 Phase shift versus drive voltage of the MOS capacitor phase shifter in Figure 6.18, at a wavelength of 1550nm for different phase shifter lengths. The symbols represent the measured phase shifts, and the solid lines are the simulated phase shifts. Results for modulators of lengths varying from 1 to 8mm are shown (From Reference [47].)

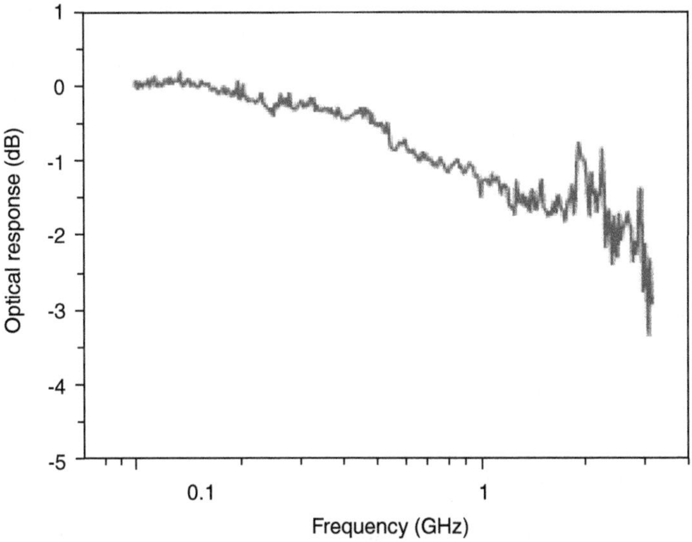

FIGURE 6.20 Frequency dependence of the optical response of a silicon MZI modulator reported in Reference [47].

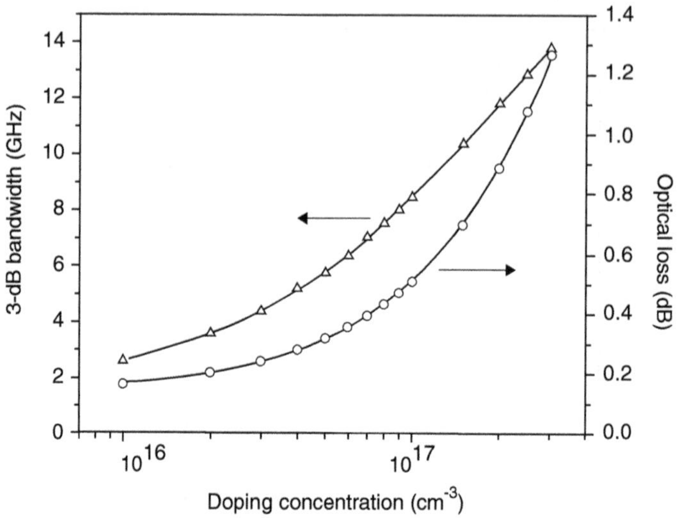

FIGURE 6.21 Modeled 3-dB bandwidth and optical loss as a function of the doping concentration for a carrier-depletion modulator similar to that in Figure 6.18. The concentration refers to the acceptors and donors on either side of the gate oxide. The gate oxide thickness is 6nm. The optical loss is for a device length that leads to a $\pi/2$ phase shift at a drive voltage of 3V. (Reproduced from Reference [48], ©IEEE2005.)

desirable that the signal is carried at a wavelength in the infrared to avoid significant, on-chip attenuation; however, this would imply almost zero responsivity for monolithically integrated detectors or optical monitors. One approach to overcome this limitation is the hybrid integration of III–V-based detectors with silicon optical circuits [34], but a truly monolithic fabrication technology is preferred. Figure 6.22 compares the absorption coefficients for Si, Ge, and (In)GaAs as a function of signal wavelength [49].

A considerable amount of recent work has been dedicated to the development of integrated optical detectors sensitive to wavelengths around 1550 nm, the fabrication of which is completely compatible with standard silicon (CMOS) processing technology. This includes the development of Ge/Si heterostructures, the incorporation of optical dopants such as erbium, into the silicon matrix [50], and the introduction of midband gap energy levels via defect incorporation into the silicon lattice. In this section, we summarize results from two of these approaches that meet the need for truly monolithic integration.

6.4.2.1 Silicon-Germanium Devices

The transparency of silicon to wavelengths around 1300 nm to 1550 nm (the property that makes it an ideal substrate for infrared optical waveguiding) prevents the straightforward fabrication of monolithically integrated detectors. The addition of Ge to the silicon matrix (i.e., the formation of $Si_{1-x}Ge_x$ alloy) shifts the absorption

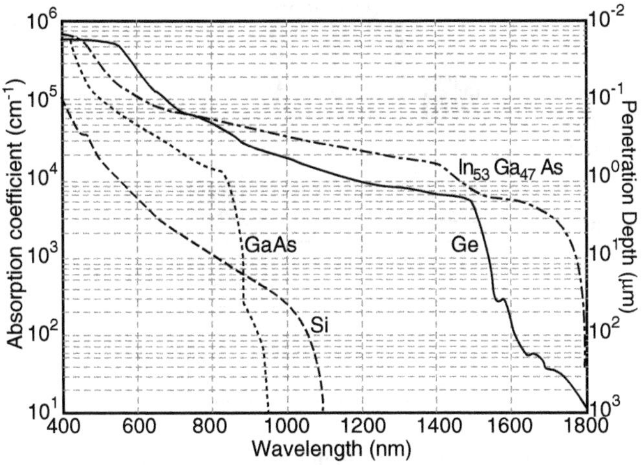

FIGURE 6.22 Absorption coefficients for Ge, Si and (In)GaAs. (Reproduced from Reference [49], ©IEEE2004.)

edge from 1100 nm, deeper into the infrared. For $x > 0.3$, absorption (and hence detection) of 1300 nm is possible; and for $x > 0.85$ even 1550 nm wavelengths can no longer traverse a $SiGe$ sample unattenuated. Of some importance is a concomitant increase in refractive index with increasing x, a property that suggests the fabrication of optical detector integration with silicon waveguides via evanescent coupling.

Several recent reports of $SiGe/Si$ heterostructures have been made. Masini et al. used a low temperature deposition technique to form a 120 nm polycrystalline Ge film on a silicon substrate [51]. Despite the poor quality film, responsivities of 16 mA/W and 5 mA/W were measured for optical wavelengths of 1300 nm and 1550 nm, respectively, with an associated dark current of 1 mA/cm^2. For a reverse heterojunction bias of 30 V, the device was shown to be capable of measuring optical source modulation at rates in excess of 2.5 Gb/s.

A significant increase in the effective absorption coefficient at 1550 nm for a thin film Ge/Si heterostructure was demonstrated by Dosunmu et al. [49]. They exploited the high refractive index contrast of Si and SiO_2 to create a resonant cavity-enhanced photodetector consisting of a Ge film of several hundred nanometers grown on a double-layer SOI (DSOI) substrate. A low temperature Ge buffer was inserted between the DSOI and Ge detection region to minimize dislocations usually introduced in hetero-layer growth. Figure 6.23 shows a cross-sectional view of a Ge-DSOI detector design that is completed by the addition of a *p-type* doped region in the upper silicon layer and a *n-type* doped region in the Ge detection volume. The detector response to wavelengths ranging from 1300 nm to 1600 nm is shown in Figure 6.24. For a detector area of 140 × 140 μm, a relatively large dark current of 54 μA was observed for a reverse bias of 1 V. This was attributed to the presence of dislocations in the Ge detection volume (despite the use of the low temperature buffer layer). The quantum efficiency (η) at 1550 nm was increased four-fold compared to the expected

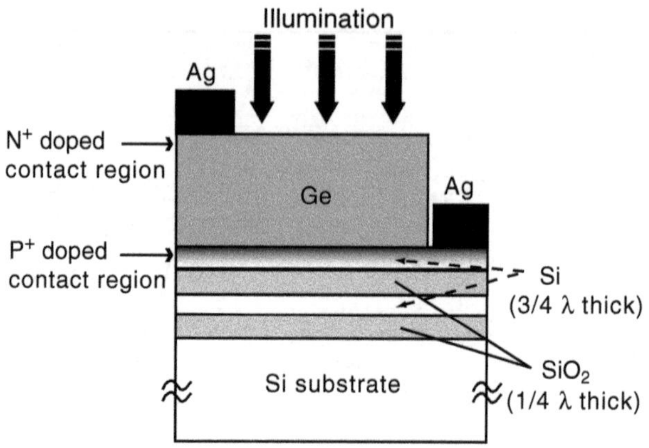

FIGURE 6.23 Cross-sectional view of a top illuminated Ge-DSOI vertical p-i-n photodetector. (Reproduced from Reference [49], ©IEEE2004.)

response of a similarly designed single-pass detector. A significant increase in η was predicted for optimized layer thicknesses, while it was proposed that cyclic annealing during layer growth would suppress dark current. Simulations suggested a bandwidth approaching 25 GHz.

The integration of Ge and a SOI waveguide was reported by El kurdi et al [52]. The significance of this work rests in the demonstrated compatibility of the growth

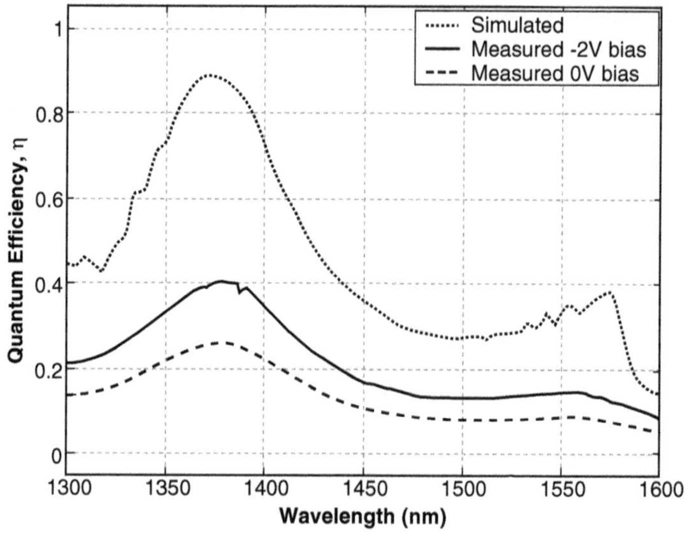

FIGURE 6.24 Ge-SOI photodetector quantum efficiency as a function of wavelength and reverse bias. (Reproduced from Reference [49], ©IEEE 2004.)

process and the formation of waveguides in SOI, where the photo-responsive element consisted of multiple layers of Ge islands. The responsivity of the detectors was relatively low at 0.25 mA/W for 1550 nm, but the device had the advantage of being in a waveguide geometry with a silicon-terminated surface. All measurements were performed at zero bias due to a significant dark current when the device was reverse biased.

6.4.2.2 Defect Mediated Detection

As early as 1959 it was realized that the absorption edge of irradiated silicon could be extended to wavelengths beyond that for intrinsic silicon. For example, a summary of the early work of Fan and Ramdas is reproduced in Figure 6.25 in which the effect of deuteron irradiation on the optical absorption of silicon is shown [53]. When crystalline silicon is irradiated with high energy ions (available via standard ion implantation), mobile defects such as interstitials and vacancies are produced. At room temperature, most of these defects rapidly reincorporate into the crystal structure. However, some combine to form complex stable defects such as the silicon divacancy — a pair of empty lattice sites.

A defect such as the divacancy disrupts the normal energy band structure of the material and introduces mid-band gap levels. The photon energy required to excite an electron from such levels is lower than the energy normally required for valence to conduction band excitation, allowing the absorption of lower energy photons. Since the 1.8 μm absorption band tail shown in Figure 6.25 passes through wavelengths around 1550 nm, the study of this excess optical absorption is of potential importance to the development of silicon-based optoelectronics in general and integrated

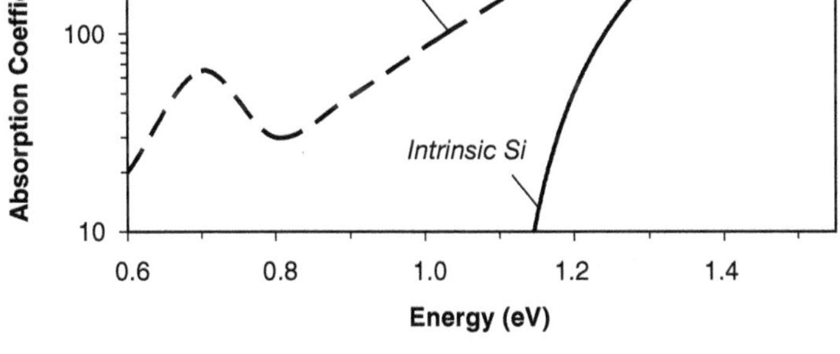

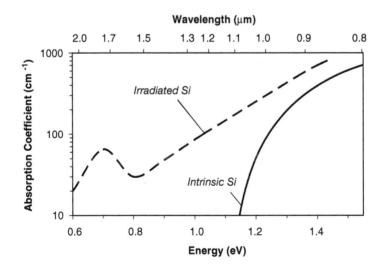

FIGURE 6.25 Absorption spectra of irradiated and intrinsic silicon. (Deduced from Reference [53].)

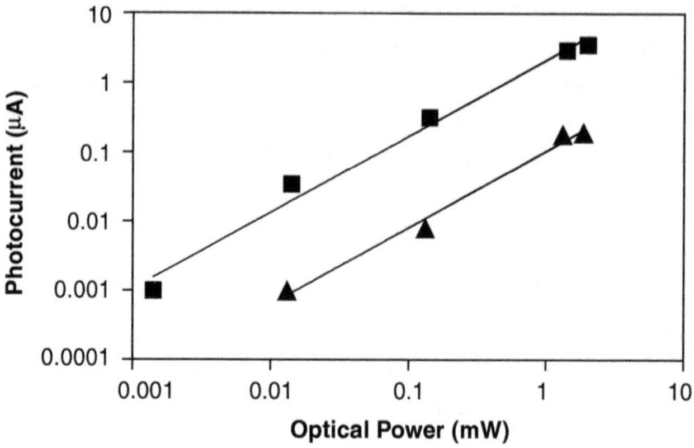

FIGURE 6.26 Unbiased photocurrent versus on-chip optical power for a proton implanted (closed squares) and unimplanted (closed triangles) waveguide detector. (Reproduced from Reference [55].)

detectors in particular. Bradley et al. exploited this infrared enhanced absorption in the fabrication of a SOI integrated detector [54]. The device was similar in design to the VOA described in Figure 6.14, with the addition of dilute concentrations of vacancy-type defects into a volume consistent with the optical mode. The defects were introduced via low energy proton irradiation. Figure 6.26 shows the response of the in-line detector to 1550 nm light. The IR sensitivity is enhanced relative to a similar device without proton irradiation by approximately two orders of magnitude, with a measured responsivity of 3 mA/W. Of some note is the capability of the fabrication process to determine the fraction of light removed from the waveguide and converted into an electrical signal. In Bradley et al. [54], this was reported as 19%, but values from less than 1% to virtually complete absorption could be absorbed simply by adjusting the proton irradiation dose. The authors suggested therefore that these detectors might find greater application as in-line monitors.

In a subsequent report the same group demonstrated monolithic integration of a carrier injection VOA and detector [55]. Results from this first demonstration of monolithic integration of variable attenuation and monitoring are reproduced in Figure 6.27. One disadvantage of defect-mediated detection is the relatively small bandwidth. For the devices described in References [54] and [55], the photosensitive region of the waveguide is required to be >2 mm in length. This leads to a bandwidth of a few megahertz suitable for monitoring thermo-optic or carrier injection modulation only.

6.4.3 Integrated Optical Sources

We have described several of the essential elements required of a highly integrated optoelectronic system and how they may be fabricated using a silicon substrate. The outstanding omission is an efficient silicon-based light source. Silicon is not well

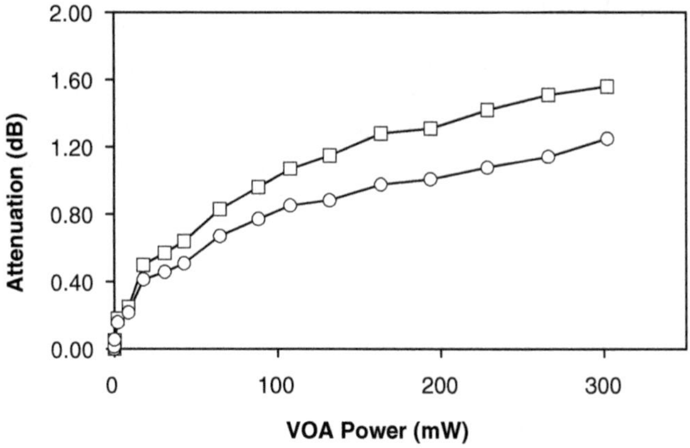

FIGURE 6.27 Attenuation of on-chip power by integrated VOA measured by external detector (open squares) and by the integrated optical power monitor (open circles). (Reproduced from Reference [55].)

suited to optical emission because of its indirect band gap. Band-edge luminescence (and absorption) in bulk Si is a three-body process involving an electron, hole, and phonon. The low probability of such an event means that the luminescence lifetime is very long — on the order of milliseconds. As the electron and hole move through the sample during this period, they typically come into contact with a defect or trapping center within a few nanoseconds and recombine nonradiatively, releasing their energy as phonons. The room temperature internal quantum efficiency of Si is thus of the order of 10^{-6}.

In terms of achieving optical amplification in Si, two further mechanisms tend to limit population inversion. The first is a nonradiative three-body process in which an electron and hole recombine but instead of creating a photon in the process, the recombination energy is instead transferred to another free carrier, exciting it to a higher energy — the so-called Auger process. Doping, current injection, and increased temperature all increase the probability of such an event because they promote population of the conduction band. The second nonradiative process is free-carrier absorption. As with the Auger process, the absorption probability increases with the Si free-carrier density. Further complicating the specifications required from a silicon optical source is the need for emission at a wavelength for which silicon is usually transparent—ideally at 1550 nm. Efficient Si light emitters operating at this wavelength require significant material engineering.

Several approaches are currently under investigation as routes toward efficient silicon optical sources: bulk Si systems, e.g., dislocation loops [56], stimulated Raman scattering in Si waveguides [57], band structure engineering via alloying with Ge [58], quantum-confined structures [59], and impurity centers, e.g., rare-earth doping [60]. Here we describe the two most popular approaches.

6.4.3.1 Silicon-Rich Oxides

Nano-sized Si structures embedded in an SiO_2 matrix represent attempts to exploit the efficient luminescence mechanism of low-dimensional Si in a structurally stable system. Ion implantation of Si into SiO_2 or plasma-enhanced chemical vapor deposition of SiH_4/O_2 is typically employed to incorporate the excess Si into a stoichiometric silicon oxide thin film. Subsequent annealing at temperatures ranging from 600 to 1200°C then determines the size and degree of crystallinity of the nano-sized Si structures. Intense, broad PL has been observed with peak wavelength dependent on the nanostructure size distribution but generally centered close to 800 nm. Luminescence is only weakly temperature-dependent because Auger recombination and free carrier absorption are limited by the wide band gap SiO_2 host.

Optical gain has been reported in this system, initially by a group led by Lorenzo Pavesi at the University of Trento [59]. Further, electrically pumped light emitting diodes have been fabricated with an external quantum efficiency (η_{ext}) approaching 0.1%; however, the operational lifetimes of these devices tend to be extremely limited due to oxide breakdown resulting from the high electric fields needed to attain current injection into the nanostructures [61].

Rare-earth doping of silicon-rich SiO_2 with Er represents an attempt to resolve the free carrier absorption and energy back-transfer problems of Er-doped Si by using a wide band gap host, while tuning the emission wavelength to a value at which silicon is transparent. Erbium luminescence shows only weak temperature quenching in stoichiometric SiO_2 but optical excitation is extremely difficult due to low Er optical absorption cross-sections and narrow line widths. The presence of Si nanoclusters in the vicinity of the Er increases the effective excitation by acting first as broadband absorbers of the pump light and then acting subsequently as sensitizers of the Er 1.54 μm emission. This system shows significantly enhanced photoluminescence compared to SiO_2:Er [62]. Electroluminescence devices based on these systems tend to exhibit the reverse trend with a SiO_2:Er MOS diode design showing $\eta_{ext} > 10\%$ at 300K, as reported by Castagna et al. [63]. Output characteristics for this diode and similar devices doped with ytterbium and terbium are reproduced in Figure 6.28. The authors showed that when biased, hot electron effects tended to degrade the devices. However the incorporation of Si nanocrystals, although reducing the signal strength, enhanced the device reliability by providing more conducting paths through the oxide.

6.4.3.2 Raman Laser

In 2004, a landmark paper reported work from the group of Bahram Jalali at UCLA [64]. It describes the construction of the first silicon laser working on the principle of the Raman effect. Jalali noted the relatively strong Raman effect in silicon (10,000 times greater than in silica fiber, for example), deducing that optical gain could be achieved over the length of a typical PLC (i.e., $\sim$1 cm). Further, it was asserted that net gain should be limited by free carrier losses resulting from two-photon-absorption (TPA) — a direct result of the high power pump laser required by a system utilizing the Raman effect.

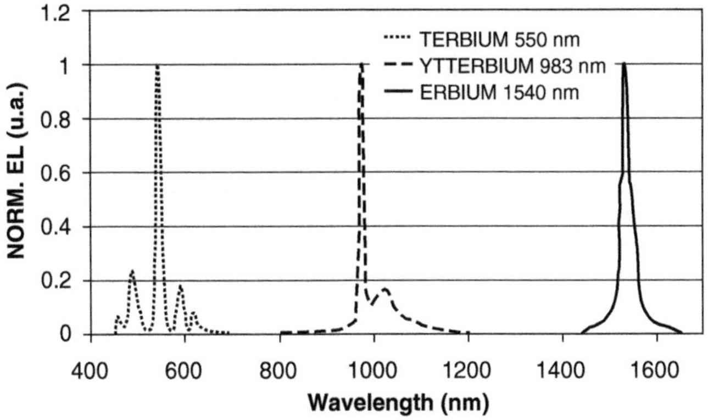

FIGURE 6.28 Normalized room temperature electro-luminescence spectra measured on MOS-type devices having Er-doped SiO_2 (solid line), Yb-doped SiO_2 (dashed line) and Tb-doped SiO_2 (dotted line) as gate dielectric. (Reproduced from Reference [63].)

To circumvent the TPA problem, these first measurements were made using a 1540 nm pump signal bunched in 30 ps pulses. The authors used a ring cavity set-up, with a SOI rib waveguide as the active volume. The laser was characterized at a wavelength of 1675 nm — that expected as the Stokes wavelength. Reproduced in Figure 6.29 from [64] is a plot of output peak power at 1675 nm versus input pump power at 1540 nm. There exists a clear lasing threshold, at 9 W, consistent with the round-trip loss and the expected Raman gain. After exceeding threshold, the output power increases linearly with pump power, with a slope efficiency of 8.5%.

A continuous wave (CW) silicon Raman laser was recently demonstrated by the Intel silicon photonics group [57]. The key addition to the previous silicon waveguide

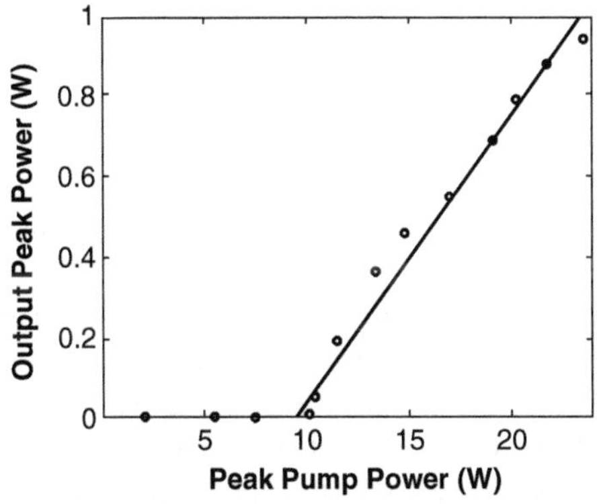

FIGURE 6.29 Measured laser output power with respect to peak pump power for silicon Raman laser. (Reproduced from Reference [64].)

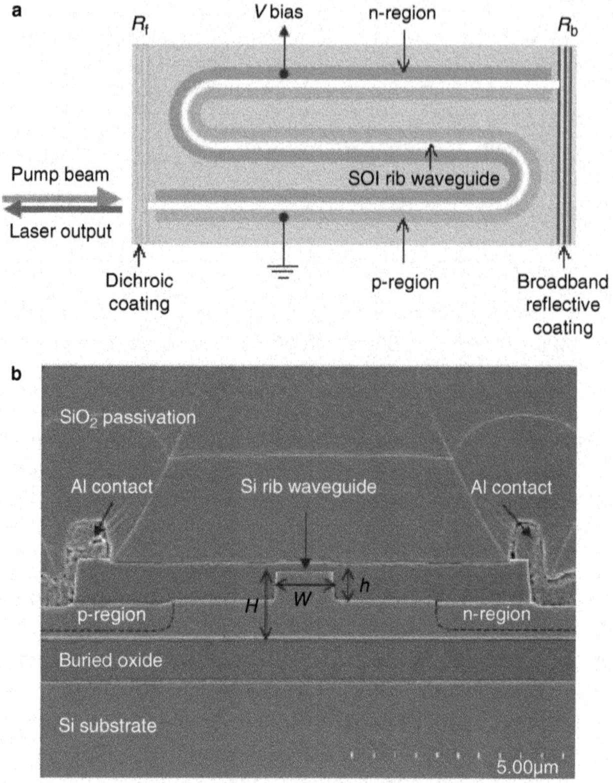

FIGURE 6.30 Intel fabricated silicon-Raman laser. a) Schematic layout of the silicon waveguide laser cavity with optical coatings applied to the facets and a p-i-n structure along the waveguide. b) Scanning electron microscope cross-section image of a silicon rib waveguide with a p-i-n diode structure. (Reproduced from Reference [57].)

Raman laser design of the UCLA group was an integrated *p-i-n* diode, very similar to that used for the carrier injection VOA and switch [43] and the defect-mediated IR-enhanced in-line detector [54] (see Figure 6.14). When reverse biased, the diode allows the fast extraction of free carriers generated by the TPA effect. A further important modification was the integration of a dichroic input facet coating and a highly reflecting back facet coating. Thus the SOI waveguide constituted the entire optical cavity. A schematic layout and scanning electron micrograph of the Intel laser [57] appear in Figure 6.30. The pump wavelength was 1550 nm, producing a Stokes wavelength of 1686 nm. The lasing threshold was dependent on both the pump power and the reverse bias applied to the diode. For a reverse bias of 5 V, the threshold was 280 mW, but for 25 V it was reduced to only 180 mW. Further, the slope efficiency was increased from 2 to 4.3% as the reverse bias was increased from 5 to 25 V. These results are summarized in Figure 6.31 reproduced from [57].

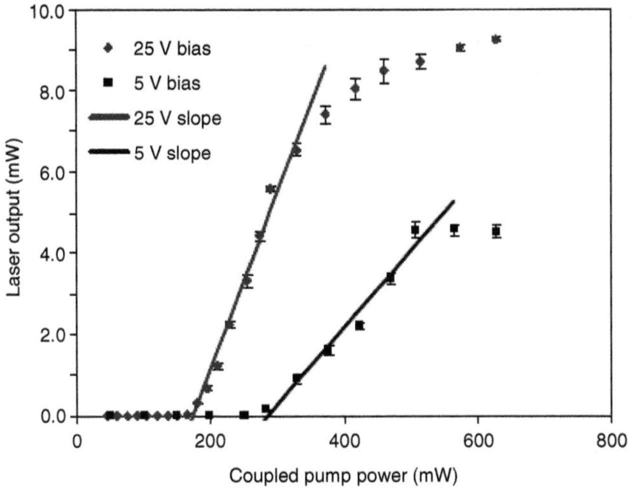

FIGURE 6.31 Intel fabricated silicon Raman laser output power as a function of the input pump power at a reverse bias of 25 and 5V. The pump wavelength is 1550 nm and the laser wavelength is 1686 nm. The slope efficiency (single side output) is 4.3% for 25V bias and 2% for 5V bias. (Reproduced from Reference [57].)

6.5 CONCLUSION

In this chapter, we have attempted to provide insights into the development and vast potential application of silicon-based waveguides and devices derived therefrom. The recurring message of those working in this field is the ease with which one may integrate photonic functionality with electronic functionality on the same substrate in one seamless process flow. Further, this process flow would be compatible with fabrication technologies already in place in the highly developed silicon microelectronics industry. Although not yet dominant as a material for optoelectronic fabrication, it is difficult to imagine highly integrated devices of any kind not based on silicon technology. The next few decades will likely witness the increased migration of silicon photonics from the research laboratory to the manufacturing facility with benefit to areas as diverse as microelectronics, telecommunications, and the biological, environmental, and medical sensor industries.

REFERENCES

1. Orton, J. W., *The Story of Semiconductors*, Oxford University Press (2004).
2. Kilby, J. S., Invention of the integrated circuit, IEEE Trans. on Electr. Dev., ED-23, 648 (1976).
3. International Technology Roadmap for Semiconductors, 2005. Available at: http://www.itrs.net/Common/2005ITRS/Home2005.htm.
4. Cocorullo, G., et al., Silicon-on-silicon rib waveguides with a high-confining ion-implanted lower cladding, *IEEE J. Sel. Top. Quantum Electron.*, 4, 983 (1998).

5. Sauer, H., Paraire, N., Koster, A., and Laval, S., Optimization of a silicon-on-sapphire waveguide device for optical bistable operation, *J. Opt. Soc. Am. B*, 5, 443 (1988).
6. Janz, S., et al., Optical properties of pseudomorphic Si_1xGex for si-based waveguides at the $\lambda = 1300$-nm and 1550-nm telecommunications wavelength bands, *IEEE J. Sel. Top. Quantum Electron.* , 4, 990 (1998).
7. Celler, G.K. and Cristoloveanu, S., Frontiers of silicon-on-insulator, *J. Appl. Phys.*, 93, 4955 (2003).
8. Soref, R.A., Schmidtchen, J., and Petermann, K., Large single-mode rib waveguides in GeSi-Si and Si-on-SiO_2, *IEEE J. Quantum Electron.*, 27, 1971 (1991).
9. Pogossian, S.P., Vescan, L. and Vonsovici, A., The single-mode condition for semiconductor rib waveguides with large cross section, *J. Lightwave Technol.*, 16, 1851 (1998).
10. Powell, O., Single-mode condition for silicon rib waveguides, *J. Lightwave Technol.*, 20, 1851 (2002).
11. Bogaerts, W., et al., Nanophotonic waveguides in silicon-on-insulator fabricated with CMOS technology, *J. Lightwave Technol.*, 23, 401 (2005).
12. Tsuchizawa, T., et al., Microphotonics devices based on silicon microfabrication technology, *IEEE J. Sel. Top. Quantum Electron.*, 11, 232 (2005).
13. Lee, K.K., et al., Fabrication of ultralow loss Si/SiO_2 waveguides by roughness reduction, *Opt. Lett.*, 26, 1888 (2001).
14. Yablonovitch, E. Photonic band-gap structures, *J. Opt. Soc. Am. B*, 10, 283, (1993).
15. McNab, S., Moll, N., and Vlasov, Y., Ultra-low loss photonic integrated circuit with membrane-type photonic crystal waveguides, *Opt. Express.*, 11, 2927 (2003).
16. Notomi, M., et al., Waveguides, resonators and their coupled elements in photonic crystal slabs, *Opt. Express.*, 12, 1551 (2004).
17. Pearson, M.R.T., et al., Arrayed waveguide grating demultiplexers in silicon-on-insulator, in *Silicon-Based Optoelectronics II*, Proc. *SPIE* 3953, 11 (2000).
18. Murphy, T.E., Hastings, J.T., and Smith, H.I., Fabrication and characterization of narrow-band bragg reflection filters in silicon-on-insulator ridge waveguides, *J. Lightwave Technol.*, 19, 1938 (2001).
19. Tidmarsh, J. et al., A narrow linewidth laser for WDM applications using silicon waveguide technology, *IEEE Lasers and Electro-Optics Society 12th Annual Meeting*, Vol. 2, 497 (1999).
20. Jones, R., Paniccia, M.J., and Merritt, S., Tunable waveguide based external cavity laser using a silicon/poly-silicon Bragg grating, *Integrated Photonics Research*, Paper IThE4 (2004).
21. Irace, A., Breglio, G., and Cutolo, A., All-silicon optoelectronic modulator with 1 GHz switching capability, *Electron. Lett.*, 39, 232 (2003).
22. Tsao, S.L., Tien, J.H., and Tsai, C.W., Simulations on an SOI grating-based optical add/drop multiplexer, *IEEE J. Sel. Top. Quantum Electron.*, 8, 1277 (2002).
23. Ang., T. W. et al., Highly efficient unibond silicon-on-insulator blazed grating couplers, *Appl. Phys. Lett.*, 77, 4214 (2000).
24. Csutak, S. M., Dakshina-Murthy, S., and Campbell, J.C., CMOS-compatible planar silicon waveguide-grating coupler photodetectors fabricated on silicon-on-insulator (SOI) substrates, *IEEE J. Quantum Electron.*, 38, 477 (2002).
25. Tsao, S.L. and Peng, P.C., An SOI Michelson interferometer sensor with waveguide Bragg gratings for temperature monitoring, *Microwave Opt. Technol. Lett.*, 30, 321 (2001).
26. Dai, D. and He, S., Analysis of the birefringence of a silicon-on-insulator rib waveguide, *Appl. Opt.*, 43, 1156 (2004).

27. Kiyat, I., Aydinli, A., and Dagli, N., A compact silicon-on-insulator polarization splitter, *IEEE Phot. Technol. Lett.,* 17, 100 (2005).
28. Simova, E., Delage, A., and Golub, I., Study of polarization splitter/combiner in high index contrast Bragg reflector waveguides, *Integrated Photonics Research*, Paper IFC3 (2004).
29. Deng, H. et al., Design rules for slanted-angle polarization rotators, *J. Lightwave Technol.,* 23, 432 (2005).
30. Brooks, C. et al., Polarization rotating waveguides in silicon on insulator, *Integrated Photonics Research*, Paper IFG4 (2004).
31. Soref, R. A., Silicon-based optoelectronics, *Proc. IEEE*, 81, 1687 (1993).
32. Pavesi, L., and Lockwood, D.J., *Silicon Photonics*, Springer-Verlag, Berlin (2004).
33. Reed, G., and Knights, A.P., *Silicon Photonics: An Introduction,* John Wiley & Sons, Chichester (2004).
34. Bestwick, T., ASOC^{TM} — A Silicon-based Integrated Optical Manufacturing Technology, *Proc. IEEE Electronic Components and Technology Conference*, 566 (1998).
35. Lin, W. and Smith, T., Silicon opto-electronic integrated circuits: bringing the excellence of silicon into optical communications. Available at www.kotura.com.
36. Paniccia, M., Krutul, V., and Koehl, S., Intel's research in silicon photonics could bring high-speed optical communication to silicon. Available at www.intel.com.
37. Soref, R.A. and Bennett, B.R., Electro-optical effects in silicon, *IEEE J. Quant. Electron.*, QE-23, 123 (1987).
38. Fischer, U., Zinke, T., Shüppert, B., and Petermann, K., Single-mode optical switches based on SOI waveguides with large cross-section, *Electron. Lett.*, 30, 406 (1994).
39. Treyz, G.V., Silicon Mach–Zehnder waveguide interferometers operating at 1.3 μm, *Electron. Lett.*, 27, 118 (1991).
40. Treyz, G.V., May, P.G., and Halbout, J.M., Silicon optical modulators at 1.3 μm based on free-carrier absorption, IEEE *Electron. Dev. Lett.*, 12, 276 (1991).
41. Treyz, G.V., May, P.G., and Halbout, J.M., Silicon Mach–Zehnder waveguide interferometers based on the plasma dispersion effect, *Appl. Phys. Lett.*, 59, 771 (1991).
42. Zhao, C.Z. et al., Silicon on insulator Mach–Zehnder waveguide interferometers operating at 1.3 μm, *Appl. Phys. Letts.*, 67, 2448 (1995).
43. Hewitt, P.D. and Reed, G.T., Improving the response of optical phase modulators in SOI by computer simulation, *Lightwave Technology J.*, 18, 443 (2000).
44. SILVACO International, Santa Clara, CA 94054.
45. Xu, Q. et al., Micrometre-scale silicon electro-optic modulator, *Nature* 435, 325 (2005).
46. Giguere, S.T. et al., Simulation studies of silicon electro-optic waveguide devices, *J. Appl. Phys.*, 68, 4964 (1990).
47. Liu, A.S. et al., A high-speed silicon optical modulator based on a metal-oxidesemiconductor capacitor, *Nature*, 427, 615 (2004).
48. Liu, A.S. et al., Scaling the modulation bandwidth and phase efficiency of a silicon optical modulator, *IEEE J. Sel. Topics Quant. Electron.*, 11, 367 (2005).
49. Dosunmu, O.I., et al., Resonant cavity enhanced Ge photodectors for 1550nm operation on reflecting Si substrates, *IEEE J. Sel. Topics Quant. Electron.*, 10, 694 (2004).
50. Kik, P.G. et al., Design and performance of an erbium-doped silicon waveguide detector operating at 1.5μm, *J. Lightwave Technol.*, 20, 862 (2002).
51. Masini, G., Colace, L., and Assanto, G., 2.5 Gbit/s polycrystalline germanium-on-silicon photodetector operating from 1.3 to 1.55 μm, *Appl. Phys. Letts.*, 82, 2524 (2003).
52. El kurdi, M., et al., Silicon-on-insulator waveguide photodetector with Ge/Si self-assembled islands, *J. Appl. Phys.*, 92, 1828 (2002).

53. Fan, H.Y. and Ramdas, A.K., Infrared absorption and photoconductivity in irradiated silicon, *J. Appl. Phys.*, 30, 1127 (1959).
54. Bradley, J.B.D., Jessop, P.E., and Knights, A.P., Silicon-waveguide integrated optical power monitor with enhanced sensitivity at 1550 nm, *Appl. Phys. Lett.*, 86, 241103 (2005).
55. Knights, A.P. et al., Monolithically Integrated Photodetectors for Optical Signal Monitoring in Silicon Waveguides, *Proceedings of Photonics West*, SPIE, San Jose, CA (2006).
56. Ng, W.L. et al., An efficient room-temperature silicon-based light-emitting diode, *Nature*, 410, 192 (2001).
57. Rong, H. et al., A continuous-wave Raman silicon laser, *Nature*, 433, 725 (2005).
58. Presting, H., et al., Room-temperature electroluminescence from Si/Ge/$Si_{1-x}Ge_x$ quantum well diodes grown by molecular beam epitaxy, *Appl. Phys. Lett.*, 69, 2376 (1996).
59. Pavesi, L. et al., Optical gain in silicon nanocrystals, *Nature*, 408, 440 (2000).
60. Franzo, G. et al., Mechanism and performance of forward and reverse bias electroluminescence at 1.54 μm from Er-doped Si diodes, *J. Appl. Phys.*, 81, 2784 (1997).
61. Franzo, G. et al., Electroluminescence of silicon nanocrystal in MOS structures, *Appl. Phys. A*, 74, 1, (2002).
62. Priolo, F. et al., Role of the energy transfer in the optical properties of undoped and Er-doped interacting Si nanocrystals, *J. Appl. Phys.*, 89, 264, (2001).
63. Castagna, M.E. et al., High efficiency light emitting devices in silicon, *Mat. Sci. Eng. B*, 105, 83 (2003).
64. Boyraz, O. and Jalali, B., Demonstration of a silicon Raman laser, *Opt. Express.*, 12, 5269 (2004).

7 Enabling Fabrication Technologies for Planar Waveguide Devices

Boris Lamontagne

CONTENTS

7.1 INTRODUCTION

How are planar waveguides fabricated? The published literature on devices based on planar waveguides rarely provides details on fabrication. In general, more emphasis is placed on modeling and testing, unless the devices failed to work on schedule or were not ready before the intended conference or publication deadline! Fabrication know-how is often hidden; sometimes scientists feel more pressure to make devices than to publish their fabrication expertise. Moreover, the recent economic downturn in the photonics industry has resulted in a workforce quake and a net loss of fabrication know-how.

This chapter covers the fundamental fabrication steps for making passive planar waveguides and discusses some issues and challenges. We will discuss the three main materials: III–V semiconductors, silica, and silicon-on-insulators (SOIs). Silica is sometimes referred to as SOS (silica-on-silicon) but could also refer to silica-on-glass substrates. Because they are based on similar material systems, several fabrication steps used to make planar waveguides are similar to those used to make microelectronic or micro-electro-mechanical systems (MEMS) devices. Figure 7.1 illustrates the basic process flow for the fabrication of a passive planar waveguide. The first step is the growth or deposition of the optical layers, after which the patterning of the various waveguide structures using various combinations of mask and lithography techniques takes place. The transfer of the pattern from the mask to the optical layers usually involves wet and/or dry etching.

Once the waveguide structures have been patterned, cleaning, passivation, upper cladding deposition, and annealing may be performed. Rather trivial but critical processes are the final dicing, cleaving, and polishing steps required to obtain high quality optical facets prior to packaging and testing. These steps will ultimately influence the efficiency of the final coupling to optical fibers.

Table 7.1 describes some optical and fabrication properties for the materials discussed in this chapter. These three materials represent the majority of planar

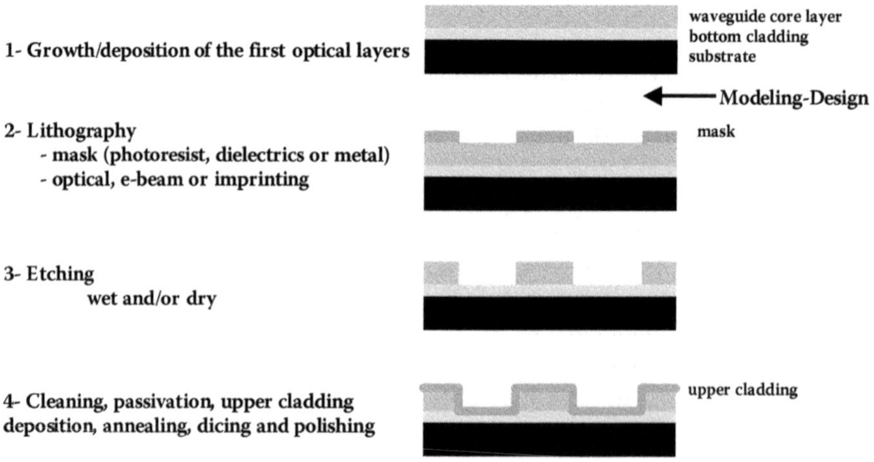

FIGURE 7.1 Basic process flow for fabrication of passive planar waveguide.

TABLE 7.1
Fabrication of Planar Waveguides in Different Materials

Material	Refractive Index (n)	Index Contrast (Δn)	Growth	Wet Etching	Dry Etching	Dicing/Optical Facet
III–V	3.1 to 3.4	0 to 15%	Epitaxy MBE-MOCVD	Iso/anisotropic	Halogenated and/or methane chemistry	Dry etching and leaving
Silica	1.44 to 1.47	0 to 1.5%	FHD-PECVD (amorphous)	isotropic	Low selectivity/ fluorocarbon chemistry	Sawing & polishing
SOI	3.47	70%	SIMOX, Smart Cut™	Iso/anisotropic	Vast range of possibilities	Dry etching and cleaving

waveguides available on the market or undergoing research and development (R&D). This table presents typical values of the optical properties as well as some fabrication aspects. Other researchers have compared various optical material platforms[1,2] without addressing the fabrication aspects.

The various material systems are characterized by comparable propagation losses ~0.1 dB/cm, even for III–V materials,[3] although silica waveguides tend to be less lossy. The propagation losses for each material were not included in the table because the waveguide losses often depend on the waveguide geometry (size, refractive index contrast, mode confinement) rather than on the material itself. The typical wafer size for III–V materials (especially InP) is usually smaller than for silica and SOI materials. Moreover the silica optical layers tend to be more complex to deposit and anneal than the other optical materials. The silica materials could also include silicon oxynitride and silicon nitride, which are characterized by higher refractive indices and contrasts. In Table 7.1, the coarse comparison of index contrast depends on waveguide geometry and in-plane and out-of-plane index contrast. In general, the index contrast increases from silica to III–V to SOI; and this determines the confinement level and minimum bending radius and hence device compactness.[1,4]

The silica layers tend to be characterized by rather high intrinsic stress levels.[1,4] Silica planar waveguides present a clear advantage over coupling with optical fibers because their size and refractive indices are close to those of optical fibers. However, higher integration (hybrid or monolithic) levels favor the SOI or III–V for future development. There are always exceptions to the rule, for example a CVD glass called Hydex^{TM} with a relatively high index contrast up to 20% and low stress has been reported.[5] The fabrication maturity of silicon seems to favor SOI, and it is expected that SOI wafers will become less expensive in the future. SOI waveguides may also have the shortest fab cycle times.

The applications of planar waveguides are diversified. No longer concentrated on telecommunications, an intense R&D effort in biophotonics now demands various types of silica and SOI planar waveguides. The various materials platforms discussed may retain their specific niche applications and markets; this is true also for other optical materials not covered in this chapter such as polymers, sol-gels, lithium niobate, ion-exchange materials, etc.[6]

7.2 OVERVIEW OF FABRICATION TECHNOLOGIES

This section will cover the basic fabrication steps: growth, deposition, lithography, etching, and finally, dicing and cleaving.

7.2.1 Growth and Deposition of Optical Materials

7.2.1.1 Growth of III–V Optical Layers

III–V compound semiconductors are the materials of choice for designing lasers, amplifiers, detectors, modulators, and other active devices. They are also obvious choices for monolithic integration. There are a variety of compounds such as GaAs, InP, GaP, GaN, as well as alloys (ternary, quaternary, and further). The substrates (bulk) are usually obtained by vertical Bridgman or Czochralski growth techniques. Subsequent growth of optical layers often takes place by epitaxial techniques to obtain varying layer properties such as bandgap or refractive index. The epitaxy is the oriented overgrowth of a thin crystalline layer on the surface relative to the crystalline substrate.

Over the years, improvements in growth control have allowed us to progress from a few thick layers (tens of micrometers) of growth to hundreds of nanometer-size layers of growth for quantum wells and quantum dots. Several techniques may be used for the epitaxy of III–V materials:

1. Liquid phase epitaxy (LPE) is the oldest technique. It was used for the preparation of the first generation of lasers. It is inexpensive but gives rather poor uniformity and control of thin layers.
2. Vapor phase epitaxy (VPE) in which the various elements including the dopants are brought to the substrates in a gaseous phase. A new version of this technique is called metal–organic chemical vapor deposition (MOCVD); it is one of the two leading techniques for growing III–V materials. It was first reported in 1968.[7] One of the chief advantages of MOCVD is its ability to grow AlGaAs, which is difficult using other VPEs; however, its utilization of very toxic arsine makes it more difficult to implement. In a large flow of hydrogen, the reactant gases such as trimethylgallium and arsine are brought in contact with the hot GaAs substrate. The gases decompose to form GaAs and methane according to the reaction:

$$(CH_3)3Ga(g) + AsH_3(g) \xrightarrow{\text{Heat}} GaAs(s) + 3CH_4(g)$$

3. Molecular beam epitaxy (MBE) is a sophisticated evaporation or physical deposition technique. The various atomic elements are evaporated from ovens called effusion cells. Highly controlled growth can take place, assuming the critical shutter operation and the vacuum quality (base pressure: 10^{-10} to 10^{-11} Torr) have been optimized. Such growth is rather slow (atomic layer/sec) and is often monitored by a sophisticated characterization technique such as reflection high energy electron diffraction (RHEED) to evaluate the crystal structure. Chemical beam epitaxy (CBE) is a variant of MBE that involves a chemical reaction on the substrate with a metal–organic source.

7.2.1.2 Deposition of Silica Optical Layers

Silica is the most common material for planar waveguides.[1,4,8] Its refractive index and index step allow easy coupling to optical fibers. Silica optical layers are typically deposited on a Si substrate, sometimes on quartz substrates for stress reasons or on III–V materials for integration with III–V devices. The Si substrate is preferably already thermally oxidized (5 to 15 μm thick) to form a buffer oxide or bottom cladding.

The two main techniques of deposition are flame hydrolysis deposition (FHD)[9–16] and plasma-enhanced chemical vapor deposition (PECVD).[17–31] Other techniques such as CVD (low pressure or atmospheric), evaporation, ion implantation, and sputter deposition are less common and will not be covered here.

For both FHD and PECVD techniques, the refractive indices of the resulting films can be controlled by varying gas flows; adding small amounts of a dopant (usually Ge) is the favored method. A high temperature annealing step is commonly required to improve the structures of the layers and the optical propagation. For the top cladding over the etched waveguide, boron and phosphorus dopants are also added to improve the flowing behavior of the silica during the temperature annealing, resulting in lower intrinsic stress and better planarization. The control of intrinsic stress build-up in these optical layers is often critical for waveguide devices.[4,32,33]

Flame hydrolysis deposition was developed mainly at NTT and is based on the process of fabrication of optical fiber preforms. For doping, it uses $SiCl_4$ vapors as well as $GeCl_4$, PCl_3, and BCl_3 burned in an oxygen/hydrogen torch, resulting in soot particles deposited on a rotating substrate. The technique has given good results for thick layers, up to 100 μm. The soot requires a consolidation step at 1000 to 1300°C.

The PECVD technique was developed for the semiconductor industry. Initially it targeted low deposition rates and thicknesses but can now reach deposition rates up to 300 μm/min and layers tens of micrometers thick. Several equipment manufacturers tried to enter the photonic market (STS, Oxford, Trikon, Unaxis, and Novellus) and many research papers reported studies of hundreds of PECVD process variables. The basic technology involves a vacuum system (Figure 7.2) in which a radio-frequency plasma is generated between parallel plates with a gas mixture typically of SiH_4 and N_2O. Carrier gases (He, N_2, or Ar) and dopant gases (GeH_4, PH_3, BH_4, or CF_4) may also be added. The basic reaction can be described by:

$$SiH_4\ (\text{gas}) + 4\ N_2O\ (\text{gas}) \rightarrow SiO_2\ (\text{solid}) + 2\ H_2O\ (\text{gas}) + 4\ N_2\ (\text{gas})$$

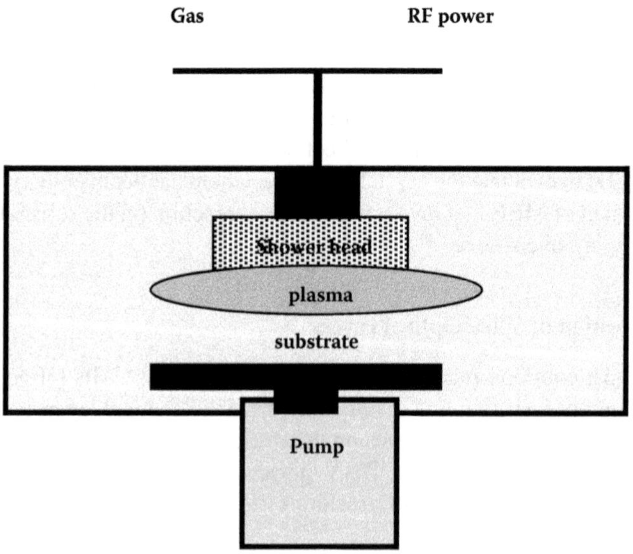

FIGURE 7.2 Basic PECVD system.

The deposition usually takes place at a temperature of 250 to 350°C. Such optical layers could be obtained at lower temperatures compared to FHD, making it possible to integrate the layers with temperature-sensitive materials such as III–V.

Extensive research efforts on PECVD layers led to a very flexible and controllable technique. Moreover, several configurations of PECVD systems have been developed as illustrated by Figure 7.3.[28] New PECVD systems using dual frequency or high density plasmas (ECR, Helicon, ICP, etc.) bring new possibilities for optimizing optical layers for low absorption, low stress, and low process temperature. PECVD layers usually require thermal annealing (~1000°C) to reduce the concentrations of Si–H and N–H bonds, and their associated absorption lines near 1480 and 1510 nm close to the C band (1525 to 1565 nm).[30,34–38] The doped P and/or B cladding layers usually require multiple deposition and anneal steps to optimize glass flow.

The advantages of high-contrast optical waveguides brought increased attention to PECVD SiN_x and SiO_xN_y layers.[39–57] They provide the possibility of higher integration, but issues such as increased absorption losses and film stress must still be resolved.

7.2.1.3 Fabrication of Silicon-on-Insulators (SOIs)

SOI technology has evolved rapidly, particularly to meet the needs of the microelectronics and MEMS industries. The simple act of placing a thin crystalline Si layer over a thin buried oxide (BOX) layer opened many new possibilities for device design such as high electron and photon confinement close to the surface. The resulting high contrast composite material is also promising for the fabrication of planar waveguides. The three main techniques used to produce SOI wafers are (1) bond and etch-back SOI (BESOI), (2) separation by implanted oxygen (SIMOX), and (3) Soitec's Smart Cut™ (Unibond™) process.[58–60]

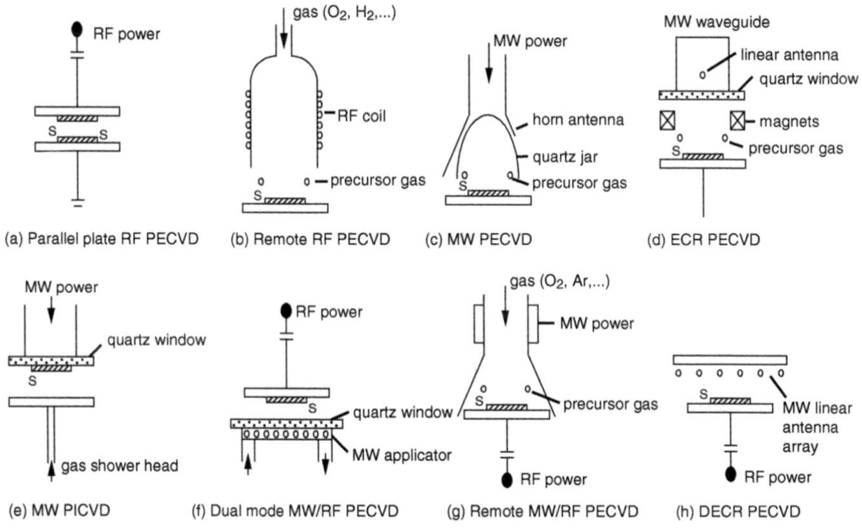

FIGURE 7.3 Various PECVD configurations.[28]

The BESOI technique involves bonding two oxidized Si surfaces, and then thinning down one of the wafers until it reaches the desired Si thickness. Such a process limits the Si thickness uniformity to 0.5 to 1.0 μm, resulting in rather thick Si layers. Moreover, the process consumes two wafers for every SOI wafer produced. The BESOI technique is appropriate only for waveguides of rather large dimension (over 10 μm).

The SIMOX technique involves implanting high energy (~200 keV) oxygen ions. The energy and dose of the ions determine the silicon and BOX thicknesses. The wafer is then thermally annealed to reduce the dislocation levels and form a distinct BOX underlayer. Usually the SOI and BOX layer are only a few hundred nanometers thick, which is often not sufficient for planar waveguides. It is always possible to overgrow Si by epitaxy to thicken the Si layer, but this step increases the cost.

Smart Cut™ is a combination of the SIMOX and BESOI techniques developed a decade ago at France's Laboratoire d'Electronique de Technologie de l'Information. Figure 7.4 (from SOITEC's website) describes the Smart Cut™ technique. A thermally oxidized Si wafer is implanted with hydrogen ions in order to introduce structural

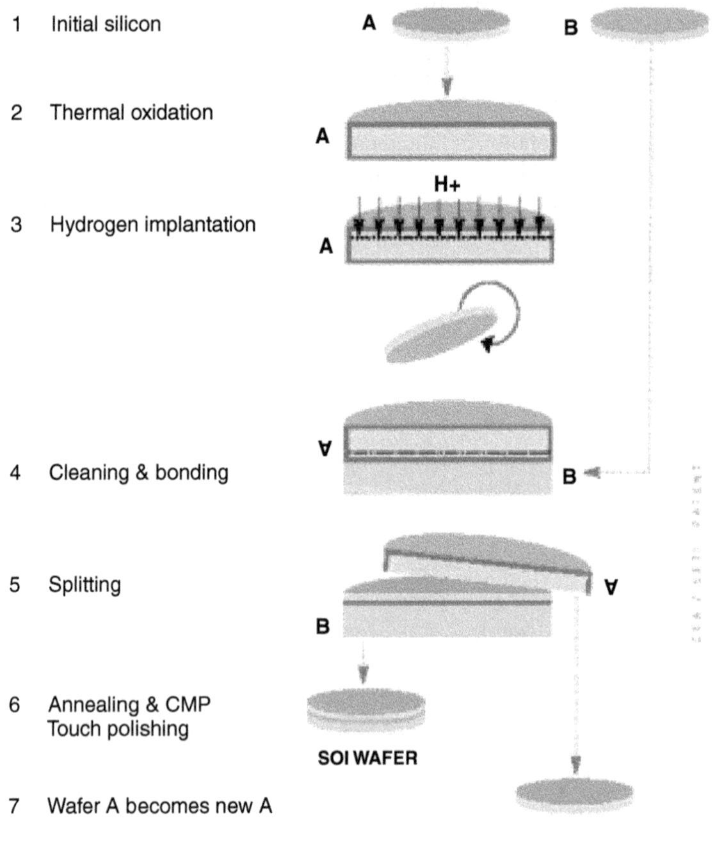

FIGURE 7.4 Schematic illustration of the Smart Cut™ technology for SOI manufacturing. (With permission from SOITEC.)

defects. The implanted wafer is then bonded to a regular Si wafer and thermally annealed to separate the wafer at the hydrogen underlayer and complete the bonding process. A final polishing leaves a good quality SOI wafer with relatively low dislocation levels and a wide range of Si and BOX layer thicknesses. The SmartCut seems to be the optimum technique for the preparation of SOIs for planar waveguides.

7.2.2 Lithography

Lithography is the process of transferring a pattern from a mask or design to the surface of a sample. It is a critical fabrication step. It is one of the most limiting factors of the semiconductor roadmap,[61] often determining the sidewall roughness, minimum feature size, and other characteristics that can ultimately be achieved. The lithographic techniques to be discussed here use polymers to be patterned and then used as masks for pattern transfers into substrates. The three main parameters that define lithographic techniques are resolution, registration, and throughput. Resolution is the size of the minimum feature that can be defined in the mask. Registration is the ability to align or overlay different mask levels over each other accurately. Throughput is the speed at which wafers can be patterned.

This section covers three different lithography techniques: optical lithography, electron-beam lithography, and imprinting lithography; they are compared in Table 7.2. This section will not cover other techniques such as laser patterning, LIGA, x-ray and ion beam lithography, or scanning probe writing.

7.2.2.1 Optical Lithography

Optical lithography is the most common patterning technique. This section briefly describes the two major optical lithography techniques: contact and projection (Figure 7.5). Optical lithography uses photons to modify a photosensitive polymer called a photoresist that typically has a thickness ranging from 0.2 to 100 μm. In order to ensure uniform coverage, the photoresist is usually spun onto a wafer, although some versions can be sprayed or evaporated. Depending on chemical composition and post-exposure process, photoresists are classified as either positive or negative.

The exposed areas of a positive photoresist will dissolve when immersed in a liquid developer, while it is the contrary for the negative resist. The mask inserted between the light source and the sample is usually made of quartz or fused silica patterned with chromium-coated opaque areas. The patterns on the high resolution masks are written by electron-beam lithography. Lower resolution masks (>1 μm)

TABLE 7.2
Coarse Comparison of Three Lithography Techniques

Technique	Resolution (nm)	Registration	Throughput
Optical	~ 150	Good	High
Electron-beam	~ 10	Very good	Low
Imprinting	~ 10	Good	Medium

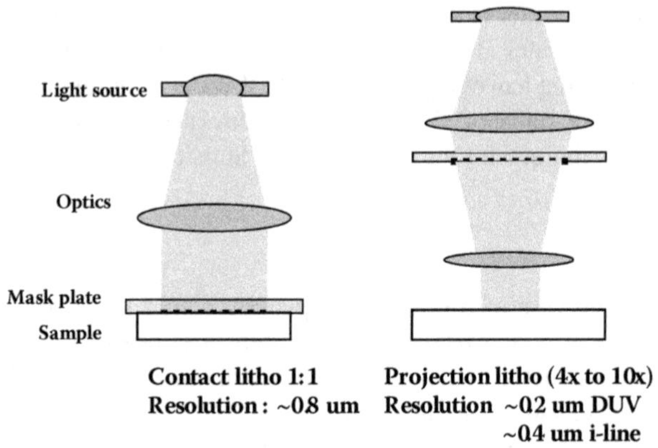

FIGURE 7.5 Two main optical lithography techniques: contact and projection lithographies.

are written by optical pattern generators. The resolution of the optical lithography process is a function of the wavelength of the light source used.

The two main exposure schemes in optical lithography are represented in Figure 7.5. Contact lithography implies that the mask plate is in mechanical contact with the photoresist spun on the sample. The typical resolution of a contact mask aligner is 0.8 μm for a common Hg lamp (i-line, 365 nm). This type of mask aligner is commonly observed in R&D or low volume production environments. In projection lithography, an image of the mask is projected onto the sample surface. The image is usually 4 to 10 times smaller than the object (mask). The resolution is typically better for projection lithography. Steppers (projection aligners) are commonly found in high volume production environments. They pattern the wafers by steps, repeating the mask unit several times on each wafer. Contact aligners are less expensive than steppers but lead to more mask contamination and sample damage during the contact step.

Projection aligners can refocus on each field to optimize the exposure locally while with contact aligners, a wafer may have areas where there are gaps (losses of mechanical contact) between the mask and the sample. Figure 7.6 presents SEM images of retro-reflectors patterned by contact lithography in photoresists on different areas of the same sample. A gap, probably related to defects or particles, induced diffraction effects.

7.2.2.2 Electron Beam Lithography

Instead of using photons, electrons can also be used to expose the photoresist. The resolution can be very high, around 10 nm. Two other advantages of electron-beam lithography are the large depth of field compared to optical lithography and very low levels of contamination damage. High resolution optical lithography suffers from small depth of field, which makes it more difficult to process non-ideal wafers from warping and already patterned (topography) wafers.

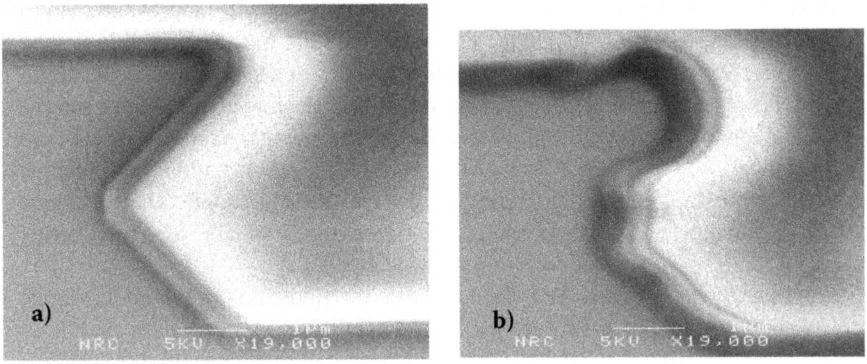

FIGURE 7.6 SEM top view images of retro-reflectors in photoresist on the same sample at different locations showing (a) good to (b) bad contact lithography. The bad contact was most probably induced by localized defects.

The two main types of electron-beam lithography are direct writing and projection (for example, SCALPEL). Direct writing systems use different spot shapes and writing modes but generally yield very low throughputs. Projection systems have been in development for many years with little success; the specialized mask needed is a difficult challenge. During the exposure process, back-scattered and secondary electrons result in proximity effects that complicate the patterning of dense features.

7.2.2.3 Imprinting Lithography

Moore's law and the International Technology Roadmap for Semiconductors (ITRS)[61] call for sub-100-nm resolution (gate length) in the near future. Extreme UV lithography (EUVL), immersion lithography, and phase shifting masks may yield feature sizes as small as 50 to 100 nm. However, for smaller dimensions, a new technology may be required. ITRS considers imprinting a possible solution for the 32-nm node or evolution step.[62] Moreover, imprinting lithography may be a less expensive choice than the complex and very expensive high resolution optical steppers. In contrast to exotic lithography techniques like x-ray and SCALPEL, several commercial imprinting systems are already available.[63]

Imprinting lithography involves a mechanical contact between a mold and a soft (polymer) material to transfer a pattern from the mold to the sample. Among the different versions of imprint lithography[64] are nanocasting, displacement printing, and chemical imprinting. The three basic versions are nanoimprinting,[65–70] soft-lithography,[71–74] and step-flash lithography.[75] They allow very high resolution ($\sim$10 nm) and three-dimensional patterns. They can pattern flat or curved and small or large (8 to 12 inches) substrates. The mask (template mold) is usually fabricated using electron-beam lithography.

Nanoimprint lithography involves a solid mold, for example, silicon or nickel. The process requires heating the surface polymer over its glass transition temperature, then embossing the pattern with the mold. Soft lithography involves the transfer of self-assembled monolayers using a flexible template. In order to improve the alignment process, an imprinting technique known as step-and-flash imprint lithography (S-FIL) using a rigid and transparent template was developed. Figure 7.7 illustrates the S-FIL process. A transparent mold of quartz is brought into contact with a sample surface previously coated with a monomer. Before removing the mold, the patterned monomer is exposed to UV light through the mold and the residual monomer is etched away (where pressed) before transferring the pattern onto the sample. This technique does not require hot and hard embossing, and is particularly suitable for good overlay accuracy.

Extensive research efforts have been dedicated to finding an optimum technique for very high resolution. S-FIL seems a good candidate, especially for photonics applications, where the required throughput is at least an order of magnitude lower than for microelectronics. None of these imprint techniques aims for very high throughputs. One challenge facing imprint lithography is the development of good anti-adhesion layers required for efficient and high yield release of templates.

7.2.3 Etching

The etching process may be defined as the selective removal of material subsequent to its patterning by a lithographic process. Because etching is highly material-dependent, each material system will be discussed separately. Etching processes can be classified according to directionality and selectivity. The directionality can range from isotropic

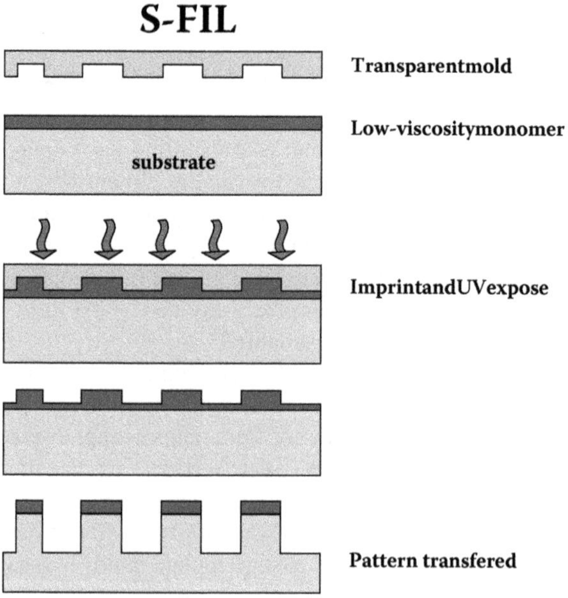

FIGURE 7.7 Step and flash imprint lithography (S-FIL).

(etch rates comparable in the vertical and lateral directions) to anisotropic (vertical etch rate much higher than that of the lateral direction). The etching selectivity may be defined as the ratio of the etch rates of the material being etched to the etch rate of the mask or another adjacent layer. Table 7.3 compares some aspects of the wet and dry etching processes.

7.2.3.1 Wet Etching

The reaction of a wet etching process takes place in a liquid medium and yields liquid or gaseous by-products. Reproducibility of the process may be a concern; it depends on the surface state preparation, cleanliness, and stability of the solution. Uniformity and control of the usual dipping in a bath are not very good, but new automatic systems such as the Wavetch^{TM} improve these characteristics. The key ingredients to a wet etch process are (1) an oxidizer (for example, H_2O_2, HNO_3), (2) an acid or base to dissolve the oxidized surface (H_2SO_4, KOH), and (3) a diluent medium (H_2O, CH_3COOH).

III–V semiconductors are usually crystalline materials obtained by epitaxy. A III–V device can contain an etch stop layer, that is a specific composition layer allowing to stop the etch at a specific depth. Several types of anisotropic etchants can be used based on sulfuric, phosphoric, and citric acids or ammonium and sodium hydroxides.[7,75–78] The concentrations of chemical compounds will influence the etch profile, selectivity, and reaction type. Isotropic etchants based on a mixture of bromine and methanol are sometimes used to smooth the surface of a sample.

Because silica is an amorphous material, it will undergo isotropic etching. The etch rate depends on the doping and the silica density, which in turn depend on the deposition techniques and conditions (PECVD, LPCVD, thermal oxide, thermal anneal). The wet etching of silica is based on mixtures containing hydrofluoric acid.

SOI materials (in fact Si layers) benefit from the vast experience of the microelectronic and MEMS industries.[80–82] In particular, the MEMS sector has developed and

TABLE 7.3
Comparison of Wet and Dry Etching

Parameter	Wet Etching	Dry Etching
Principle	Liquid chemical solution	Mixture of physical and chemical processes
Equipment	Simple bath at atmospheric pressure, manual or robotic system	Vacuum chamber, manual loading or cassette
Directionality	Usually isotropic, except for crystalline substrates	Anisotropic (good control)
Selectivity	High	Low to medium
Smallest features	$\sim 1\mu m$	$\sim$ 10nm
Damage	No	Yes
Reproducibility	Fair	Good
Cost	Inexpensive	Expensive

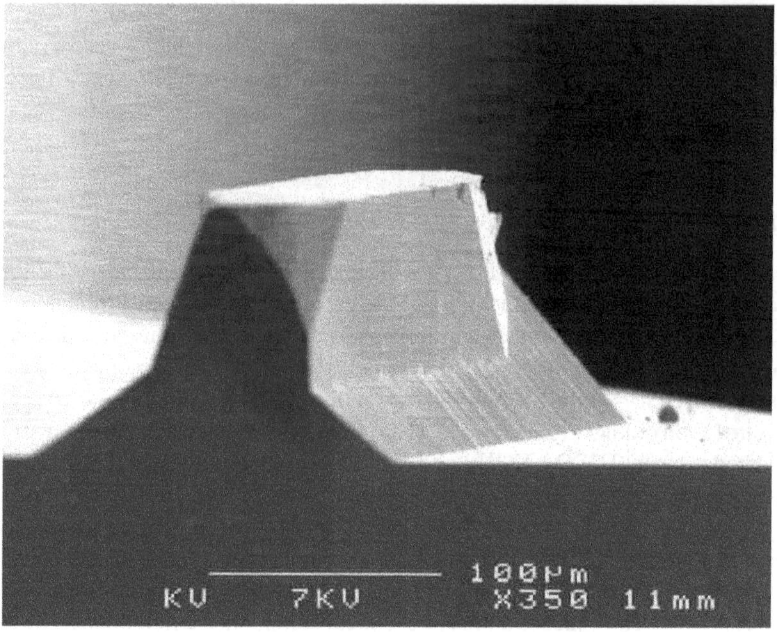

FIGURE 7.8 SEM image of wet-etched Si ridge.

tested a variety of anisotropic etchants with different aims, for example, to achieve compatibility with the CMOS process. The anisotropic etching processes have etch rates (ERs) dependent on the crystallographic plane ER (100) > (110) > (111). The roughness of the etched surface depends on the KOH concentration. The best etching mask for KOH seems to be SiN_x. Figure 7.8 shows a SEM image of a wet etched Si ridge. Table 7.4 summarizes the most common wet etching solutions for various materials.

TABLE 7.4
Main Etching Solutions

Material	Typical Wet Etchant Anisotropic	Isotropic
III–V:		
InP-based	H_3PO_4:HCl	Br:methanol
GaAs-based	$H_2SO_4:H_2O_2:H_2O$	Br:methanol
Silica		HF
SOI (silicon)	KOH with or without isopropanol Alkali-OH, EDP, TMAH, N_2H_4-H_2O	HF:HNO_3:CH_3COOH:H_2O

TABLE 7.5
Coarse Comparison of Dry Etching Processes

Process or Technique	Process Pressure (mTorr)	Ion Energy (eV)	Anisotropy	Selectivity
Plasma etching	~ 100	~ 10	Low	Very good
Reactive ion etching (RIE)	~ 10 to 100	~ 100	Good	Good
High density plasma (HDP) etching (ICP, ECR, Helicon)	~ 10	~ 10 – 100	Good	Good
CAIBE–RIBE	~ 1	~ 100 to 1000	Good	Low
IBE–ion milling	~ 1	~ 1000	Good	Low

7.2.3.2 Dry Etching

The dry etching process could be described as the selective removal of materials in the gas phase, usually under partial vacuum and plasma assisted. Most dry etching systems are based on specific proportions of chemical and physical processes. The chemical aspect involves the concentration of radical (reactive species), while the physical aspect is related to the bombardment of energetic ions. There are numerous types of dry etching processes depending, for example, on the pressure. Table 7.5 lists a few examples of dry etching systems and some of their characteristics.

Lower process pressures generally are coupled to higher ion energies. Figure 7.9 illustrates a common, versatile, high density plasma etching system with an inductively coupled plasma (ICP) source and antenna coil. The ICP source generates a high density plasma by radio-frequency excitation, creating reactive species. A second

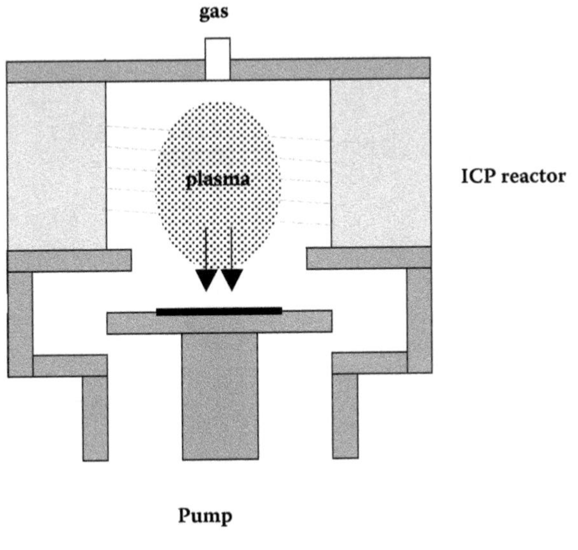

FIGURE 7.9 High density plasma (HDP) etcher based on ICP.

radio-frequency source coupled to the platen (wafer holder) creates an electric field gradient attracting the ions toward the sample. This type of system allows the decoupling of the chemical and physical aspects of dry etching and thus gives a better control of etching results (profile, selectivity, etc.). Dry etching processes are complex. Reactions take place at the sample surface and also in the plasma and on the walls of the vacuum chamber. Reproducibility requires a good knowledge and control of the numerous parameters involved in the dry etching process.

At the surface of the etched areas of the sample, several etching and passivation mechanisms govern the etch rate and resulting profile. The etching processes usually involve ion bombardment enhancing the reaction of the surface with reactive species created by the plasma. The ion bombardment is generally perpendicular to the sample surface and thus leads to anisotropic (vertical) etching. The etched trenches commonly result from the competition of etching and passivation phenomena. Figure 7.10 illustrates some of the processes. Depending on the etching chemistry, pressure, and ion bombardment, various proportions of mask charging, sidewall passivation or inhibition, ion bowing, and scattering modify the etch profile.

Figure 7.11 illustrates the importance of selecting the optimum proportions of physical and chemical aspects of dry etching. Several SEM images of etched InP ridges by chemically assisted ion beam etching, also known as CAIBE (Cl_2/Ar) are shown. The ratio of ion current to Cl_2 flow was reduced from the top image to the bottom. The top image (Figure 7.11a) shows a typical physical etch, characterized by an overcut profile (slanted sidewalls), microtrenches, mask erosion, and smoothness. Slightly more chlorine gives an optimum etching profile (Figure 7.11b). The roughness of the etched surface is related to the preferential etching of the phosphorus, leading to In clusters and thus micro-masking. This micro-masking, revealed by rough areas induced by etch-resistant particles, actually helps the vertical etching. Adding more chlorine results in chemically dominated etching revealed by the spontaneous etching

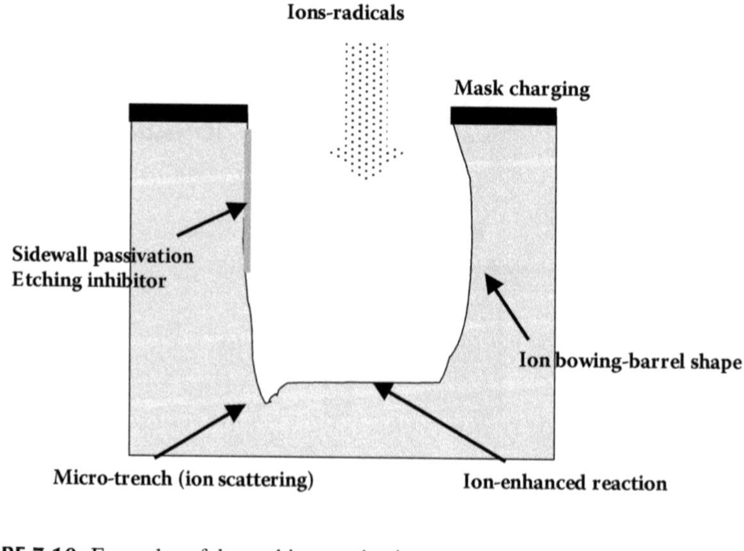

FIGURE 7.10 Examples of dry etching mechanisms.

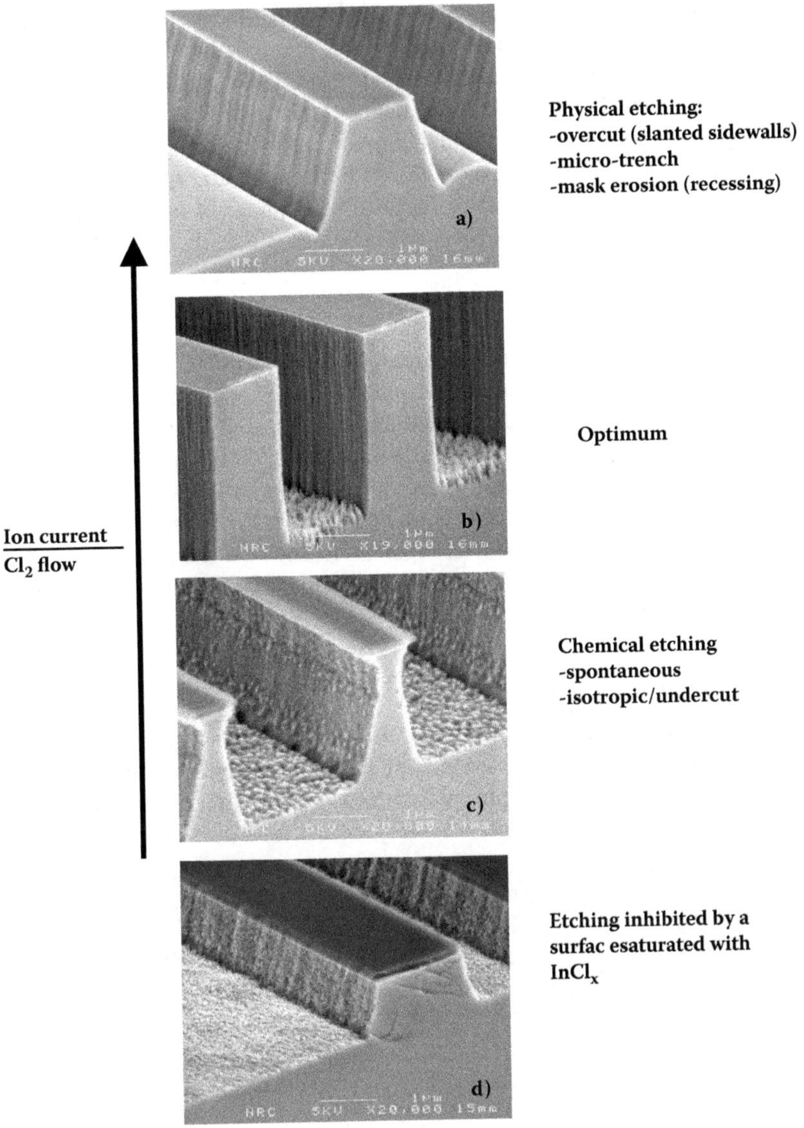

FIGURE 7.11 Physical- versus chemical-dominant etching. SEM images of InP etched by CAIBE (Cl_2/Ar). (From Lamontagne, B. et al., *J. Electrochem. Soc.*, 146, 1918, 1999. With permission.)

(undercut) under the SiO_2 mask. Increasing the chlorine concentration even further saturates the surface and inhibits further etching (Figure 7.11d).

7.2.3.2.1 Dry Etching of III–V

GaAs-based materials are often dry etched by a mixture containing BCl_3, Cl_2, HBr, and/or $SiCl_4$.[7,84–86] InP-based materials traditionally have been etched using two recipes: methane–hydrogen[87–92] and chlorine,[84,93–96] but in recent years numerous complex chemistries have also been studied and developed.[97–100] The methane–hydrogen process gives a smooth and slow etch. Nevertheless it also produces thick polymer on the sidewalls and H-induced damage for active devices. The chlorine chemistry gives a good etch profile and forms no polymer on the sidewalls, but usually requires a high temperature (~200°C) to promote desorption of the $InCl_x$ formed during the etch process. Figure 7.12 is a SEM image of InP trenches etched by CAIBE (Cl_2/Ar) with a mask of SiO_2 PECVD.

7.2.3.2.2 Dry Etching of Silica

Silica is a relatively inert material and requires a significant amount of ion energy to promote reactions between plasma species such as CF_2 with SiO_2. Traditionally, RIE systems have etched silica using mixtures of CF_4, CHF_3, and oxygen.[102–104] The more recent high density plasma systems such as ICP often use a mixture of C_4F_8 (or less frequently C_2F_6) with oxygen or argon, with a mask of photoresist, Si, or Al.[105–113] This etching process is based on the passivation of the sidewalls by polymerization of etch by-products. Figure 7.13 shows a silica ridge etched by ICP-RIE with a mixture of C_4F_8/Ar with an aluminum mask. The etch is vertical and provides smooth sidewalls.

7.2.3.2.3 Dry Etching of SOI

Over the years, several methods have been developed to dry etch silicon, depending on the device requirements. Gas phase etching using XeF_2 was developed to release MEMS structures. It does not require plasma and needs only a reactor chamber to

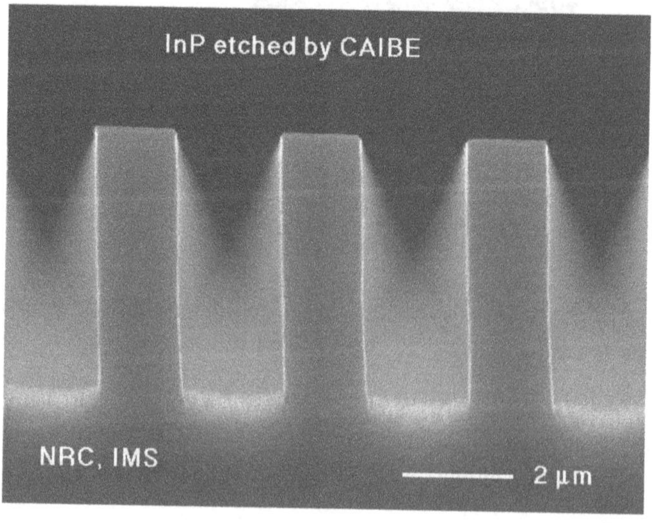

FIGURE 7.12 SEM image of InP trenches etched by CAIBE (Cl_2/Ar). (From He, J.J. et al., *J. Lightwave Technol.*, 16, 631, 1998. With permission.)

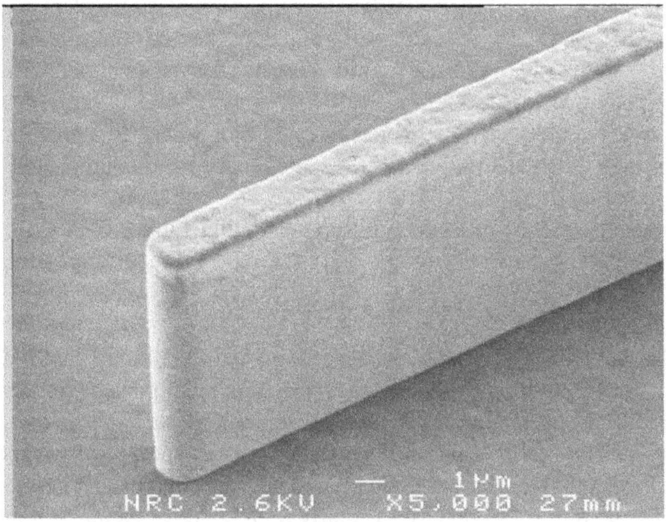

FIGURE 7.13 Silica ridge etched by ICP-RIE using a C_4F_8/Ar chemistry with aluminum mask.

bring the XeF_2 gas (or sometimes BrF_3 or ClF_3) into contact with the Si. The process is purely chemical and isotropic.

Chlorine- and bromine-based plasmas have been used to etch Si devices, mainly for microelectronics.[113–121] This method tends to be characterized by a slower etch rate than fluorine-based etching. However, it has the virtue of being less prone to aspect ratio-dependent etching (ARDE), i.e., etching profiles and depths that are dependent on the sizes of the features to be etched.

A process known as deep reactive ion etching (DRIE) is commonly used to etch Si for MEMS and MOEMS (micro-optical-electro-mechanical systems) applications. The two DRIE techniques are commonly known as the Bosch and cryogenic processes. Both can quickly etch vertical trenches through a wafer. The patented Bosch process[122–123] uses cycles of etching (SF_6) and passivation (C_4F_8), at room temperature. Horizontal striations, scallops or ripples are usually observed on the sidewalls. The cryogenic process uses SF_6 with a small amount of oxygen[124–134] at a temperature of about –120°C. Thick photoresist masks may crack at such low temperatures. Oxygen containing by-products inhibit etching on the sidewalls. Vertical and smooth sidewalls can be achieved. For both DRIE techniques, the ability to ramp parameters as the etch progresses can be important and the results may be dependent on geometry. Both DRIE processes are characterized by very high mask selectivity.

For a SOI planar waveguide, the etch depth typically ranges from 0.3 to 3 μm. The sidewalls should be as smooth as possible to reduce waveguide propagation loss due to light scattering on rough sidewalls. The DRIE techniques described above are not the optimum methods for this application. A more adequate etching process has been developed and involves a mixture of SF_6 and C_4F_8 (etching plus passivation) at room temperature.[31,135] This technique combines the advantages of the cryogenic and the Bosch processes without their disadvantages. By tuning parameters such as the proportions of SF_6 and C_4F_8, the etch profile and depth can be controlled accurately. The etch rate is in the range of 0.5 μm/min, adequate for the small and critical etch depth of SOI planar waveguides.

Recent work[135] on dry etching optical facets in SOI described a process capable of etching through the Si layer as well as the buried oxide layer continuously without a notch.

The various dry etching processes are summarized in Table 7.6.

7.2.4 Dicing, Cleaving, and Polishing

After samples have been processed (patterned, etched, metallized, etc.), a critical step still remains: separation of the samples into individual pieces (die) for optical characterization and packaging. This involves operations such as dicing, cleaving and often also polishing. III–V planar waveguides devices are typically lapped, back polished, scribed, and then cleaved. Silica-based devices are typically sawn and then polished along the facets. No process is clearly preferable for dicing SOI materials; the operations are characterized by low reproducibility and low yield.

Another possibility for obtaining good quality waveguide optical facets is the dry etching used on III–V materials and more recently on SOI materials.[135] It offers high reproducibility and accurate positioning of the facet relative to the waveguide. It involves two etching steps: (1) etching of the cladding, the Si waveguide core layer, and the buried oxide cladding in one step, and then (2) making a deep etched trench for cleaving and then coupling to the optical fiber, as such etched facets can be integrated with V-groove etches for optical fiber alignment. Figure 7.14 shows a SEM image

TABLE 7.6
Dry Etching Processes

Material	Chemistry	Comments
GaAs	BCl_3, Cl_2, HBr, $SiCl_4$	
InP	CH_4/H_2	Smooth, slow, room temperature passivation, damage
	Cl_2	No polymer formation, fast, high temperature required
Silica	CHF_3, CF_4 (RIE) C_4F_8 (high density plasma)	
SOI	XeF_2 or sometimes BrF_3, ClF_3	Chemical dry etching, isotropic, no plasma
	Cl_2, BCl_3, HCl, HBr	Slow, microelectronics
	DRIE–Bosch cycles of SF_6 then C_4F_8	Deep, rough sidewalls
	DRIE–cryogenic SF_6-O_2	Deep, smooth, very selective, no thick PR
	Fluorocarbon mixture like C_4F_8/SF_6	Medium etch rate, smooth
	$C_4F_8/SF_6/O_2$	Etching through the SOI and the box without notch

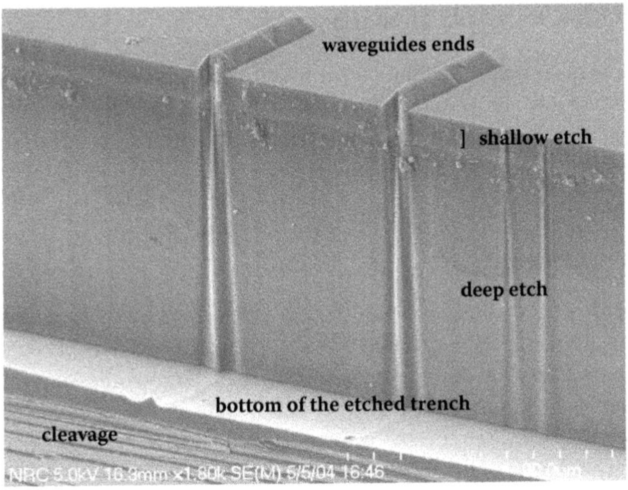

FIGURE 7.14 SEM images of etched and cleaved SOI planar waveguide facets.

of such a SOI sample with waveguides, a 40-μm etched groove, and a mechanical cleave.

Other possibilities for coupling planar waveguides to other devices include diffractive optical elements (DOEs)[136–137] and out-of-plane mirrors.[138] Out-of–plane mirrors or inclined facets have been developed in III–V,[139–143] polymer-silica,[144–147] and Si[148–49] platforms. The fabrication of inclined facets is not trivial; techniques like wet etching along preferential crystal planes, reactive ion etching using a Faraday cage, focused ion beam (FIB), ion milling, sawing, and CAIBE have been used. Figure 7.15 shows out-of-plane mirrors fabricated in SOI by tilted CAIBE.[149] The roughness visible at the bottom of the mirrors is induced by the Ni mask sputtered off residues; it does not induce significant optical loss.

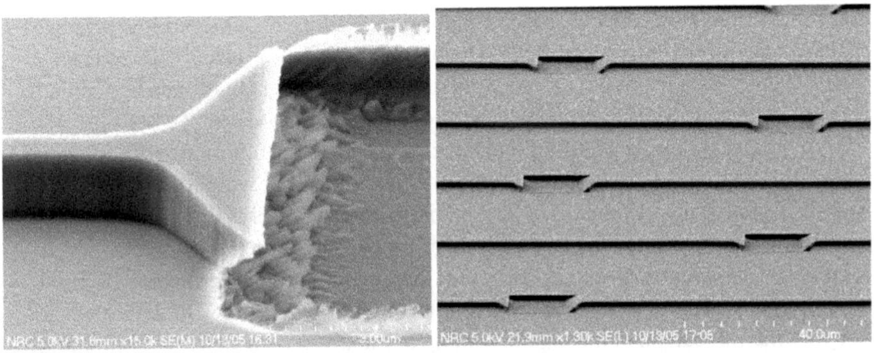

FIGURE 7.15 SEM images of SOI out-of-plane mirrors fabricated by tilted CAIBE.

7.3 FABRICATION CHALLENGES

This section covers some of the challenges found in the fabrication of planar waveguides. Because of limited space, some important aspects such as wafer bonding, process control, laser trimming, erbium doping, planarization, and removal of polymer passivation cannot be addressed here.

7.3.1 Roughness

As the dimensions of waveguides shrink and their index steps increase, sidewall roughness becomes a significant source of light scattering and thus propagation loss. The roughness associated with a waveguide is usually the roughness around its core. The propagation loss due to the roughness below and on top of the core resulting from the growth deposition processes is generally considered negligible.

The major roughness is usually located on the vertical sides of the core; terms like side wall roughness (SWR) and line edge roughness (LER) are often used. The roughness is typically defined by the root mean square (RMS) deviation and its correlation length. Both are important to consider in order to minimize losses. Furthermore, roughness dominated by vertical striations on the sidewalls also induces polarization-dependent loss (PDL), while the induced waveguide width variations may introduce crosstalk between waveguides. The semiconductor roadmap (ITRS) considers roughness reduction one of the most important outstanding issues.

The main characterization techniques to determine the sidewall roughness are SEM (top view, then extracting trace line) and AFM (tilted sample). Sidewall roughness is typically generated during the patterning process. Optimization of the masking scheme has been studied on III–V waveguides.[150–155] For nanoscale devices, it is believed that the roughness is mainly induced during the patterning of the photoresist. Some authors have attempted to modify the exposure process or use photoresists with lower molecular weights or smaller aggregates[156–159] with some success. Barwicz and Smith[160] noted that adjusting fabrication aspects such as polymerization during dry etching and a sharp resist profile for lift-off reduced the roughness.

After etching the waveguides, several attempts have been made to smooth the waveguide sidewalls of silica (flowing) and silicon-based devices by thermal treatment. Smoothing induced by thermal oxidation of SOI[161–162] achieved low propagation losses: an improvement from 32 dB/cm to 0.8 dB/cm in one case.[163] In addition to oxidation, thermal annealing–smoothening of Si waveguides in hydrogen atmosphere has been reported to yield promising results (LER = 1.4 nm).[164]

In summary, assuming the lithography technique has been optimized, the optimum smoothness may be approached by using (1) a low molecular weight photoresist, (2) a multilayer masking scheme, (3) a dry etching involving polymer passivation, and (4) a subsequent thermal anneal (oxidation or flow).

Different devices will require different approaches to the reduction of losses caused by sidewall roughness. As an example, straight waveguides in devices based on crystalline InP can benefit from combining dry and wet etch steps. A combination of dry etching (chlorine- or methane-based) and wet etching is typically used to etch an InP-based waveguide with an etch stop layer (InGaAsP). The dry etching is performed and stopped before reaching the etch stop layer; this step defines the position of the waveguide, but the roughness induced by the mask can be noticeable.

The subsequent wet etching employs a mixture of H_3PO_4 and HCl (anisotropic) and a subsequent optional polishing step using HCl and water. This wet etching step will stop on the quaternary layer, defining accurately the etch depth but also smoothing out the sidewall and the bottoms of the etched areas. Such wet etching also provides the advantage of removing contaminants and damages induced by the dry etching steps. Very low propagation losses (< 0.1 dB/cm) have been measured on InP waveguides using such methodology.[3]

7.3.2 Aspect Ratio-Dependent Etching (ARDE)

The aspect ratio is the ratio of the depth to the width of an etched area. ARDE describes the etching characteristics (rate and profiles) dependent on the etched features sizes and depths. Figure 7.16 presents a basic case in which the etch rate is dependent on the opening size. This effect is caused by the differential transport of ions and neutral species into etched openings. ARDE may have adverse effects when devices contain fine features such as photonic crystals and/or wires, micro-ring resonators, and Y splitters. Designers of photonic devices must keep the ARDE effect in perspective, avoiding etch-sensitive fine and large openings on the same mask level. Several papers have addressed the ARDE effect on Si[165–169] and other materials.[109,170–172]

ARDE can be reduced by making the ions (rather than the radicals) control the etching rate; i.e., using ion-limited process conditions. For example, an ion-induced process such as chlorine chemistry to etch Si (HBr chemistry is even better) is less sensitive to ARDE than an ion-inhibitor etching process such as the SF_6-based chemistry. Moreover, reducing the etching pressure seems to alleviate ARDE; hence the CAIBE technique is less prone to ARDE than conventional RIE. A patterning technique such as gray-scale lithography was also suggested to compensate for ARDE.[173]

FIGURE 7.16 SEM image of trenches of various widths (0.5 to 5 μm) etched in InP by CAIBE.

7.3.3 Stress

Stress in optical materials is an important parameter for designing and fabricating photonic devices, especially when birefringence properties are critical or the alignment of tens or hundreds of waveguides is required with minimum bowing induced by stress.[4] The several sources of stress in optical layers include different coefficients of thermal expansion for the substrate and optical layers, lattice mismatch, and intrinsic (growth) stress. For example, the stress resulting from the deposition of silica layers has been a critical issue.

The characterization techniques for stress include the measurement of the wafer curvature or by micro-raman (Si). It is possible to reduce stress levels by etching stress-release grooves along the waveguides or by depositing an overlayer with appropriate stress to compensate. It is also possible to take advantage of such stress to make new types of photonic devices or improve existing ones, such as AWG compensated with stress.[174]

7.3.4 Surface Damage Induced by Dry Etching

Energetic particles used in dry etching processes may disturb the optical properties of a waveguide; this is especially true for crystalline materials like III–Vs [175–178] and SOI. The density and depth range of the damages depend on the ion density and energy as well as the proportion of chemical versus physical etching. A chemically dominated etching process would remove (etch) the damage at a faster rate than a physically dominated process. The damages on the sidewalls are shallower than those located at the bottom of the etch. Damage depths reported are between 20 and 200 nm;[90,179–180] the main technique to quantify the depth of damage is photoluminescence using quantum wells. Obviously, highly confined small waveguides and those patterned with gray-scale lithography are more sensitive than larger waveguides, because the optical density is higher near the surface for the former.

To limit the damage low ion energy processes like ICP-RIE with low DC bias are used. It has also been reported that a thermal anneal of the etched devices reduces the extent of the damage.[181–182] Further mild wet chemical etching also helps to reduce the effects of damage.[183]

7.3.5 Trench Filling

After etching the core of a silica waveguide, an upper cladding layer is often required. SOI and III–V waveguides often require a coating layer to act as an optical cladding, a passivation, an insulator, a protection, or for symmetry reasons. This coating may be a polymer, spin-on-glass (SOG), but is typically PECVD SiO_2 or SiN_x. For aspect ratios higher than one, trenches or holes are difficult to fill properly. Figure 7.17 illustrates a keyhole or gap formed by the trench filling process with conventional PECVD SiO_2.

Various options can avoid gap formation during trench filling. One is to use several thin cladding layers doped with phosphorus and boron. Intermediate thermal

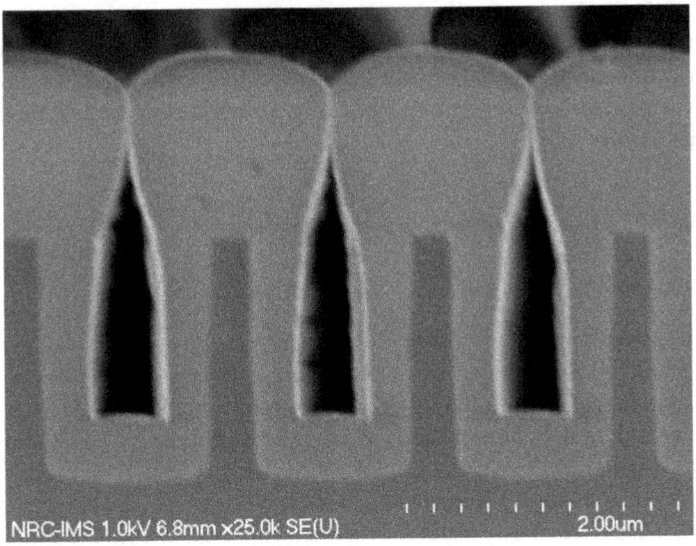

FIGURE 7.17 SEM image of etched Si trenches partially filled with PECVD SiO_2.

annealing steps are required to make the thin layers flow and fill the trench. Another possibility is to use successive deposition and sputter etch cycles to avoid the mushroom shapes associated with the conventional PECVD process.[184] Yet another option is a recent approach using a high-density PECVD system allowing one to control (and increase) the ion energy arriving at the surface of the sample during the PECVD process.[185–186] The growth mechanisms are highly affected by ion bombardment. The technique known as atomic layer deposition (ALD) would also be suitable for narrow trenches.

7.4 MODELING VERSUS FABRICATION

This section addresses some small knowledge gaps that exist between the people who design and model planar waveguide devices and the people who fabricate them. Many planar waveguides devices are fabricated in research environments where yield and reproducibility are not priorities. Tolerances for critical dimensions (CDs), thicknesses, etching depth, etching profiles, and overlay accuracy (alignment between different levels) need to be clarified. Designers should keep in mind that scratches may appear on devices, so several similar sets of devices and a back-end protective coating are desirable.

The process flow should stay simple and robust, thus leading to short fabrication runs. ARDE effects may give rise to fabrication challenges, particularly if they were not planned for in the design phase. PECVD layers are usually well characterized, but modelers should be aware that PECVD layers grown on vertical surfaces like the sidewalls of waveguides usually have poorer performances than those grown on horizontal surfaces. Designers should incorporate test structures for characterizing the waveguides (geometry), optical properties, etching, metallization, and cladding steps. Fabrication people should measure and report the actual values of the waveguide dimensions and also report any fabrication issues experienced while making the photonic devices.

7.5 NOVEL FABRICATION TECHNOLOGIES

7.5.1 Hollow Waveguides

Hollow waveguides represent a novel type of planar waveguide[187–191] that combines free space and integrated optics. They are of interest for new applications of planar waveguides in MOEMS, micro-fluidic, and bio-photonics applications. Hollow waveguides allow guiding light in a specific gas or liquid media present in a hollow core usually coated with reflective metal or dielectrics. The two main hollow waveguide fabrication techniques are (1) deep etching to form the core and then bonding a cover plate and (2) using a previously patterned sacrificial material to form the hollow core. It is possible to tune the optical properties by changing the dimensions of the core. Dicing hollow waveguides can be difficult because dicing debris can enter the core region.

7.5.2 Gray-Scale Lithography

Standard lithography processes use photoresist and mask in a binary digital mode in which the photoresist is either fully exposed or not exposed. Gray-scale lithography[192–193] allows the fabrication of curved surfaces and three-dimensional (3-D) shapes by allowing a continuous gradient between exposed and not exposed areas; this is an analog version of lithography. Gray-scale requires specific photomasks and photoresists, or E-beam lithography. It is particularly useful for fabricating microlenses and 3-D tapers for coupling waveguides to optical fibers.[194] Figure 7.18 shows SEM images of 3-D structures patterned by gray-scale lithography.

7.5.3 New Plasma Systems

Pulsed plasma etching systems seem to be the next generation of dry etching systems. They permit the control of ARDE effects and mask charging effects that cause sidewall bowing or notches.

High density PECVD systems incorporate sample biasing sources (already implemented on HDP-RIE systems). These new systems allow the control of the ion energy fluxes arriving at the sample during the PECVD process and provide much better control of the structural and chemical properties of the deposited layers. Optical layers can be obtained with much lower losses at reduced deposition temperatures. Moreover, the planarization properties of the HD-PECVD layers are markedly improved.

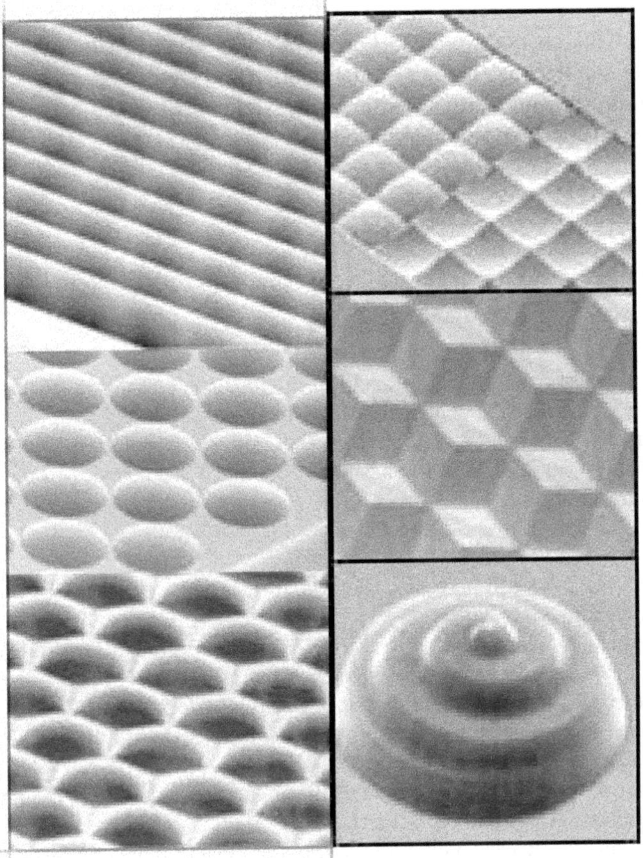

FIGURE 7.18 SEM images showing various three-dimensional structures obtained by gray-scale lithography. (From MEMS Optical. with permission.)

7.5.4 Fabrication Processes and Global Warming

Plasma processes use gases such as CF_4, NF_3, C_4F_8, and SF_6 that have global warming potential (GWP, based on their infrared absorption and lifetimes) ranging 6,000 to 30,000 times higher than CO_2.[195] The processes in question are dry etching and PECVD chamber cleaning processes. Several promising approaches have been investigated to reduce the levels of GWP emissions: (1) more efficient processing by increasing the amount of reacted–dissociated etchant gases,[196–197] (2) introducing additive gases like N_2 or NO to C_4F_8 plasma,[198–199] (3) substituting etchant gases with less GWP or more efficient gases[200–202] such as C_4F_6, FNO, and CF_3CH_2F, and (4) implementing an efficient gas abatement system.

ACKNOWLEDGMENTS

The writing of this chapter would not have been possible without the valuable expertise of the author's colleagues in the Nanofabrication and Optoelectronic Devices Group at the Institute for Microstructural Sciences, National Research Council of Canada and former colleagues at Optenia, Inc.

REFERENCES

1. Eldada, L., Optical networking components, Communications Design Conference, 2002.
2. Communications Technology Roadmap, Microphotonics: hardware for the information age: silicon microphotonics, MIT Microphotonics Center, 2005.
3. Grabtchak, S. et al., Very low loss shallow ridge waveguides for InP-based monolithically integrated WDM components, presented at Photonics North Conference, Montreal, 2003.
4. Janz, S., Silicon-based waveguide technology for wavelength division multiplexing, *Top. Appl. Phys.*, 94, 323, 2004.
5. Little, B.E., A VLSI photonics platform, invited talk. Optical Fiber Communication Conference, 2003.
6. Nishihara, H., Haruna, M. and Suhara, T., *Optical Integrated Circuits,* Fischer, R. and Smith, W., Eds, McGraw-Hill, 1989, ch. 6 and 7.
7. Williams R., *Modern GaAs Processing Methods*, Artech House, Boston, 1990.
8. Miya, T., Silica-based planar lightwave circuits: passive and thermal active devices, *IEEE J. Sel. Top. Quant. Electron.*, 6, 38, 2000.
9. Kawachi, M., Silica waveguides on silicon and their application to integrated optic components, *Opt. Quant. Electron.*, 22, 391, 1990.
10. Kawachi, M., Recent progress in silica-based planar lightwave circuits on silicon, *IEEE Proc. Optoelectron.*, 143, 257, 1996.
11. Schneider, H.W., Realization of SiO_2–B_2O_3–TiO_2 waveguide and reflectors on Si substrate, *Mater. Res. Soc. Symp. Proc.*, 244, 337, 1992.
12. Clemens, P.C. et al., Flat-field spectrograph in SiO_2/Si, *IEEE Photonics Technol. Lett.*, 4, 886, 1992.
13. Kazarinov, R., Silica waveguides and waveguide devices on silicon: technology and applications, *Proc. EFOC N,* 1993, p.21.
14. Sun, C.J. and Schmidt, K.M., Building passive components with silica waveguides, *Proc. SPIE*, 3795, 313, 1999.
15. Kim, Y.T. et al., Influence of hydrogen on SiO_2 thick film deposited by PECVD and FHD for silica optical waveguide, *Cryst. Res. Technol.*, 37, 1257, 2002.
16. Okuno, M. et al., Recent advances in optical switches using silica-based PLC technology, *NTT Tech. Rev.,* 1, 20, 2003.
17. Imoto, K. and Hori, A., High refractive index difference and low loss optical waveguide fabricated by low temperature processes, *Electron. Lett.*, 29, 1123, 1993.
18. Lai, Q. et al., Simple technologies for fabrication of low-loss silica waveguides, *Electron. Lett.*, 28, 1000, 1992.
19. Lai, Q. et al., Formation of optical slab waveguides using thermal oxidation of SiO_x, *Electron. Lett.*, 29, 714, 1993.

20. Pereyra, I. and Alayo, M.I., High quality low temperature DPECVD silicon dioxide, *J. Noncrystal Sol.*, 212, 225, 1997.
21. Bazylenko, M. et al., Fabrication of low-temperature PECVD channel waveguides with significantly improved loss in the 1.50–1.55 μm wavelength range, *IEEE Photon. Technol. Lett.*, 7, 774, 1995.
22. Bazylenko, M. et al., Pure and fluorine-doped silica films deposited in a hollow cathode reactor for integrated optic applications, *J. Vac. Sci. Technol.*, A14, 336, 1996.
23. Durandet, A. et al., Silica buried channel waveguides fabricated at low temperature using PECVD, *Electron. Lett.*, 32, 326, 1996.
24. Boswell, R.W. et al., Deposition of silicon dioxide films using the helicon diffusion reactor for integrated optics applications, *NATO ASI Series E*, 336, 433, 1997.
25. Grand, G. et al., Low-loss PECVD silica channel waveguides for optical communications, *Electron. Lett.*, 26, 2135, 1990.
26. Wensley, P.R. et al., Improved waveguide technology for silica-on-silicon integrated optics, *Proc. ECIO*, 1993, p. 9–8.
27. Ojha, S.M. et al., Simple method of fabricating polarisation-insensitive and very low crosstalk AWG grating devices, *Electron. Lett.*, 34, 78, 1998.
28. Martinu, L. and Poitras, D., Plasma deposition of optical films and coatings: a review, *J. Vac. Sci. Technol.*, A18, 2619, 2000.
29. Wosinski, L., Technology for photonic components in silica/silicon material structure, doctoral dissertation, Royal Institute of Technology, Stockholm, 2003.
30. Dainese, M., Plasma assisted technology for Si-based photonic integrated circuits, doctoral dissertation, Royal Institute of Technology, Stockholm, 2005.
31. Kim, T.H. et al., Effective silicon oxide formation on silica-on-silicon platforms for optical hybrid integration, *ETRI J.*, 25, 73, 2003.
32. Wörhoff, K. et al., Birefringence compensation applying double-core waveguiding structures, *IEEE Photon. Technol. Lett.*, 11, 206, 1999.
33. Kilian, A. et al., Birefringence free planar optical waveguide made by flame hydrolysis deposition (FHD) through tailoring of the overcladding, *J. Lightwave Technol.*, 18 , 193, 2000.
34. Dominguez, C., Rodriguez, J.A., and Zine, N., Effect of thermal annealing on mechanical stress in PECVD silicon oxides, *Quim. Analyt.*, 18, 55, 1999.
35. Dominguez, C. et al., Plasma-enhanced CVD silicon oxide films for integrated optic applications, *Vacuum*, 52, 395, 1999.
36. Rodriguez, J.A., Llobera, A., and Dominguez, C., Evolution of the mechanical stress on PECVD silicon oxide films under thermal processing, *J. Mater. Sci. Lett.*, 19, 1399, 2000.
37. Cramer, J.K. and Murarka, S.P., Stress-temperature behavior of electron cyclotron resonance oxides and their correlation to hydrogenous species concentration, *J. Appl. Phys.*, 77, 3048, 1995.
38. Kato, K., Ishii, M., and Inoue, Y., Packaging of large-scale planar lightwave circuits, *IEEE Trans. Compon. Packag. Manufac. Technol.*, B21, 121, 1998.
39. Lam, D.K.W, Low temperature plasma chemical vapor deposition of silicon oxynitride thin-film waveguides, *Appl. Opt.*, 23, 2744, 1984.
40. Bruno, F. et al., Plasma-enhanced chemical vapor deposition of low-loss SiON optical waveguides at 1.5-μm wavelength, *Appl. Opt.*, 30, 4560, 1991.
41. Tu, Y.K., Chou, J.C., and Cheng, S.P., Single-mode SiON/SiO_2/Si optical waveguides prepared by plasma-enhanced chemical vapor deposition, *Fiber Integ. Opt.*, 14, 133, 1995.

42. Yokoyama, S. et al., Optical waveguides on silicon chips, *J. Vac. Sci. Technol.*, A13, 629, 1995.
43. Sun, Z.J., McGreer, K.A., and Broughton, J.N., Demultiplexer with 90 channels and 0.3 nm channel spacing, *IEEE/LEOS Summer Mtgs.*, 1997, p. 52.
44. Alayo, M.J., Pereyra, I., and Carreno, M.N.P., Thick SiO_xN_y and SiO_2 films obtained by PECVD technique at low temperatures, *Thin Solid Films*, 332, 40, 1998.
45. Wörhoff, K. et al., Plasma enhanced chemical vapor deposition silicon oxynitride optimized for application in integrated optics, *Sensors Actuators*, 74, 9, 1999.
46. Hoffmann, M., Kopka, P., and Voges, E., Low-loss fiber-matched low-temperature PECVD waveguides with small-core dimensions for optical communication systems, *IEEE Photonics Technol. Lett.*, 9, 1238, 1997.
47. Bulat, E.S. et al., Fabrication of waveguides using low-temperature plasma processing techniques, *J. Vac. Sci. Technol.*, A11, 1268, 1993.
48. Wörhoff, K., Lambeck, P.V., and Driessen, A., Design, tolerance analysis, and fabrication of silicon oxynitride based planar optical waveguides for communication devices, *J. Lightwave Technol.*, 17, 1401, 1999.
49. Gorecki, C., Optimization of plasma-deposited silicon oxinitride films for optical channel waveguides, *Optics Lasers Eng.*, 33, 15, 2000.
50. Bona, G.L., Germann, R., and Offrein B.J., SiON high-refractive-index waveguide and planar lightwave circuits, *IBM J. Res. Dev.*, 47, 239, 2002.
51. De Ridder, R.M. et al., Silicon oxynitride planar waveguiding structures for application in optical communication, *IEEE J. Sel. Top. Quant. Electron.*, 4, 930, 1998.
52. Hilleringmann, U. and Goser, K., Optoelectronic system integration on silicon: waveguides, photodetectors, and VLSI CMOS circuits on one chip, *IEEE Trans. Electron. Dev.*, 42, 841, 1995.
53. Agnihotri, O.P., Tyagi, R., and Kato, I., Silicon oxynitride waveguides for optoelectronic integrated circuits, *Jpn. J. Appl. Phys.*, 36, 6711, 1997.
54. Haus, H.A., Kimerling, L.C., and Romagnoli, M., Application of high index contrast technology to integrated optical devices, *Proc. ECOC-IOOC*, 3, 354, 2003.
55. Larsen, B.H. et al., A low-loss, silicon-oxynitride process for compact optical devices, *Proc. ECOC-IOOC*, 3, 366, 2003.
56. Lambeck, P.V. and Wörhoff, K., SiON-technology for integrated optical sensors, *Proc. SPIE*, 4944, 195, 2003.
57. Sabac, A. et al., Silicon oxinitride waveguides developed for opto-mechanical sensing functions, *Proc. SPIE*, 4944, 214, 2003.
58. Reed, G.T. and Knights, A., *Silicon Photonics: An Introduction*, John Wiley & Sons, 2004.
59. Aalto, T. et al., Fabrication and characterization of waveguides structures on SOI, *Proc. SPIE*, 4944, 183, 2003.
60. Ang, T.W. et al., 0.15 dB/cm loss in Unibond SOI waveguides, *Electron. Lett.*, 35, 977, 1999.
61. International Technology Roadmap for Semiconductor, Sematech, 2004.
62. Hand, A., One on one: a closer look at nanoimprinting, *Semiconductor Int.*, 27, 40, 2004.
63. Hand, A., Nanoimprint lithography and its predicted liftoff, *Semiconductor Int.*, December 2005, p. 26.
64. Resnick, D.J. et al., Imprint lithography for integrated circuit fabrication, *J. Vac. Sci. Technol.*, B21, 2624, 2003.
65. Li, Y. et al., Fabrication of sub-micron gratings based on embossing, *Proc. Conf. Design, etc. MEMS/MOEMS*, Cannes, 2003, p. 350.

66. Belotti, M. et al., Investigation of SOI photonic crystals fabricated by both electron-beam lithography and nanoimprint lithography, *Microelectron. Eng.*, 73–74, 405, 2004.
67. Hirai, Y., et al., Imprint lithography for curved cross-sectional structure using replicated Ni mold, *J. Vac. Sci. Technol.*, B20, 2867, 2002.
68. Chao, C.Y. and Guo, L.J., Polymer microring resonators fabricated by nanoimprint technique, *J. Vac. Sci. Technol.*, B20, 2862, 2002.
69. Gourgon, C. et al., Nanoimprint lithography on 8 inch wafers, *Proc. EIPBN,* 2003, p. 59.
70. Lei, X. et al., Fabrication of low-loss polymeric waveguides with reduced edge roughness using nanoimprint lithography, *Proc. EIPBN,* 2003, p. 131.
71. Michel, B., Printing meets lithography: soft approaches to high-resolution patterning, *IBM J. Res. Dev.*, 45, 697, 2001.
72. Paul, K.E., Prentiss, M., and Whitesides, G.M., Patterning spherical surfaces at the two-hundred-nanometer scale using soft lithography, *Adv. Funct. Mater.*, 13, 259, 2003.
73. Wolfe, D.B., Fabrication of planar optical waveguides by electrical micro-contact printing, *Appl. Phys. Lett.*, 84, 1623, 2004.
74. Xia, Y. and Whitesides, G.M., Soft lithography, *Angew. Chem. Int. Ed.*, 37, 550, 1998.
75. Le, N.V. et al., Selective dry etch process for step and flash imprint lithography, *Microelectron. Eng.*, 78–79, 464, 2005.
76. Clawson, A.R., Guide to references on III–V semiconductor chemical etching, *Mater. Sci. Eng.*, 31, 1, 2001.
77. Seto, M. et al., Fabrication of submillimeter-radius optical waveguide bends with anisotropic and isotropic wet chemical etchants, *J. Ligthwave Technol.*, 8, 264, 1990.
78. Elias, P. et al., Wet-etch bulk micromachining of (100) InP substrates, *J. Micromech. Microeng.*, 14, 1205, 2004.
79. Buchmann, P. and Houghton, J.N., Optical y-junctions and s-bends formed by preferentially etched single-mode rib waveguides in InP, *Electron. Lett.*, 18, 850, 1982.
80. Williams, K.R. and Muller, R.S, Etch rates for micromachining processing, *J. Microelectromechan. Syst.*, 5, 256, 1996.
81. Williams, K.R., Gupta, K., and Wasilik, M., Etch rates for micromachining processing, part II, *J. Microelectromechan. Syst.*, 12, 761, 2003.
82. Seidel, H. et al., Anisotropic etching of crystalline silicon in alkaline solutions: orientation dependence and behavior of passivation layers, *J. Electrochem. Soc.*, 137, 3612, 1990.
83. Lamontagne, B. et al., InP etching using chemically assisted ion beam etching (Cl_2/Ar), *J. Electrochem. Soc.*, 146, 1918, 1999.
84. Shul, R.J. et al., High-density plasma etching of compound semiconductors, *J. Vac. Sci. Technol.*, A15, 633, 1997.
85. Daleiden, J. et al., Chemical analysis of a $Cl_2/BCl_3/IBr_3$ chemically assisted ion-beam etching process for GaAs and InP laser-mirror fabrication under cryo-pumped ultrahigh vacuum conditions, *J. Vac. Sci. Technol.*, B13, 2022, 1995.
86. Thomas, S., III and Pang, S.W., Atomic force microscopy study of III–V materials etched using an electron cyclotron resonance source, *J. Vac. Sci. Technol.*, B13, 2350, 1995.
87. Sendra, J.R. and Anguita, J., Reactive ion beam etching of indium phosphide in electron cyclotron resonance plasma using methane/hydrogen/nitrogen mixtures, *Jpn. J. Appl. Phys.*, 33, L390, 1994.
88. Chakrabarti, U.K., Pearton, S.J., and Ren, F., Sidewall roughness during dry etching of InP, *Semicond. Sci. Technol.*, 6, 408, 1991.
89. Burness, A.L. et al., Low-loss mirrors in InP/InGaAsP waveguide, *Electron. Lett.*, 29, 520, 1993.

90. Carlström, C.F., Ion beam etching of InP based materials, thesis, Royal Institute of Technology, Sweden, 2001.
91. Carlström, C.F., Landgren, G., and Anand, S., Low energy ion beam etching of InP using methane chemistry, *J. Vac. Sci. Technol.*, B16, 1018, 1998.
92. Etrillard, J. et al., Sidewall and surface induced damage comparison between reactive ion etching and inductive plasma etching of InP using a $CH_4/H_2/O_2$ gas mixture, *J. Vac. Sci. Technol.*, A14, 1056, 1996.
93. Ren, F. et al., High rate dry etching of InGaP in BCl_3 plasma chemistries, *Appl. Phys. Lett.*, 67, 2497, 1995.
94. Daleiden, J. et al., Low-temperature CAIBE process for InP-based optoelectronics, *Proc. Int. Conf. InP Rel. Mater.*, 1995.
95. Hong, J. et al., Dry etching of InGaAlP alloys in Cl_2/Ar high ion density plasmas, *J. Electron. Mater.*, 25, 1428, 1996.
96. Dzioba, S., et al., High temperature operation of InGaAsP/InP heterostructure lasers and integrated back facet monitors fabricated by chemically assisted ion beam etching, *Appl. Phys. Lett.*, 62, 2486, 1993.
97. Murad, S.K. et al., Selective reactive ion etching of InGaAs and InP over InAlAs in $SiCl_4/SiF_4$/HBr plasmas, *J. Vac. Sci. Technol.*, B13, 2344, 1995.
98. Eisele, K.M., Daleiden, J., and Ralston, J., Low temperature chemically assisted ion-beam etching process using Cl_2, CH_3I, and IBr_3 to etch InP optoelectronic devices, *J. Vac. Sci. Technol.*, B14, 1780, 1996.
99. Lai, L.S., Kao, H.C., and Chan, Y.J., Low damage and selective gate recess RIE etching of InAlAs/InGaAs HEMTs using fluorine and chlorine gas mixtures, *Proc. Intern. Conf. Indium Phosphide Rel. Mater.*, Tsukuba, 1998, p. 179.
100. Pearton, S.J., Hobson, W.S., and Levi, A.F.J., Comparison of plasma chemistries for patterning InP-based laser structures, *Plasma Sources Sci. Technol.*, 3, 19, 1994.
101. He, J.J. et al., Monolithic integrated wavelength demultiplexer based on a waveguide Rowland circle grating in InGaAsP/InP, *J. Ligthwave Technol.*, 16, 631, 1998.
102. Kim, B. Kwon, K.H., and Park, S.H., Characterizing metal-masked silica etch process in a CHF_3/CF_4 inductively coupled plasma, *J. Vac. Sci. Technol.*, A17, 2593, 1999.
103. Rueger, N.R. et al., Role of steady state fluorocarbon films in the etching of silicon dioxide using CHF_3 in an inductively coupled plasma reactor, *J. Vac. Sci. Technol.*, A15, 1881, 1997.
104. Jung, S.T. et al., Inductively coupled plasma etching of SiO_2 layers for planar lightwave circuits, *Thin Solid Films*, 341, 188, 1999.
105. Cho, S.B. et al., Improved etch characteristics of SiO_2 by the enhanced inductively coupled plasma, *J. Vac. Sci. Technol.*, A19, 1308, 2001.
106. Matsui, M., Tatsumi, T., and Sekine, M., Observation of surface reaction layers formed in highly selective SiO_2 etching, *J. Vac. Sci. Technol.*, A19, 1282, 2001.
107. Gaboriau, F., et al., Selective and deep etching of SiO_2: comparison between different fluorocarbon gases (CF_4, C_2F_6, CHF_3) mixed with CH_4 or H_2 and influence of the residence time, *J. Vac. Sci. Technol.*, B20, 1514, 2002.
108. Li, X. et al., Effects of Ar and O_2 additives on SiO_2 etching in C_4F_8-based plasmas, *J. Vac. Sci. Technol.*, A21, 284, 2003.
109. Bretoiu, S. et al., Inductively coupled plasma etching for arrayed waveguide gratings fabrication in silica on silicon technology, *J. Vac. Sci. Technol.*, B20, 2085, 2002.
110. LeLuyer, Y., Reactive ion etching of alumna-doped silica thin films for integrated optics applications, *Vide*, 284, 65, 1997.
111. An, K.J. et al., A study of the characteristics of inductively coupled plasma using multidipole magnets and its application to oxide etching, *Thin Solid Films*, 341, 176, 1999.

112. Hayashi, S. et al., SiO_2 etching using inductively coupled plasma, *Electron. Commun. Jpn*, 81, 21, 1998.
113. Schaepkens, M., Oehrlein, G.S., and Cook, J.M., Effect of radio frequency bias power on SiO_2 feature etching in inductively coupled fluorocarbon plasmas, *J. Vac. Sci. Technol.*, B18, 848, 2000.
114. Oehrlein, G.S. and Rembetski, J.F., Plasma-based dry etching techniques in the silicon integrated circuit technology, *IBM J. Res. Develop.*, 36, 140, 1992.
115. Tian, W.C., Weigold, J.W., and Pang, S.W., Comparison of Cl and F-based dry etching for high aspect ratio Si microstructures etched with an inductively coupled plasma source, *J. Vac. Sci. Technol.*, B18, 1890, 2000.
116. Sato, M. and Arita, Y., Etched shape control of single-crystal silicon in reactive ion etching using chlorine, *J. Electrochem. Soc.* , 134, 2856, 1987.
117. Lee, J.H. et al., Study of shallow silicon trench etch process using planar inductively coupled plasmas, *J. Vac. Sci. Technol.,* A15, 573, 1997.
118. Rangelow, I.W. And Löschner, H., Reactive ion etching for micromechanical system fabrication, *J. Vac. Sci. Technol.,* B13, 2394, 1995.
119. Juan, W.H. and Pang, S.W., Control of etch profile for fabrication of Si microsensors, *J. Vac. Sci. Technol.*, A14, 1189, 1996.
120. Rakshandehroo, M.R et al., Dry etching of Si field emitters and high aspect ratio resonators using an inductively coupled plasma source, *J. Vac. Sci. Technol.*, B16, 2849, 1998.
121. Panda, S. et al., Effect of rare gas addition on deep trench silicon etch, *Microelectron. Eng.*, 75, 275, 2004.
122. Hopkins, J., DRIE of silicon for MEMS inkjet heads, *Semic. Intl.*, November 2004, p. 83.
123. Volland, B.E., Profile simulations of gas chopping etching processes, Ph.D. thesis, University of Kassel, 2004.
124. Jansen, H.V., Plasma etching in microtechnology, Ph.D. thesis, Twente University, 1996.
125. Blauw, M.A., Deep anisotropic dry etching of silicon microstructures by high-density plasmas, Ph.D. thesis, Delft University of Technology, 2004.
126. Esashi, M. et al., High-rate directional deep dry etching for bulk silicon micromachining, *J. Micromech. Microeng.*, 5, 5, 1995.
127. Jansen, H. et al., A survey on the reactive ion etching of silicon in microtechnology, *J. Micromech. Microeng.*, 6, 14, 1996.
128. De Boer, M.J. et al., Guidelines for etching silicon MEMS structures using fluorine high-density plasmas at cryogenic temperatures, *J. Microelectromech. Sys.*, 11, 385, 2002.
129. Boufnichel, M. et al., Profile control of high aspect ratio trenches of silicon. I. Effect of process parameters on local bowing, *J. Vac. Sci. Technol.*, B20, 1508, 2002.
130. Figueroa, R.F. et al., Control of sidewall slope in silicon vias using SF_6/O_2 plasma etching in a conventional reactive ion etching tool, *J. Vac. Sci. Technol.*, B23, 2226, 2005.
131. Marcos, G., Rhallabi, A., and Ranson, P., Topographic and kinetic effects of the SF_6/O_2 rate during cryogenic etching process of silicon, *J. Vac. Sci. Technol.*, B22, 1912, 2004.
132. Boufnichel, M. et al., Origin, control and elimination of undercut in silicon deep plasma etching in the cryogenic process, *Microelectron. Eng.*, 77, 327, 2005.
133. Dussart, R. et al., Passivation mechanisms in cryogenic SF_6/O_2 etching process, *J. Micromech. Microeng.*, 14, 190, 2004.

134. Cracium, G. et al., Temperature influence on etching deep holes with SF_6/O_2 cryogenic plasma, *J. Micromech. Microeng.*, 12, 390, 2002.
135. Yap, K.P. et al., Fabrication of lithography-defined optical coupling facets for SOI waveguides by ICP etching, *J. Vac. Sci. Technol.*, A24, 812, 2006.
136. Taillaert, D. et al., An out-of-plane grating coupler for efficient butt-coupling between compact planar waveguides and single-mode fibers, *IEEE J. Quant. Electron.*, 38, 949, 2002.
137. Cheben, P. et al., A broad-band waveguide grating coupler with a subwavelength grating mirror, *IEEE Photon. Technol. Lett.*, 18, 13, 2006.
138. Haramatsu S. and Kinoshita, M., Optical path redirected multichannel waveguide connectors for surface mount technologies, *IEEE Photon. Technol. Lett.*, 16, 2281, 2004.
139. Whitaker, T., Etched facets produce reliable long-wavelength surface-emitting lasers, *Compound Semicond.*, 10, 14, 2004.
140. Frateschi, N.C. et al., Low threshold InGaAs/GaAs 45° folded cavity surface-emitting laser grown on structured substrates, *IEEE Photon. Technol., Lett.*, 5, 741, 1993.
141. Chao, C.P. et al., Low-threshold, high-power, 1.3-μm wavelength, InGaAs-InP etched-facet folded-cavity surface-emitting lasers, *IEEE Photon. Technol. Lett.*, 7, 836, 1995.
142. Lee, H.P. et al., 1.57-μm InGaAsPInP surface emitting lasers by angled focus ion beam etching, *Electron. Lett.*, 28, 580, 1992.
143. Janiak, K. et al., 1.3 μm BH-FP laser with integrated monitor photodiode, 45 deg reflector for bottom side emission employing full on-wafer fabrication, *Proc. IPRM*, Sweden, 2002, p. 31.
144. Kagami, M., Kawasaki, A., and Ito, H., A polymer optical waveguide with out-of-plane branching mirrors for surface-normal optical interconnections, *J. Lightwave Technol.*, 19, 1949, 2001.
145. Terui, H. and Shutoh, K., Novel micromirror for vertical optical path conversion formed in silica-based PLC using wettability control of resin, *J. Lightwave Technol.*, 16, 1631, 1998.
146. Bazylenko, M.V. et al., Fabrication of light-turning mirrors in buried-channel silica waveguides for monolithic and hybrid integration, *J. Lightwave Technol.*, 15, 148, 1997.
147. Liu, Y. et al., Optoelectronic integration of polymer waveguide array and metal–semiconductor–metal photodectector through micromirror couplers, *IEEE Photon. Technol. Lett.*, 13, 355, 2001.
148. Hwang, S.H. et al., VCSEL array module using (111) facet mirrors of a V-grooved silicon optical bench and angled fibers, *IEEE Photon. Technol. Lett.*, 17, 477, 2005.
149. Lamontagne, B. et al., Fabrication of out-of-plane micro-mirrors in silicon-on-insulator planar waveguides, *J. Vac. Sci. Technol.* A, A24, 718, 2006.
150. Bae, J.W. et al., Characterization of sidewall roughness of InP/InGaAsP etched using inductively coupled plasma for low loss optical waveguide applications, *J. Vac. Sci. Technol.*, B21, 2888, 2003.
151. Chakrabarti, U.K., Pearton, S.J., and Ren, F., Sidewall roughness during dry etching of InP, *Semicond. Sci. Technol.*, 6, 408, 1991.
152. Zhao, W. et al., Effect of mask thickness on the nanoscale sidewall roughness and optical scattering losses of deep-etched InP/InGaAsP high mesa waveguides, *J. Vac. Sci. Technol.*, B23, 2041, 2005.
153. Jang, J.H. et al., Direct measurement of nanoscale sidewall roughness of optical waveguides using an atomic force microscope, *Appl. Phys. Lett.*, 83, 4116, 2003.
154. Jang, J.H. et al., Study of the evolution of nanoscale roughness from the line edge of exposed resist to the sidewall of deep-etched InP/InGaAsP heterostructures, *J. Vac. Sci. Technol.*, B22, 2538, 2004.

155. Deichsel, E. and Unger, P., High-brightness semiconductor lasers fabricated with improved dry-etching technology for ultra-smooth laser facets, *Dig. Papers Microprocesses Nanotechnol.*, 2001, p. 218.
156. Yasin, S. et al., Correlation of surface roughness with edge roughness in PMMA resist, *Microelectron. Eng.*, 78–79, 484, 2005.
157. Namatsu, H. et al., Influence of edge roughness in resist patterns on etched patterns, *J. Vac. Sci. Technol.*, B16, 3315, 1998.
158. Yagura, K., Niu, H., and Kotera, M., Dependence of linewidth and its edge roughness on EB exposure dose, *Proc. EIPBN*, Orlando, 2005, p. 287.
159. Yamaguchi, A. et al., Spectral analysis of line-edge roughness in polyphenol Eb-resists and its impact on transistor performance, *Proc. EIPBN*, Orlando, 2005, p. 231.
160. Barwicz, T. and Smith, H.I., Evolution of line-edge roughness during fabrication of high-index-contrast microphotonic devices, *J. Vac. Sci. Technol.*, B21, 2892, 2003.
161. Takahashi J.I., et al., Oxidation-induced improvement in the sidewall morphology and cross-sectional profile of silicon wire waveguides, *J. Vac. Sci. Technol.*, B22, 2522, 2004.
162. Juan, W.H. and Pang, S.W., Controlling sidewall smoothness for micro-machined Si mirrors and lenses, *J. Vac. Sci. Technol.*. B14, 4080, 1996.
163. Lee, K.K. et al., Fabrication of ultralow-loss Si/SiO_2 waveguides by roughness reduction, *Opt. Lett.*, 26, 1888, 2001.
164. Fritze, M. et al., Fabrication of three-dimensional mode converters for silicon-based integrated optics, *J. Vac. Sci. Technol.*, B21, 2897, 2003.
165. Jansen, H. et al., A survey on the reactive ion etching of silicon in microtechnology, *J. Micromech. Microeng*. 6, 14, 1996.
166. Gottscho, R.A., Jurgensen, C.W., and Vitkavage, D.J., Microscopic uniformity in plasma etching, *J. Vac. Sci. Technol.*, B10, 2133, 1992.
167. Richards, D.F. et al., Modeling plasma processes in microelectronics, *Vacuum,* 59, 168, 2000.
168. Rickard, A. and McNie, M., Characterisation and optimisation of deep dry etching for MEMS applications, *Proc. SPIE*, 4407, 78, 2001.
169. Blauw, M.A., Zijlstra, T., and Van der Drift, E., Radical transport in deep silicon structures during dry etching, *Proc. SeSens Workshop*, 2000, p. 617.
170. Misaka A. and Harafuji, K., Simulation study of micro-loading phenomena in silicon dioxide hole etching, *IEEE Trans. Electron Dev.*, 44, 751, 1997.
171. Abdollahi-Alibeik, S. et al., Modeling and simulation of feature-size-dependent etching of metal stacks, *J. Vac. Sci. Technol.,* B19, 179, 2001.
172. Xie, R.J., Kava, J.D. and Siegel, M., Aspect ratio-dependent etching on metal etch: modeling and experiment, *J. Vac. Sci. Technol.*, A14, 1067, 1996.
173. Morgan, B., Waits, C.M., and Ghodssi, R., Compensated aspect ratio-dependent etching (CARDE) using gray-scale technology, *Microelectron. Eng.*, 77, 85, 2004.
174. Xu, D.X. et al., Eliminating the birefringence in silicon-on-insulator ridge waveguides by use of cladding stress, *Opt. Lett.*, 29, 2384, 2004.
175. Sendra, J.R., Armalles, G., and Anguita, J., Optical study of InP etched in methane-based plasmas by reactive ion beam etching, *Semicond. Sci. Technol.*, 11, 238, 1996.
176. Several short papers, *Proc. EIPBN*, 1997–1998.
177. Tamura, M. et al., Surface damage in GaInAs/GaInAsP/InP wire structures prepared by substrate potential-controlled reactive ion beam etching, *Jpn. J. Appl. Phys.*, 34, 3307, 1995.

178. Heinbach, M., Kaindl, J., and Franz, G., Lattice damage in III/V compound semiconductors caused by dry etching, *Appl. Phys. Lett.*, 67, 2034, 1995.
179. Iber, H. et al., Characterization of surface damage in dry-etched InP, *Semicond. Sci. Technol.*, 12, 755, 1997.
180. Dienelt, J. et al., Roughness and damage of a GaAs surface after chemically assisted ion beam etching with Cl_2/Ar^+, *Microelectron. Eng.*, 78–79, 457, 2005.
181. Wuu, D.S. et al., High-density plasma-induced etch damage of wafer-bonded AlGaInP/mirror/Si light-emitting diodes, *J. Vac. Sci. Technol.*, A20, 766, 2002.
182. Goubert, L. et al., A study of electrically active defects created in p-InP by CH_4:H_2 reactive ion etching, *J. Appl. Phys.*, 82, 1696, 1997.
183. Qian, Y.Y.H. et al., Techniques for achieving low leakage current in dry etched InGaAs waveguide PIN detectors, *J. Vac. Sci. Technol.*, A22, 1062, 2004.
184. Schwartz, G.C. and Johns, P., Gap-fill with PECVD SiO_2 using deposition/sputter etch cycles, *J. Electrochem. Soc.*, 139, 927, 1992.
185. Barr-Ilan, A.H. and Gutmann, N., A comparative study of sub-micron gap filling and planarization techniques, *Proc. SPIE*, 2636, 277, 1995.
186. Yota, J. et al., Integration of ICP high-density plasma CVD with CMP and its effects on planarity for sub-0.5 μm CMOS technology, *Proc. SPIE*, 2875, 265, 1996.
187. Madkour, K. et al., Silicon hollow waveguide for MEMS applications, *Proc. ECOC*, 3, 650, 2003.
188. Jenkins, R.M. et al., Hollow waveguides for integrated optics, *Proc. ECOC*, 2, 162, 2003.
189. Miura, T. et al., Hollow waveguide with variable air core for tunable planar waveguide devices, *Proc. ECOC*, 2, 160, 2003.
190. Miura, T., Koyoma, F., and Matsutani, A., Modeling and fabrication of hollow waveguide for photonic integrated circuits, *Jpn. J. Appl. Phys.*, 41, 4785, 2002.
191. Schmidt, H. et al., Hollow-core waveguides and 2-D waveguide arrays for integrated optics of gases and liquids, *IEEE J. Sel. Top. Quant. Electron.*, 11, 519, 2005.
192. Waits, C.M., Modafe, A., and Ghodssi, R., Investigation of gray-scale technology for large area 3D silicon MEMS structures, *J. Micromech. Microeng.*, 13, 170, 2003.
193. Waits, C.M. et al., Microfabrication of 3D silicon MEMS structures using gray-scale lithography and deep reactive ion etching, *Sensors Actuators,* A119, 245, 2005.
194. Dillon, T., Process development and application of grayscale lithography for efficient three-dimensionally profiled fiber-to-waveguide couplers, *Proc. SPIE*, 5183, 123, 2003.
195. www.epa.gov
196. Johnson, A.D., Ridgeway, R.G., and Maroulis, P.J., Reduction of PFC emissions to the environment through advances in CVD and etch processes, *IEEE Trans. Semicond. Manufac.*, 17, 491, 2004.
197. Raju, R. et al., Warming potential reduction of C_4F_8 using inductively coupled plasma, *Jpn. J. Appl. Phys.*, 42, 280, 2003.
198. Kim, K.J. et al., Global warming gas emission during plasma cleaning process of silicon nitride using c-C_4F_8O/O_2 chemistry with additive Ar and N_2, *J. Vac. Sci. Technol,.* B22, 483, 2004.
199. Oh, C.H. et al., Increase of cleaning rate and reduction in global warming effect during C_4F_8O/O_2 remote plasma cleaning of silicon nitride by adding NO and N_2O, *Thin Solid Films,* 435, 264, 2003.

200. Yonemura, T. et al., Evaluation of FNO and F_3NO as substitute gases for semiconductor CVD chamber cleaning, *J. Electrochem. Soc.*, 150, G707, 2003.
201. Goyette, A.N., Wang, Y., and Olthoff, J.K., Inductively coupled plasmas in low global warming potential gases, *J. Phys. D Appl. Phys.*, 33, 2004, 2000.
202. Fracassi, F., d'Agostino, R., and Fornelli, E., SiO_2 etching with perfluorobutadiene in a dual frequency plasma reactor, *J. Vac. Sci. Technol.*, A21, 638, 2003.

[illegible]

8 Biomedical Fiber Optics

Vadakke Matham Murukeshan

CONTENTS

8.1 INTRODUCTION

With the recent advances in normal and tunable coherent (lasers) and ultra-broad bandwidth light sources, the technology of fiber optics in medical applications has taken a big leap. Such light sources have found tremendous applications in the biomedical arena for a wide range of electromagnetic spectra such as x-ray, ultraviolet (UV), visible, near infrared (NIR) and extending to the mid-IR region. This wide range of wavelengths transmitted from such coherent and incoherent sources to the target tissues by flexible devices such as optical fibers enables easy manipulation of the probe light beams in a medical chair-side setting.

If we recall the early stages of lasers employed in medicine, we found that laser beam energy is transmitted by a series of tubes and reflecting mirrors, with varying degrees of freedom, to form an articulated arm. Early devices were cumbersome and posed many challenges for efficient use until the incorporation of rigid and flexible image and light conduits such as optical waveguides. With the introduction of endoscope concepts, surgical procedures moved toward minimally invasive from the conventional open procedures, as indicated by the new approach: "open hole, key hole, NO HOLE surgery!" The use of such approaches for diagnostics, surgical, and post-operative procedures actually reduce hospitalization time, post-operative pain, and discomfort and cause fewer complications. The variety of novel concepts and complex instruments that researchers have devised to exploit the properties of light in combination with optical fibers for applications range from imaging abnormalities to detection and sensing of diseases, to cures.

This chapter, dedicated to these aspects of fiber optics (waveguide) concepts, provides an overview of past, current, and future developments in the challenging area of biomedical fiber optics. Following this brief introduction, the chapter overviews the different types of conventional and specialty optical fibers employed in biomedical applications. Generally, fiber optics applications in the biomedical area are mainly focused on (i) optical spectroscopy and (ii) bio-imaging applications. While many reviews on optical spectroscopy applications are available the same cannot be said about imaging. This chapter focuses mainly on these aspects, particularly on *in vivo* imaging, and concludes with discussion of the recent combined imaging and sensing configurations in biomedical fiber optics and illustrative examples.

8.2 WHY FIBER OPTICS FOR MEDICAL APPLICATIONS?

Optical fibers are currently available for transmitting and guiding different wavelengths. They can be tailored for various biosensing applications by properly tailoring the configurations and exploiting the properties of optical fibers and target biochemical agents. This can make minimally invasive procedures even more common. Fiber optics can be tailored for the transmission and reflection modes of operation. Signals such as fluorescence from tissues can be guided toward detectors and specialty cameras through image or signal conduits such as specialty coherent optical fiber bundles. Novel instrumentation concepts are being designed with and without the incorporation of endoscopic components for various biomedical applications.

8.2.1 Optical Fibers in Medical Applications

The basic concepts of optical fibers and their cross sectional view diagrams will be covered briefly. An optical fiber consists of a core, a cladding, and a protective jacket [1]. A light beam is guided through the fiber core through total internal reflection (TIR) and hence the refractive index of core is slightly higher than that of the cladding.

The four special types of fibers used in biomedical applications are (a) conventional solid core fibers [2], (b) hollow waveguides [3], (c) custom-made image fiber bundles [4] and (d) liquid light guides [5]. Each type of fiber guides optical radiation by different phenomena. For example, solid silica core fibers guide optical radiation by TIR, whereas the light guiding principle of hollow waveguides is pure reflection.

8.2.2 Conventional and Specialty Optical Fibers

8.2.2.1 Conventional Solid Core Fibers

There are two main types: the step-index type and the graded-index type. While the refractive index profile is a step function of a step-index fiber, it generally varies along the core radius for the graded-index fiber. These fibers are also classified as single mode and multimode with respect to guides. Only one mode is guided through the single mode fibers whereas many modes are transmitted through multimode fibers, thus satisfying the general V-number equation [6]. Attenuation is a factor that is difficult to avoid when one deals with solid core fibers. It generally depends on waveguide (fiber) dimensions. It increases when the fiber length increases or when the inner radius decreases [7]. The research leading to subduing these attenuation mechanisms in solid and hollow core fibers has led to the development of multilayer waveguides (generally known as photonic crystals) [8].

8.2.2.2 Multilayer Waveguides

The general structure of a multilayer fiber is shown in Figure 8.1. The fiber is made from an alternating structure combining two dielectric materials and is typical of a photonic crystal. Wave propagation through such structures follows Bloch theory, according to which there exist forbidden band gaps through which photons (such as electrons in solid state physics) with certain energies can not propagate [9]. In such a scenario, photons are perfectly guided accordingly, thus attaining idealistic minimum attenuation, and eventually allowing the generation of photonic crystal fibers. Several groups have reported successful implementation [10–12].

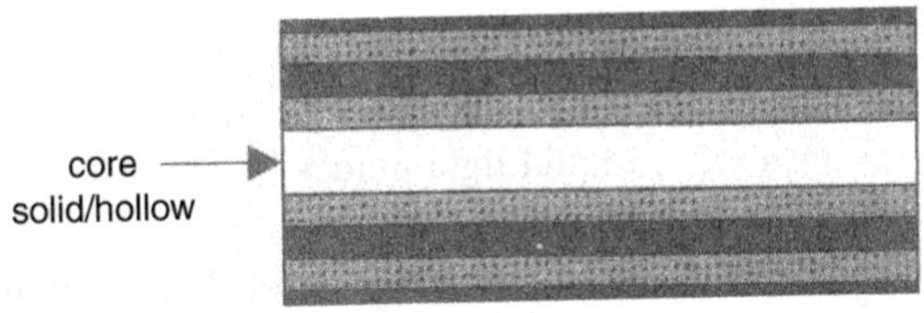

FIGURE 8.1 Schematic of multilayer waveguides.

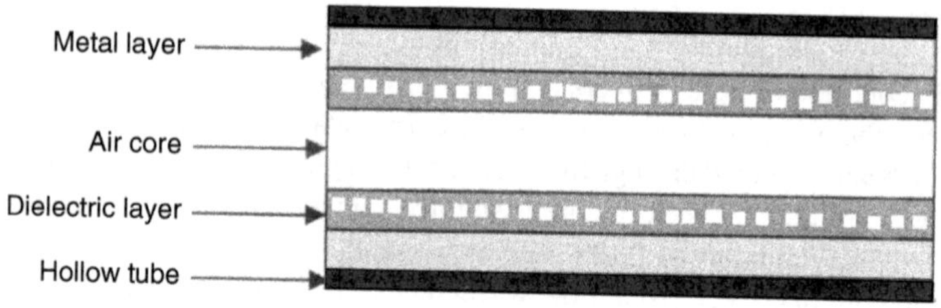

FIGURE 8.2 Schematic of hollow waveguide.

The use of x-ray waveguides with solid cores must be mentioned in this context. x-ray radiation has been used in medical imaging for a long time now [13]. Generally, it is difficult to manipulate x-radiation and different methodologies were adopted and reported such as those that employ Fresnel lenses, coated fibers, etc. However, such approaches including the generation of large spot sizes were too cumbersome to employ for minimally invasive applications. This problem can be subdued by intelligent use of specialty waveguides such as the multilayer version proposed by Pfeiffer et al. [14].

8.2.2.3 Hollow Waveguides

A hollow waveguide consists of a general structure as shown in Figure 8.2. It consists of a hollow tube coated internally with a thin metal layer and a dielectric layer. The optical beam is guided through the reflection at the inner layers. The two types of such waveguides are characterized by (i) leaky waveguides and (ii) attenuated total reflection waveguides involving hollow tubes and guiding mechanisms similar to solid core fibers. These fibers were used for the mid-IR region used for many medical applications [15–16].

8.2.2.4 Glass Core Fibers

Detailed information and schematic diagrams of cross-sections of these fibers can be found in standard textbooks. Glass and silica core fibers have applications at a wide range of wavelengths from UV-visible to the NIR [17–18]. Certain specialty fibers such as silver halide and thallium halide are useful for IR beam delivery applications [19–20] and can find applications in medical diagnostics and therapeutics.

8.2.2.5 Liquid Core Fibers

A liquid core fiber is a hollow tube filled with a liquid that is transparent in the wavelength region of interest. Liquid light guides in the visible and IR regions are available in the commercial market and have found many applications as laser and light beam delivery conduits. The advantage of these fibers over solid core fibers is the possibility of obtaining a purer liquid than solid material. For example, liquids such as $C_2Cl_4$and CCl_4 find good waveguide core applications in the IR region [21–22].

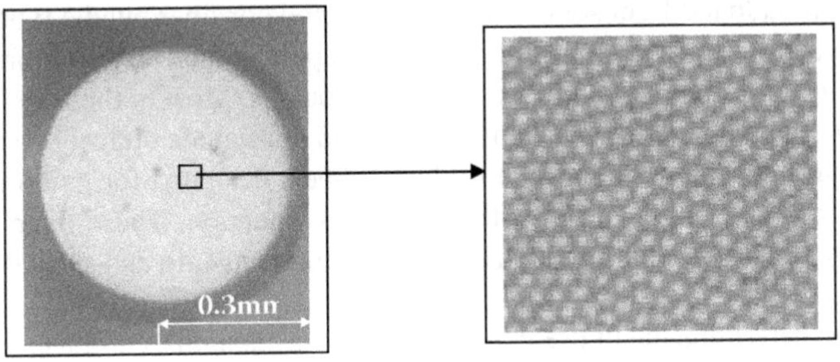

(a) 100X (b) 200X

FIGURE 8.3 End face of the image fiber bundle (Fujikura FIGH-15-600N). (b) is the enlarged view.

Specialty fibers for beam delivery and excitation of test samples will be explained in Sections 8.3.4 and 8.3.5.

8.2.2.6 Image Fiber Bundle

The designs of all fiber optic probes require the image or return light to be guided through an image fiber to the probe proximal end. The selection of image fiber in the scheme is intended to increase the flexibility of the probe and miniaturize the whole image collection scheme. More information on the image fibers and their applications can be found in the literature [23–24]. The image fiber bundle has a double role to perform here, i.e., as an image conveying device and also as a fluorescence signal transmitting element. Image fibers differ in the number of individual fibers (fiberlets) they contain and thus differ in their image resolution levels. An image fiber consisting of a higher number of pixels has better resolution than an image fiber with a lesser number of pixels. However, the flexibility of the image fiber decreases as the number of individual fibers increases. Thus, a compromise between flexibility and image resolution requires selecting the most appropriate image fiber for the flexible fiberscope. A microscopic image of the end face of one such fiber that has 15,000 pixels (Fujikura FIGH-15-600N) is shown in Figure 8.3. The image of the fiber bundle was taken under an Axiotech reflected light microscope with a SONY Hyper HAD camera at magnifications of 100× and 200×.

8.3 LIGHT SOURCES AND DETECTORS

8.3.1 Broad Band and Coherent Light Sources

There are different types of optical fibers rated for each wavelength or wavelength band, depending on the bandwidth of the light sources such as broadband or coherent narrowband sources. Broadband light sources include white light, mercury and xenon arc lamps, light emitting diodes (LEDs), superluminescent diodes (SLDs) that are specially employed in biomedical instrumentation. Most of the fibers employed for guiding such broadband light sources are of multimode or fiber bundle types unless otherwise specially configured for certain specialty applications such as tailored single mode fibers in combination with broadband light sources. Most of the coherent light sources such as lasers, employ both single mode and multimode optical fibers as well as custom-made special configuration (tailored) fibers designed for guiding certain wavelengths or wavelength bands without much attenuation. These fibers include SM (single mode), MM (multimode), HiBi (highly birefringent), and specialty image types.

In order to determine the fluorescence emission wavelengths, certain test samples that include biochemical agents (natural as well as exogenous fluorophore contents) are excited with a broadband light source followed by detecting the emission spectra. For this purpose, a poly light (for example, a broad band light source such as the Omnichrome Spectrum 9000, Melles Griot) can be employed. This uses a xenon arc lamp with specific filters at the emission side to tune the wavelength. The tunable mode feature of this light source allows for the tuning of the wavelength (300 to 700 nm) and bandwidth (up to 100 nm). A liquid light guide is used as the transmission cable.

8.3.2 Monochromators and Spectrometers

A monochromator scans one wavelength of the input light at a time and presents from its exit slit, whereas a spectrometer presents a range of wavelengths simultaneously at the exit focal plane. The key components of a monochromator or spectrometer are an entrance slit, mirror M1 as a collimator, a grating based dispersing unit for wavelength selection, mirror M2 as a focusing element, and an exit slit (see Figure 8.4). For a fluoroscope system, the input to the monochromator will be the output light from the endoscope proximal end as shown in Figure 8.4. The grating can be either ruled (where the groove density will be 50 to 3600 grooves/mm) or holographic (where the groove density will be higher than the ruled one).

An important parameter relevant to the selection of a monochromator or spectrometer is the reciprocal linear dispersion (RLD) which is inversely related to the grating groove density, diffraction order, and focal length of the focusing element. RLD when multiplied by the entrance slit width provides the spectral resolution. For fluorescence studies, generally a spectral resolution of 10 nm is sufficient [25]. Stray light rejection is another important factor in the selection of monochromators; it can be reduced by

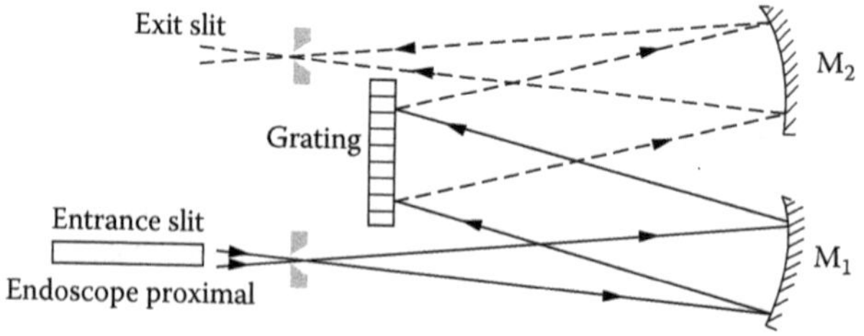

FIGURE 8.4 Schematic of a monochromator.

using holographic gratings compared to ruled gratings. Transmission efficiencies of both holographic and ruled gratings are comparable to a certain percentage in the VIS spectral region, with a maximum efficiency at the blazed wavelength.

8.3.3 Fluorescence Spectrometer

Generally, fluorescent spectrometers are used to record the fluorescent spectrum. One such spectrometer that employs fiber optics is shown in Figure 8.5. This fiber optic spectrometer (AVS-2000, Avantes) is used to record the fluorescence intensity as a function of wavelength [26]. It has an entrance slit of 200 μm and a grating resolution of 600 lines/mm. The spectrometer covers a wavelength range from 200 to 850 nm. It uses a linear 2048-element charged-coupled device (CCD) array as the detector. The fluorescence signal is collected using the fiber, which has a collimating lens attached to it. The fluorescence spectrum is displayed on the monitor of the computer associated with the related software. Figure 8.5 is a schematic of a fluorescence spectrometer set-up [27].

8.3.4 Fluorescence Lifetime Analyzer

Knowledge of the fluorescence lifetimes of test samples is highly desirable for dealing with bio-imaging and sensing analysis. The measurement of the lifetime of fluorescence emission is always a requirement and it is generally carried out using a lifetime measurement system. In this section, a lifetime measurement system that employs a dual bifurcated optical fiber system using spectrofluorometer is explained [28–29].

Lifetime measurements of the test samples can be made using a time-correlated single photon counting (TCSPC) spectrofluorometer (for example, the Mini-Tau, Edinburgh Instruments) that uses a microwatt power pulsed light emitting diode (LED) as a light source with a picosecond pulse width. The detector of this instrument is

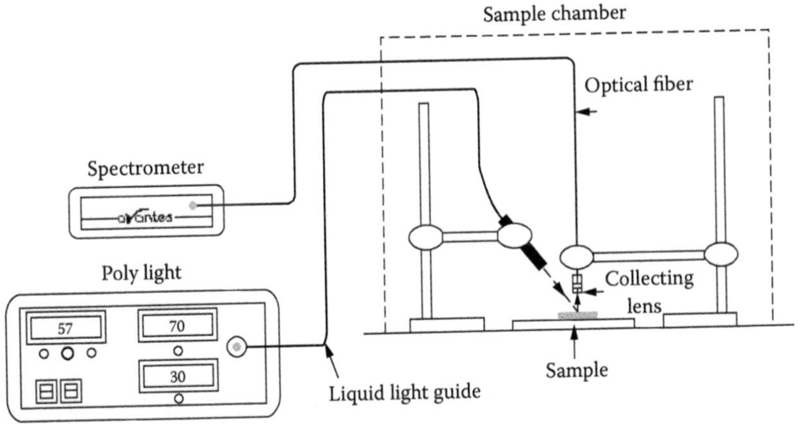

FIGURE 8.5 Schematic of fiber optic spectrometer used for determination of fluorescence emission wavelengths. Liquid light guide is used as the illumination/excitation arm and optical fiber is used as the collection arm.

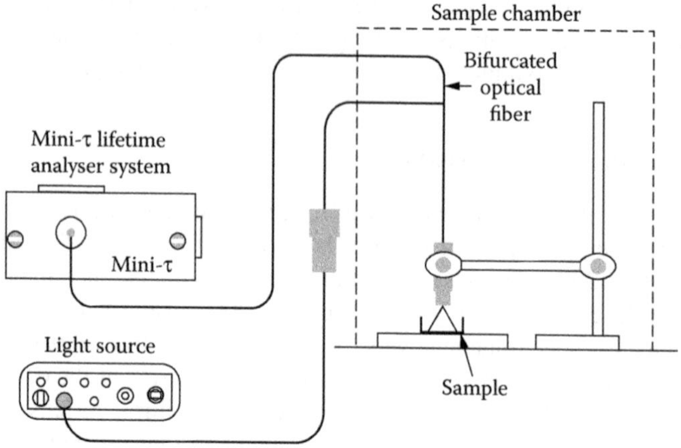

FIGURE 8.6 Schematic of fluorescence lifetime measurement system.

a photo multiplier tube (PMT). The light from the pulsed LED is coupled into an optical fiber that has two ports. One of the ports acts as the excitation arm and the fluorescence signal collected using the same fiber is then fed to the second port that acts as the detection port. Figure 8.6 is a schematic of the bifurcated fiber-optic set-up used for lifetime measurement. The excitation and collection of the signal are carried out with the help of a bifurcated optical fiber.

Two LEDs at wavelengths of 380 nm and 498 nm can be used as the excitation sources. While samples that fluoresce under UV excitations are excited with a 380-nm LED, those that fluoresce in the visible band are excited with blue-green light with 498 nm.

8.3.5 Detectors

Generally, a photodetector system is placed after the spectrometer or monochromator to measure the emitted fluorescence intensity for all emission wavelengths. If measurements are being made at only one or several wavelengths, a single-channel photoemissive tube, i.e., a photomultiplier tube (PMT) can be used [25], as explained in the previous sections. It is important to note that the PMT must detect emitted photons of all wavelengths with equal efficiency [30]. Fluorecence spectroscopy can be performed using a monochromator coupled to a PMT.

As far as the detector performance is concerned, other considerations include quantum efficiency and sources of noise. The detector quantum efficiency (QE) is defined as the ratio of induced current to the induced flux (in electrons/photons). QE depends on the wavelength of light used, the material type and shape, and other physical parameters. Different types of noises during the detection of light can be classified as shot noise, dark noise/dark signal, and read-out noise [25]. A proper selection of the PMT used for fluorescence analysis should ensure maximum QE and minimum noise factors so as to ensure a higher signal-to-noise ratio.

8.4 ROLE OF FIBER OPTICS IN BIOMEDICAL ARENA: DIFFERENT CONFIGURATIONS FOR ILLUMINATION AND COLLECTION

The role of fiber optics is not limited only to the illumination and collection of target excitation or targeted fluorescent light collection, as we have seen in the previous sections. A multitude of configurations have been researched by various researchers over the years for specific biomedical applications. This section briefly overviews some of them with focus on *in vivo* applications.

Fluorescence techniques in endoscopy are of great diagnostic value, for example, the qualitative differences in tissue fluorescence intensity are used to identify the presence of cancer during endoscopy [31]. Fluorescence images of cancerous tissue will show reduced intensity as compared to images of normal tissue and thus such images can normally highlight regions of suspicious tissue to provide biopsy guidance.

Normally, the reported and commercial fluoroscopes make use of a light source (coherent and incoherent) to provide the necessary excitation for inducing fluorescence. In fiberscopes, this excitation is delivered to the tissue through single mode or multimode fibers. The collection of the emitted signal is also achieved using optical fibers. Different arrangements of the excitation and collection fibers for the fluorescence analysis of colonic tissues making use of several geometries have been reported in the literature.

8.4.1 Single Fiber for Excitation and Collection

In this configuration, a single optical fiber is used to deliver the excitation light and collect the emitted fluorescence. An example of such a system is given in Figure 8.7 [32].

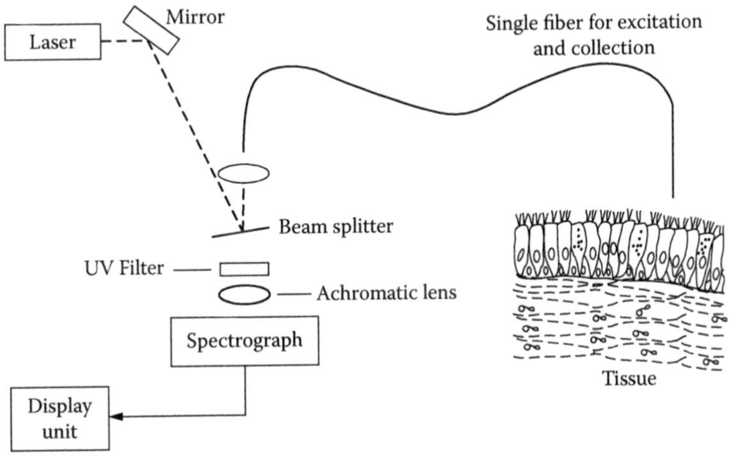

FIGURE 8.7 Fluorescence analysis with single fiber excitation and emission collection. Adapted from Kapadia et al. [32].

In such a set-up, the excitation source is coupled to an optical fiber (with diameter $\approx 400\mu$ m) that in turn induces endogenous tissue fluorescence. By the use of a custom beam splitter, the simultaneous excitation and emission collection are carried out. But the fiber was positioned perpendicular to and in contact with the tissue surface. This type of contact method suffers from the disadvantage of distortion in the collected signal by the variable pressure on the tissue. Also, the area of tissue from which the emission is collected will be small. This type of set-up is useful only for biopsy examinations outside the body and not in a real endoscope set-up.

8.4.2 Single Fiber for Excitation and Multiple Fibers for Collection

An example of fluorescence analysis with single fiber excitation and multiple fibers for emission collection is reported in the literature [33]. In this set-up, the fluorescence analysis probe consists of a circular array of collection fibers surrounding a single excitation fiber, as shown in Figure 8.8.

Having the collection fibers in contact with the tissue, can lead to the disadvantage of signal distortion. Also, as the collection fibers are fewer in number and arranged surrounding the excitation fiber, such an arrangement may not be maximized for emission collection and definitely cannot be used for fluorescence imaging. The concept of fluorescence imaging along with other imaging modalities will be discussed in Section 8.5.

8.4.3 Fiber Bundles for Excitation and Collection

Fluorescence analysis using a standard colonoscope is reported in the literature [34]. The excitation is achieved by the illumination of the optical fiber bundle of the endoscope (Figure 8.9). The emission is collected via an optical probe inserted through the

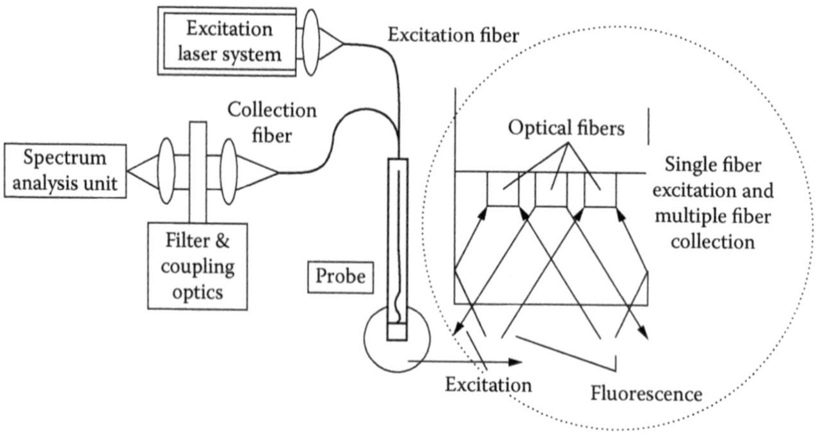

FIGURE 8.8 Fluorescence analysis with single fiber excitation and multiple fibers for emission collection. Adapted from Cothren et al. [33].

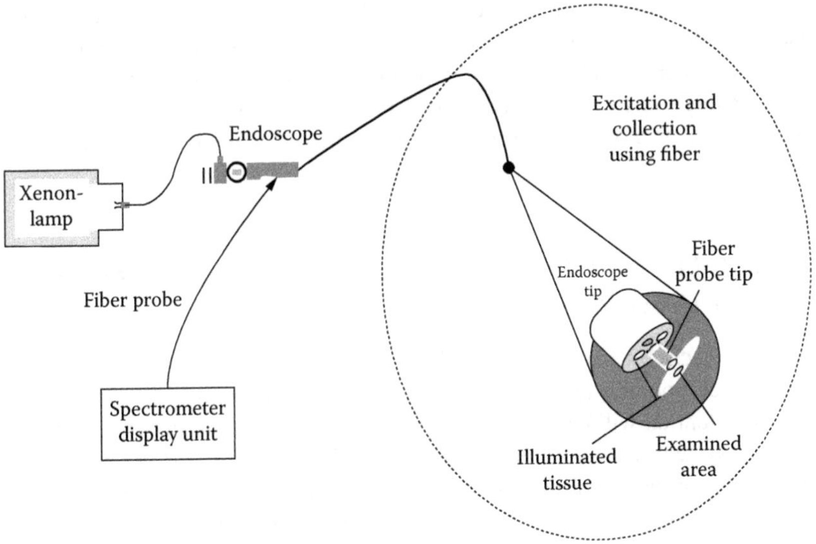

FIGURE 8.9 Fluorescence analysis with excitation and emission collection using fiber bundles. Adapted from Mayinger et al. [34].

accessory channel of the same endoscope. The collection probe consists of a bundle of fibers as shown in the figure.

The above configuration, like all standard colonoscopes, also allowed white light imaging modality. Although, the use of collection fiber bundles increased the examined tissue area as compared with collection using a single fiber, the above method is definitely complicated and involves two separate probes (commercial endoscope and custom-made collection probe) for the fluorescence analysis. Also, the study mainly concentrates on biopsy samples taken from the human body.

All the fluoroscope designs discussed above deal with single fiber excitation. A symmetrical configuration of two illumination fiber ports can give flexibility in enlarging the interrogating area, providing uniform specimen excitation at the desired angle. Also, the collection of emissions depends on the imaging lens used, the imaging lens collection fiber coupling, and the collection fiber transmission efficiency. A single collection fiber can only transmit the emission, whereas with the use of a collection fiber bundle, the fluorescence image of the tissue can be transmitted and can provide an alternative to spectrum analysis. An attempt to collect such fluorescence images using a standard white light colonoscope with laser excitation provided by an external fiber has been reported in the literature [35]. The study involved examination of biopsy samples. The fluorescence excitation and image collection were not carried out using a single probe, and thus could not be carried out inside body cavities. Since the fluorescence spectrum of cancerous tissue shows a decrease in intensity compared to normal tissue the cancerous regions are seen with decreased intensity in fluorescence images.

Because the emitted fluorescence intensity from the live tissue is weak in autofluorescence studies, the selection and use of proper imaging elements in the endoscope for fluorescence analysis is very important for collecting sufficient amounts of the emitted energy. Also, the emitted signal from the endofluoroscope passes through the spectrometer or monochromator where the reflected and emission wavelengths are scanned.

8.5 SPECIALTY APPLICATION PROBES AND CURRENT RESEARCH AND DEVELOPMENT

This section discusses one of the novel designs using conventional and specialty fiber optics incorporating various concepts and techniques for imaging with multiple modalities. Attention will be given to the recently developed multimodality imaging and sensing system known as the Endo-Speckle-Fluoroscope (E-S-F) probe [36]. Finally, the use of the E-S-F system in identifying the presence of growth or cancer will be illustrated by using a phantom model of the colon as the test specimen.

8.5.1 Design Considerations of E-S-F System

The design of the E-S-F system was based on the concept of multiple modality imaging that can provide a more accurate diagnosis of colon cancer. The whole system consists of four main elements namely (i) the E-S-F probe, (ii) the collection lens, (iii) the optical component selection unit (OCSU), and (iv) the associated image processing and detection systems (the details of which are given in the following sections). Figure 8.10 represents the E-S-F system.

Light from the laser source is coupled into a sheathed single mode fiber (standard single mode fiber with protective covering) that is then split into two sheathed single mode fibers via a 1 × 2 fiber splitter. The two sheathed fibers are connected to two bare fibers (with 1 mm removable sheath) via connectors known as finger splices (FSs). These bare fibers are then fed into the illumination channels of the E-S-F probe. This type of arrangement allows the easy plug-in and plug-out of the E-S-F probe and the source. The source light is delivered to the colon surface from the fiber ports at the distal end of the E-S-F probe. The return light from the colon surface is collected by an imaging lens and is transmitted through an image fiber (Fujikura 15-600N). An objective lens placed at the proximal end of the probe collects the return image or light from the image fiber and directs it to the OCSU, where different constituent elements are employed to facilitate the multimodality imaging. The output from the OCSU is directed to a normal imaging speckle correlation analysis unit or fluorescence spectrum analysis unit, depending on the employed imaging modality. Initially, the colon walls will be checked for visible abnormalities on the surface (normal imaging modality). Secondly, the colon walls will be subjected to the speckle correlation analysis modality where surface abnormalities of millimeter order and subsurface abnormalities are diagnosed. If such abnormalities are found, the system is switched

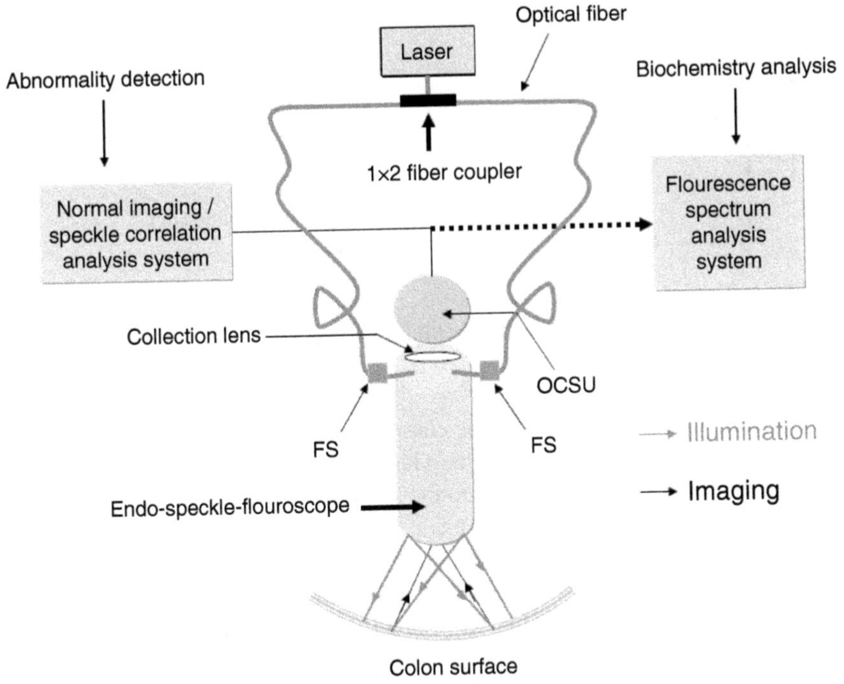

FIGURE 8.10 Schematic of E-S-F system. Adapted from Murukeshan et al. [24].

to the final modality of fluorescence analysis where the intensity of the emitted light will be checked to diagnose the presence of cancer.

A green diode pumped solid state (DPSS) laser ($\lambda = 532$ nm) was selected as the source for the E-S-F system in order to generate biospeckles from the colon tissue as well as excite some of the fluorophores contained in the colon tissue. The difference in emitted intensities of these fluorophores makes possible a diagnosis of cancer, as is evident from the literature.

8.5.2 The E-S-F Probe

An all-fiber optic approach is adopted here for the E-S-F probe in order to facilitate the working of the system in multiple modalities. Single mode optical fibers were used to guide the light from the laser source through the E-S-F body. The E-S-F probe and its constituent elements are shown in Figure 8.11.

The probe consists of two illumination fibers for providing necessary illumination or excitation. An imaging lens–image fiber combination is used as the imaging unit in the probe. The probe has rigid distal and proximal ends and a flexible part to facilitate

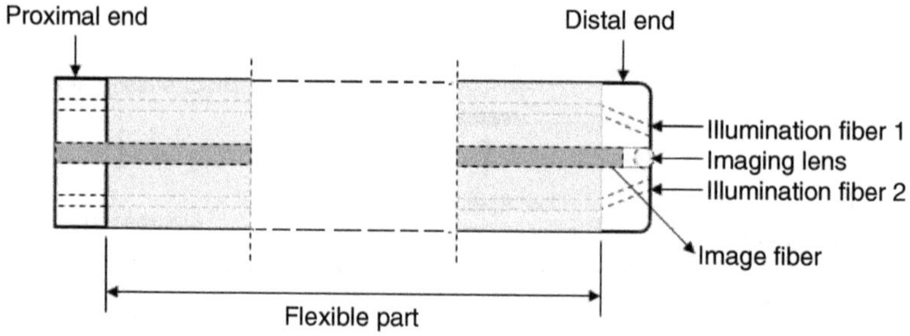

FIGURE 8.11 The E-S-F probe. Adapted from Murukeshan et al. [37, 38].

inspection inside body cavities such as the colon. In order to facilitate the complete inspection of the inner colon walls, a certain length of the probe can be bent. This helps to keep the probe far from and near to the colon tissue wall. When the probe is located far from the colon wall, the interrogated and imaged area will be greater as compared to the probe position near to the tissue wall. The latter will provide a close view of the colon and allow a more accurate inspection. The main considerations for designing the E-S-F probe were the selection of source, the design aspects of the probe distal end, the choice of the image fiber, and the selection of imaging lenses. They will be discussed in the sections below.

8.5.3 E-S-F Probe Distal End

The E-S-F probe distal end consists of optical elements selected and integrated to meet target objectives such as normal imaging, analysis with different speckle interferometric configurations, fluorescence emission collection, and subsequent analysis.

The distal end of the E-S-F probe consists mainly of two single mode fiber ports and an imaging port for carrying an imaging unit that consists of an imaging lens–image fiber combination. The two fiber ports are intended to deliver the source light to the test target. In normal imaging modality, the use of dual illumination ports will enhance the total illumination light, thus resulting in a brighter image. For speckle correlation modality, the use of dual illumination ports will enable the probe for in-plane sensitive configuration. For out-of-plane and shear configurations, illumination from one of the fiber ports is required; the other port is disconnected accordingly. For the fluorescence spectrum analysis modality, the use of double illumination will enhance the interrogation area and also the excitation light delivered to the tissue so as to obtain a considerable amount of emission from the concerned tissue. In order to make the center of the illuminated or excited area in line with the imaging unit, care must be taken to orient the two illumination fibers at equal angles from their respective optic axes.

8.5.4 Selection of Image Fiber

The choice of imaging elements forms an important part of an endoscope probe design. The design of an all-fiber optic probe requires the images or return light to be guided through an image fiber to the probe proximal end. The selection of image fiber in the scheme is intended to increase the flexibility of the probe and miniaturize the whole image collection scheme. More information on image fibers and their applications can be found in the literature [38]. The image fiber bundle has a double role to perform here, i.e., it acts as an image-conveying device and also as a fluorescence signal transmitting element. Image fibers differ in the number of individual fibers (fiberlets) and thus differ in their image resolution. An image fiber consisting of a greater number of pixels has better resolution compared to an image fiber with fewer pixels. However, the flexibility of the image fiber decreases as the number of individual fibers increases. Thus, a compromise between flexibility and image resolution is required when selecting a suitable image fiber for a flexible fiberscope. The selected image fiber should have sufficient flexibility to bend as well. A microscopic image of the end face of a Fujikura FIGH-15-600N fiber bundle is shown in Figure 8.3.

8.5.5 Illustration

A phantom colon model (purchased from Buy-A-Mag Corporation [39]) that resembles the human colon was used as the test specimen. At different locations inside this phantom colon, layers of phantom tissues obtained from Simulab Corporation were pasted [40]. Each tissue layer had a thickness equal to the layer thickness of a human colon.

For normal imaging modality, the OCSU output is directed toward a CCD camera connected to a PC installed with a frame grabber (EDC-2000N computer camera system). The images of the test target (phantom colon surface) may be viewed on the PC monitor with the help of appropriate software. For speckle correlation analysis, the software allowed subtractive correlation of frames corresponding to the different states of the test object.

The E-S-F system was first operated in the normal and speckle correlation analysis modalities to detect the presence of abnormalities or suspicious cancerous growths. In case of such findings, the whole system was switched to the final diagnostic modality, i.e., fluorescence spectrum analysis. In this modality, the collection lens output is directed toward a monochromator system in which the OCSU output is scanned according to the set resolution of the monochromator. This scanned output is allowed to pass through a PMT. The PMT output signal is connected to a system interface card installed in a PC. The fluorescence spectrum (emission intensity versus emission wavelength) is displayed on the PC monitor with the help of related software. The diagnosis of cancerous regions in the colon tissue is made by determining the difference in emitted intensities of normal and cancerous regions. The experimental investigation using the developed probe as per the design was carried out on the phantom colon model.

Phantom tissue material representing cancerous growth was simulated for experiments with E-S-F probe. The abnormal colon surface was prepared to represent real cancerous tissues having emissions in the same wavelength range.

8.6 E-S-F SYSTEM IN MULTIMODALITY IMAGING

8.6.1 Normal Imaging Modality

In the normal imaging modality, the two illumination fibers in the E-S-F probe are used for providing necessary illumination of the inner walls of phantom tissue model and the imaging unit in the probe images different parts of the phantom colon model to check for the presence of cancerous growths. In this modality, all the visible cancerous growths protruding from the colon wall could be imaged for identification. Figure 8.12(a) shows an image of a normal phantom colon taken using the E-S-F system. The phantom colon image with cancerous growth imaged with the probe is illustrated in Figure 8.12(b). The protrusions from the colon wall image indicate the presence of cancerous growths (polyps).

8.6.2 E-S-F System in Speckle Correlation Modality

The E-S-F system developed as per the design was able to perform all three configurations of speckle interferometry: in-plane, out-of-plane, and shear configurations. In this way, the analysis of the phantom colon wall in the in-plane and out-of-plane directions was made possible. Also, abnormality growth detection was possible in all the configurations. For all these configurations, the reflected speckle pattern from the tissue wall is to be guided and transmitted through the image fiber confined within the E-S-F probe before it is fed to the CCD for further processing. We explain only one of the three possible displacement sensitive configurations: in-plane, out-of-plane, and shear [36, 41]. All three configurations can be realized by employing one or two

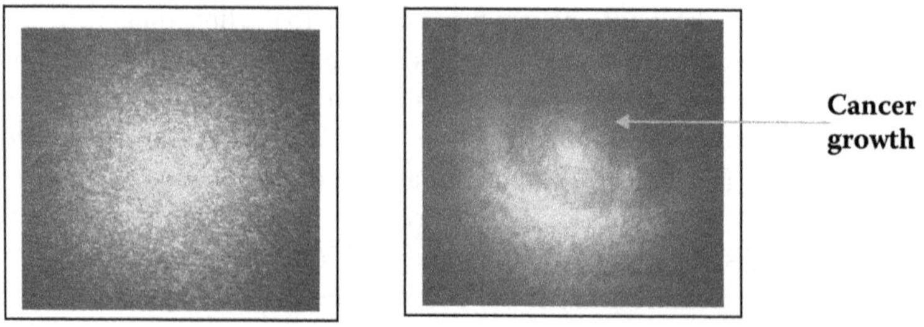

(a) Normal colon (b) Cancerous colon

FIGURE 8.12 Images of colon wall obtained using E-S-F system in normal imaging modality [41]. (a) Normal colon. (b) Cancerous colon.

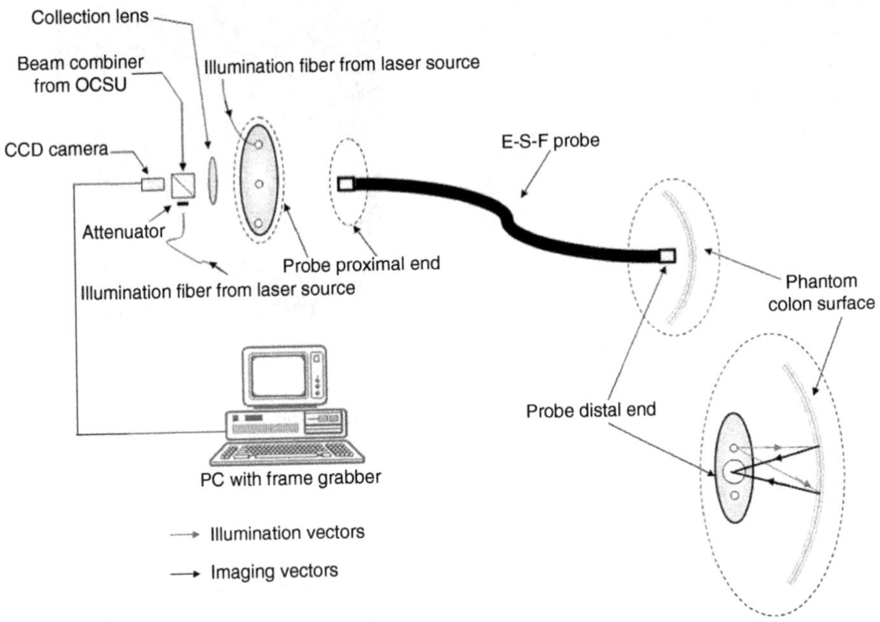

FIGURE 8.13 E-S-F system in out-of-plane configuration.

illumination fibers at the distal end and suitable optical elements from the OCSU at the proximal end. Experiments were carried out with the same set-up for normal and cancerous colon phantoms. The E-S-F system in the out-of-plane configuration is shown in Figure 8.13.

In this configuration, only one of the fiber ports is used for illuminating the curved colon phantom surface. The reflected speckle pattern is collected by the ball lens and transmitted through the image fiber. The collection lens after the probe proximal end collects the speckle pattern transmitted through the image fiber. This collected speckle pattern is combined with the reference beam (supplied by the other illumination fiber connected to the laser) at the beam combiner. The reference beam is attenuated before it reaches the beam combiner using an attenuator (variable ND filter). The reference beam combined with the reflected image is then captured by the CCD camera placed behind the beam combiner.

Experimental results of E-S-F system in the out-of-plane configuration with normal and abnormal colon phantoms are shown in Figures 8.14(a) and 8.14(b). Figure 8.14(a) represents the fringe pattern for a normal colon phantom with the E-S-F system in the out-of-plane configuration, where the fringes followed concentric oval shape.

Figure 8.14(b) shows the fringe pattern obtained from a sample with cancerous growth (abnormality) simulated with a size of 3 mm. It can be noted that the presence of simulated cancer growth is indicated by a kink in the fringe pattern.

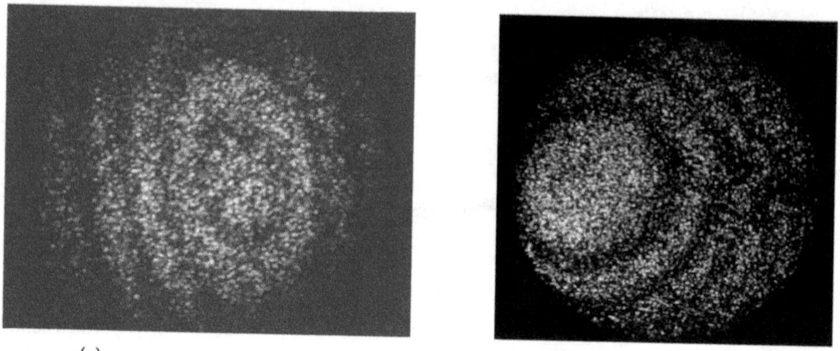

(a) (b)

FIGURE 8.14 Out-of-plane fringes obtained from (a) normal colon and (b) cancerous phantom model.

In this configuration, growth size smaller than 3 mm was difficult to identify. Similar to the case of in-plane configuration, smaller abnormalities required more of out-of-plane fringes to be formed in order to identify them as kinks. With shear configuration, even smaller size abnormalities can be identified.

If the presence of an abnormality is detected in any of the in-plane, out-of-plane, or shear fringes, the E-S-F system is switched into the third and final diagnostic modality: fluorescence spectrum analysis. In the next section, a comprehensive theoretical analysis of the performance of the E-S-F system in fluorescence spectrum analysis modality is discussed.

8.6.3 E-S-F System in Fluorescence Spectrum Analysis Modality

The analysis of the performance of the E-S-F system in the fluorescence spectrum collection modality requires information regarding parameters such as excitation energy, distance between the E-S-F distal end to the tissue surface, spatial distribution of the image intensity, etc. for better quantitative imaging of the biological tissue or surface. Exact information regarding the excitation energy is needed for deciding its optimum level because too much energy causes tissue damage and too little energy gives only undetectable signals. Energy level also determines the closeness of the distal end to the tissue wall. Since absorption coefficient is a dependent parameter with wavelength, use of different excitation wavelengths requires different optimum excitation energy levels for better performance.

For the E-S-F system, the effect of the illumination beams and the geometry of the object surface must also be considered. For example, when the DSPI configurations are confined in endoscope form to illuminate a large object area, the illuminations should be divergent. Even for oblique illumination, the use of small collimating lenses

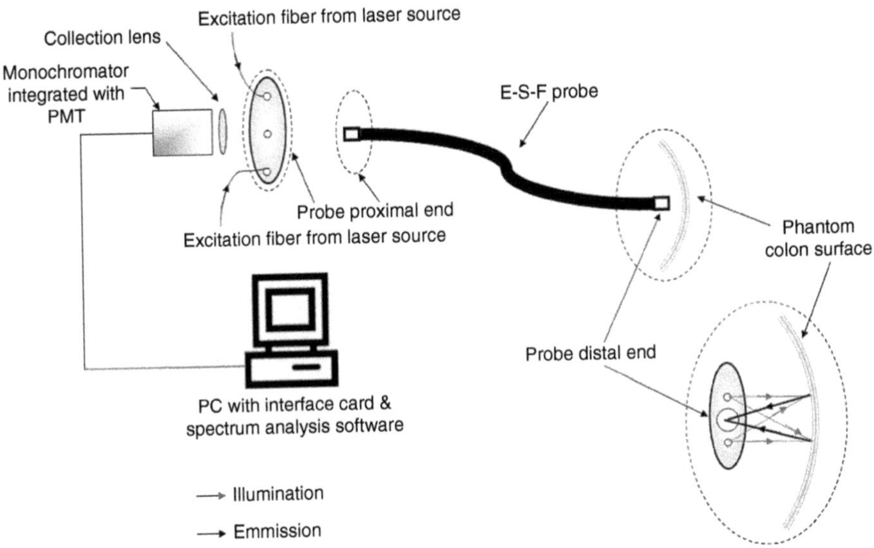

FIGURE 8.15 E-S-F system in fluorescence spectrum analysis modality [36].

(restricted by the probe maximum diameter) will result in illuminating only a small area. Thus the area imaged via normal imaging will be small and this makes the process of diagnosis more time-consuming. Also, in the case of speckle correlation analysis, only fractions of a fringe can be seen with this reduced imaged area, which again makes the process of diagnosis difficult. Hence, the significance of employing divergent illumination beams only.

During the experimental studies, the E-S-F system was operated in the fluorescence spectrum analysis modality for collecting the fluorescence emission from the test tissue phantom. The schematic diagram of E-S-F system in the fluorescence spectrum analysis modality is given in Figure 8.15.

The excitation laser light is delivered to the test tissue by the two fiber optic ports in the distal end of the E-S-F probe. The emission from the test tissue is imaged by means of the ball lens–image fiber combination. The collected emission is transmitted to the proximal end of the E-S-F probe via the same image fiber. The transmitted emission is collected by a collection lens placed after the probe proximal end. The collection lens output is directed to the fluorescence spectrum analysis unit (monochromator integrated with PMT).

Figure 8.16 shows the excitation–emission spectrum from normal and cancerous phantom tissues collected using the E-S-F system. In the case of simulated cancerous tissue, the collected fluorescence emission showed a decrease in emission intensity at a similar emission wavelength range, as expected.

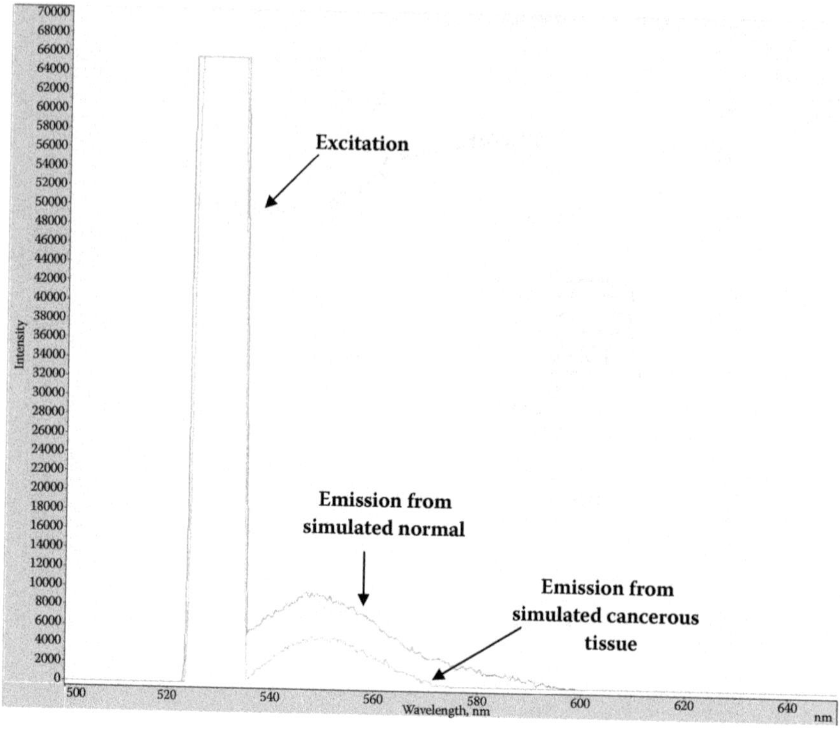

FIGURE 8.16 Excitation emission spectra collected from simulated normal and cancerous tissues using E-S-F probe [36].

8.7 CONCLUSION

This chapter provides an overview of various optical fiber configurations for illumination, fluorescence excitation, and imaging with a focus on *in vivo* diagnosis. A comprehensive list of literature references will be useful for practicing biomedical engineers, students, and researchers.

ACKNOWLEDGMENTS

This chapter cannot be concluded without acknowledging the support of my students, especially Dr. N. Sujatha and Dr. U.S. Dinish, who worked in our lab to pursue their Ph.D. programs. Acknowledgments are also due for the support and encouragement received from A/P L.S. Ong and A/P L.K. Seah. Many of the details included in this chapter on biomedical fiber optics were taken from the author's research and development data from various projects funded by SDS, AcRF (NTU), etc. since 2001 and their support is duly acknowledged here. Finally, I thank my wife and son who were very patient when I was involved in this project and burned the midnight oil on many weekends holidays, and evenings.

REFERENCES

1. Yeh, C., *Handbook of Fiber Optics: Theory and Applications*, Academic Press, New York, 1989.
2. Gloge D., Weakly guiding fibers, *Appl. Optic.*, 10, 2252, 1971.
3. Ben-David, A. et al., Theoretical and experimental studies of infrared radiation propagation in hollow waveguides, *Opt. Eng.*, 39, 1384, 2000.
4. Murukeshan, V.M. et al., Imaging considerations in fiber optic endoscopy system for the gastrointestinal endoscopy. Proceedings of SPIE, 5143, 170, 2003.
5. Takahashi, H. et al., Optical transmission loss of liquid core silica fibers in the infrared region, *Opt. Commun.*, 53, 164, 1985.
6. Smith, W.J., *Modern Optical Engineering*, McGraw-Hill, New York, 2000.
7. Kawano, K. and Kitoh, T., *Introduction to Optical Waveguide Analysis*, John Wiley & Sons, New York, 2001.
8. Harrington, J., A review of IR transmitting hollow waveguides, *Fibers Integ. Opt.*, 19, 211, 2000.
9. Joannoppoulos, J.D., Meade, R.D., and Winn, J.N., *Photonic Crystals: Molding the Flow of Light*, Princeton University Press, Princeton, NJ, 1995.
10. Knight. J. et al., Photonic band gap guidance in optical fibers, *Science*, 282, 1476, 1998.
11. Bjarklev, A. et al., Photonic crystal fibers: a variety of applications, in optical fibers and sensors for medical applications, *Proc. SPIE*, 4616, 73, 2002.
12. Gannot, I., Optical fibers and sensors for medical applications, *Proc. SPIE*, 4253, 42, 2001.
13. Feng, Y.P. et al., X-ray flux enhancements in thin-film waveguides using resonant beam coupler, *Phys. Rev. Lett.*, 71, 537, 1993.
14. Pfeiffer, F.T. et al., X-ray waveguides with multiple guiding layer, *Phys. Rev B*, 62, 16939, 2000.
15. Ben-David, M. et al., The effect of scattering on the transmission of IR radiation through hollow waveguides, *J. Optoelectronic. Adv. Mater.*, 3, 23, 1999.
16. Gannot, I. et al., Mid-IR optimized multilayer hollow waveguides, *Bios 2001*, 4253, 11, 2001.
17. Sanghera, J.S. and Agarwall, I.D., *Infrared Fiber Optics*, CRC Press, Boca Raton, 1998.
18. Chen, C.L., *Elements of Optoelectronics and Fiber Optics*, Irwin, 1996.
19. Saito, M., Takizawa, M., and Miyagi M., Optical and mechanical properties of infrared fiber, *J. Lightwave Technol.*, 6, 233, 1988.
20. Alimpiev, S.S. et al., Polycrystalline IR fibers for laser scalpels, *Int. J. Optolectron.*, 3, 333, 1988.
21. Takahash, H. et al., Optical transmission loss of liquid core silica fibers in the infrared region, Opt. Commun., 53, 164, 1985.
22. Klein, S. et al., High power laser waveguide with a circulating liquid core for IR applications, in *Specialty Fiber Optics for Biomedical and Industrial Applications*, Katzir, A. and Harrington, J.A., Eds., *Proc. SPIE*, 2977, 155, 1997.
23. Holder, L., Okamoto, T., and Asakura, T., A digital speckle correlation interferometer using an image fiber, *Meas. Sci. and Technol.*, 4, 746, 1993.
24. Murukeshan, V.M., et al., Effect of image fiber on speckle fringe pattern in image fiber guided DSPI endoscopy, *Opt. Laser Technol.* , 2005.
25. Ramanujam, N., *Fluorescence spectroscopy in Vivo: Encyclopedia of Analytical Chemistry*, John Wiley & Sons Chichester 2000, pp. 20–56.
26. Avasoft Full Version 5.1 User's Manual, Avantes, Netherlands.

27. Dinish, U.S. et al., Fluorescence optimisation and lifetime studies of fingerprints treated with magnetic powders, *J. Forens. Sci. Int.*, 152, 249, 2005.
28. Dinish, U.S. et al., Nanosecond resolution in fingerprint imaging using optical technique, *Int. J. Nanosci.*, 4, 695, 2005.
29. Mini-Tau Miniature Fluorescence Lifetime Analyser, Operating Instructions, Edinburgh Instruments, U.K.
30. Lakowicz, J. R., *Principles of Fluorescence Spectroscopy*, Kluwer/Plenum, New York, 1999, ch. 1.
31. Marcon, N.E. and Wilson, B.C., The value of fluorescence techniques in gastrointestinal endoscopy: better than an endoscopist's eye? II. The North American experience, *Endoscopy*, 30, 419, 1998.
32. Kapadia, C.R., et al., Laser-induced fluorescence spectroscopy of human colonic mucosa: detection of adenomatous transformation, *Gastroenterology*, 99, 150, 1990.
33. Cothren, R.M. et al., Detection of dysplasia at colonoscopy using laser-induced fluorescence: a blinded study. *Gastrointestinal Endoscopy*, 44, 168, 1996.
34. Mayinger, B. et al., Endoscopic light-induced autofluorescence spectroscopy for the diagnosis of colorectal cancer and adenoma, *J. Photochem. Photobiol. B*, 70, 13, 2003.
35. Wang, T.D. et al., Fluorescence endoscopic imaging of human colonic adenomas, *Gastroenterology*, 111, 1182, 1996.
36. Murukeshan, V.M. and Sujatha, N., An integrated simultaneous dual modality imaging endo-speckle fluoroscope system for early colon cancer diagnosis, *Opt. Eng. Lett.*, 2005.
37. Murukeshan, V.M., Min, L., and Sujatha, N., Design and development of fiber optic probe for cavity imaging, internal report, 2004.
38. Murukeshan, V.M. et al., An all-fiber optic endoscope probe distal end for disease diagnosis in body cavities, *Progr. Biomed. Opt. and Imaging*, *Proc. SPIE*, 5651, 5, 321, 2004.
39. Home page of Buy-A-Mag corporation (2004), www.buyamag.com.
40. Home page of Simulab Corporation (2004), www.simulab.com.
41. Murukeshan, V.M. et al., A flexible endoscope system for dual mode intracavity investigations, *Proc. SPIE*, 5686, 6, 353, 2005.

9 Neutron Waveguides and Applications

Ramón F. Alvarez-Estrada and María L. Calvo

CONTENTS

9.1 INTRODUCTION

The existence of the neutron was experimentally established by J. Chadwick and announced in 1932.[1] That fundamental discovery can be regarded as the culmination of other important research by W. Bothe, H. Becker, I. Curie, H.C. Webster, P.I. Dee, F. Joliot, F. Perrin, N. Feather, E. Majorana, W. Heisenberg and D. Iwanenko.[2] In 1936, W.M. Elsasser suggested that neutrons evolve according to quantum mechanics.[3] Two experiments (by H. Halban and P. Preiswork[4] and by D.P. Mitchell and P.N. Powers[5]), also in 1936, provided adequate indication that crystalline materials diffracted slow neutrons. It has been regarded[6] that systematic research establishing neutron optics on firm ground started by 1944 with the works by E. Fermi and W.H. Zinn. Their experiments provided evidence of mirror reflection of thermal neutrons.[6] Since then, a whole variety of experiments have established several optical-like effects for slow neutrons (with associated de Broglie wavelength in an interval about $1\mathring{A}$). A number of researchers (G.E. Bacon, B. Brockhouse, L.M. Corliss, J.M. Hastings, W.C. Koehler, C.G. Shull, M.K. Wilkinson, and E.O. Wollan) at various institutions made important contributions, developing and employing neutron diffraction in order to analyze the atomic structure of matter (e.g., locations of the atoms and their dynamics). In particular, B. Brockhouse and C.G. Shull were awarded the 1994 Nobel Prize in Physics for their works on those subjects.

No attempt will be made here to offer a broad view of neutron optics: for comprehensive presentations, the interested reader should consult other sources.[2,6–8] We shall limit discussion to a few general aspects of neutron physics and neutron optics (including, occasionally, discussions of a few recent topics), and we shall concentrate on one specific topic, namely, the confined propagation of slow neutrons along waveguides of decreasingly small transverse cross sections. In so doing, we shall keep in mind well known similar phenomena for electromagnetic radiation with decreasing wavelengths.

Section 9.2 will outline various general aspects of neutron physics.[2,7,9] Section 9.3 will summarize wave propagation of slow neutrons, using quantum mechanics. Section 9.4 will deal briefly with total internal reflection and hollow neutron guides with large cross sections. Section 9.5 will discuss the first conjectures regarding confined propagation of slow neutrons in thin waveguides (neutron fibers). Section 9.6

will outline the quantum mechanical treatment of slow neutron propagation in neutron fibers: propagation modes, quasi-classical estimates of their number, examples for circular cross-section, statistical mixtures, and bending or curvature losses. Section 9.7 will deal with experimental implementations of neutron fibers by means of polycapillary glass fibers, in comparison with the use of the latter for x-ray optics. Section 9.8 will be devoted to the possibility of confined thermal neutron propagation in carbon nanotubes. In Sections 9.7 and 9.8, we also include short discussions about various possible loss mechanisms in the neutron flux: nuclear absorption and diffuse scattering, tunneling effects across the clad, and bending losses. At this stage, and as a detour, we remind readers that boron neutron capture therapy (BNCT) and, more generally, neutron capture therapy (NCT) have emerged as relatively recent and very interesting interdisciplinary applications of slow neutrons. Section 9.9 will present an overview of BNCT and NCT in order to provide some updated accounts of this very interesting application of slow neutron physics, even if their connection to neutron optics and to neutron fibers may seem somewhat loose. As an attempt to proceed one step beyond and to argue about the latter connection, Section 9.10 will speculate about possible applications of neutron optics and neutron fibers to BNCT and discuss some preliminary attempts. Section 9.11 will discuss the possibility of employing neutron fibers for condensed matter analysis. Section 12 contains the conclusions and some discussions.

9.2 NEUTRON PHYSICS: SOME GENERAL ASPECTS

9.2.1 Properties of Neutrons

The neutron (denoted by n) is one of the constituents of atomic nuclei. Its mass is $m_n = 939.57\ MeV/c^2$, its spin is $\hbar/2$ (thereby being half integral), its magnetic moment is $\mu_n = -1.913\mu_N$ and it is electrically neutral. We recall that $1\ MeV = 10^6 eV$ ($1\ eV$ denoting one electron-volt), c $(= 2.998 \times 10^{10}\ cm.s^{-1})$ is the speed of light in vacuum, $\hbar(= 6.58 \times 10^{22}\ MeV.s)$ is Planck's constant and $\mu_N = \hbar \mid e \mid / (2m_p c)$ is one nuclear magneton. In turn, s denotes one second, $\mid e \mid = 1.602 \times 10^{-19}$ $Coulomb$ is the proton electric charge and $m_p = 938.27\ MeV/c^2$ is the proton mass. Many experiments appear to indicate that the neutron has a vanishing electric dipole moment: more specifically, experimental upper bounds on its magnitude are of the order $10^{-25} \mid e \mid .cm$.[2,9]

The neutron is not strictly elementary: rather, it is composed of three more elementary or fundamental entities named quarks. Quarks have fractional electric charges. Specifically, the neutron is a bound system of one certain type of quark with electric charge $+(2/3) \mid e \mid$ (named quark u) and of two other quarks, each with an electric charge of $-(1/3) \mid e \mid$ (named quark d).[2,9] The overall electric charge of the neutron as a bound system (due to strong interactions) of those three charged entities vanishes consistently. All phenomena to be dealt with in this article take place at ranges of energies and at length scales such that each of the neutrons involved behaves as an elementary particle (its composedness not becoming manifest); this is the point of view to be followed here, that is, we shall disregard completely and consistently the fact that the neutron is composed of other more fundamental entities.

The neutron is unstable, its lifetime being 886.7 s. All neutron phenomena (in particular, the travels of slow neutrons along various, eventually macroscopic, distances) treated in this article occur during time intervals shorter than its lifetime. Accordingly, one can disregard neutron instability as a reliable approximation and hence treat the neutron as a stable particle (as if it had an infinite lifetime).

A discussion about the influence of gravity on slow neutrons, moving on the surface of the Earth with differences of height not exceeding one m (which will always be the case here) will be given in Section 9.3.

Let us consider a non-relativistic neutron with kinetic energy E_K and de Broglie wavelength λ_{dB}. They are related through $E_K = (2\pi\hbar)^2/(2m_n\lambda^2)$. In a rather strict sense, a neutron is designated thermal if it has $E_K \simeq 0.025\ eV$ and $\lambda \simeq 1.8\AA$. If a non-relativistic neutron moves through a material medium at thermodynamical equilibrium at absolute temperature $T = 293\ K$ (20 °C), then it is just strictly thermal, as $0.025\ eV = (3/2)K_BT$, K_B being Boltzmann's constant.

It is convenient to attribute various specific names to non-relativistic neutrons according to the ranges of their kinetic energy, although there does not seem to exist a unique and precise convention that is unanimously accepted (as the reader may realize just by comparing different sources and by what follows). Thus, in a less strict sense, neutrons with E_K ranging from about $0.003\ eV$ up to about 0.4 or $0.5\ eV$ are also thermal. Optical-like phenomena with slow neutrons having E_K smaller than about $0.03\ eV$ are also of considerable interest; such neutrons are denoted generically as cold. In a somewhat more strict sense, neutrons also receive the following more specific names: cold ($5 \times 10^{-5}\ eV \leq E_K < 0.025\ eV$), very cold ($5 \times 10^{-5}\ eV > E_K \simeq 2 \times 10^{-7}\ eV$) and ultra-cold ($E_K < 2 \times 10^{-7}\ eV$).[2] Neutrons with E_K in the range $0.4 - 0.5\ eV$ to 0.1 and, even, to a few tens of keV are usually called epithermal ($1\ keV = 10^3\ eV$). In a general sense, we shall refer to ultra-cold, cold, thermal, and epithermal neutrons as slow. Neutrons with E_K in the range just above the epithermal one, up to about 200 keV, are named intermediate energy neutrons. Neutrons with kinetic energy in the range 200 keV up about 10 MeV are referred to as fast. One usually refers to neutrons with E_K larger than 10 MeV as high energy (or even relativistic). Throughout this article, we shall keep in mind thermal neutrons (in both strict and non-strict senses). However, various results for other slow (epithermal) and even fast neutrons will also be mentioned. On the other hand, the basic features of the quantum mechanical analysis for thermal neutrons will also hold for slow neutrons (in particular, for epithermal ones), except for a few specific details. For this reason, we shall frequently replace thermal neutrons with slow neutrons in the theoretical wave-mechanical studies and give values and orders of magnitude for various phenomena for those different kinds of slow neutrons.

9.2.2 Interest in Slow Neutrons

We shall comment very briefly on a few reasons supporting the specific interest in slow neutrons.[2,7,9] Of course, they are enormously important due to their key role in nuclear reactors, but we shall not cover this aspect (except for a few comments in the next subsection).

The de Broglie wavelength for (strictly) thermal neutrons with $E_K \simeq 0.025\ eV$ is $\lambda_{dB} = 1.8\AA$, which has the same order of magnitude as the spacing between two neighboring atoms in a typical crystal. This allows for a beam of (strictly) thermal neutrons, going through a crystal lattice, to give rise to interference and diffraction phenomena, as already displayed in research at an early stage.[4,5] By analyzing the resulting diffraction patterns, such (strictly) thermal neutrons allow us to explore inter-atomic spacings and structures in crystals. In this application, (strictly) thermal neutrons do a job similar (or, as qualified below, complementary) to the one x-rays do. More generally, slow neutrons with $0.1\AA \leq \lambda_{dB} \leq 20\AA$ have proved to be quite adequate and very useful in order to study correlations among atoms and among magnetic sub-domains in condensed matter. See Subsection 9.4.2.

Slow neutrons with energies E_K in the range 10^{-2} to $10^{-1}\ eV$ enable to analyze molecular vibrations, while those with E_K about $10^{-6}\ eV$ allow for exploring certain slow dynamics of polymers. Slow neutrons can penetrate into various materials without being absorbed appreciably along a certain limited depth and without producing significant modifications in the former. In comparison with slow neutrons, and on the opposite side, x-rays turn out to be more strongly absorbed and rather destructive. That (relatively) non-destructive character of slow neutrons makes them quite convenient for (1) the experimental study of several properties and behaviors of biological matter and (2) the determinations of the compositions of small samples of certain materials without producing appreciable damage in them (neutron activation analysis).

Slow neutrons that have penetrated into matter beyond some depth will at a later stage be eventually absorbed through suitable nuclear reactions occurring at certain localized domains. In certain cases, those nuclear reactions can be relatively non-destructive but in other cases, they turn out to be certainly destructive. In the case of biological matter (human tissue), the damage produced by those nuclear reactions can be beneficial by killing tumor cells (therapies based on neutron capture). We shall give an account of this rather recent and very active interdisciplinary application of slow neutron physics in Section 9.9.

9.2.3 Sources

It may be useful to provide a short survey of various small, medium, and large sources of slow neutron beams.[2,9] Recall that ${}^A_Z X$ denotes a nucleus of chemical species X, with atomic number Z and mass number A (that is, with Z protons and $A-Z$ neutrons). For the sake of brevity, our notations will not distinguish between the atomic nucleus and the atom that contains it; the former or the latter will be understood and easily deduced from the context.

A source of α particles (say, the fully ionized nucleus ${}^4_2He$, helium) produced, for instance, in α decay of another nuclei and a beryllium target, through the nuclear reaction $\alpha + {}^9_4Be \rightarrow n + {}^{12}_6C$, gives rise to neutrons with E_K about $5\ MeV$. ${}^{12}_6C$ denotes a carbon nucleus. This reaction, which gives rise to a neutron source at a laboratory scale, was precisely the one that led to the discovery of the neutron.[1] See Figure 9.1.[1]

Photons (γ), previously emitted in the decay of some adequate radioisotope (like ${}^{124}Sb$, antimony), give rise to photo-nuclear reactions such as $\gamma + {}^9_4\ Be \rightarrow n + {}^8_4\ Be$. The neutrons so produced have E_K smaller than $1\ MeV$.

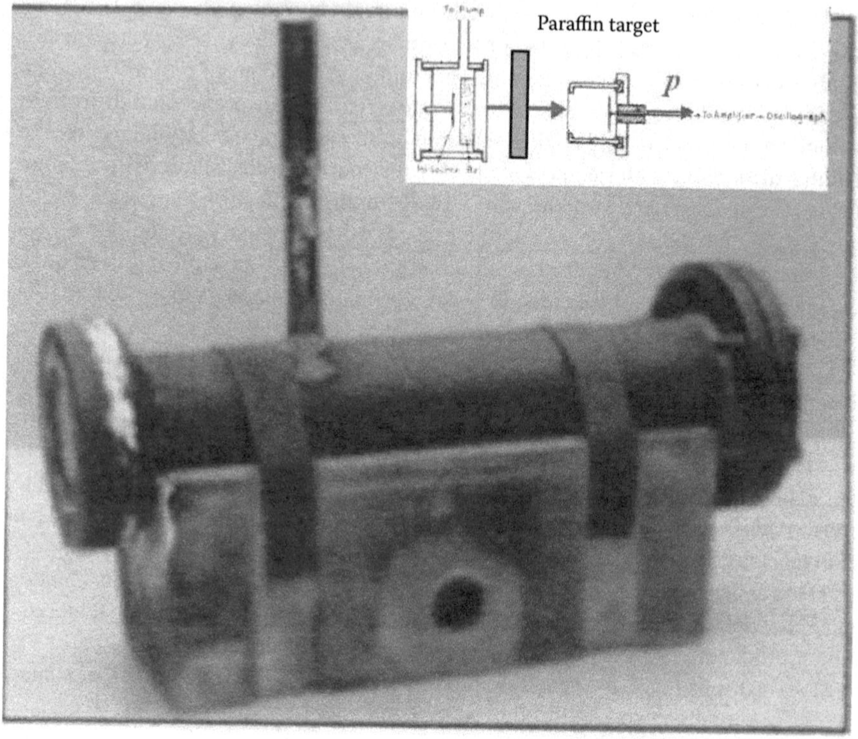

FIGURE 9.1 Chadwick's device and experiment provide illustrative examples of a neutron source, a converter, and a neutron detector. A source containing Po (polonium) emits α particles which, by means of nuclear reactions in the Be target, become, in turn, sources of an unknown neutral radiation (solid arrow). The unknown neutral radiation goes across a paraffin target and enters an ionization chamber where further nuclear reactions (also generated by that unknown neutral radiation) occur, giving rise to protons that in turn are detected. The unknown neutral radiation emitted by the Be target is finally and consistently interpreted as a neutron beam. The paraffin target acts as a converter; it contains H, Li, Be, B, and N. It is where the neutrons undergo elastic collisions. The detections of protons are accomplished by means of an amplifier and an oscillograph.

Californium ($^{252}_{98}Cf$) is a source of neutrons. $^{252}_{98}Cf$ suffers spontaneous nuclear fission, which gives rise to fast neutrons. The half-life of $^{252}_{98}Cf$ is 2.6 years, which implies that, as a neutron source, it may require a relatively frequent replacement. We shall comment later in Subsection 9.2.5 on one recent application of this type of source.

Small accelerators also play an important role (even if neutrons cannot be directly accelerated by externally applied electric fields). Thus, suitable nuclear reactions produced by incoming charged particles that have been accelerated previously in small accelerators constitute other sources of neutrons. An example of those reactions is $p +^7_3 Li$(Lithium) $\rightarrow n +^7_4 Be$ (with protons that have acquired energies about $2\ MeV$ or a bit higher in an accelerator).

Spallation sources. Let us consider a suitable charged particle beam (protons, α-particles, deuterons) with energies about 100 MeV or higher (coming from some accelerator). Actually, such beams do not have constant intensities and are not stationary but rather constitute successive pulses. If those pulsed beams collide with a target containing some suitable heavy metal, the resulting nuclear reactions (generically called spallation reactions) give rise copiously to successive pulses or pulsed beams of neutrons. For instance, let the target contain uranium $^{238}_{92}U$. Then, the resulting fission of the latter, originated by an incoming beam of protons with energies up to about 800 MeV, may generate about 30 neutrons per proton on the average. A pulse of neutrons produced in a spallation source may contain 10^{16} neutrons/s and even more.[2,10]

Nuclear reactors. A nuclear reactor contains in its inner part (core) fissile material (fuel) and other elements, to be commented on shortly. Typical fuels are natural or enriched uranium U (containing, respectively, 0.7 percent of $^{235}_{92}U$ or more) or $^{233}_{92}U$ or $^{239}_{94}Pu$ (plutonium). Neutrons propagating through the fuel are captured by the specifically fissile atomic nuclei $^{235}_{92}U$ (or $^{239}_{94}Pu$ or $^{233}_{92}U$), each of which suffers nuclear fission, breaks into two smaller nuclei and emits fast neutrons (having a few MeV).

The neutrons produced after a fission can give rise to new fission processes with other fissile atomic nuclei in the fuel and so on. In many reactors, the fast neutrons after a fission must first be slowed down (moderated) from MeV energies down to the thermal energy range, in order to be able to produce new fission processes effectively. The process of moderation of fast neutrons (to be discussed shortly in the next subsection) is performed by other material (the moderator), also located in the core of the reactor. In certain reactors (fast neutron reactors) the fast neutrons just produced after a fission and without need of moderation give rise directly to the subsequent nuclear fission processes. For the purposes of this chapter, the most interesting feature is that beams of slow or thermal neutrons (after having suffered moderation) can be extracted from the core of the reactor, be subject to adequate physical manipulations (collimation, monochromatization), commented in Subsection 9.2.5 and employed at a later stage for various purposes.

Small nuclear reactors (a few megawatts of power) can give rise to stationary neutron fluxes about 10^9 epithermal $neutrons/(cm^2.s)$.

Large nuclear reactors produce neutron beams with high constant flux. For instance, the nuclear reactor (with 57-megawatt power) at the Institut Laue Langevin (Grenoble, France) produced fluxes about 10^{15} $neutrons/cm^2.s$.

It is a fact that new programs and budgets for building nuclear reactors for nuclear plants (specifically, for producing energy at large scale) were strongly reduced or even eliminated in the last decades of the 20th century in many advanced countries. Those nuclear reactor power plants were perceived by society as sources of pollution and potential dangers (nuclear accidents), and thus their operations should have been closed down or not continued. In principle, nuclear reactors as sources of neutrons for various applications (therapies, condensed matter research) unrelated to energy production should not be confused with nuclear reactors for power plants that produce much larger amounts of nuclear residues. However, given the above unfavorable perception of the nuclear reactor power plants by society, the possibility that nuclear reactors (with power above certain thresholds) as neutron sources are not affected (e.g., their levels of acceptance and criticism by society) should not be entirely excluded. In such case, the future availability of the nuclear reactors as neutron sources (in particular, the larger ones) could be subject to increasing criticism or reduced. However, in recent times, other sources of energy production at large scale have been shown to either give rise to other important problems for society and/or to suffer from limited availability. It appears that society in advanced countries, unavoidably and progressively, accepts the necessity of allowing and employing several different sources of large scale energy production and, in particular, of not excluding nuclear power plants and allowing them to a certain extent. This change of perception has also been facilitated by considerable advances and improvements in recent decades in performance, security, and treatment of nuclear residues. Thus, we have recently witnessed new studies and the rebirth of new budgets and programs for building nuclear reactors for power plants in several countries (China, Finland, France, India, Iran, Japan, Russia, South Africa, and United States). Consequently, one may foresee and expect improved acceptance of nuclear reactors as sources of neutrons as well.

9.2.4 DETECTORS

Because the neutron is electrically neutral, its detection is not direct but it proceeds indirectly as follows.[2,9] Arrangements are made so that the neutron to be detected gives rise to some nuclear reaction. The detection of some suitably chosen final product (a charged particle or a photon, γ) in that nuclear reaction constitutes an indirect detection of the initial neutron. See Figure 9.1.

Three different detection procedures are based, respectively, upon the nuclear reactions $n +^{10}_{5} B$ (boron)$\rightarrow$ $\alpha +^{7}_{3} Li$ (in proportional counters containing boron trifluoride), $n +^{6}_{3} Li \rightarrow \alpha +^{3}_{1} H$ (in spark counters), and $n +^{3}_{2} He \rightarrow p +^{3}_{1} H$ (in proportional counters). $^{3}_{1}H$ denotes tritium. The subsequent detections of the outgoing charged particles (the α particle or the proton, p) in the counters provide the indirect detection of the neutron that triggered the corresponding nuclear reaction.

Another procedure with important applications is based on activation reactions induced by incoming beams of slow neutrons on suitable targets. Examples of such reactions are: $n +^{55}_{25} Mn$ (manganese)$\rightarrow$ $\gamma +^{56}_{25} Mn$, $n +^{197}_{79} Au$ (gold)$\rightarrow$ $\gamma +^{198}_{79} Au$, and $n +^{59}_{27} Co$ (cobalt)$\rightarrow$ $\gamma +^{60}_{27} Co$. The $^{55}_{25}Mn$, $^{197}_{79}Au$ and $^{59}_{27}Co$ in the targets are the isotopes found in nature with abundance about 100 percent in each case. Each of those reactions gives rise to isotopes ($^{56}_{25}Mn$, $^{198}_{79}Au$ and $^{60}_{27}Co$, in the above examples) that are radioactive, with lifetimes ranging from minutes to days. These isotopes decay, returning to their ground states (or to less excited ones), and, in so doing, they emit various radiations (electrons, and photons) that are detected. The activity $\tilde{A}$ (that is, the number of those products emitted by the radioactive isotopes in the targets and counted by suitable detectors, per unit time) is measured. A useful approximate formula for $\tilde{A}$ (in counts per second) is:[9]

$$\tilde{A} \simeq \frac{m\sigma F_0}{A}[1 - \exp(-\lambda t)] \tag{9.1}$$

F_0 is the flux of incoming slow neutrons (measured in $neutrons/(cm^2.s)$). See Subsection 9.6.4 for discussions and interpretations of neutron fluxes. m and A are the mass (in grams) and the mass number of the element (isotopes found in nature) subject to neutron bombardment, respectively, σ is the cross-section for slow neutron capture by that element, λ is the decay constant for the decay of the isotope formed by the neutron capture, and t is the duration of the neutron bombardment.

In a typical application of activation reactions to neutron detection, such reactions allow for measuring specifically the flux (F_0) of the incoming neutron beam, without introducing significant perturbations in the latter. Thus, a thin target of a suitable material is exposed to the neutron beam to be measured. After some reasonable time, the irradiated thin target is removed from the beam and the γ radioactivity of the corresponding radioactive isotopes induced by the captured neutrons is analyzed subsequently in suitable counters. Specifically, the activity $\tilde{A}$ is measured. The value of F_0 follows from Eq. (9.1), if the remaining quantities in it are known. The technique is applied typically to beams of thermal and epithermal neutrons. A wide variety of materials have been employed for the thin target, in particular, ${}^{55}_{25}Mn$, ${}^{197}_{79}Au$, ${}^{59}_{27}Co$, ${}^{113}_{48}Cd$ (cadmium), ...

These activation detections are related to the so-called neutron activation analysis to be discussed in Subsection 9.2.5 which has important applications.

9.2.5 Miscellaneous Properties

We summarize here some useful facts and items related to neutron physics.[2,7,9] ${}^{6}_{3}Li$ (with abundance 7.4 percent) captures and absorbs thermal neutrons with high probability. Thus, reasonable amounts (say, widths of about several mm) of ${}^{6}_{3}Li$ may be sufficient to provide protection from slow neutrons. On the other hand, natural lithium (say, the mixture of isotopes of this element naturally occurring and found) has an absorption probability for thermal neutrons about an order of magnitude smaller. Natural boron (B) and, in particular, cadmium (Cd) absorb thermal neutrons strongly.

Natural beryllium (Be) and carbon (C) have a very small probability to capture and absorb thermal neutrons. The same is true for deuterium (containing the deuteron nuclei ${}^{2}_{1}H$) and for natural bismuth (Bi).

Graphite (say, C) and Bi absorb photons with large probability and hence contribute to eliminating γ's produced when neutron beams propagate through matter and interact with atomic nuclei in the latter.

Neutron beams coming from nuclear reactors also typically include fast neutrons and hard photons. Achieving an adequate degree of suppression of both fast neutrons and hard photons in the beams is very desirable in general. The hard photon content can be reduced by employing adequate devices (filters) containing lead (Pb) or bismuth (Bi). On the other hand, Bi also reduces fast neutron contaminations. Pb and Bi are relatively transparent to thermal neutrons. Outside of areas in which neutron beams are propagating, high-density concrete (say, mixed with minerals containing iron) can be employed in order to reduce hard photons.

Moderators. The atomic nuclei of certain materials (with low atomic mass) have small probability to capture and absorb slow neutrons. Neutrons propagating in those materials lose energy and are slowed down (moderated) through successive collisions with their atomic nuclei, but they are not absorbed by the latter. In particular, hydrogen is a good moderator and so are suitable (non-absorbing) materials containing an adequate proportion of the former. Graphite, heavy water (containing deuterium), Be, and paraffin (due to its hydrogen content) are widely used moderators. Other possibilities include Al (aluminum), Al_2O_3, AlF_3, and Teflon^{TM}. Typically, moderators are devices located inside nuclear reactors.

Collimation. It is very important to dispose of neutron beams formed by neutrons with momenta having approximately similar directions (although still with rather different absolute values) or, at least, with velocities not diverging much (in a transverse plane, orthogonal to some average direction of propagation). The devices producing beams in which the directions of motion of all neutrons are more or less analogous (to within some solid angle) are named collimators. Holes in the shields of nuclear reactors and walls of steel coated with Cd (which absorbs neutrons) constitute approximate or partial collimators. Those located inside the shielding reflect neutrons back into the beam. Collimators placed near the beam exit are beam delimiters, so that they should absorb rather than reflect neutrons. The magnitudes of the velocities of all neutrons in those (partially) collimated beams typically sweep an interval (or spectrum): they are said to be non-monochromatic.

Monochromatization. It is very important that an already (partially) collimated beam be formed by neutrons also with similar velocities in absolute value (be almost monochromatic). Monochromators are devices that give rise to neutron beams with all momenta in a narrow interval (in direction and magnitude) out of incoming partially collimated non-monochromatic beams. Single crystals (for instance, calcium fluoride) constitute monochromators. Incoming partially collimated non-monochromatic neutron beams suffer Bragg reflections in suitably chosen crystal planes, with typical lattice spacing (distance between neighboring atoms) equal to d (a few $\mathring{A}$). We recall that the reflected neutrons with appreciable probability fulfill Bragg's law: $2d \sin\varphi = n\lambda_{dB}$, where $\varphi = 2^{-1}(\pi - \theta)$, θ being the angle between the momenta of the incoming neutron and the outgoing reflected one, λ_{dB} is de Broglie's wavelength, and $n = 1, 2, \ldots$ For given d and small values of n, Bragg's law selects values of θ and λ_{dB} and, so, outgoing sub-beams of neutrons that are better collimated and are approximately monochromatic.

Converters. Let us consider an initial beam, formed by neutrons having energies mostly concentrated in a certain interval. In a broad sense, a converter is a device that creates (or regenerates) out of that initial beam another neutron beam with energy spectrum concentrated in another range, for various specific purposes (say, to improve the detection efficiency of the resulting beam or to increase the probability that the latter beam gives rise to some subsequent nuclear reaction). See Figure 9.1. A typical (fission) converter is a row of fuel elements located in the beam line, but adequately away from the reactor core. We shall limit ourselves to illustrate that concept by describing succinctly a fission converter put into operation at the Massachusetts Institute of Technology some years ago.[11] This converter contains fissionable material (fuel), moderators, and filters (aluminum, teflon and cadmium in this case), a suitable shield (lead) in order to absorb photons, and a large collimator (with walls made of lead). In this converter, a (large area) thermal neutron beam coming from the nuclear reactor at the institution impinges upon the fissionable material. The resulting fission processes give rise to production of neutrons of higher energies which, after moderation and filtering, generate a new beam of epithermal neutrons (about 10^{10} $neutrons/(cm^2 \times s)$) with high purity. In particular, the latter epithermal neutron beam has been designed for advanced research on neutron capture therapy.

Neutron activation analysis. This is an important non-destructive technique for the determination of the composition of small samples of certain materials without producing appreciable damage in them. Specifically, it is based upon the capture of slow neutrons in the nuclear reaction $n + B \rightarrow C + \gamma$. B denotes a stable nucleus (contained in the small sample) which will become radioactive by virtue of the irradiation with slow neutrons, and C is the radioactive isotope. The slow neutrons belong to an incoming beam generated in, say, a nuclear reactor, and the incoming flux F_0 is supposed to be known with adequate accuracy (contrary to what happened in the activation detection considered in Subsection 9.2.4). After removing the irradiated sample from the neutron beam (e.g., extracting the sample from the reactor), the radioactivity of C is measured. Typically, the radioactive nucleus C emits first one β-decay electron and subsequently one or several photons until it gives rise to a stable nucleus. The specific energies of the photons so emitted characterize the decay. The precise measurement of the energies of those photons allows us to determine the radioactive isotope C and hence its parent B. Moreover, the measurement of the activity $\tilde{A}$ enables us to estimate, using Eq. (9.1), the amount m of the element B present in the small sample. In fact, $\tilde{A}$, the knowledge of F_0, t and of other quantities in Eq. (9.1) yield m. There are other interesting and useful applications (forensic science and archaeological research) of neutron activation analysis.[9]

Further possible applications. The number and variety of possible applications of thermal neutron beams have extended and grown through the years, and they continue to do so. Just as an example, we recall their possible use to detect buried mines and hidden explosives.[12] In general, identification of buried explosives can be based on the amounts of light elements (carbon, nitrogen, oxygen and hydrogen) they contain, for instance, the measurements of concentration ratios of carbon to nitrogen and carbon to oxygen are quite effective in discriminating hidden explosives from the medium in which they are located. In particular, an anomalous concentration of nitrogen appears to be characteristic of most explosives. Having this in mind, some ideas in the possible application of thermal neutrons to explosive identification are the following.

The detection device contains a small radioactive source (specifically, californium $^{252}_{98}Cf$) which produces fast neutrons, and a moderator which thermalizes them. A beam of thermalized neutrons is sent from the device onto the inspected area in the soil. After penetration into the latter, the neutrons are eventually captured by various elements in the investigated volume. The capture of the neutrons by those nuclei gives rise to emission of photons (gamma rays) that are characteristic of the nuclear transitions involved. Suitable detectors (large scintillation ones) above the inspected area allow detection of the characteristic gamma rays so emitted and, in particular, those arising from thermal neutron capture by nitrogen nuclei. This allows investigation of a possible anomalous concentration of chemicals containing nitrogen and thus the possible existence of buried explosives. This identification technique appears to be effective if a sample of about 800 grams of explosives is buried at a depth not larger than 15 *cm*. The latter provides a rough but essentially correct estimate of the penetration depth for thermal neutrons in matter. The overall size of a prototype[12] does not exceed tens of centimeters.

9.3 WAVE PROPAGATION OF SLOW NEUTRONS

Wave and optical phenomena for slow neutrons in three-dimensional space are based upon the following approximate Schrödinger equation ($\bar{x} = (x, y, z)$):[2,6]

$$\left[-\frac{\hbar^2}{2m_n}\Delta + V(\bar{x})\right]\Psi(\bar{x};t) = i\hbar\frac{\partial\Psi(\bar{x};t)}{\partial t} \tag{9.2}$$

$\Psi(\bar{x};t)$ is the time (t)-dependent coherent (two-component) wave function for the neutron, which describes how it propagates through either vacuum or some given material media. As usual, $\Delta = \partial^2/\partial x^2 + \partial^2/\partial y^2 + \partial^2/\partial z^2$.

$V(\bar{x})$ is the approximate (optical) potential on the neutron due to a given material medium, while $V(\bar{x}) \equiv 0$ in vacuum. The neutron is subject to the strong interaction due to the atomic nuclei of the material medium through which the former propagates. Coulomb forces due to either electrons or atomic nuclei do not act on neutrons, as the latter are electrically neutral. In magnetic materials, a neutron can be subject to magnetic interactions of its magnetic moment with those of atoms (due to their electrons). There have been detailed studies about coherence and incoherence phenomena in the propagation of matter waves associated to slow neutrons through material media.[13–20] In this chapter, we shall concentrate on non-magnetic materials and, except for a short discussion later in this section, we shall disregard those magnetic interactions due to atomic magnetic moments. We shall accept that $V(\bar{x})$ provides an effective description of neutron optics phenomena arising from strong interactions with atomic nuclei. We shall limit discussion to neutron propagation through a chemically pure material medium in which the nuclei are fixed in space (as a first approximation) at the positions $\bar{a}$. Chemical purity still allows for different isotopes of the same element to be present at various locations, as discussed below. Then, an approximate representation for $V(\bar{x})$ is:[2,6]

$$V(\bar{x}) = \frac{2\pi\hbar^2}{m_n}\sum_{\bar{a}}[b_{\bar{a}} + c_{\bar{a}}\bar{s}\bar{I}_{\bar{a}}]\delta^{(3)}(\bar{x} - \bar{a}) \tag{9.3}$$

$\delta^{(3)}(\bar{x} - \bar{a})$ is the three-dimensional delta function and the summation over $\bar{a}$ is carried out over all nuclei in the material medium. $\bar{s}$ is the spin operator for the neutron: the j-th component of $\bar{s}$ equals $(\hbar/2)\sigma_j$, σ_j being the standard j-th Pauli matrix,[2] $j = 1, 2, 3$. $\bar{I}_{\bar{a}}$ and $I_{\bar{a}}$ are the spin operator and the spin value, respectively, for the atomic nucleus fixed at $\bar{a}$. $b_{\bar{a}}$ and $c_{\bar{a}}$ are suitable parameters describing the low energy (purely nuclear) scattering of a neutron by an atomic nucleus at the position $\bar{a}$, due to strong interactions. We allow for the possibility that different isotopes of the same stable nucleus are located at different positions $\bar{a}$. If the atomic nuclei located at different $\bar{a}$ are identical (say, they are not different isotopes) and if they have non-zero spin $I_{\bar{a}} = I \neq 0$, then there are incoherence effects in the propagation and scattering of neutrons through the material medium (spin incoherence). Moreover, the possible occurrence of different isotopes at different positions $\bar{a}$ in the material medium may give rise to further incoherence effects (isotope incoherence).

We can approximate Eq. (9.3) by the following:[2,6–8]

$$V(\bar{x}) = \frac{2\pi\hbar^2\rho b}{m_n} \tag{9.4}$$

Eq. (9.4) can be regarded as some sort of spin and spatial average of Eq. (9.3) and corresponds to homogeneous medium. ρ is the number of nuclei per unit volume (say, per cm^3) in the homogeneous material medium. For the material media to be considered here, ρ is about 10^{22} nuclei per cm^3. b now stands for the average (over nuclear spins, when $I_{\bar{a}} \neq 0$, and over isotopes) coherent amplitude for the low energy purely nuclear scattering of a neutron by the nucleus of an atom in the medium. Specifically, let us consider a material medium formed by a mixture of L nuclear isotopes (each with its own b_l and c_l at the position $\bar{a}$) with fractions f_l ($f_l > 0$, $\sum_{l=1}^{L} f_l = 1$). In particular, at each $\bar{a}$ we replace $b_{\bar{a}}$ by b_l and the average coherent amplitude b turns out to be $\sum_{l=1}^{L} f_l b_l$.[2,6,7]

Throughout this chapter, we shall work with Eq. (9.4) (and, hence, with b), rather than with Eq. (9.3): such an approximation will suffice for our purposes and it still allows for the possibility that the neutron could be absorbed by atomic nuclei. If that absorption can be neglected, then b is real ($b = Re(b)$, Re denoting the real part). However, if nuclear absorption is not negligible, then b is complex: $b = Re(b) + iIm(b)$, with imaginary part $Im(b) < 0$. b has been experimentally measured for many naturally occurring elements in the Periodic Table and for many isotopes as well. Various compilations of data for b exist.[2,7,21] The order of magnitude of the real part $Re(b)$ is about $10^{-13}cm$.[7] The sign of the (average coherent) amplitude $Re(b)$ is positive for almost all nuclei, except for a few: $Re(b) < 0$ for the naturally occurring H, Li, Ti, V, Mn. On the other hand, $Re(b) < 0$ holds individually for several isotopes of certain nuclei (say, without averaging over isotopes), for instance, for ${}^{62}_{28}Ni$ and ${}^{64}_{28}Ni$. However, upon averaging over isotopes, one has $Re(b) > 0$ for natural Ni.[2,7] One also has $Re(b) < 0$ for ${}^{152}_{62}Sm$ (samarium). In other cases, $Re(b)$ has opposite signs for different isotopes, but the contribution of one isotope having $Re(b) < 0$ dominates in the (isotope averaging yielding the) naturally occurring element. This is the case for Li and Ti, where both ${}^{7}_{3}Li$ (abundance 92.6 percent) and ${}^{48}_{22}Ti$ (abundance 73.8 percent) have $Re(b) < 0$. On the other hand, the isotope ${}^{55}_{25}Mn$ has an abundance about 100 percent, so that it coincides, in practice, with natural Mn.

The order of magnitude of the right-hand side of Eq. (9.4) for typical material media of physical interest is about 10^{-8} to 10^{-7} eV.

The approximate Eq. (9.4) omits any sort of spin-dependent (or magnetic) interaction of the neutron. Upon employing Eqs. (9.4) and (9.2), both components of $\Psi(\bar{x}; t)$ (corresponding to the two possible and opposite projections of the neutron spin on the z-axis) have similar behavior. In such a case, we shall safely restrict our study to just one component, that is, $\Psi(\bar{x}; t)$ will be regarded as a one-component wave function (thereby omitting the neutron spin).

We shall now turn briefly to the influence of gravity on slow neutrons. The difference of potential energies of a neutron, subject to gravity, between two different positions with heights z_1 and $z_0(< z_1)$, vertically on the surface of the Earth, is $V_{grav,0} = m_n.g.(z_1 - z_0)$ ($g = 9.8\ m.s^{-2}$, m denoting one meter). For $z_1 - z_0$ about one meter, $V_{grav,0}$ is about $10^{-7}\ eV$, which is comparable to typical values for Eq. (9.4). We shall regard z_0 and z_1 as reference heights. If the slow neutron is at the arbitrary height z, the difference of potential energies between z and z_0, also subject to gravity, is $V_{grav}(z) = V_{grav,0}[1 + (z - z_1)/(z_1 - z_0)]$. For $\mid z - z_1 \mid$ not exceeding $1\ cm$ which, in practice, will be the case for the (essentially or almost) horizontal motions of neutrons, to be considered here, $V_{grav}(z)$ will be approximately equal to $V_{grav,0}$. In turn, a constant $V_{grav,0}$ cancels out in physically interesting quantities. Then, for a slow neutron moving on the surface of the Earth with differences of height not exceeding $1\ cm$, and as a zeroth-order approximation, the effects of gravity on it will be neglected and so shall we proceed in our study. If more accuracy in the treatment of gravity effects were required, one could start with Eq. (9.2), $V(\bar{x})$ being replaced by $V(\bar{x}) + V_{grav}(z)$ and $V(\bar{x})$ being now given through either (9.3) or (9.4). Then, for differences of height not exceeding $1\ cm$, one could regard $V_{grav}(z)$ as a small perturbation of $V(\bar{x})$.

The scattering and the absorption of a slow neutron by an individual atomic nucleus in the medium are relevant processes. We need to consider σ_s and σ_{abs}, which are the cross-sections (integrated over all angles) for those processes, respectively. In turn, one has $\sigma_s \simeq \sigma_{coh} + \sigma_{inc}$, where σ_{coh} and σ_{inc} stand for the coherent and incoherent scattering cross-sections, respectively, and the effects of diffuse scattering[6] have been disregarded in a zeroth-order approximation. One has $\sigma_{coh} = 4\pi \mid b \mid^2$, b being the average coherent amplitude discussed above. Recall that, at different positions $\bar{a}$, there may be l nuclear isotopes, each with its own spin I_l and parameters b_l and c_l, with fractions f_l, fulfilling $f_l > 0$, $\sum_l f_l = 1$. One has:

$$\sigma_s = 4\pi \sum_{l=1}^{L} f_l \left[\frac{I_l + 1}{2I_l + 1} \mid b_l + 2^{-1}c_l I_l \mid^2 + \frac{I_l}{2I_l + 1} \mid b_l - 2^{-1}c_l(I_l + 1) \mid^2 \right] \quad (9.5)$$

where suitable averages over nuclear spins and isotopes are performed. σ_{inc} describes both spin and isotope incoherence. The total cross-section is $\sigma_{tot} = \sigma_s + \sigma_{abs}$. Table 9.1 summarizes some useful data[2,7] for $b, \sigma_s, \sigma_{abs}$ for various isotopes (omitting, for simplicity, their atomic numbers and certain data in some cases) and natural elements. In many cases of physical interest, σ_{coh} is larger than σ_{inc}, but not always (as happens for $H, Li, Na, V, \ldots$),[2,6] The order of magnitude of both σ_s and σ_{abs} is about $10^{-24}\ cm^2$ (= 1 barn), with appreciable variations.[2,7]

For atomic number Z below 45, σ_s is usually (but not always) larger than σ_{abs}, while for $Z \geq 45$ the typical situation seems to be the opposite one.[7]

The linear coefficient μ for slow neutrons in the medium is defined as $\mu \equiv \rho\sigma_{tot}$.[7] μ characterizes the attenuation of a neutron beam propagating in a medium, due to the interaction of any neutron with the atomic nuclei. Let us consider a slab of thickness z_0, made up of certain homogeneous material with linear coefficient μ, and some neutron beam which has approached that slab from outside (along a direction orthogonal to

TABLE 9.1
Useful Data

Nucleus	$b(10^{-12}\text{cm})$	$\sigma_s(10^{-24}\text{cm}^2)$	$\sigma_{abs}(10^{-24}\text{cm}^2)$
${}^{1}_{1}H$	−0.378	81.5	0.19
${}^{2}_{1}H$	0.65	7.6	0.0005
${}^{7}_{3}Li$	−0.22	1.4	—
${}^{6}_{3}Li$	0.2	—	570
Li	−0.18	1.2	40
${}^{9}_{4}Be$	0.774	7.54	0.005
B	—	4.4	430
${}^{12}_{6}C$	0.661	5.51	0.003
${}^{14}_{7}N$	0.940	11.4	1.1
${}^{16}_{8}O$	0.577	4.24	0.0001
${}^{23}_{11}Na$	0.351	3.4	0.28
${}^{27}_{13}Al$	0.35	1.5	0.13
Si	0.42	2.2	0.06
K	0.35	2.2	1.2
${}^{48}_{22}Ti$	−0.58	—	—
Ti	−0.34	4.4	3.5
V	−0.05	5.1	2.8
Mn	−0.37	2.0	7.6
Fe	0.96	11.8	1.4
${}^{62}_{28}Ni$	−0.87	—	—
Ni	1.03	18.0	2.7
Cd	0.36	—	2650
In	0.38 + i0.12	—	115
Sm	—	—	11700
Gd	—	—	19200
Pb	0.96	11.4	0.1
${}^{238}_{92}U$	0.85	—	2.1

one of its surfaces) and entered into it through the surface at $z = 0$. Let $F_0(z)$ be the flux corresponding to that neutron beam inside the slab at a distance z from that entrance surface ($0 \leq z \leq z_0$); z does not represent vertical height here. Discussions and interpretations of neutron fluxes will be deferred to Subsection 9.6.4. Standard arguments yield $F_0(z) \simeq F_0(z = 0)\exp[-z\mu]$,[6,7] which displays the physical interest of μ. Typical values for μ in many cases are about 10^{-1} cm. μ is about 5×10^{-4} cm for Be, C and Bi and air, but about 26, 60 and 121 cm for Li, B and Cd, respectively.

Through various analyses in some of the following sections, we shall realize the convenience and the necessity of the wave (quantum mechanical) description for slow neutrons, specifically as one investigates the possibility of employing them to explore short distances.

9.4 TOTAL INTERNAL REFLECTION AND HOLLOW NEUTRON GUIDES

9.4.1 REFRACTIVE INDEX AND TOTAL INTERNAL REFLECTION

In the geometrical optics approximation, the index of refraction (or refractive index), n, for slow neutrons with total energy E moving through a medium (and hence under the action of the potential $V(\bar{x})$ due to the latter) is defined as:[2,8]

$$n^2 = 1 - \frac{V(\bar{x})}{E} \tag{9.6}$$

In turn, by using Eq. (9.4):

$$n^2 = 1 - \frac{2\pi b\rho\hbar^2}{m_n E} \tag{9.7}$$

If a slow neutron moves in vacuum with wave vector $\bar{k}$, then $V(\bar{x}) \equiv 0$, $n = 1$ and $E = E_K = \hbar^2\bar{k}^2/(2m_n)$. Let such a slow neutron, having moved initially in vacuum, approach a material medium with $V(\bar{x}) \neq 0$ and real $b > 0$ (that is, nuclear absorption being neglected). Consequently, the refractive index $n(< 1)$ is given in Eq. (9.7). Then, from the standpoint of neutron optics (and upon recalling the analogy with ordinary optics of light, where the refractive index is usually larger than 1) such a medium could be regarded as less dense than vacuum. Recall that, for most materials, one has $Re(b) > 0$ and, if $Im(b) \simeq 0$ also holds, it follows that $n < 1$. Let φ be the angle between $\bar{k}$ and the normal to the limiting surface separating vacuum from the medium. Then, according to standard geometrical optics-like arguments, the neutron suffers total reflection back into vacuum and it does not penetrate into the medium if $\varphi \geq \varphi_{cr}$, where the critical angle φ_{cr} fulfills $\sin\varphi_{cr} = n(< 1)$. See Figure 9.2. More precisely (going beyond the geometrical optics approximation and employing the solution of the Schrödinger equation), in such a case, the probability

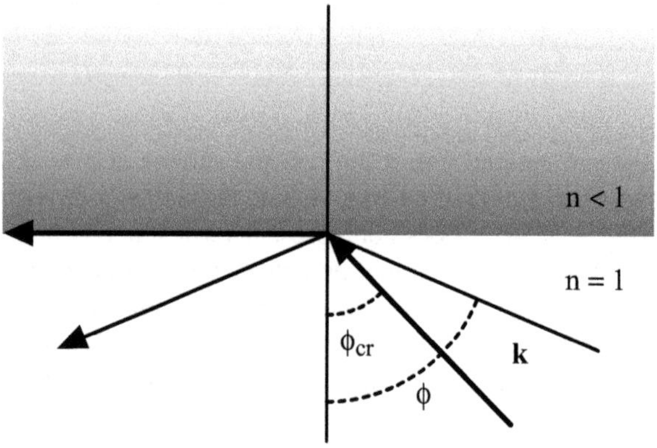

FIGURE 9.2 Vacuum corresponds to $n = 1$, while $n < 1$ in the medium. The incoming neutron propagates in vacuum with wavevector $\bar{k}$. φ is the angle between $\bar{k}$ and the normal to the medium at the entrance surface. If $\varphi \geq \varphi_{cr}$, total internal reflection occurs.

for the neutron to penetrate a distance d into the medium decreases exponentially with d. Let us introduce the glancing angle $\theta = \pi/2 - \varphi$, and the critical glancing angle $\theta_{cr} = \pi/2 - \varphi_{cr}$. Then, total reflection occurs for $\theta \leq \theta_{cr}$. with $b > 0$. For typical values of $b(> 0)$ and ρ, n and φ_{cr} are a bit smaller than $+1$ and $\pi/2$, respectively. Then, one has $\theta_{cr} \simeq \lambda_{dB}[b\rho/\pi]^{1/2}$. See Section 9.5 for a related discussion.

9.4.2 Hollow Neutron Guides with Large Cross-Sections

The geometrical optics-like phenomenon of total reflection has provided the basis for the hollow neutron guides (with suitably large transverse dimension about several cm), enabling the channeling of a slow neutron beam along relatively long distances:[22–26] the neutrons propagate through the inner empty space of the guide and, if $\theta \leq \theta_{cr}$ holds, they suffer multiple total reflections on the walls made by some suitable material with $b > 0$.

In practice, one deals with neutron beams that are not monochromatic but have a spectrum of de Broglie wavelengths. Typically, the walls of hollow neutron guides are made up of nickel-coated boron glass. Their cross-sections may be rectangular, having sides about 10 cm, and their lengths may be about 80 m. They are not straight, but have some curvature radius R_{cu}. Those designed to transport thermal neutron beams have R_{cu} about one to several hundred meters[2]. Since typical transverse dimensions of these guides are much larger (by several orders of magnitude) than the de Broglie wavelengths for thermal neutrons, the latter have been described through geometrical optics approaches.

One important feature (and application) of such hollow neutron guides that makes them technologically interesting is that they can be connected directly to nuclear reactors. The guides are employed to extract slow neutron beams from the latter and to channel and transport such beams along relatively long distances for various purposes (investigation of condensed matter and biological matter samples).[2,8] See Figure 9.3.

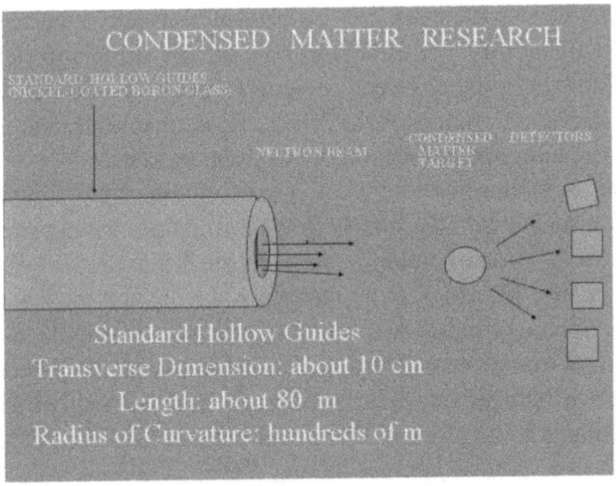

FIGURE 9.3 Application of hollow neutron guides for the investigation of condensed matter.

Microguides for thermal neutrons have been realized experimentally (using a suitable evaporation technique).[27] In that work, one microguide consisted of a sandwich of alternate Ni and Al layers: it contained 100 double layers, with total thickness 0.037 mm and radius of curvature 15 cm. Thus, the distance between walls in the microguide was about 320 nm. The microguides were also employed to study the deflection of neutrons.[27]

There are further and interesting developments on neutron guides (using devices called super-mirrors).[28,29]

9.5 NEUTRON FIBERS: ANALOGIES AND CONJECTURES

Light can be transmitted along optical fibers, namely thin solid (glass-like or plastic) dielectric waveguides.[30–32] Interesting physics follows from the fact that the electromagnetic wave aspects of light have to be taken into account (a geometrical optics description does not suffice, in general). Light is transmitted along the fiber only in the form of some specific distributions of the electromagnetic field (solving Maxwell's equations), named propagation modes. In order that the waveguiding effect be manifest (which is the physically interesting feature for us), the transverse dimensions of the fiber should be about or larger than a few average wavelengths of the transmitted light. On the other hand, such transverse dimensions should not be exceedingly larger (by several orders of magnitude) than the average wavelength because wave aspects would become masked and the propagation of light would be described by geometrical optics (with increasing accuracy as the transverse dimensions increase). A very important waveguiding phenomenon is the propagation of ordinary light along the retinal photoreceptors in the eyes.[33,34] Each photoreceptor works, regarding light transmission, as a (biological) waveguide, the transverse dimensions of which are only a few wavelengths of light.

Based upon an analogy with light propagation along optical fibers, the possible confined propagation of slow neutrons along waveguides of small transverse cross-sections (neutron fibers) and some speculations about possible applications (to neutron radiotherapy) were discussed earlier.[35,36] Before entering into a discussion of those proposals, we notice that for a few isotopes and naturally occurring elements (recall Section 9.3 and Table 9.1), the coherent low-energy neutron–nucleus scattering amplitude b fulfills $Re(b) < 0$, that is, $Re(n^2) > 1$. Those materials can be regarded as more dense than vacuum, from a neutron optics point of view.

One proposal[35] was the possible existence of confined propagation modes of thermal neutrons in fully solid (say, non-hollow) waveguides having small transverse cross-sections (fibers), made of suitable materials such that the real part of the corresponding b is negative ($Re(b) < 0$, so that $Re(n^2) > 1$) and nuclear absorption is adequately small. Such a proposal was based on the wave phenomena associated to thermal neutrons and on various properties of suitable (propagation mode) solutions of the associated Schrödinger Eq. (9.2). Among the various nuclei with $Re(b) < 0$ mentioned in Section 9.3, the proposal[35] concentrated on Ti (73.8 percent of natural Ti is ${}^{48}_{22}Ti$, which is a spinless nucleus having $Re(b) = -5.84 \times 10^{-13}$ cm), with some additional discussion of ${}^{62}_{28}Ni$ (which is also spinless, but with an abundance of 3.7 percent only in natural Ni). In their simplest version, those neutron fibers could

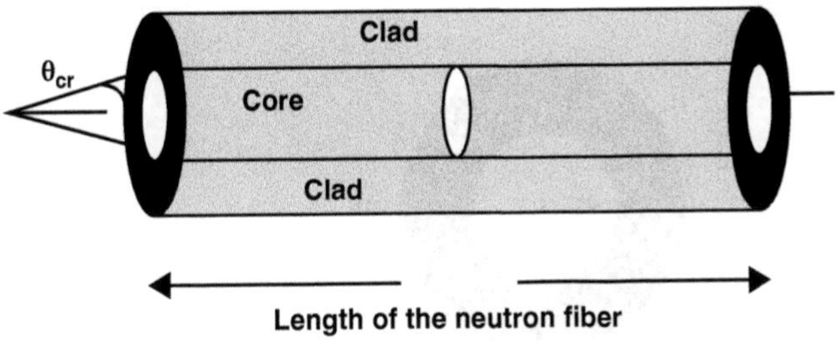

FIGURE 9.4 A possible neutron fiber is displayed. The core could be Al, Si or air and the clad could be made of Ni or Fe. The length of the fiber could be about 10 cm and the radius R of the core could be $\ll$ 100 μm. The acceptance angle θ_{cr} (for a thermal neutron beam to enter into the fiber and propagate confined along it) is about 10^{-2} and 10^{-3} radians.

be unclad and surrounded by air, due to the fact that $Re(n^2) > 1$ (and that $\mid \rho b \mid$ for air is about three orders of magnitude smaller than for the core). By employing the approximate formula $F_0(z) \simeq F_0(z = 0)\exp[-z\mu]$ (Section 9.3), the thermal flux $F_0(z)$ transmitted in the case of a Ti neutron fiber with length about $z = 5$ cm was estimated[36] to be about 10 percent of $F_0(z = 0)$ (while that for a $^{62}_{28}Ni$ one can be estimated to be smaller). Upon formulating such proposals, an attempt was made to exploit the possible analogies between those neutron fibers and the known optical ones, commonly employed to transport light. We also entertained,[35] among others, another possibility: a cladded neutron fiber, namely a cylindrical solid waveguide (core) of an element like those mentioned above, surrounded by an outer coaxial cylinder (the clad) of finite thickness which, in turn, would be surrounded by air. See Figure 9.4. The element making up the clad would be required to have positive nuclear scattering amplitude ($Re(b) > 0$) and small nuclear absorption. The possible confined propagation of slow neutrons along magnetized waveguides (e.g., those containing manganese) was also discussed briefly.[35] Spin (and birefringence) effects for neutrons would play a role. Fibers consisting of the compounds Mn_2N (ferrimagnetic), Mn_2Sb (ferrimagnetic), and MnS (antiferromagnetic) were considered.[37] In particular, in a Mn_2N fiber, magnetized along the fiber (z-) axis, thermal neutrons would propagate confined only if their spins were anti-parallel to the net magnetization.

Subsequently,[36] some improving alternatives for cladded neutron fibers were analyzed. The aim was to allow confined neutron propagation along reasonable lengths longer (with smaller scattering and absorption losses) than those for a Ti fiber. In those improved cladded fibers (see also Figure 9.4), the core (subscript co) could be either Al or Si or air, while the clad (of course, also of finite thickness) could be either Ni or Fe. Now, $\rho_{co}Re(b_{co}) \geq 0$ and $\rho_{cl}Re(b_{cl}) > 0$, with $\rho_{co}Re(b_{co}) < \rho_{cl}Re(b_{cl})$. Again, the actual cladded fiber is surrounded by air, for which $\rho Re(b)$ is also positive and much smaller. The refractive index profile for one of these cladded neutron fibers is displayed in Figure 9.5. Recall that the linear coefficient for air is about 5×10^{-4} cm.

Even if those proposals rely and focus on waveguide features of slow neutron propagation, let us carry out a few estimates based upon geometrical optics. Let us consider one single straight hollow cladded neutron fiber having a suitable clad (see Figure 9.4) and small overall cross-section. For convenience, the axis of the fiber is

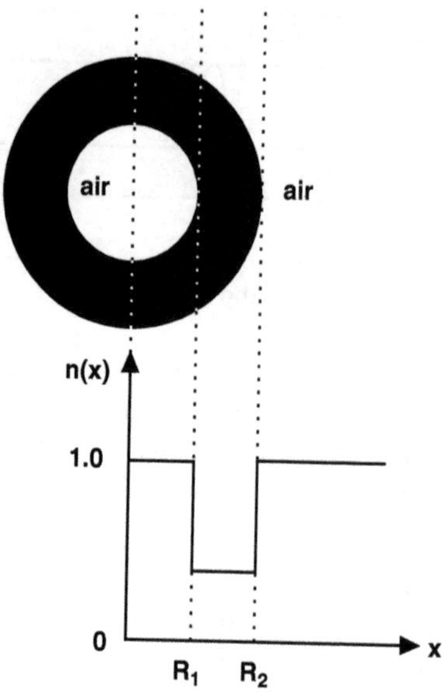

FIGURE 9.5 Transverse cross-section of a cladded neutron fiber surrounded by air. The core is also air ($n \simeq 1$), and the clad has $b_{cl} > 0$ and, then, $n_{cl} < 1$ (with very small nuclear absorption, regarded as negligible). The profile of the refractive index is also displayed, as $x = r$ varies along some direction.

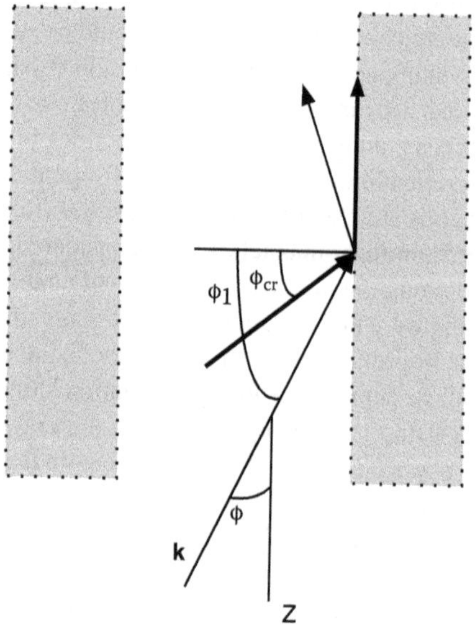

FIGURE 9.6 The incoming neutron propagates outside the entrance surface of the fiber with wavevector $\bar{k}$. φ is the angle between $\bar{k}$ and the z-axis (orthogonal to the entrance surface). $\bar{k}$ forms an angle φ_1 with the normal to the internal surface of the clad. One has $\varphi + \varphi_1 = \pi/2$. For $\varphi_1 \geq \varphi_{cr}$, total internal reflection occurs (compare with Figures 9.2 and 9.4).

parallel to the z-axis. Let a neutron moving in vacuum (air) with wave-vector $\bar{k}$ outside the entrance end of the fiber enter and propagate confined along the empty core (with $n \simeq 1$) of the latter, by successive total internal reflections in the inner surface of the clad (without penetrating into it). Let φ be the angle between $\bar{k}$ and the z-axis. Then, as in Section 9.4 and as shown in Figure 9.6, the neutron propagating along the empty core suffers total internal reflection from the clad, if $\varphi \leq \theta_{cr} = 2^{-1}\pi - \varphi_{cr}$, where φ_{cr} satisfies $\sin\varphi_{cr} = n_{cl}$, The index of refraction of the clad, n, is given in Eq. (9.7) (with $\rho b = \rho_{cl} b_{cl} > 0$), and nuclear absorption in the former is assumed to be negligible. Consequently, θ_{cr} is also the maximum angle for neutrons to enter into the fiber and to propagate confined along the empty core (acceptance angle). θ_{cr} is small and the approximation for it given in Subsection 9.4.1 can be recast as:

$$\theta_{cr} \simeq \frac{\hbar[4\pi\rho_{cl}b_{cl}]^{1/2}}{[2m_n E]^{1/2}} \tag{9.8}$$

with $E = E_K = \hbar^2\bar{k}^2/(2m_n)$. The numerical values of θ_{cr} lie typically between 10^{-2} and 10^{-3} radians.

9.6 NEUTRON FIBERS: QUANTUM MECHANICAL ANALYSIS

9.6.1 PROPAGATION MODES

There are some interesting differences between the propagation of ordinary light[30–32] and that of slow neutrons: (i) the interaction giving rise to the phenomena (electromagnetic for light, versus strong nuclear force for neutrons) and (ii) the ranges of wavelengths (about some thousands of angstroms for light, versus an interval about $1 Å$ for thermal neutrons, or even shorter for epithermal ones). Let us compare the treatment of the propagation modes for slow neutrons in neutron fibers, to which we now turn, with that of those for optical fibers, specifically, regarding the role of the possible polarization states. The vector character of the electromagnetic field in the propagation modes along optical waveguides is usually taken into account.[32] On the other hand, and as commented in Section 9.3, the use of Eqs. (9.2) and (9.4) (in the absence of interactions of the spin of the neutron) amounts to omitting the neutron spin. This simplification may suffice as a zeroth-order approximation, for slow neutrons (provided that magnetized fibers are excluded).

Let us consider, in three-dimensional space, a very lengthy and straight neutron fiber with zero curvature and finite transverse cross-section T. The latter lies in the (x, y)-plane, so that the z-axis is parallel to any axis of the fiber. T may have an arbitrary shape. See Figure 9.7. Three- and two-dimensional vectors will be denoted by over-bars and boldface symbols, respectively, so that $\bar{x} = (x, y, z) = (\mathbf{x}, z)$. The confined propagation of slow neutrons along the fiber can be modelled through Eq. (9.4), where $V(\bar{x}) = V(\mathbf{x})$ fulfills: $V(\mathbf{x}) = V_{co} = 2\pi\hbar^2 b_{co}\rho_{co}/m_n$ for $\mathbf{x}$ inside the inner part of T (the core, represented by the subscript co), while $V(\mathbf{x}) = V_{cl} = 2\pi\hbar^2 b_{cl}\rho_{cl}/m_n$ for $\mathbf{x}$ outside T (the infinite outer medium or clad, with subscript cl). Here, b_{co} and b_{cl} stand for the average coherent amplitudes for the low-energy scattering of a neutron by an atomic nucleus belonging to the core and the clad, respectively. The possible absorptions of slow neutrons by both the core and the clad

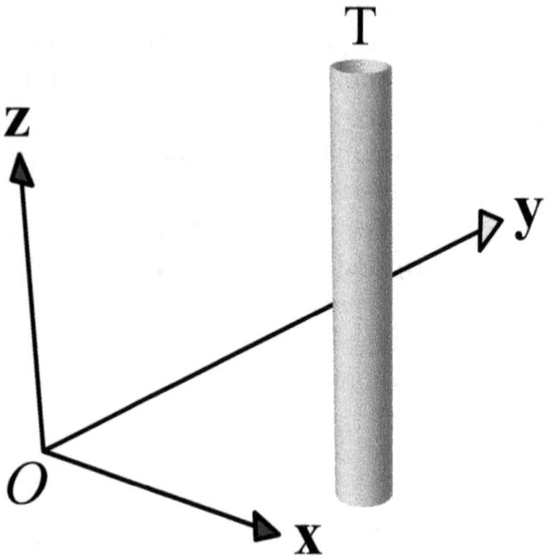

FIGURE 9.7 Unbent three-dimensional fiber having (small) transverse cross-section T.

are taken as negligible, so that both b_{cl} and b_{co} are assumed to be real. The inclusion of nuclear absorption has also been taken into account.[35,36] The numbers of nuclei per cubic centimeter in the core and in the clad are ρ_{co} and ρ_{cl}, respectively. It is necessary that $V_{co} < V_{cl}$, in order that confined neutron propagation along the fiber occurs.

The neutron propagation modes, with total energy E and propagation constant β_α (in an interval about one $\AA^{-1}$ or about 10 nm^{-1}), are the physically relevant solutions of the Schrödinger equation. They bear the following structure:

$$\Psi = \phi(\boldsymbol{x})_\alpha . \exp i\beta_\alpha z . \exp(-iEt/\hbar) \tag{9.9}$$

with $\int d^2\boldsymbol{x} \mid \phi(\boldsymbol{x})_\alpha \mid^2 = 1$ (normalization). α denotes a set of additional quantum numbers, also required in order to specify uniquely, together with either E or β_α, the solution of the Schrödinger equation. As for optical waveguides, a neutron fiber is named monomode if there is only one possible choice of values for α. Otherwise, the neutron fiber is called multimode.

As both b_{cl} and b_{co} are real, then so is β_α, and we take $\beta_\alpha > 0$. As the kinetic energy is positive, one has $E > V_{cl}$. The time-dependent Schrödinger Eq. (9.2) and (9.9) yield:

$$[-\frac{\hbar^2}{2m_n}\Delta_T + V(\boldsymbol{x}) - V_{cl}]\phi(\boldsymbol{x})_\alpha = -\frac{\hbar^2}{2m_n}\chi_\alpha^2\phi(\boldsymbol{x})_\alpha \tag{9.10}$$

$$E = V_{cl} + \frac{\hbar^2}{2m_n}\beta_\alpha^2 - \frac{\hbar^2}{2m_n}\chi_\alpha^2 \tag{9.11}$$

where $\Delta_T = \partial^2/\partial x^2 + \partial^2/\partial y^2$. Moreover, since both b_{cl} and b_{co} are real, χ_α is also real and fulfills $\chi_\alpha > 0$. In particular, these properties hold for a hollow fiber with a clad in which absorption of neutrons is negligible. Eq. (9.10) has and fully determines a finite set of real negative eigenvalues $-\chi_\alpha^2$. Then, for given E and for the finite set of allowed values of χ_α^2, Eq. (9.11) determines a finite set of values of β_α only. Notice that $V(\boldsymbol{x}) - V_{cl}$ vanishes or equals a negative constant, depending on whether $\boldsymbol{x}$ lies outside or inside T, respectively. Both ϕ_α and κ_α can be obtained by solving the time-independent Schrödinger Eq. (9.10) with the following standard boundary conditions: both ϕ and its normal derivative have to be continuous across all walls, and $\phi(\boldsymbol{x})$ has to vanish exponentially as $\boldsymbol{x} \to +\infty$. Moreover, ϕ_α has to be finite for any $\boldsymbol{x}$. As the Schrodinger Eq. (9.10) refers to two spatial dimensions and involves an attractive potential,[35,36] the fundamental mode (or ground state) always exists, but there may or may not be further propagation modes (see later). In particular, $\kappa_\alpha/\beta_\alpha \leq 10^{-2}$ holds for thermal neutrons in typical cases. The following bound holds:[38]

$$\chi_\alpha \leq [4\pi(b_{cl}\rho_{cl} - b_{co}\rho_{co})]^{1/2} \equiv \chi_{max} \tag{9.12}$$

The largest value of $\chi_\alpha(> 0)$ allowed by Eq. (9.10) and compatible with the bound in Eq. (9.12) corresponds precisely to the fundamental mode. The values of $\chi_\alpha(> 0)$ associated to higher modes decrease in magnitude.

9.6.2 Number of Propagation Modes: Quasi-Classical Estimate

For an unbent fiber, the number of allowed propagation modes, $N_{pm}(\chi_0^2)$ in the range $0 \leq \chi_0^2 \leq \chi_\alpha^2 \leq \chi_{max}^2$ is the number of independent solutions $\phi(\boldsymbol{x})_\alpha$ of Eq. (9.10) as α varies, having χ_α^2 in that range (excluding higher modes with $0 \leq \chi_\alpha^2 \leq \chi_0^2$). Provided that $N_{pm}(\chi_0^2)$ is appreciably larger than unity (the neutron fiber being multimode) the former can be estimated easily in quasi-classical approximation. In turn, the previous condition will be fulfilled provided that transverse dimension of the fiber (e.g., the square root of the area $A(T)$ of T) is much larger than β_α^{-1}. General studies of quasi-classical approximations in quantum mechanics[39–42] provide the foundations for the approximate formula for $N_{pm}(\chi_0^2)$ to be employed. It reads:[36,38]

$$N_{pm}(\chi_0^2) = \frac{A(T)}{2\pi}[4\pi(b_{cl}\rho_{cl} - b_{co}\rho_{co}) - \chi_0^2] \tag{9.13}$$

The total number of propagation modes having χ_α^2 in the range $0 \leq \chi_\alpha^2 \leq \chi_{max}^2$ (with $\chi_0^2 = 0$) is also in quasi-classical approximation:

$$N_{pm} = N_{pm}(\chi_0^2 = 0) = 2A(T)(b_{cl}\rho_{cl} - b_{co}\rho_{co}) \tag{9.14}$$

We anticipate that the quantum mechanical description will be necessary for slow neutrons propagating confined in the short length scales corresponding to the hollow interior of a carbon nanotube. In the latter case, the quasi-classical approximations in Eqs. (9.13) and (9.14) will be unreliable in principle.

9.6.3 Fiber with Circular Cross-Section

As a first example,[43] we shall consider a very long and straight absorptionless fiber, its cross-section T being a circle of radius R in the (x, y) plane. The origin $\bar{x} = (0, 0, 0)$

is chosen to be the center of one of those cross-sections. Let the standard polar coordinates r and φ ($x = r\cos\varphi$, $y = r\sin\varphi$) be introduced. The clad and the core correspond to $r > R$ and $r > R$, respectively. In cylindrical coordinates (r, φ, z), the solution of Eq. (9.10), which is finite at $r = 0$ and continuous at $r = R$, is:

$$\phi(\boldsymbol{x})_\alpha = \exp iM\varphi . J_{|M|}(K_1 r), r < R \tag{9.15}$$

$$\phi(\boldsymbol{x})_\alpha = \frac{J_{|M|}(K_1 R)}{H^{(1)}_{|M|}(i\chi_\alpha R)} . \exp iM\varphi . H^{(1)}_{|M|}(i\chi_\alpha r), r > R \tag{9.16}$$

$$K_1^2 + \chi_\alpha^2 = 4\pi[b_{cl}\rho_{cl} - b_{co}\rho_{co}] \tag{9.17}$$

with $M = 0, 1, 2, \ldots$ Here, α denotes M and the K's, collectively. $H^{(1)}_{|M|}$ and $J_{|M|}$ are the standard Hankel and Bessel functions.[44] One still has to impose one additional boundary condition, namely that $\partial\phi(\boldsymbol{x})_\alpha/\partial r$ be continuous, at $r = R$ for any φ. It reads:

$$[J_{|M|}(K_1 R]^{-1}[\frac{dJ_{|M|}(K_1 r)}{dr}]_{r=R} = [H^1_{|M|}(i\chi_\alpha R)]^{-1}[\frac{dH^1_{|M|}(i\chi_\alpha r)}{dr}]_{r=R} \tag{9.18}$$

The allowed values of χ_α (for given M) turn out to be the solutions of Eq. (9.18). The allowed values of K_1 follow from Eq. (9.17). The ground state of Eq. (9.10) corresponds to its lowest eigenvalue, that is, to the largest χ_α (which is close to but slightly smaller than χ_{max}); the corresponding K_1 is very small. That ground state is independent of φ (as $M = 0$), does not vanish for any finite r, and decreases as $r^{-1/2}\exp[-\chi_\alpha r]$ if $r \to +\infty$. The remaining propagation modes are excited ones; they correspond to decreasing values of χ_α (with either $M = 0$ or increasing integer $\mid M \mid$), decrease as $r^{-1/2}\exp[-\chi_\alpha r]$ if $r \to +\infty$, and develop an increasing number of zeroes for finite values of r. Further results and details for those propagation modes appear in precedent works.[35,36,43] In particular, some approximate formulas[36] for K_1 or χ_α (for both $M = 0$ and $M \neq 0$), constitute the counterparts of related formulas for optical waveguides. For $M = 0$, other useful approximate formulas together with some detailed numerical solutions for the 16 lowest eigenvalues of Eq. (9.10) (except for the ground state) have been obtained.[45]

9.6.4 Linear Superpositions and Statistical Mixtures

We shall treat a slow neutron beam that in general has an energy spectrum. In typical reactor beams, the average separations among the centers of the wavepackets associated to different thermal neutrons are appreciably larger than their average de Broglie wavelengths; for a simple estimate, see Subsection 9.7.3. It is reasonable to neglect overlaps, interferences, and interactions among different neutrons in the beam as a first approximation.

In a bit more detail, let us consider a system containing more than one neutron (e.g., a thermal neutron beam). As the neutron has half-integral spin, the neutrons contained in that system or beam behave as fermions (described by antisymmetric wave functions), in principle: that is, their wave functions should change sign under the simultaneous exchange of the coordinates and spin projections of any two neutrons.

However, in the neutron beams to be considered, the overlaps among the wavepackets associated to any two or more different neutrons are negligible (again, because the typical average separations between those wavepackets are appreciably larger than the average de Broglie wavelengths). Consequently, one may safely disregard that neutrons are fermions, their wavefunctions need not be antisymmetric, and they can be treated as independent particles.

The neutrons in any typical beam treated in this article are regarded as independent particles, each of which propagates through and interacts with various material media. Moreover for a typical reactor beam, the different neutrons in it can be regarded to constitute an ensemble of identical copies of the same quantum system (formed by just one neutron) prepared in the same quantum state.[2] It may be adequate to discuss qualitatively some conceptual aspects regarding such beams.

Let us consider the case in which relative phases are not destroyed and incoherence effects are negligible for single neutrons, say, for single copies (even if there may be an energy spectrum). The quantum state describing the slow neutrons is represented by one time-dependent wave function $\Psi(\bar{x};t)_0$ fulfilling Eq. (9.2). $\Psi(\bar{x};t)_0$ is normalized $(\int d^3\bar{x} \mid \Psi(\bar{x};t)_0 \mid^2 = 1)$ by assumption.

Let F_0 be the quantum mechanical probability flux (per unit area and unit time interval)[39,42] of the wave function $\Psi(\bar{x};t)_0$ across a small surface $dS = dxdy$ in the (x, y)-plane about some (x, y, z), at time t:

$$F_0 = \frac{\hbar}{m_n} Re \left[\Psi^*(\bar{x};t)_0 (-i) \frac{\partial}{\partial z} \Psi(\bar{x};t)_0 \right] \tag{9.19}$$

The wave function $\Psi(\bar{x};t)_0$ provides a statistical description of all neutrons in the beam. One could regard all neutrons in the beam (N_n) as an ensemble. In fact, all those neutrons, being independent from one another approximately, can be regarded to be in the same quantum state and behave as N_n identical copies of one and the same system. Thus, let us imagine that N_n identical idealized experiments (independent from one another) were carried out on that ensemble of N_n neutrons. In each experiment, a measurement is performed, aimed to detect whether one neutron has gone across the small surface $dS = dxdy$ in the (x, y)-plane, about (x, y, z) during the small time interval $(t - dt/2, t + dt/2)$. The detections are carried out by adequate counters (all being entirely analogous in all measurements). Let $N_{n,c}(\bar{x}, t)$ be the total number of those counts (that is, the total number of times in which the arrival of one neutron is detected). The statistical interpretation of quantum mechanics[46,47] states the following. For large N_n, $N_{n,c}(\bar{x}, t)/(N_n \times dS \times dt)$ (namely the probability for one neutron to have reached the counter, per unit area in the (x, y) plane about (x, y, z) per unit time) approaches F_0, as given in Eq. (9.19). See Figure 9.8. The wave function $\Psi(\bar{x};t)_0$ represents that ensemble and encodes the largest amount of information possible about the ensemble, and this appears to suffice for practical purposes.

Trying to go beyond the above practical viewpoint, and as stated more or less explicitly, the wave function $\Psi(\bar{x};t)_0$ also describes each single neutron (say, each individual copy in the ensemble). Each single neutron, independently of the other copies in the ensemble, would be represented by $\Psi(\bar{x};t)_0$. The flux F_0, as given in Eq. (9.19), could be regarded as some measure of (or quantitative value for) the tendency of that single neutron to go across the small surface $dS = dxdy$ in the

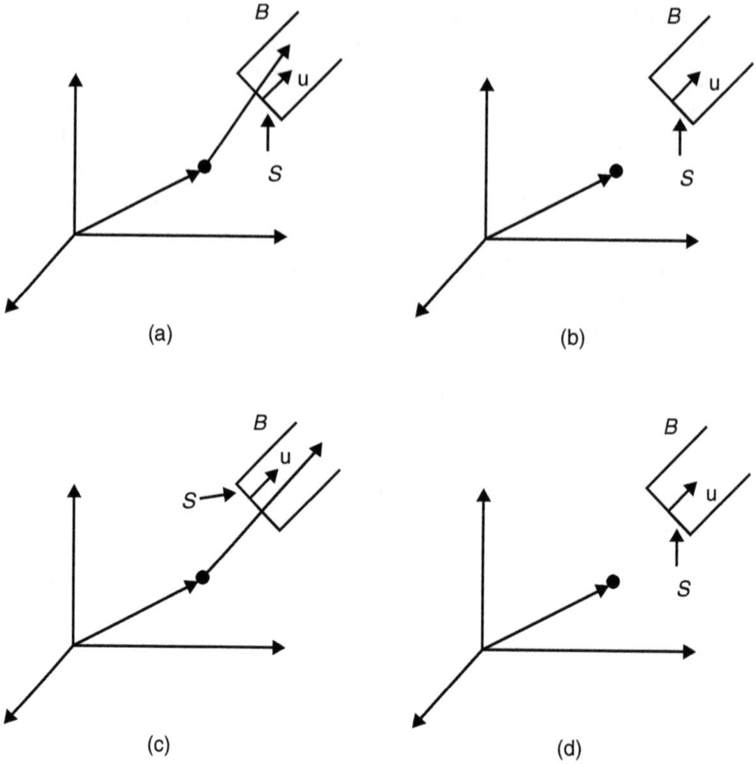

FIGURE 9.8 Probabilistic interpretation of the quantum mechanical probability flux at some (x, y, z) at time t. We consider a small surface S, orthogonal to the unit vector $\bar{u}$ (generalizing the case of a small surface $dS = dxdy$ with $\bar{u}$ along the z axis, treated in the text). There are $N_n = 4$ identical copies (each containing one neutron) and $N_n = 4$ identical experiments (with identical counters, B), aimed to detect whether one neutron has gone across S during the small time interval $(t - dt/2, t + dt/2)$. The total number of counts in which the arrival of one neutron is detected is $N_{n,c}(\bar{x}, t) = 2$. The interpretation applies for both cases, namely for the pure case (described by $\Psi(\bar{x}; t)_0$) and for a statistical mixture.

(x, y)-plane about (x, y, z) during the small time interval $(t - dt/2, t + dt/2)$. That quantification of the tendency would try (in a rather loose way) to give meaning to the idea of probability (or probability flux) for one single neutron. Various physicists, however, do not take that step forward, but they argue that, from a unique experiment performed upon one single particle (from only one copy in the ensemble!), it is impossible to infer statistical information, that is, the probability (understood as a frequency or as the number of counts, $N_{n,c}(\bar{x}, t)$ over the total number, N_n, of repeated experiments) per unit area and per unit time, associated to the wave function $\Psi(\bar{x}; t)_0$. For this purpose, one would need many repeated and similar experiments, performed on identically prepared systems, that is, an ensemble. What $\Psi(\bar{x}; t)_0$ (or more precisely, its flux F_0, as given in Eq. (9.19)) does is represent that ensemble. There exist various interesting discussions about that conceptual aspect.[46,47]

Let $F_{0,tot}$ be the total quantum mechanical probability flux (per unit time)[39,42] of $\Psi(\bar{x};t)_0$ across the whole (x, y) plane, for any z. One has:

$$F_{0,tot} = \int_{-\infty}^{+\infty} dx \int_{-\infty}^{+\infty} dy \frac{\hbar}{m_n} Re\left[\Psi^*(\bar{x};t)_0(-i)\frac{\partial}{\partial z}\Psi(\bar{x};t)_0\right] \tag{9.20}$$

$F_{0,tot}$ turns out to be not constant, but time-dependent.

Under various circumstances, typical slow neutron beams generated directly in nuclear reactors (according to some Maxwellian distribution) do not correspond to a unique wave function. In such cases, relative phases are destroyed and taking into account incoherence effects (for single neutrons or copies) becomes essential.[2,7] Such beams appear to be the analogue for neutrons of incoherent beams for ordinary light. As in the above case in which relative phases were not destroyed (pure case), all neutrons (N_n) in the actual incoherent beam, being in the same quantum state and approximately independent from one another, are also regarded as an ensemble of identical copies. The distinguishing feature now is that a statistical mixture of wave functions[2,6,39] is what provides the statistical description of all neutrons in the beam. In the actual case, the beam is represented by a statistical mixture of the (normalized) wave functions $\Psi(\bar{x};t)_{0,j}$ with certain probabilities p_j ($p_j \geq 0, \sum_j p_j = 1$), the index (or set of indices) j varying in some discrete or continuous set. What is attributed to any individual neutron in the beam is not one single wave function $\Psi(\bar{x};t)_{0,j}$ (in some sort of one-to-one correspondence), but the whole statistical mixture of wave functions precisely due to incoherence. The ensemble of all N_n thermal neutrons in the beam (the ensemble of N_n copies) is represented by the whole statistical mixture of all wave functions $\Psi(\bar{x};t)_{0,j}$. In turn, the statistical mixture is represented by the density matrix:

$$\rho(\bar{x};\bar{x}';t)_0 = \sum_j p_j \Psi(\bar{x};t)_{0,j} \Psi^*(\bar{x}';t)_{0,j} \tag{9.21}$$

As in the above pure case, let us imagine that N_n identical idealized experiments (independent from one another) were carried out on that ensemble of N_n neutrons, aimed to detect whether one neutron has gone across the small surface $dS = dxdy$ in the (x, y) plane about (x, y, z) during the small time interval $(t - dt/2, t + dt/2)$. Let $F_{0,j}$ be given in Eq. (9.19), for each $\Psi(\bar{x};t)_{0,j}$. Let $N_{n,c}(\bar{x}, t)$ be, as before, the number of those experiments in which the counters do detect the arrival of one neutron. In the actual incoherent case, for large N_n, $N_{n,c}(\bar{x}, t)/(N_n \times dS \times dt)$ (the probability for one neutron to have reached the counter, per unit area in the (x, y)-plane about (x, y, z), per unit time) approaches $\sum_j p_j F_{0,j} =< F_0 >$. $< F_0 >$ is the quantum-mechanical probability flux (per unit area and unit time interval) for the statistical mixture. See Figure 9.8. One could formulate statistical averages, say, the values for the p_j's for neutron beams, with probability fluxes which are non-monochromatic and time-dependent. Those statistical averages, omitted here, could perhaps be adequate for describing neutron beams coming from pulsed (spallation) sources. It is instructive to compare different sources of thermal neutrons (spallation sources versus nuclear reactors) and their associated fluxes.[10]

Next, we shall consider a slow neutron beam (also with an energy spectrum), that propagates confined along the fiber. First, we treat the case in which relative phases are not destroyed and incoherence effects are negligible. The quantum state describing the confined propagation of slow neutrons is represented by one time-dependent wave function $\Psi(\bar{x};t)_{in}$. The latter is a linear superposition of the propagation modes $\phi(\boldsymbol{x})_\alpha \exp(i\beta_\alpha z)\exp(-iEt/\hbar)$, given in Eq. (9.9). As there is an energy spectrum ($V_{cl} < E_{min} \leq E \leq E_{max}$), the linear super-position referred to above reads, for any $\boldsymbol{x}$:

$$\Psi(\bar{x};t)_{in} = \sum_\alpha \int_{E_{min}}^{E_{max}} dE c(E)_\alpha \phi(\boldsymbol{x})_\alpha \exp(i\beta_\alpha z)\exp(-iEt/\hbar) \qquad (9.22)$$

where $c(E)_\alpha$ is a given amplitude characterizing that incoming beam. The set of quantum numbers, α, over that the summation in Eq. (9.22) runs, corresponds to all modes that effectively propagate. $\Psi(\bar{x};t)_{in}$ is also normalized by assumption.

Let F_{in} be the quantum mechanical probability flux of $\Psi(\bar{x};t)_{in}$ (Eq. (9.22)) across a small surface $dS = dxdy$ in the (x, y)-plane, about some (x, y, z) at time t. F_{in} is given by the right-hand side of Eq. (9.19) with $\Psi^*(\bar{x};t)_0$ replaced by $\Psi(\bar{x};t)_{in}$. The total quantum mechanical probability flux, $F_{in,tot}$, of the incoming wave $\Psi(\bar{x};t)_{in}$ across the whole (x, y)-plane for any z, is given by the right-hand side of Eq. (9.20), with $\Psi^*(\bar{x};t)_0$ replaced by $\Psi(\bar{x};t)_{in}$:

$$F_{in,tot} = \frac{\hbar}{m_n} \sum_\alpha \int_{E_{min}}^{E_{max}} dE \int_{E_{min}}^{E_{max}} dE' \beta_\alpha \exp[-i(E' - E)t/\hbar] c^*(E)_\alpha c(E')_\alpha \quad (9.23)$$

as the $\phi(\boldsymbol{x})_\alpha$'s are orthonormalized. This $F_{in,tot}$ is time-dependent.

In the actual case of confined propagation, let the beam be represented by a statistical mixture of the (normalized) wave functions $\Psi(\bar{x};t)_{in,j}$ given in Eq. (9.22) (each with some amplitude $c(E)_{\alpha,j}$), with certain probabilities p_j ($p_j \geq 0, \sum_j p_j = 1$). The statistical mixture is now represented by a density matrix $\rho(\bar{x};\bar{x}';t)_{in}$ given by the right-hand side of Eq. (9.21), with $\Psi(\bar{x};t)_{0,j}$ replaced by $\Psi(\bar{x};t)_{in,j}$. Eqs. (9.22) and (9.21) yield:

$$\rho(\bar{x};\bar{x}';t)_{in} == \sum_{\alpha\alpha'} \int_{E_{min}}^{E_{max}} dE dE' < c(E)_\alpha c(E')^*_{\alpha'} >$$
$$\times \phi(\boldsymbol{x})_\alpha \phi^*(\boldsymbol{x}')_{\alpha'} \exp(i\beta_\alpha z)\exp(-i\beta_{\alpha'} z')\exp[-i(E - E')t/\hbar] \qquad (9.24)$$

$$< c(E)_\alpha c(E')^*_{\alpha'} > \equiv \sum_j p_j c(E)_{\alpha,j} c(E')^*_{\alpha',j} \qquad (9.25)$$

Typically, the physically interesting situations for us here (namely, slow neutron beams generated in nuclear reactors) correspond to quantum mechanical probability fluxes that are both non-monochromatic and, on the average, approximately constant (that is, time-independent, at least approximately and in some average sense). In those cases, $< c(E)_\alpha c(E')^*_{\alpha'} >$ can be taken to be negligibly small for $(\alpha, E) \neq (\alpha', E')$ but very large for $(\alpha, E) = (\alpha', E')$. Such cases can be formulated through the assumption:[38]

$$< c(E)_\alpha c(E')^*_{\alpha'} > = \delta(E - E') e(E)_{\alpha,\alpha'} \qquad (9.26)$$

Here, $\delta(E - E')$ denotes the Dirac delta function and $e(E)_{\alpha,\alpha'}$ is certain (non-negative) spectral density, characterizing the beam (very strongly peaked for $\alpha = \alpha'$).

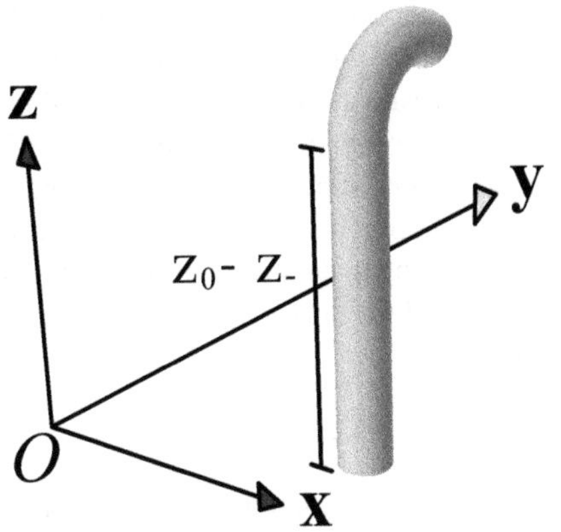

FIGURE 9.9 Curved three-dimensional fiber. The radius of curvature R_{cu} is the distance from the origin O up to the center of the (small) cross-section T of the fiber in the curved part.

The relevant quantum mechanical probability flux for the incoherent beam, $< F_{in} >$, will be understood and evaluated as the statistical average $\sum_j p_j F_{in,j}$, where $F_{in,j}$ is given by the right-hand side of Eq. (9.19) for each $\psi(\bar{x};t)_{in,j}$. In the statistical mixture case for confined propagation, Eqs. (9.23) and (9.26) imply readily for the total quantum-mechanical probability flux $< F_{in,tot} >$ across the whole (x, y)-plane:[38]

$$< F_{in,tot} > = \frac{\hbar}{m_n} \sum_{\alpha} \int_{E_{min}}^{E_{max}} dE \beta_\alpha e(E)_{\alpha,\alpha} \tag{9.27}$$

The interpretations of F_{in} and $< F_{in} >$ for the confined beam are omitted, as they are entirely analogous to those for F_0 and $< F_0 >$ (indicated in Figure 9.8).

9.6.5 Curved Neutron Fibers

On physical grounds, let us now consider a very lengthy three-dimensional fiber and let us suppose that: (i) both the z-axis and the origin $\bar{x} = (0, 0, 0)$ lie far outside the fiber, (ii) the fiber is perfectly straight from some $z = z_-$ up to some $z = z_0$, and (iii) the fiber has a large curvature radius R_{cu}, for $z_0 < z < z_1$. Both z_- and z_1 are finite: the straight part of the fiber starts at z_-, while the curved one ends at z_1. The length $z_0 - z_- = l$ is assumed to be quite large, so that, as a first approximation, one could set $z_- = -\infty$. See Figure 9.9. R_{cu} will be supposed to be much larger than any neutron wavelength and larger than the transverse dimensions of the fiber (for instance, $R_{cu} \geq 0.1\ m$). $z_1 - z_0 = l$ can be regarded as the length of the curved part of the fiber.

We suppose a confined beam of slow neutrons propagating initially along the straight part of the fiber and represented by some statistical mixture like that described in Subsection 9.6.4. Let $F_{in,tot}$ be the incoming quantum mechanical probability flux across the transverse section of the straight part. Typically, each neutron is represented by a statistical mixture of wave functions, so that $F_{in,tot}$ should be interpreted as (and replaced by) $< F_{in,tot} >$ (Subsection 9.6.4). As those neutrons enter into the curved part of the fiber, there is a non-vanishing probability for them to have no longer confined propagation but to escape outside the fiber. The tendency of the initially confined neutrons to propagate unconfined toward increasing separations from the fiber is named curvature or bending loss. Let $F(z_1)$ be the quantum mechanical probability flux for neutrons to propagate still confined across the transverse cross-section of the fiber, by the end z_1 of the curved part of the fiber. One has the following three-dimensional formula:[38]

$$F(z_1) \simeq F_{in,tot} \Upsilon \tag{9.28}$$

Υ can be interpreted as its transmission coefficient which assesses the importance of bending losses. Arguments have been given to justify the following approximate formula displaying how Υ depends on R_{cu} and on the length l of the curved part of the fiber:[38]

$$\Upsilon \simeq \exp\left[-a_{cu} l \exp[-b_{cu} R_{cu}]/R_{cu}^{1/2}\right] \tag{9.29}$$

Eq. (9.28), which holds in three spatial dimensions, describes the decrease of the probability flux of the confined neutron propagating along the curved fiber, due to curvature losses. The constants a_{cu} and b_{cu} depend, in principle, on whether the waveguide is multimode or monomode. It can be justified[38] that $b_{cu} = (2/3)\beta_0[\chi_0/\beta_0]^3(> 0)$. χ_0 is the smallest in the set of all χ_α that are excited effectively in the unbent part of the fiber; hence, the corresponding $\beta_\alpha \equiv \beta_0$ is maximum. Thus, we expect that all modes such that $\chi_0 \leq \chi_\alpha \leq \chi_{max}$ are excited effectively. With this χ_0^2, the number of modes that propagate effectively can be estimated as $N_{pm}(\chi_0^2)$ (Eq. (9.13)). To find a_{cu} is more difficult, and its determination will proceed through different routes, depending on whether the waveguide is multimode or monomode; see Subsections 9.7.4 and 9.8.3. In order to realize the physical consistency of the approximate Eq. (9.29), we notice the following. Let R_{cu} become very large: Υ tends quickly toward unity. As R_{cu} tends to zero, Υ goes quickly to zero.

Previously, a theoretical study of bending losses for a curved neutron fiber in the simpler case of two spatial dimensions had been carried out.[36] There exist various analyses of bending losses for electromagnetic radiation, based both on geometrical optics and Maxwell equations.[30,32,48–50] Another treatment of bending losses using geometrical optics has been applied for x-rays and neutrons.[51]

9.7 CONFINED PROPAGATION IN MULTICAPILLARY GLASS FIBERS

9.7.1 MULTICAPILLARY GLASS FIBERS: CASE FOR X-RAYS

In principle, polycapillary glass fibers could be obtained by stretching glass tubes adequately at suitably high temperatures. A typical polycapillary fiber may have a diameter in the millimeter range (a bit smaller eventually) and a length about a few tens of centimeters. Each polycapillary glass fiber contains many (about 10^3) individual, essentially parallel, hollow capillary channels. Single hollow capillary channels with internal diameters about a few microns have indeed been achieved experimentally, and we shall concentrate on them. The former can be regarded potentially as hollow waveguides for the confined propagation of radiation with wavelengths adequately smaller than the internal diameter. Consequently, x-rays can propagate confined along each single hollow capillary channel. Bundles of such polycapillary glass fibers are currently employed for the focusing of x-rays. Thus, those polycapillary glass fibers serve as certain basic devices for performing experiments in capillary optics of x-rays.

A survey of the experimental development of capillary optics for x-rays together with short comments about their possible applications for slow neutrons was made in the 1980s by M.A. Kumakhov and F.F. Komarov.[51] Other reviews[8] deal with neutron capillary optics more briefly. Kumakhov developed a patent for experimental capillary optics in 1984.[51,52] Since then, the field of experimental capillary optics has advanced considerably and applied to x-rays. We cite the I.V. Kurchatov Institute in Moscow, the Center for X-ray Optics at State University of New York at Albany, the Institute for Roentgen Optics (based on the Kurchatov Institute), Unisantis S.A. (www.unisantis.com/) in Geneva, and the Institute for Advanced Instruments (IfG) GmbH (www.ifg-adlershof.de) in Berlin. They are specifically devoted to developing various analytical x-ray instruments (spectrometers, microscopes, diffractometers) based upon capillary optics, mostly for x-rays. X-ray capillary optics constitutes a very active research field with important and expanding applications. Interesting research and advances in x-ray capillary optics have been made in recent times.[53–62]

We shall consider single hollow capillary channels with inner diameters far larger than the micrometer scale only briefly in Subsection 9.10.2.

9.7.2 Multicapillary Glass Fibers for Neutron Optics: Early Experiments

An important experimental implementation of neutron fibers was achieved in 1992.[63,64] Bundles of polycapillary glass fibers, as described in the previous subsection, were employed specifically for the confined propagation and focusing of slow neutrons.[63,64] The polycapillary fibers employed in the experiments[63,64] typically have a diameter $d_{pf} = 0.4$ mm and a length between 150 mm and 200 mm and contain more than $N_{pf} = 1,000$ parallel individual hollow capillary channels. In turn, each single hollow capillary channel in a polycapillary glass fiber with an internal diameter $d_{hcc} = 6\mu m$ could be regarded reasonably as a hollow waveguide of very small cross-section along which neutrons propagate. See Figure 9.10.

Each polycapillary fiber employed in the early experiments[63,64] was made of lead-silica glass. Accordingly, each of the above hollow capillary channels was surrounded by a lead-silica clad (with $Re(n^2) < 1$) and had a width about 5 μm, so that beyond such a distance the cladding region ended and another neighboring parallel hollow channel was met.

The fibers may be straight or bent; in the latter case, they have a curvature radius $R_{cu} \geq 0.1$ m (meters).[63,64] See Figure 9.11.

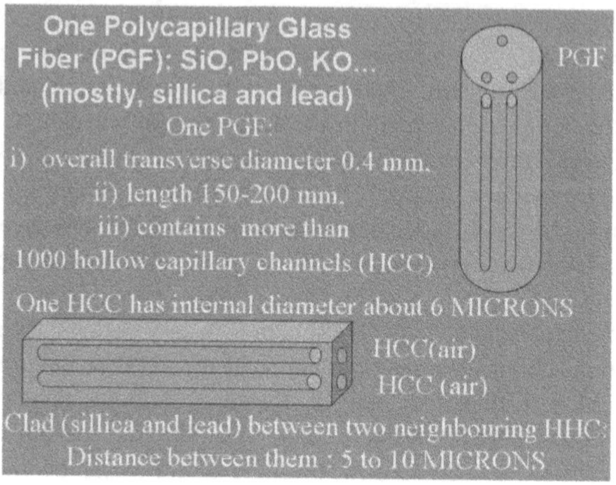

FIGURE 9.10 One single polycapillary glass fiber.

Those devices in principle enabled researchers to increase the flux of confined neutrons that emerge from the exit end of a parallel assembly of fibers and concentrate onto a small region (focal spot) located at the focal distance. Specifically, a parallel assembly of 721 polycapillary fibers with an overall diameter of 15 *mm* at the exit end has provided an amplification of the neutron flux by a factor of 7, at a focal distance equal to 104 *mm*.[64] The diameter of the focal spot was 1 *mm*. See Figure 9.12.

The 1992 experiments[63,64] were analyzed by means of geometrical optics. Some related and quite detailed studies of neutron transmission have been carried out[65–67] using geometrical optics.

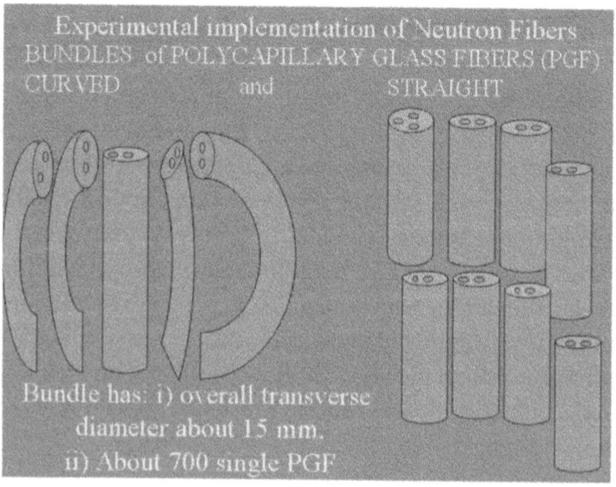

FIGURE 9.11 Bundle of polycapillary glass fibers.

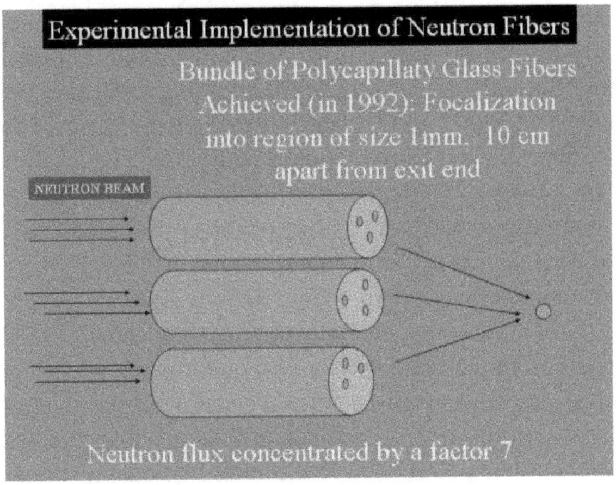

FIGURE 9.12 Focalization of a neutron beam by means of a bundle of polycapillary glass fibers.

9.7.3 Multicapillary Glass Fibers for Neutron Optics: Estimates and Analysis

It appears[38] that a single individual hollow capillary channel with small cross-section is conceptually analogous to and can be regarded as a single neutron fiber conjectured theoretically.[35,36] In order to avoid confusion, neither a single polycapillary glass fiber nor bundles thereof (which have much larger cross-sections) will be regarded to be a single neutron fiber here, consistent with the convention adopted initially[38] (although a somewhat different convention appeared recently.[68])

Let the average energy of a thermal neutron in a typical beam coming from a nuclear reactor be about $10^{-2}\ eV$, so that the average neutron velocity v is about $10^3\ ms^{-1}$. Let F_0 denote some average thermal neutron flux (per unit surface and time). Typically, each neutron is represented by a statistical mixture of wave functions, so that F_0 should be interpreted as and replaced by $< F_0 >$ (Subsection 9.6.4). The average smallest separation between two neighboring thermal neutrons propagating in the beam is about $d_{n-n} = (v/F_0)^{1/3}$. For the highest fluxes, F_0 is about 10^{15} neutrons $cm^{-2}s^{-1}$. Then, d_{n-n} is of the order of $10^4 Å$ to $10^5 Å$. The thermal neutron beam is typically non-monochromatic, so that a spectrum of wavelengths is met; for instance, it ranged between $2Å$ and $9Å$ in the first experiments.[64] Since the wavelengths in such intervals are smaller than d_{n-n}, one can reasonably neglect the overlaps between the wave packets associated to different neutrons in the beam and regard each confined neutron as propagating independently. Recall the comments in Subsection 9.6.4.

The average distance, d_{nhc}, between two neighboring (parallel) hollow capillary channels can be estimated from $d_{pf}^2 \simeq d_{nhc}^2 N_{pf}$, with the fiber diameter $d_{pf} = 0.4\,mm$. One finds $d_{nhc} \simeq 5$ to $10\ \mu m$.

A priori, one could entertain the possibility that one slow neutron, propagating initially confined along certain hollow capillary channel (denoted by $hcc1$) could escape by transmission through the surrounding clad due to quantum-mechanical tunnel effect to one of the neighboring parallel hollow capillary channels and eventually escape finally toward the externally surrounding medium (air). Let d be the radial distance from the internal surface limiting the $hcc1$ to some point in its surrounding clad, so that $0 \leq d \leq 5\ \mu m$. According to some estimates,[36] the probability for such a tunnel effect is exponentially small and hence negligible for almost all thermal neutrons for values of the number of nuclei per cubic centimeter and of the neutron–nucleus scattering amplitude typical of those in the lead-silica clad, provided that the clad thickness is about or larger than $d \geq 0.5\mu m = d_{tu}$. Thus, d_{tu} is some characteristic tunnel effect length. Consequently, since d_{nhc} is about one order of magnitude larger than d_{tu}, one concludes that such a tunnel effect is negligible (absence of cross-talk in terms typical of optical communications or optical fibers). Stated in other terms, the time required for a thermal neutron to escape across the cladding toward air is several orders of magnitude larger than the time needed for that neutron to travel confined along 1 m in the core. Equivalently, one could simply say that neutrons propagate confined along any individual hollow capillary channel as if the channel were surrounded by a clad of infinite width (as if $d_{nhc} = \infty$), which will simplify the analysis; this can be the practical point of view to be adopted in the study of thermal neutron propagation in polycapillary glass fibers. The importance of tunnel effects will have to be reconsidered again and rather critically when dealing with confined thermal neutron propagation in carbon nanotubes; see Subsection 9.8.3.

The description of thermal neutrons propagating confined along the above hollow capillary guides with small transverse cross-sections appears to require coherent wave functions and the associated Schrödinger Eq. (9.2) in principle.[38] Based upon that physical standpoint, we shall discuss various qualitative aspects of the quantum mechanical approach below.

Let us concentrate on a typical hollow capillary channel, like those in the early experiments guiding thermal neutrons.[64] E can be taken to be about 0.025 eV, as $V(\boldsymbol{x})$ and V_{cl} are much smaller. For such a channel ($b_{co}\rho_{co} = 0$), and for values of the number (ρ_{cl}) of nuclei per cm^3 and of the neutron–nucleus scattering amplitude $b_{cl}(> 0)$[7] of a typical clad, one expects that the right-hand side of Eq. (9.12) ranges from about $10^{-3}Å$ to about $10^{-2}Å$. Then, for typical values of β_α ranging from, say, $0.7Å^{-1}$ to $3Å^{-1}$, χ_α/β_α for the highest modes (say, the fundamental one and the first few above it) varies from 10^{-2} to 10^{-4}. The hollow capillary channel behaves naturally as a multimode waveguide and the total number of propagation modes N_{pm} allowed in principle varies between 4.5×10^2 and 4.5×10^4. Of course, for a given incoming slow neutron beam, it is not warranted *a priori* that all allowed propagation modes will be effectively excited. A reasonable expectation seems to be that as neutrons enter into the hollow capillary channel, they find it more favorable to propagate into lower modes — the fundamental one and those having smaller values of χ_α. An interesting problem is to provide an estimate of the number of propagation modes that get excited effectively; see Subsection 9.7.4.

Even if the inner diameter of a hollow capillary channel ($d_{hcc} = 6\mu m$) in a polycapillary glass fiber is three to four orders of magnitude larger than the average de Broglie wavelength of a thermal neutron and the total number of allowed propagation modes for the latter is rather large, one should not rely entirely on a geometrical optics description. In fact, a quantum mechanical description for slow neutrons propagating confined in a hollow capillary channel in a polycapillary glass fiber should be entertained, as least for a suitable fraction of those neutrons. Thus, for certain effective propagation modes (namely the fundamental one and a subset of higher modes above it) $\chi_\alpha(d_{hcc}/2)$ may be of order unity and even larger and hence, quantum effects may be important. In fact, for values typical of the actual hollow capillary channel, Eq. (9.12) implies that $\chi_\alpha d_{hcc}$ has an upper bound that varies between 2.8 and 28.

Slow neutrons propagating confined along capillary channels with small transverse cross-sections can give rise to interference phenomena. This has been the point of view adopted by Rohwedder.[45] The detailed and very interesting analysis[45] on neutron propagation along capillary guides (under certain specific conditions which, in turn, are met in various experiments) has provided further arguments supporting the physical necessity of a wave optical treatment of that phenomenon. Specifically, it has been stated[45] that "At first sight, this ray-optical point of view seems appropriate, since even in the case of capillary guides with core diameters of only a few microns, neutron de Broglie wavelengths are still smaller by a factor of roughly 1000. But this in itself is not a correct criterion for disregarding diffraction phenomena." Then, for slow neutrons and under those conditions, it is proposed[45] that (i) treatments based upon geometrical optics, in which interference effects are not taken into account, give rise to inconsistencies, (ii) quantum mechanical wave functions are required for a proper description, and (iii) the observation of single and multiple self-images in neutron pipes with circular or rectangular profile may represent an unambiguous means to demonstrate the relevance of waveguide-internal diffraction phenomena. Self-imaging distances could be macroscopically magnified by the accumulation of a large number of (microscopic) Goos–Hänchen shifts for neutrons.[45]

Let us consider an incoming thermal neutron flux that approaches one end of the polycapillary glass fiber from outside. The flux of the confined neutrons that have entered into and propagate confined along the polycapillary glass fiber is substantially smaller. We shall estimate this important reduction in Subsection 9.8.1 for the case of carbon nanotubes and in Subsection 9.10.1 for polycapillary glass fibers.

9.7.4 Bending Losses in Multicapillary Glass Fibers

Polycapillary fibers are not straight. When initially confined beams of slow neutrons propagate in curved multicapillary glass fibers, the flux decreases due, in particular, to bending losses. Those bending losses have been experimentally measured (that is, the associated transmission coefficient Υ).[64] Those experimental data have also been analyzed theoretically,[64] by means of a formula for Υ (based upon geometrical optics), which had been obtained previously.[51]

A new theoretical analysis of bending losses in three dimensions, based upon quantum mechanics and specifically addressed to the experimental results,[64] has been carried out.[38] We shall now outline the latter analysis. This proceeds by applying Eq. (9.29) to the analysis of those curvature losses, accepting naturally that the hollow capillary channels behave as multimode waveguides. An appealing fact was that the dependence of Υ on R_{cu}, as given by Eq. (9.29), appeared to allow for an approximate description of the experimental data. For multimode waveguides, a_{cu} is rather difficult to evaluate theoretically, and it has been estimated through a fit to the

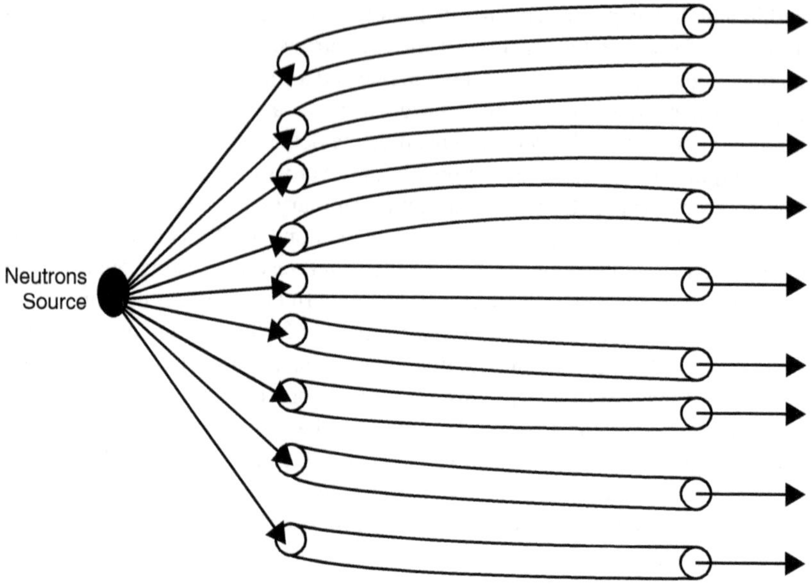

FIGURE 9.13 Part of a multicapillary fiber with $N_{pf} = 9$ individual parallel hollow capillary channels. The neutrons previously emitted by the source propagate confined along the individual capillary channels. Each individual capillary channel with small cross-section can be regarded as a neutron fiber. The part not displayed of the multicapillary fiber is more or less symmetric of the one shown. All individual capillary channels converge toward some small region, onto which the propagating neutrons are focused.

experimental data.[38] The behavior in Eq. (9.29), with $a_{cu}l \simeq 0.95\ m^{1/2}$ and $b_{cu} \simeq 0.51\ m^{-1}$ is approximately consistent[38] with the experimental data[64] for the multimode case, as R_{cu} varies in, say, $0.1\ m < R_{cu} < 13\ m$. The order of magnitude of b_{cu} seems physically reasonable and essentially consistent with the above arguments (and with estimates of $(2/3)\beta_\alpha[\chi_\alpha/\beta_\alpha]^3$) for typical average values of β_α about $1\mathring{A}^{-1}$ and of χ_α/β_α between 10^{-3} and 10^{-4}). For $b_{cu} \simeq 0.51\ m^{-1}$ and a maximum propagation constant for the effectively propagating modes (namely β_0) about $3\mathring{A}^{-1}$, one gets $\chi_0/\beta_0 \simeq 2.9 \times 10^{-4}$. If the right-hand side of Eq. (9.12) is about $10^{-3}\mathring{A}$ so that $N_{pm} \simeq 4.5 \times 10^2$, the total number of effectively excited modes is estimated to be $N_{pm}(\chi_0^2) \simeq 108$. If, on the other hand, the right-hand side of Eq. (9.12) is about $10^{-2}\mathring{A}$ so that $N_{pm} \simeq 4.5 \times 10^4$, the total number of effectively excited modes is $N_{pm}(\chi_0^2) \simeq 4.2 \times 10^4$ (which is much larger than in the previous case where $N_{pm} \simeq 4.5 \times 10^2$). The behavior of the experimental data[64] changes at $R_{cu,0} \simeq 3\ m$, which seems physically consistent. More detailed discussions about this are available.[38,43]

The case with N (more or less) parallel hollow capillary channels (with $N >> 1$) that occurs in reality with polycapillary fibers (so that $N = N_{pf}$) and with bundles thereof can be treated similarly, and leads approximately to the same Υ.[38] See Figure 9.13.

9.8 NANOTUBES AS NEUTRON FIBERS

9.8.1 Some General Aspects

Carbon nanotubes (CNTs) have played an increasingly important role in nanotechnology since their discovery in 1991 by means of high-resolution microscopy observations.[69] CNTs are very interesting by virtue of several specific properties: their very small transverse sizes at the nanometer or nm scale ($1\ nm = 10^{-3}\ \mu m = 10^{-9}\ m$), their unique structure (to be discussed below), their possibility of being metallic or superconducting (depending on their geometric structures), their remarkable capabilities for allowing for ballistic transport, their extremely high thermal conductivity and optical polarizability, and their possibilities for reaching high structural perfection. They attract extensive research motivated by their potential applications, particularly in composite materials, battery electrode materials, field emitters, nanoelectronics, and nanoscale sensors.[70–72] Many important aspects of CNTs that have generated great interest lie outside the scope of this chapter.

Here, we shall concentrate on the possible confined propagation of thermal neutrons along CNTs behaving as waveguides with transverse dimensions at nanometer (nm) scales. In so doing, we shall follow a proposal made originally by G.F. Calvo (unpublished) and developed later.[73] This section will constitute a summary of the previous work,[73] to which we refer for a detailed presentation and discussion. Such an analysis will play, for thermal neutrons, a role similar to that already carried out for focusing x-rays at nanometer scales. The x-ray case has been covered in a number of interesting works.[74–84] Since the focusing of x-rays at nanoscales may be relevant, it may be natural and adequate to investigate the confined propagation of thermal neutrons at those scales. Thus, the important analogies between x-rays and thermal neutron beams and their complementary roles[7] will be stressed and confirmed. The experimental assessment of the behavior of slow neutrons and its comparison with quantum mechanical predictions at decreasing scales constitutes a subject of considerable interest in fundamental research. Thus, the basic interactions of neutrons at the micron scale have been investigated recently.[85] Other possible reasons for the interest in the confined propagation of thermal neutrons at nanoscales will be discussed in Subsection 9.11.2.

There exist accounts about CNTs at various levels.[70–72] We consider a CNT with length L ($1\ \mu m \leq L \leq 100\ \mu m$). A single wall CNT (SWCNT) has one ($N = 1$) single layer composed of carbon atoms folded into a hollow cylinder (the hollow or empty interior containing air). A multiwall CNT (MWCNT) has $N(\geq 2)$ carbon layers wrapped coaxially. The MWCNT resembles a SWCNT coated by $N - 1$ additional concentric walls, also made by carbon atoms, so that there are $N - 1\ (> 0)$ additional empty interiors (also filled by air) between successive walls. See Figure 9.14. Regarding the preparation of CNTs, we shall limit discussion to a few short comments; detailed accounts are available at other sources.[72] Three main methods of producing SWCNTs are arc-discharge, laser ablation, and chemical vapor deposition (CVD). The first two methods are suitable mostly at laboratory scales, while CVD appears to be more adequate for producing larger amounts of SWCNTs. CVD is also employed to produce MWCNTs.

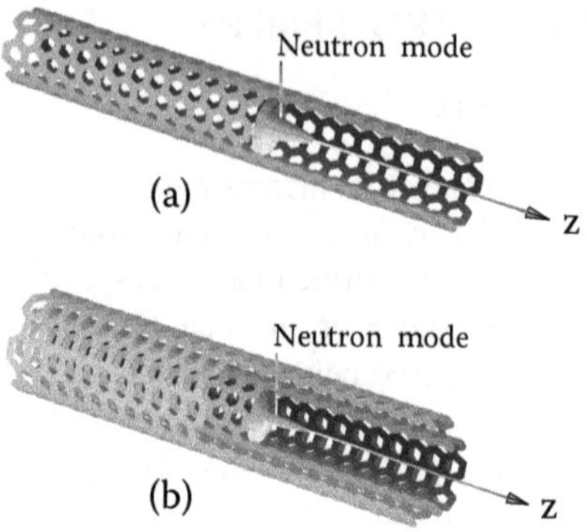

FIGURE 9.14 SWCNT (a) and MWCNT (b). The possible confined propagation on one neutron in the fundamental (Bessel) mode is also displayed for a SWCNT (a) and for a MWCNT (b).

For a SWCNT, typical dimensions are 10 to 17 nm of overall transverse diameter and diameter d of the internal empty cylinder about 5 to 7 nm. The thickness w of the wall of a SWCNT is about 2.5 to 5 nm. For a MWCNT, we shall take 50 nm as an average or typical overall transverse diameter and the distance Δ between two successive coaxial walls to be about 0.35 nm. As a first approximation, we shall assume that the diameter of the innermost central empty cylinder and the thickness of any wall in a typical MWCNT are of the same order as in a SWCNT. Then, we may estimate that N varies between 4 and 8. See Figure 9.15. CNTs with transverse diameters as small as 0.42 nm have been reported.[72] We shall concentrate on essentially straight CNTs (with very small curvatures) so as to reduce bending losses as much as possible. We shall disregard details of the discrete structures of walls, say, the locations of the carbon atoms, in the same way we approximated Eq. (9.3) by Eq. (9.4).

Let A_{tot} be the area of the total transverse cross-section of the nanotube (in square nanometers). Also, let A ($< A_{tot}$ and in nm^2 as well) be the total area of all transverse cross-sections of all N empty interiors of the CNT (effective area). For an SWCNT

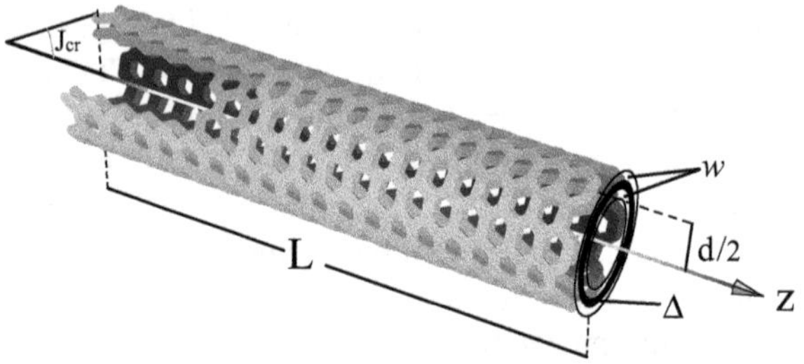

FIGURE 9.15 Simplified representation of a straight MWCNT with $N = 2$ walls. L is the length of the MWCNT. w is the width of each of the carbon walls. Δ is is the distance between adjacent walls. The shaded ring represents an empty region. The acceptance angle for a neutron to penetrate from outside into the MWCNT is denoted by θ_{cr} (also given by the right-hand side of Eq. (9.8)).

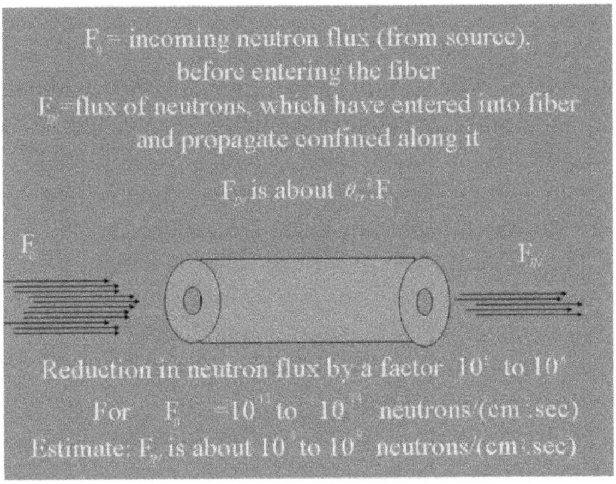

FIGURE 9.16 Reduction of the flux of confined neutrons that have entered from outside (vacuum) into a fiber and do propagate confined along it.

($N = 1$) that has a unique empty interior, $A = A(1) \simeq (\pi/4)d^2$. On the other hand, for a MWCNT having its innermost central empty interior similar to that for a SWCNT and $N - 1$ successive empty interiors with areas $A_n \simeq \pi\Delta[d + 2(n-1)(w + \Delta)]$ ($n = 2, \ldots, N$), one finds: $A = A(\text{N}) \simeq A(1) + \sum_{n=2}^{N} A_n$. As examples, we consider (a) a SWCNT with $d = 6\ nm$ and $A(1) \simeq 28\ nm^2$, and (b) a MWCNT with $N = 6$ and overall transverse diameter $\simeq 50\ nm$ in which $A(6) \simeq 207\ nm^2$ (the total area of its transverse cross-section being $A_{tot} \simeq 2044\ nm^2$).

Let us consider an incoming thermal neutron flux F_0 (neutrons/($cm^2 \times s$)) that approaches one end of the CNT from outside. In the usual case, each neutron is represented by a statistical mixture of wave functions, so that F_0 means $< F_0 >$ (Subsection 9.6.4). Then, the flux F_{in} of the confined neutrons (understood as $< F_{in} >$ for a statistical mixture) that have entered into and propagate confined along the CNT can be roughly estimated to be $F_{in} \simeq \theta_{cr}^2 \times F_0$, with the same θ_{cr} as in Eq. (9.8) (compare with Figure 9.4). Then, θ_{cr}^2 reduces the neutron flux by a factor about 10^{-5}. See Figure 9.16. For $F_0 \simeq 10^{14}$ neutrons/($cm^2 \times s$), one estimates $F_{in} \simeq 10^9$ neutrons/($cm^2 \times s$). The number of confined neutrons transmitted across the area A during τ hours is $\simeq 3.6 \times 10^{-2} A \times \tau$, which we shall now estimate. For a SWCNT ($A = A(1)$) we get $\simeq 1$ neutron per hour. For a MWCNT with $N = 6$ and $A(6) \simeq 207\ nm^2$, one has $\simeq 6$ neutrons per hour. It may be not unreasonable to expect that the above estimates of the number of confined neutrons transmitted across A during τ hours hold, in order of magnitude, in a quantum description of the thermal neutron propagation. The necessity for such a description will be discussed in the next subsection.

9.8.2 Quantum Mechanical Description for Neutrons in Nanotubes

The total area A of all transverse cross-sections of all empty interiors of the CNT is very small. Thus, a quantum mechanical description of the confined propagation of thermal neutrons along CNTs is necessary. The following arguments will support this expectation.

1. Let N_{pm} be the total number of propagation modes for neutrons moving confined along a waveguide, in quasi-classical approximation. The general formula for N_{pm} in Eq. (9.14) gives for a CNT (either SWCNT or MWCNT) $N_{pm} \simeq 2b\rho A$ (the factor 2 counting both spin projections for the neutron). Here, $b\rho(\equiv b_{cl}\rho_{cl})$ refers to any carbon wall (which constitutes the clad). Notice that the hollow interiors (the cores) contain air, for which we take, as a first approximation, $b_{co}\rho_{co} \simeq 0$. Using $b\rho \simeq 7.5 \times 10^{-4} nm^{-2}$ for the carbon walls and the estimates for A for both SWCNTs and MWCNTs given in the previous subsection, one finds $N_{pm} < 1$. The reliability of the quasi-classical approximation[39,40,42] for the computation of the total number of propagation modes cannot be trusted.

2. For thermal neutrons, the de Broglie wavelength is about 10^{-1} nm to 1 nm, to be compared to the transverse sizes of the CNT.

3. As stated previously,[38,45] quantum effects should not be neglected in the confined neutron propagation of thermal neutrons in the first experiments reported[63,64] that employed hollow capillary channels with transverse sizes at the micrometer scale. Such quantum effects should be even more relevant at the nanometer scale. To be more quantitative, recall that optical waveguides with transverse diameter d and refractive index n_1, surrounded by a medium with refractive index $n_2(< n_1)$ are characterized by the parameter $v = (\pi d/\lambda)(n_1^2 - n_2^2)^{1/2}$ (λ being the wavelength of light).[32] As an interesting example, recall that retinal photoreceptors in the human eye regarded as optical waveguides have transverse dimensions only a few times the wavelength λ of visible light and that they support a small number of propagation modes.[33] For retinal photoreceptors, and leaving aside the fundamental mode (which is always allowed), v for the other allowed modes lies between 2.4 and about 5.5. For slow neutrons propagating in a thin waveguide, an effective counterpart of the above v for optical waveguides is $v_{sn,eff} = [2N_{pm}]^{1/2}$, which also counts both neutron spin projections and applies, in a formal sense at least, for hollow capillary channels and for both SWCNTs and MWCNTs. For a hollow capillary channel, $v_{sn,eff}$ lies between 10 and 100.[38] For MWCNTs with $N = 6$ and overall transverse diameter $\simeq 51.5$ nm, one finds $v_{sn,eff} \simeq 0.74$, and for SWCNTs $v_{sn,eff}$ is smaller. These estimates may suggest that the possible confined propagation of thermal neutrons along MWCNTs could be some sort of analogue of the confined propagation of ordinary light along retinal photoreceptors in the human eye. The latter suggestion appears to be supported by a rather detailed analysis[73] and it will be summarized below.

The above arguments indicate that the proper description of thermal neutron confined propagation in CNTs requires quantum mechanics and, due to the smallness of A, one may anticipate that only one propagation mode (namely the fundamental one) would at best be relevant.

A study of the propagation modes in CNTs with $N(\geq 1)$ empty regions has been carried out,[73] to which we refer for details. We shall assume (i) the most external wall (that after the $n = N$-th empty space) of the CNT has infinite thickness. Assumption (i) will be removed in the next subsection, upon studying tunneling effects. It will also be assumed, for simplicity, that (ii) all walls and empty regions have circular cross-sections. Recall the structures in Eqs. (9.15), (9.16) and (9.17).

We consider a SWCNT ($N = 1$) first. According to several arguments and estimates,[73] the fundamental mode ($M = 0$) of a confined slow neutron in the SWCNT appears to spread far outside the waveguide, in the transverse plane. Loss effects (see Subsection 9.8.3) would upset quite dramatically the probability distributions determined by that propagation mode in the transverse plane and hence its confined propagation along the SWCNT. Of course, it appears to be extremely unlikely that the SWCNT could have excited propagation modes.

We now consider a MWCNT with $N > 1$ concentric coaxial empty regions, having spatial sizes as in Subsection 9.8.1 and behaving as a monomode waveguide. For this MWCNT, which has a larger effective area $A(N)$, the fundamental mode becomes spatially more concentrated in the $N(> 1)$ successive empty interiors than it was for $N = 1$ (as $A(1) < A(N)$). For the monomode MWCNT, estimates indicate that the probability distribution for the fundamental mode in the transverse plane would be essentially concentrated into the waveguide, without extending significantly beyond the most external wall.[73] Thus, $N > 1$ makes confined neutron propagation along MWCNTs more likely and less sensible to loss effects than in SWCNTs (see the next subsection). These conclusions[73] seem to hold also for other shapes of the transverse cross-sections, that is, the above assumption (ii) does not seem necessary. See Figure 9.14 in which the possible propagation on one neutron in the fundamental mode (also denoted as the Bessel mode) is represented schematically along a SWCNT (Figure 9.14, (a)) and a MWCNT (Figure 9.14, (b)).

9.8.3 Losses in Confined Neutron Flux

Several additional physical effects not taken into account in the previous subsection could either upset the confined propagation of thermal neutrons along CNTs or make their flux decrease appreciably. As for the case of polycapillary glass fibers, we shall outline some estimates[73] about the following loss effects: nuclear absorption and diffuse scattering, bending (or curvature) losses, and tunneling.

The relative decrease of the flux of thermal neutrons due to nuclear absorption and diffuse scattering by carbon nuclei (in the walls of the CNT, say, the clads) and by air (in the empty spaces, that is, the cores), is given approximately by $\exp(-\mu z)$, after propagation along a distance z (recall Section 9.3). $\mu(\geq 0)$ is the linear coefficient of the corresponding medium that accounts for both effects. For thermal neutrons, one has $\mu \simeq 0.6\ cm^{-1}$ for carbon, while $\mu \simeq 5.7 \times 10^{-4}\ cm^{-1}$ for air. Then, for typical lengths of CNTs ($L \leq 100\ \mu m$), the decrease of the neutron flux due to both effects is negligible both in the carbon walls and in the empty spaces.

If the CNT with length L is not perfectly straight but instead has some curvature radius $R_{cu}(< +\infty)$, there is a non-zero probability for the neutron to escape far from the former in the transverse directions. In such a case, the propagation mode in Eq. (9.9) cannot be an exact solution of the Schrödinger equation. A previous general analysis[38] and the above approximate Eqs. (9.28) and (9.29) also apply to the actual CNT monomode case. In the latter, the same approximate formula for the constant b_{cu} given in Subsection 9.6.5 holds and, moreover, it is possible to obtain a simple approximate formula for a_{cu} as well.[73] For the CNT monomode case, one estimates roughly that (for $A \simeq 160\,nm^2$) $a_{cu} \simeq 2\times10^{-2}\,nm^{-1/2}$, $b_{cu} \simeq 1.5\times10^{-7}\,nm^{-1}$. Then, for 30 $\mu m < L < 50$ μm and $R_{cu} = 1$ mm, one gets $0.57 > \Upsilon_{cu} > 0.39$. MWCNTs are, on the average, much straighter than SWCNTs. Bending losses are expected to be much smaller (and hence more tolerable) for MWCNTs than for SWCNTs.[73]

We shall remove assumption (i) in the previous subsection so that the most external ($n = N$) carbon wall has also finite thickness w. A thermal neutron, propagating initially confined along the CNT, could escape, upon performing tunnelings across the various carbon walls and eventually across the outermost (N-th) one, toward the external medium (air) surrounding the CNT. In principle, this loss effect is more relevant for CNTs than for polycapillary glass fibers, due to the fact that w is also in the nanoscale. We estimated the relative importance of tunneling for thermal neutrons,[73] through quasi-classical arguments. Let τ_L and τ_{tr} be the average times required, respectively, for the thermal neutron to travel the length L along the z-axis and along the diameter d of the $n = 1$ internal empty space in the transverse $\boldsymbol{x}$-plane of a straight CNT. Also, let Υ_{tu} be some probability (to be estimated in quasi-classical approximation) for a thermal neutron, moving initially in the fundamental mode, mostly in the $n = 1$ empty interior of a CNT, to escape, by tunneling across all carbon walls, toward the external medium (air). Roughly, one may regard that tunneling losses are tolerable if $(\tau_L \Upsilon_{tu})/\tau_{tr} < 1$. For $N = 6$, $d \simeq 6$ nm, $w \simeq 3.5$ nm, 30 $\mu m < L < 50$ μm and $\kappa_\alpha/\beta_\alpha \simeq 3 \times 10^{-3}$, it has been estimated[73] that $\Upsilon_{tu} \simeq 0.19$ and $0.28 < (\tau_L \Upsilon_{tu})/\tau_{tr} < 0.47$, which seems to be tolerable.

In conclusion, for confined propagation of thermal neutrons (mostly in the fundamental mode) at the nanoscale, MWCNTs would be a more favorable possibility than SWCNTs. The nanoscale seems, according to the analysis in this section, the lower limit in which neutron confined propagation could occur. At this point, we recall that the minimum transverse size for an x-ray beam (whatever the focusing device) is about 10 nm,[74] that is, the nanoscale again.[75–84]

Further devices may be needed in the process of focalizing neutrons between, say, the micron scale and the nanoscale one; for such a purpose, polycapillary glass fibers may be adequate.[73]

9.9 NEUTRON CAPTURE THERAPY

9.9.1 General Aspects

Radiations, particularly those emitted in the decays of atomic nuclei and in nuclear reactions, allow for the possibility of killing cancer tumors in the tissues of the human body (radiotherapy). The very important subject of cancer treatment by means of nuclear radiations (protons, heavy ions, fast and thermal neutrons) is as old as nuclear physics itself. The reader may review the book by J.F. Fowler (for an account up to 1981)[86] and the presentation by M. Abe,[87] among other interesting presentations and overviews.[9,88] Before turning to the specific areas of interest, namely slow neutrons (and hence BNCT and NCT), a few general comments are in order.

Typical (mammalian) cells have diameters about 10 to 15 μm. In order of magnitude, there may be about 10^9 cells per gram (g). This estimate also applies to tumors. Typical orders of magnitude of volumes and surface areas of (large) tumors containing many cancerous cells are expressed as cubic centimeters and square centimeters, respectively.

In order to assess and compare the effects produced in material media in general and in living tissue in particular by different types of propagating radiations, some physical quantities are usually employed.[9,86] Let us summarize them:

1. The linear energy transfer (LET) is the amount of energy deposited per unit distance by the radiation as it propagates through tissue. X-rays and electron beams have low LET. α particles have high LET.
2. The relative biological effectiveness (RBE) or the quality factor (QF) of a radiation can be used to measure effectiveness. The RBE is the ratio of the dose of radiation to the dose of x-rays that produces the same biological effect. For α radiation, $1 \leq RBE \leq 20$. Radiations which deposit a small amount of energy per unit path length have $QF \simeq 1$. In cases where it may be not easy to measure the RBE, it is convenient to work with QF measurement.
3. The absorbed dose, D characterizes the energy absorbed by media due to ionizing radiation. D measures the amount of energy deposited per unit mass of the material through its ionization. A very common unit for D is the $gray$ (Gy), which is defined as $1\ Joule/1\ Kg$.

Neutrons are very densely ionizing particles and, for this reason, are potentially useful for cancer radiotherapy.

9.9.2 NCT AND BNCT: SHORT HISTORICAL OVERVIEW

After the discovery of the neutron by J. Chadwick,[1] it was soon noticed that if a neutron hits an atomic nucleus and if the former is captured by the latter, an α particle is frequently released (see examples in Subsection 9.2.4). The possibility that the α particle and the energy liberated in that nuclear reaction have beneficial applications for cancer treatment was already considered in 1936 by G.L. Locher.[89] The therapy based upon the capture of neutrons by suitable atomic nuclei located in tumors (and upon the potential damage produced to the cancerous cells by the energy released by those nuclear reactions) is denoted generically as neutron capture therapy (NCT). A very suitable nucleus for that kind of radiotherapy is boron $^{10}_{5}B$ and the resulting specific radiotherapy is known as boron neutron capture therapy (BNCT).

In order to perform BNCT, the required beams of thermal neutrons are extracted typically from nuclear reactors. Some specific issues in standard BNCT are[86] (i) a drug containing a suitable concentration of $^{10}_{5}B$ should have been delivered previously to the malignant tumor of the patient, (ii) the α particles released after the bombardment of the tumor with thermal neutrons and the capture of the latter by $^{10}_{5}B$ deliver in turn a relatively large amount of energy to (and give rise to ionization in) the cells of the malignant tumor, and (iii) both effects in (ii) could produce an irreversible damage to those malignant cells (more specifically, to their DNA), thereby preventing their reproduction and hence stopping the growth of the tumor. See Figure 9.17. Due to the high LET produced, one single hit of one neutron can produce damage to both chains of one double-stranded DNA macromolecule. This kind of hit is regarded as

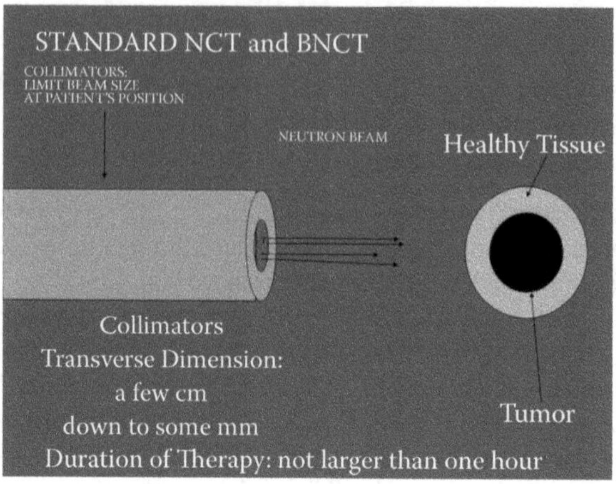

FIGURE 9.17 Standard NCT and BNCT. The use of collimators (Subsection 9.2.5) to concentrate the neutron beam.

irreversibly lethal to the cancerous cells that suffer the bombardment. If only one of the two chains is hit the DNA macromolecule in a tumor cell has a greater probability for repair and for survival; this is what may occur if low LET radiation is employed.

Clinical trials with thermal neutrons were started in the United States by 1951. However, by 1961, BNCT fell out of use after some discouraging clinical results. Two main reasons for those failures were (i) inadequacy of the boron compounds employed to the specificity of the tumors and (ii) insufficient penetration of thermal neutrons. More generally, it appears that physical drawbacks involved in the applications of BNCT are[86,90] the thermal neutron doses reaching and damaging normal (non-malignant) tissues and poor depth doses of thermal neutrons delivered effectively to the malignant cells.

In spite of those failures, various researchers (H. Hatanaka, Y. Hayakawa, Y. Nakagawa, and other scientists, in Japan) pursued, revitalized and certainly improved BNCT with thermal neutrons. More adequate boron compounds were introduced. Further studies qualitatively demonstrated the ability of BNCT to eradicate tumors with little damage to the surrounding normal tissue. The treatment was preceded by suitable surgery to allow the thermal neutron beams to reach tumors in deep parts of the brain. Those researchers initiated clinical trials in Japan by 1968. An extremely interesting paper by Hatanaka,[91] containing a wealth of information on BNCT, summarized work done up to 1991. Very active research and clinical programs on BNCT in humans continue in Japan, and have led to a variety of improvements. Clinical trials in humans have been carried out in several institutions in the United States (in particular, at the Massachusetts Institute of Technology, Harvard, and Brookhaven) through 2004. Since the mid-1990s clinical trials have also been carried out in The Netherlands (Petten), Finland (Helsinki), Sweden (Studsvik), the Czech Republic, Italy, and Argentina. Of course, those programs are affected from time to time by shortages in budgets and nuclear reactors shut-downs. There are also facilities for BNCT production under construction in Russia and Taiwan. A recent conference in Boston[92] provided an updated account of various proposals and other facilities under construction.

More recent developments on BNCT and NCT from various standpoints are well documented.[87,92–98] Detailed assessments of the results of the clinical trials (in particular, statistics of survival, duration of survival after treatment, life quality after treatments) and discussions of future proposals can also be found in those references[87,92–97] and in an article by Hatanaka.[91]

BNCT has been applied mostly to malignant brain tumors (gliomas, intra-cerebral melanomas and glioblastomas). In particular, glioblastoma (multiforme) has been one main target for BNCT clinical applications due to poor prognosis of patients after other treatments. In particular, there was some critical assessment by 2001 whether the results of BNCT therapies demonstrated significant benefits for patients compared to other treatments.[95] Independent protocols appeared to indicate that BNCT can produce median survival in patients with glioblastoma that appears equivalent to conventional photon therapy. More recent results indicate typical survival of about 23 months for glioblastoma.[98] The best survival data for BNCT are at least comparable with those obtained by current standard therapies for patients with multiform glioblastoma.[97] A very important aspect is that BNCT treatments appear to give rise to improved quality of life. Head and neck tumors and cutaneous melanomas have also been treated. BNCT has also been applied to multiple liver metastases.[92]

Future important and successful advances depend, among other issues, on the developments of better boron compounds (more selective and effective boron delivery agents) and improved methods to estimate quantitatively the boron content in tumors before treatment.[95–97]

9.9.3 NCT and BNCT: Miscellaneous Aspects (Mostly Neutron Physics)

${}^{10}_{5}B$ has a relatively large cross-section for thermal neutron capture (about $3800 \times 10^{-24}\ cm^2$). Basic energy-releasing nuclear reactions upon which BNCT is based, are (a) ${}^{10}_{5}B + n \rightarrow {}^{4}_{2}He + {}^{7}_{3}Li + 2.8MeV$ and (b) ${}^{10}_{5}B + n \rightarrow {}^{4}_{2}He + {}^{7}_{3}Li + \gamma + 2.3MeV$, where γ denotes a photon with energy $0.48\ MeV$. The reactions (a) and (b) occur with probabilities 0.6 and 0.94, respectively. The detections of the photons (γ) emitted in the reactions (b) give rise to one method to determine (at least grossly) the concentrations of ${}^{10}_{5}B$ in liquid samples, tissue, or blood (this method being known as "prompt γ neutron activation analysis").

One single capture of one neutron by one ${}^{10}_{5}B$ nucleus usually liberates enough energy to kill one cell in a tumor. For a tissue exposure of about 10^{12} neutrons$/cm^2$, in order to produce about two to three neutron capture events in a tumor cell, one would require about 10^9 atoms of ${}^{10}_{5}B$ in that cell. The ${}^{4}_{2}He$ and ${}^{7}_{3}Li$ nuclei produced have both high LET and high RBE. It can be estimated that the average ranges of the ${}^{4}_{2}He$ and ${}^{7}_{3}Li$ nuclei in tissue are about 8 and 5 μm, respectively.[95] The damage produced by the latter (in particular, to tumor cells) is concentrated in a spatial domain, having size ranging from 4 up to 15 μm about the smaller region where the capture of the neutron by ${}^{10}_{5}B$ has taken place.

As estimated by Hatanaka,[91] for a thermal neutron beam to be useful in BNCT (or NCT), its flux should be, at least, 5×10^8 neutrons/$(cm^2 \times s)$. Accordingly, neutron fluxes (coming from nuclear reactors) ranging from 5×10^8 up to 10^{12} neutrons/$(cm^2 \times s)$ are employed typically.

Diameters of transverse cross-sections of neutron beams range from millimeters to centimeters. Typically, circular apertures of 12 to 14 cm diameter are used in clinical trials. See Figure 9.17.

The so-called ratio between the total neutron current and the total neutron flux provides a measure of the fractions of neutrons that are moving approximately in the forward beam direction.[95] A high value is important for two reasons: (i) to limit divergence of the neutron beam and, so, reduce undesired irradiation to other tissues, and (ii) to permit flexibility in patient positioning. An acceptable value for that ratio is larger than 0.7.

Theoretically, the maximum depth in current BNCT with thermal neutrons is about 6 cm from the surface. In practice, the largest part of the thermal neutron flux reaches a depth of 2 to 3 cm and smaller fluxes are delivered to domains deeper than that. A thermal neutron flux that has entered into tissue falls off exponentially from the surface; it drops off below 20 percent at a depth about 4 cm. Thermal neutrons are considered to be most useful for superficial tumors, at depths less than 4 cm (for instance, for subcutaneous melanomas). They have also been extensively employed (in particular in Japan), combined with surgery, for the treatment of deeper tumors.

An epithermal neutron flux that has entered into tissue increases first up to a maximum, at a depth between 2 and 3 cm from the surface, and falls off exponentially for larger penetrations. The epithermal neutron flux drops below 25 percent (of the incoming flux at the surface) for a depth about 14 cm from the surface. Thus, epithermal neutrons (with typical energies from 0.5 eV up to 10 keV) penetrate better in tissue, but their probability of interaction with ${}^{10}_{5}B$ is lower than that for thermal neutrons. In recent times, extensive programs based upon epithermal neutron beams (with energies from about 0.5 eV up to about 20 keV) for BNCT on deeper tumors have been initiated at the Massachusetts Institute of Technology, Washington State University, and Brookhaven.[87,92,94] At other institutions, reactor-based facilities are under construction for BNCT using epithermal neutrons.[92]

The compounds (containing ${}^{10}_{5}B$) employed in BNCT clinical trials are those known as BSH (or sulfhydryl borane, with chemical formula $Na_2B_{12}H_{11}SH$) and BPA (or amino acid *p*-(dihydroxyboryl)-phenylalanine). Boron concentrations (of approximately, 20 μg per g of tumor) appear sufficient to deliver therapeutic doses of radiation to the tumor, with minimal toxicity to normal tissue. Others (like borated porphyrin, also known as BOPP) have been approved for clinical use.

Usually, overall location, shape, and boundaries of a tumor are determined by means of a computed tomography (CT) scan, nuclear magnetic resonance, or positron-emission tomography. According to Hatanaka,[91] a normal brain can tolerate a single dose of approximately 40 to 50 Gy in a current BNCT regime with thermal neutrons. A brain containing a tumor would tolerate a duration of irradiation during BNCT that should not exceed the above limit. Consistent with this, typical delivered doses in BNCT are equivalent to 10 (to a few dozen) Gy. Some typical full BNCT treatments require a total of $N_n \simeq 10^{14}$ neutrons, to be delivered to a tumor of volume V_{mc} (some tens of cubic centimeters) in a reasonable time duration τ (about an hour).

Neutrons tend to spread in tissue due to scattering interactions, specially with hydrogen. The effectiveness of NCT relies on the differential boron uptake of tumor and healthy tissues. Thus, flooding with neutrons the entire region having a potentially viable tumor is essential.

Therapies with fast neutrons seem to suffer from unsatisfactory dose localization.[86] There are also critical comments about the seriously damaging effects of fast neutrons on brain matter.[91] That is, fast neutrons can be very effective in killing tumor cells, but they can be equally lethal to normal tissue. Nevertheless, the possibility and reliability of performing BNCT based upon faster than thermal and epithermal neutron beams have been investigated and continue to be active research subjects.[92,99–102] A useful account on fast neutron therapy can be seen at www.leiomyosarcoma.infoIneutrons.htm.

Almost all BNCT research performed has been based upon nuclear reactors. Research has also been carried out in order to employ small compact accelerators for BNCT. An accelerator-based neutron source (using a Li target) is under construction in Russia.[92]

The power of the typical nuclear reactors involved in BNCT may be low (about 100 kilowatts) or medium (megawatts). There are proposals based upon fast reactors with smaller power (4 to 5 kilowatts).[92]

9.10 BNCT AND POLYCAPILLARY GLASS FIBERS

9.10.1 Some Estimates

A priori, it would seem desirable to control and reduce the spatial spread of a beam of slow neutrons to be delivered in BNCT. That reduction is what neutron fibers could perhaps accomplish. The importance for BNCT of a large ratio between the total neutron current and the total neutron flux[95] has been cited in Subsection 9.9.3. Speculations about the possible applications of neutron fibers to BNCT have been made at a very early stage.[35] At a later stage, after the experimental arrangement of a neutron lens[63] and one employed for guiding and focusing neutron beams[64] by means of multiple capillary fiber bundles, it did not seem unreasonable to speculate[43,103] whether further developments based on these devices could perhaps be useful for BNCT. Independently, succinct proposals regarding applications of polycapillary glass fibers to BNCT have also been made in connection with the first experiments.[63]

In order to make those proposals[35,43,103] less qualitative, some additional simple estimates have been carried out and reported recently.[68] We shall now summarize them in the remainder of this subsection.

Let θ_{cr} (Eq. (9.8)) be the maximum angle (acceptance angle) for neutrons that approach one end of a hollow capillary channel (HCC) from outside, to enter into and to propagate confined along that HCC. A typical estimate is: $10^{-2} \leq \theta_{cr} \leq 10^{-3}$ radians. Let F_0 be an incoming thermal neutron flux (neutrons/$(cm^2 \times sec)$) outside the fiber (say, the HCC). For a statistical mixture of wave functions, F_0 is interpreted as $< F_0 >$ (Subsection 9.6.4). Then, the flux F_{in} ($< F >_{in}$, for statistical mixtures) of the confined neutrons that have entered into and propagate confined along the HCC can be roughly estimated as $F_{in} \simeq \theta_{cr}^2 \times F_0$. Then, θ_{cr}^2 reduces the flux by a factor about 10^{-5}. For $F_0 \simeq 10^{13}$ to 10^{14} neutrons/$(cm^2 \times sec)$, one estimates $F_{in} \simeq 10^8$ to 10^9 neutrons/$(cm^2 \times sec)$. Although this F_{in} is not high, one should notice that the directions of motion for those neutrons propagating confined along the thin fiber would be controlled by the latter and could be delivered selectively to a region R of interest. For the same reason, one could avoid (or decrease) the delivery of those neutrons to regions R_1 around R.

Let us consider a bundle of some n_{PGF} parallel polycapillary glass fibers (PGF), each with transverse cross-section about 0.13 mm^2. Such a bundle would transmit $n_{PGF} \times 10^5 \times \tau$ to $n_{PGF} \times 10^6 \times \tau$ neutrons during τ seconds. For $\tau = 3.6 \times 10^3 s$ (1 hour), we would have $N_{PGF} \simeq 3.6 \times n_{PGF} \times 10^8$ to $3.6 \times n_{PGF} \times 10^9$ neutrons. As an (extreme) example, the bundle could be arranged as in Reference 63, with $n_{PGF} \simeq 720$ and transverse diameter 15 mm. One may compare the above estimates with those corresponding to thermal neutron fluxes that have been useful in BNCT, as discussed previously. Of course, a neutron flux too small would force delivery of the beam during a long period of time.

Assume that such a full BNCT neutron flux has treated a tumor with $V_{mc} \simeq 10^n cm^3$, $1 \leq n \leq 2$. Another small tumor of volume $V_{mc} \simeq 1\ mm^3$ containing malignant cells could be treated at about the same time as the above full BNCT treatment, by a total of some $N_{st} = x \times 10^9$ neutrons, $10 > x > 1$ (say, about $x \times 10^{-3}\ Gr$). N_{st} has an order of magnitude about that of N_{PGF} (for suitable n_{PGF}).

Let us consider a tumor with volume V_{mc} (cm^3) and typical size L_{mc} cm so that $V_{mc} \simeq L_{mc}^3$. It contains N_{mc} malignant cells so that $N_{mc} = V_{mc} \times \rho_{mc}$, ρ_{mc} equals the number of malignant cells per cubic centimeter in the tumor. The thin border of the tumor (healthy and malignant tissues meet at its external and internal sides, respectively) has area about L_{mc}^2 and some small width d_{mc} (say, about some fraction of a millimeter). Such a border contains $N_{mc;b} \simeq L_{mc}^2 \times d_{mc} \times \rho_{mc}$ malignant cells. We then have $N_{mc;b} \simeq \rho_{mc}^{1/3} \times d_{mc} \times N_{mc}^{2/3} \simeq 7.9 \times 10^2 \times d_{mc} \times N_{mc}^{2/3}$ malignant cells. In the last estimate, we have taken $V_{mc} \simeq 10^n cm^3$, $1 \leq n \leq 2$, and 15 μm as a typical size for a cell, so that $\rho_{mc} \simeq 5 \times 10^8$ cells per cubic centimeter. If d_{mc} is $\leq 0.5mm$ $N_{mc;b}$ is less than about $40 \times N_{mc}^{2/3}$. Let $N_{n;b}$ be the total number of neutrons required to treat also in a standard NCT duration all malignant cells $N_{mc;b}$ contained just in the thin border of a tumor having the above size. If N_n is the total number of neutrons required for the volume V_{mc} in a typical full BNCT treatment (see Subsection 9.9.3), it may be not unreasonable to expect that $N_{n;b} \simeq \alpha \times N_n^{2/3}$. α would be some numerical constant (the order of magnitude of which should not exceed some tens). Then, for $N_n \simeq 10^{14}$, one has $N_{n;b} \simeq \alpha \times 2 \times 10^9$ neutrons. The order of magnitude of $N_{n;b}$ is about that for N_{st} (for a small tumor of volume $\simeq 1\ mm^3$) and about that for N_{PGF} (for suitable n_{PGF}). These estimates could be adequate to within, say, one order of magnitude.

For treatment of the bulk of a tumor with $V_{mc} \simeq 10^n\ cm^3, n = 1, 2$, standard BNCT treatments with collimators are adequate. After such standard therapy of the bulk has been essentially completed, there may remain smaller subdomains still containing surviving malignant tissue. Let us consider such small tumors with $V_{mc} \leq 1\ mm^3$ (case (a) above). We could use a neutron beam with small transverse cross-section, like that provided by some curved parallel polycapillary glass fibers (or some bundle

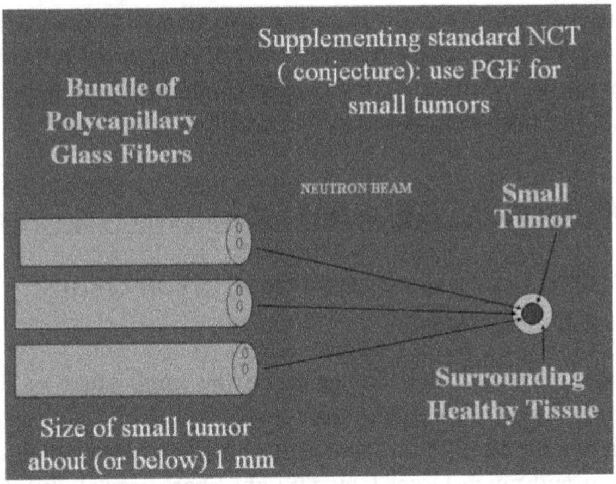

FIGURE 9.18 Possible complementary NCT (or BNCT) for small tumors by means of polycapillary glass fibers.

thereof), in order to try complementary BNCT (or NCT) for those smaller tumors in some typical BNCT duration. See Figure 9.18. One could also speculate about a similar complementary BNCT (or NCT) of thin (surviving malignant) borders of tumors (case (b) above) also using some (possibly bent) PGF or bundles thereof in a typical BNCT duration.

9.10.2 Some Preliminary Experiments

Further related research addressed toward possible biomedical applications of neutrons has been carried.[104–107] Some preliminary and very interesting trials of BNCT in experimental mice, using polycapillary glass fibers have been reported recently.[108] The inner diameter of each single hollow capillary channel was not chosen to be at the micron scale but larger, at millimeter scale (specifically 2 mm). The length of each capillary was 640 mm. The focal distance was 1000 mm and the diameter of the focal spot was 6 mm. The number of the confined neutrons[108] which, having entered each single hollow capillary channel are transmitted across its transverse cross-section per second appeared to be essentially consistent with the estimates given in the previous subsection of the total number transmitted per second across the transverse cross-section (0.13 mm^2) of one of the polycapillary glass fibers. The hollow capillary channels cited in Subsection 9.10.1 lie in the micron range. Notice the essential consistency of the number of transmitted neutrons through each single hollow capillary channel[108] with other speculation noted in Subsection 9.10.1. The size of the present hollow capillary channel lies between those for the hollow neutron guides (Subsection 9.4.2) and those considered in Section 9.7.

The motion of one single neutron inside each single hollow capillary channel[108] has been described by geometrical optics. Thus, the average number of internal reflections suffered by one single neutron on the inner walls of the single hollow capillary channel (through which the former propagates) has been been estimated to be 0.64.[108] For comparison, we remark that the average number of internal reflections suffered by one single neutron on the inner walls of one single hollow capillary channel in the first experiments[64] has been been estimated to be about 100.

9.11 CONDENSED MATTER ANALYSIS AND NEUTRON FIBERS

9.11.1 Multicapillary Glass Fibers for Neutron Optics: Further Developments

Further experimental research and advances on the application of flexible polycapillary glass fibers for neutron optics have been reported,[109] with several flexible fibers with capillary diameters 14 μm and 20 μm and successive radii of curvatures equal to 8 m, 10 m and much larger (say, straight fibers practically). Other experimental studies (regarding halo backgrounds of both slow and fast neutrons and hard photons, and their reduction or suppression) deal with further related effects.[52,110]

At least some of the institutions cited in Subsection 9.7.2 even if mostly dedicated to the development of x-ray polycapillary optics, also appear to conduct additional research activities and development on neutron polycapillary optics. Experiments on neutron focusing based on polycapillary glass fibers have also been carried out recently at IfG and reported.[111]

9.11.2 Possible Applications to Condensed Matter Analysis

Recall the so-called neutron activation analysis, discussed in Subsection 9.2.5, in connection with an important application, namely the determination of the composition of materials.

Possible applications of confined propagation of thermal neutrons by means of polycapillary glass fibers, due to their greater spatial resolutions, may be not restricted to BNCT.[63] They may also provide improvements to traditional neutron activation analysis:[64]

"Neutron focusing will give neutron absorption techniques greater spatial resolution, allowing them to complement other techniques for non-destructive analysis of trace elements in materials. It will be straightforward to raster a large specimen, such as a weld or a solder joint, across the focused neutron beam, and obtain spatially resolved elemental composition at selected areas. This would allow two-dimensional prompt radiation analysis with sub-millimeter resolution of elements such as H, B, Li, N and Na. The technique should also lead to three-dimensional imaging for use in mapping elements in microelectronic devices and grain boundaries in materials. Applications to neutron-scattering instrumentation, such as small-angle scattering, may become possible."

This is intimately connected to the importance of achieving a large ratio between the total neutron current and the total neutron flux; recall Subsection 9.9.3.

As early applications of the developments arising from the first experiments,[63,64] converging lenses (employing thousands of such polycapillary fibers) have been used in order to focus cold neutrons by increasing their current density for performing materials analysis.[112–114] Synchrotron radiation is increasingly employed for investigations and authentications of art and of objects of historical interest. Specifically, by exploring the micron scales, synchrotron radiation allows us to investigate compositions and ancient production techniques. We are reminded of the complementary roles played by short wavelength electromagnetic radiation and thermal neutrons, at decreasing scales.

Another possible application of confined propagation of thermal neutrons using CNTs is the characterization of nanostructured materials of importance for biotechnology.[115]

9.12 CONCLUSION AND DISCUSSION

Several general aspects of neutron optics have been outlined. We have focused on one aspect of it, namely the possibility of letting thermal neutrons propagate along suitable guides (and waveguides) with decreasingly small transverse cross-sections (neutron fibers). A quantum mechanical three-dimensional treatment of the confined propagation of slow neutrons (increasingly necessary, as transverse sizes decrease) has been summarized.

Our renewed interest in this subject has been triggered by the development of technology based upon polycapillary glass fibers and by experiments at several institutions. Slow neutrons have been guided along the hollow channels formed in the polycapillary glass fibers. The hollow channels then behaved as thin waveguides for the neutrons. We have summarized some recent work on the possible confined propagation of thermal neutrons for the smallest transverse dimensions which could presumably allow activity at the nanometer scale. We have included some short general discussion of BNCT as another interdisciplinary application of slow neutron physics. We have speculated that further developments based on the confined propagation of thermal neutrons in polycapillary glass fibers could perhaps result in improvement of standard BNCT. Possible applications of confined propagation and focusing of slow neutrons at small scales may not be restricted to BNCT. They may be employed to perform materials analysis. All in all, it appears that neutron guides and waveguides have become active and interesting research subjects.

ACKNOWLEDGMENTS

R.F. Alvarez-Estrada acknowledges the support of CICYT (Proyecto FPA2004-02602), Ministerio de Educacion y Ciencia, Spain. M.L. Calvo wishes to thank the CICYT (Proyecto TEC2005/02180), Ministerio de Educacion y Ciencia, Spain. We are grateful to Drs. A. A. Bjeoumikhov, G. I. Borisov, J. F. Crawford, W. M. Gibson, S. V. Kukhlevsky, D. F. R. Mildner, J. R. Venhuizen, and R. Wedell for interesting correspondence, information, and discussions. We are also grateful to Dr. G. F. Calvo, for analogous reasons and for providing very valuable help with several figures and with the LaTeX files.

REFERENCES

1. J. Chadwick, Nature 129, 312 (1932).
2. J. Byrne, *Neutrons, Nuclei and Matter: An Exploration of the Physics of Slow Neutrons* (Institute of Physics Publishing, Bristol, 1994).
3. W.M. Elsasser, C. R. Acad. Sci. Paris 202, 1029 (1936).
4. H. Halban and P. Preiswerk, C. R. Acad. Sci. Paris, 203, 73 (1936).
5. D.P. Mitchell and P.N. Powers, Phys. Rev. 50, 486 (1936).
6. V.F. Sears, *Neutron Optics* (Oxford University Press, Oxford, 1989).
7. G.E. Bacon, *Neutron Diffraction* (Clarendon Press, Oxford, 1962).
8. D.F.R. Mildner, *Neutron Optics*, Chapter 36 in *Handbook of Optics*, Volume III, 2nd edition, (McGraw-Hill, New York, 2001).
9. K.S. Krane, *Introductory Nuclear Physics* (John Wiley & Sons, New York, 1988).
10. J. Finney and U. Steigenberger, Physics World, December (1997).
11. W.S. Kiger III, S. Sakamoto and O.K. Harling, Nuclear Sci. Eng., 131, 1 (1999).
12. G. Nebbia and J. Gerl, Europhysics News 36, 119 (2005).
13. V.F. Sears, Phys. Reports 82, 1 (1982).
14. D.J. Hughes and M.T. Burgy, Phys. Rev. 81, 498 (1951).
15. M. Lax, Rev. Mod. Phys. 23, 287 (1951).
16. M. Lax, Phys. Rev. 85, 621 (1952).
17. L.L. Foldy, Phys. Rev. 67, 107 (1945).
18. M.L. Goldberger and F. Seitz, Phys. Rev. 71, 294 (1947).
19. R. Weinstock, Phys. Rev. 65, 1 (1944).
20. O. Halpern, M. Hamermesh and M.H. Johnson, Phys. Rev. 59, 981 (1941).
21. C.G. Shull and E.O. Wollan, Phys. Rev. 81 527, (1951).
22. J. Crist and T. Springer, Nukleonik 4, 23 (1962).
23. H. Maier-Leibnitz and T. Springer, J. Nuclear Energy A/B17, 217 (1963).
24. B. Alefeld, J. Crist, D. Kukla, R. Scherm and W. Schmatz, *Berichte der Kernforschungsanlage*, Jülich Ju 1-194-NP (1965).
25. B. Jacrot, *Proc. Symp. Instrumentation for Neutron Inelastic Scattering Research*, Vienna 1969, (IAEA, Vienna 1970).
26. O. Schärpf and D. Eichler, J. Phys. E: Sci. Instrum. 6, 774 (1973).
27. D. Marx, Nucl. Instr. Meth. 94, 533 (1971).
28. A. Schebetov, A. Kovalev, B. Peskov, N. Pleshanov, V. Pusenkov, P. Schubert-Bischoff, G. Shmelev, Z. Soroko, V. Syromyatnikov, V. Ulianov and A. Zaitsev, Nucl. Instr. Meth. Phys. Res. *A* 432, 214 (1999).
29. M.Rossbach, O. Schärpf, W. Kaiser, W. Graf, A. Schirmer, W. Faber J. Duppich and R. Zeisler, Nucl. Instr. and Meth. in Phys. Res. B 35, 181 (1988).
30. D. Marcuse, *Light Transmission Optics* (Van Nostrand Reinhold, New York, 1972).
31. D. Marcuse, *Theory of Dielectric Optical Waveguides*, Academic Press, New York (1974).
32. A.W. Snyder and J.D. Love, *Optical Waveguide Theory* (Chapman & Hall, London, 1983).
33. J.M. Enoch and F.L. Tobey, Editors, *Vertebrate Photoreceptor Optics* (Springer-Verlag, Berlin, 1981).
34. V. Lakshminarayanan and J.M. Enoch, Biological waveguides optics, in *Handbook of Optics*, Volume III, 2nd edition, (McGraw-Hill, New York, 2001).
35. R.F. Alvarez-Estrada and M.L. Calvo, J. Phys. D: Appl. Phys. 17, 475 (1984).
36. M.L. Calvo and R.F. Alvarez-Estrada, J. Phys. D: Appl. Phys. 19, 957 (1986).
37. R.F. Alvarez-Estrada and M.L. Calvo, Journal de Physique Colloque C3, Suppl. au n. 3, Tome 45, C3-243 (1984).

38. M.L. Calvo, J. Phys. D: Appl. Phys. 33, 1666 (2000).
39. A. Messiah, *Quantum Mechanics*, Vol. I, (North-Holland, Amsterdam, 1961).
40. M.V. Berry and K.E. Mount, Rep. Prog. Phys. 35, 315 (1972).
41. A. Martin, Helv. Phys. Acta 45, 140 (1972).
42. L. Landau and E. Lifchitz *Mecanique Quantique*, Section 48 (Editions Mir, Moscow, 1967).
43. M.L. Calvo and R.F. Alvarez-Estrada, *Neutron Optics, Neutron Waveguides and Applications* in *International Trends in Applied Optics*, Volume V, (The International Society for Optical Engineering Bellingham, 2002).
44. M. Abramowitz and I.A. Stegun, Editors, *Handbook of Mathematical Functions* (Dover, New York, 1965).
45. B. Rohwedder, Phys. Rev. A 65, 043619-1 (2002).
46. L.E. Ballentine, *Quantum Mechanics. A Modern Development* (World Scientific, Singapore, 1998).
47. D. I. Blokhintsev, *Mecanique Quantique* (Masson et Cie., Paris, 1967).
48. E.A.J. Marcatili, Bell Syst. Tech. J. 48, 2103 (1969).
49. D. Marcuse, J. Opt. Soc. Am. 66, 216 (1976).
50. M.L. Calvo and R.F. Alvarez-Estrada, J. Opt. Soc. Am. A4, 683 (1987).
51. M.A. Kumakhov and F.F. Komarov, Phys. Rep. 191, 289 (1990).
52. M.A. Kumakhov, Nucl. Instr. Meth. Phys. Res. A 529, 69 (2004).
53. V.A. Arkadiev, V.I. Beloglasov, A.A. Bjeoumikhov, H.-E. Gorny, N. Langhoff and R. Wedell Surface x-ray, Synchrotron Neutron Res. 1, 48-54 (2000).
54. H. Bronk, S. Rohrs, A.A. Bjeoumikhov, N. Langhoff, J. Schmalz, R. Wedell, H.-E. Gorny, A. Herrold and U. Waldschlager, Fresenius J. Anal. Chem. 371, 307 (2001).
55. A. Erko, N. Langhoff, A.A. Bjeoumikhov and V.I. Beloglasov, Nucl. Instr. Meth. Phys. Res. A467, 832 (2001).
56. A.A. Bjeoumikhov, N. Langhoff, R. Wedell, V.I. Beloglasov, N. Lebed'ev and N. Skibina, x-ray Spectrometry 32, 172 (2003).
57. T. Wroblewski and A.A. Bjeoumikhov, Nucl. Instr. Meth. Phys. Res. A 521, 571 (2004).
58. A.A. Bjeoumikhov, N. Langhoff, J. Rabe and R. Wedell, x-ray Spectrometry 33, 312 (2004).
59. J. Bartoll, S. Rohrs, A. Erko, A. Firsov, A.A. Bjeoumikhov and N. Langhoff, Spectrochimica Acta Part B 59, 1587 (2004).
60. A.A. Bjeoumikhov, N. Langhoff, S. Bjeoumikhova and R. Wedell, Rev. Sci. Instr. 76, 063115 (2005).
61. A.A. Bjeoumikhov, S. Bjeoumikhova, N. Langhoff and R. Wedell, Appl. Phys. Lett. 86, 144102 (2005).
62. T. Wroblewski and A.A. Bjeoumikhov, Nucl. Instr. Meth. Phys. Res. A 538, 771 (2005).
63. M.A. Kumakhov and V.A. Sharov, Nature 357, 390 (1992).
64. H. Chen, R.G. Downing, D.F.R. Mildner, W.M. Gibson, M.A. Kumakhov, I. Yu. Ponomarev and M.V. Gubarev, Nature 357, 391 (1992).
65. D.F.R. Mildner and H. Chen, J. Appl. Cryst. 27, 316 (1993).
66. D.F.R. Mildner, H. Chen and V.A. Sharov, J. Appl. Cryst. 28, 793 (1995).
67. D.F.R. Mildner, V.A. Sharov and H.H. Chen-Mayer, J. Appl. Cryst. 30, 932 (1997).
68. R.F. Alvarez-Estrada and M.L. Calvo, Appl. Rad. Isot. 61, 841 (2004).
69. S. Iijima, Nature 354, 56 (1991).
70. P.M. Ajayan and T.W. Ebbesen, Rep. Prog. Phys. 60, 1025 (1997).
71. H. Dai, J. Kong, C. Zhou, N. Franklin, T. Tombler, A. Cassell, S. Fan and M. Chapline, J. Phys. Chem. B 103, 11246 (1999).
72. *Issue Devoted to Advances in Carbon Nanotubes* Materials Research Society Bulletin 29, Number 4, April (2004).

73. G.F. Calvo and R.F. Alvarez-Estrada, Nanotechnology 15, 1870 (2004).
74. C. Bergemann, H. Keymeulen and J.F. van der Veen, Phys. Rev. Lett. 91, 204801-1 (2003).
75. S.V. Kukhlevsky and G. Nyitray, Phys. Lett. A 291, 459 (2001).
76. S.V. Kukhlevsky, G. Nyitray and L. Kantsyrev, Opt. Commun. 192, 225 (2001).
77. S.V. Kukhlevsky, F. Flora, A. Marinai, G. Nyitray, Zs. Kozma, A. Ritucci, L. Palladino, A. Reale and G. Tomassetti, x-ray Spectr. 29, 354 (2000).
78. S.V. Kukhlevsky, F. Flora, A. Marinai, G. Nyitray, Zs. Kozma, A. Ritucci, L. Palladino, A. Reale and G. Tomassetti, in *Selected Research Papers on Kumakhov Optics and Application of 1998–2000*, M.A. Kumakhov, Editor, SPIE, Russia Chapter, 4155, 61 (2000).
79. S.V. Kukhlevsky, x-ray Spectr. 32, 223 (2003).
80. S.V. Kukhlevsky and M. Mechler, J. Opt. A Pure Appl. Opt. 5, 256 (2003).
81. S.V. Kukhlevsky, M. Mechler, L. Csapo and K. Janssens, Phys. Lett. A 319, 439 (2003).
82. S.V. Kukhlevsky and G. Nyitray, J. Opt. A Pure Appl. Opt. 4, 271 (2002).
83. L. Vincze, S.V. Kukhlevsky and K. Janssens, *in Comput. Meth.* x-rays *Neutron Opt., 49th Annual Meeting*, SPIE, 5536, 81 (2004).
84. S.V. Kukhlevsky, G. Lupkovics, K. Negrea and L. Kozma, J. Opt. A Pure Appl. Opt. 6, 102 (1997).
85. H. Abele, S. Baessler and A. Westphal, arXiv:hep-ph/0301145 v1 (2003).
86. J.F. Fowler, *Nuclear Particles in Cancer Treatment*, (Adam Hilger, Bristol, 1981).
87. *Proc. Ninth Int. Symp. on NCT for Cancer* (Osaka, Japan, 2000).
88. Courrier CERN, Numero Special, *Applications des Accelerateurs* 35, (1995).
89. G. Locher, A.J.R. 36, 1 (1936).
90. D. Drollette, Biophotonics International, January–February, 46 (2000).
91. H. Hatanaka, in *Glioma* (Springer-Verlag, Berlin, 1991).
92. *Topics on Neutron Capture Therapy, Proc. 11th World Congress on NCT*, Boston (October 11–15, 2004), Appl. Rad. and Isot. 61, Issue 5 (2004).
93. H. Hatanaka, in *Advances in Neutron Capture Therapy* (Plenum Press, New York, 1993).
94. W. Sauerwein, R. Moss and A. Wittig. *Research and Development in NCT* (Monduzzi Editore, Bologna, 2002).
95. V. Levin, P. Andreo and B. Dodd, *Current Status of Neutron Capture Therapy* International Atomic Energy Agency, IAEA-TECDOC-1223 (2001).
96. J.A. Coderre, J.C. Turcotte, K.J. Riley, P.J. Binns, O.K. Harling and W.S. Kiger III, *Technology in Cancer Research and Treatment*, (Adenine Press, 2003).
97. R.F. Barth, J.A. Coderre, M.G. Vicente and T.E. Blue, Clin. Cancer Res. 11, 3987 (2005).
98. T. Yamamoto, A. Matsumura, K. Nakai, Y. Shibata, K. Endo, F. Sakurai, T. Kishi, H. Kumada, K. Yamamoto and Y. Torii, Appl. Rad. Isot. 61, 1089 (2004).
99. J.F. Crawford, B. Larsson, H. Reist, S. Teichmann and R. Weinreich, in *International Symposium on Hadron Therapy* (Como, Italy, 1993), Institute for Medical Radiobiology, Paul Scherrer Institute (Villigen, Switzerland, 1993).
100. J. Burmeister et al., Med. Phys. 32, 666 (2005).
101. D. Nigg et al., Med. Phys. 27, 2 (200).
102. A.F. Thornton and G.E. Laramore, *Particle Radiation Therapy*, in *Clinical Oncology*, L. Gunderson and J. Tepper Eds. (Churchill Livingston, New York, 2000).
103. M.L. Calvo, *Neutron Capture Therapy* (in Spanish), Investigacion y Ciencia (Spanish Edition of Scientific American) 298, 35 (2001).
104. G.I. Borisov, Atomnaya Energiya 60, 341 (1986).

105. G.I. Borisov, A.M. Demidov, Atomnaya Energiya 66, 408 (1989).
106. G.I. Borisov, M.M. Komkov, V.F. Leonov, Atomnaya Energiya 63, 404 (1987).
107. G.I. Borisov, L.I. Govor, A.M. Demidov, M.M. Komkov, Atomnaya Energiya 74, 394 (1993).
108. G.I. Borisov, R.I. Kondratenko and M.A. Kumakhov, Communication No. 13, Poster Session, Nuclear Engineering, *11th World Congress on Nuclear Capture Therapy* (2004).
109. G.I. Borisov, M.A. Kumakhov, Nucl. Instr. Meth. Phys. Res. A 529, 102 (2004).
110. G.I. Borisov, M.A. Kumakhov, Nucl. Instr. Meth. Phys. Res. A 529, 129 (2004).
111. International Meeting COST P7, Instituto de Fisica Aplicada, CSIC, Madrid, 5–7, May (2005).
112. Q.F. Xiao, H. Chen, V.A. Sharov, D.F.R. Mildner, R.G. Downing, N. Gao, and D.M. Gibson, Rev. Sci. Instr. 65, 3399 (1994).
113. H. Chen, V.A. Sharov, D.F.R. Mildner, R.G. Downing, R.L. Paul, R.M. Lindstrom, C.J. Zeissler and Q.F. Xiao, Nucl. Instr. Meth. Phys. Res. B 95, 107 (1995).
114. H.H. Chen-Mayer, V.A. Sharov, D.F.R. Mildner, R.G. Downing, R.L. Paul, R.M. Lindstrom, C.J. Zeissler and Q.F. Xiao, J. Radioanal. Nucl. Chem. 215, 141 (1997).
115. C.-K Loong, P. Thiyagarajan and A.I. Kolesnikov, Nanotechnol. 15, S664 (2004).

Index

B

C

D

E

M

N

O

P

Q

R

S

T

U

V

W

X

Y

Z